25th International Conference on Vacuum Technique and Technology 2018

IOP Conference Series: Materials Science and Engineering
Volume 387

St. Petersburg, Russia
5 – 7 June 2018

ISBN: 978-1-5108-6783-3
ISSN: 1757-8981

Printed from e-media with permission by:

Curran Associates, Inc.
57 Morehouse Lane
Red Hook, NY 12571

Some format issues inherent in the e-media version may also appear in this print version.

Printed by Curran Associates, Inc. (2018)

For permission requests, please contact the Institute of Physics at the address below.

Institute of Physics
Dirac House, Temple Back
Bristol BS1 6BE UK

Phone: 44 1 17 929 7481
Fax: 44 1 17 920 0979

techtracking@iop.org

Additional copies of this publication are available from:

Curran Associates, Inc.
57 Morehouse Lane
Red Hook, NY 12571 USA
Phone: 845-758-0400
Fax: 845-758-2633
Email: curran@proceedings.com
Web: www.proceedings.com

Table of contents

Volume 387

25th International Conference on Vacuum Technique and Technology

5–7 June 2018, St. Petersburg, Russian Federation

Accepted papers received: 4 July 2018
Published online: 17 July 2018

Preface

Papers

IOP Conf. Series: Materials Science and Engineering **387** (2018) 011001 doi:10.1088/1757-899X/387/1/011001

Overview of the 25th International Conference on Vacuum Technique and Technology

D K Kostrin

Department of electronic instruments and devices, Saint Petersburg Electrotechnical University "LETI", 197376, Saint Petersburg, Russia

E-mail: dkkostrin@mail.ru

Abstract. In this overview key features and main results of the 25th International Conference on Vacuum Technique and Technology that was held on 5–7 June 2018 in Saint Petersburg, Russia are discussed.

From 5 to 7 June 2018 the 25th anniversary International Conference on Vacuum Technique and Technology was held in the Saint Petersburg Electrotechnical University "LETI". The Conference was organized by the Saint Petersburg Electrotechnical University "LETI" together with the Peter the Great Saint Petersburg Polytechnic University, All-Russian Research Institute of Metrology n. a. D. I. Mendeleev, Euro-Asian Cooperation of National Metrological Institutions (COOMET) and Russian Scientific and Technical Vacuum Society n. a. Academician S. A. Vekshinskii.

The aim of the Conference was to review the results of research in the field of vacuum physics, vacuum measurement, mass spectrometry and tightness control. The Conference was attended by over 80 specialists from leading universities, research institutes and enterprises of Russia, as well as representatives of the scientific community from Ukraine and Turkey.

The Conference began with a moment of silence in memory of the recently deceased scientists: the chairman of the Conference, D. Sc., professor of the Department of electronic instruments and devices of Saint Petersburg Electrotechnical University "LETI" Alexander A. Lisenkov, D. Sc., professor, head of the research department of the Research Institute of Electrophysical Equipment n. a. D. V. Efremov Georgy L. Saksaganskii; general director of LLC "Vacuum and cryogenic systems" and head of the cryogenic and vacuum laboratory of the Peter the Great Saint Petersburg Polytechnic University Marxen P. Larin. These well-known scientists made a significant contribution to the development of vacuum technology and the creation of the Conference.

Almost all of the life and work of Alexander A. Lisenkov (figure 1) was associated with Saint Petersburg Electrotechnical University "LETI". After graduating from school and service in the Soviet Army, he entered the evening Department of radio electronics at Leningrad Electrotechnical Institute "LETI" and worked at the Department of gas discharge devices as a laboratory assistant. In 1991 he entered the full-time graduate school. In 1995 he defended his Ph. D. thesis, in 2007 – D. Sc. thesis. For work "Creation of scientific bases, development and introduction of new technologies, equipment and materials in the production of powerful generator lamps" in 2000 he was awarded the title of Laureate of the state prize of the Russian Federation in the field of science and technology.

Working in the field of vacuum and plasma technologies, he was able to solve any technical and scientific problems. During his research, significant progress has been made in the field of special coating technology. In scientific and professional activities Alexander A. Lisenkov was distinguished by the highest professionalism, hard work and the desire to do any work with maximum quality. Alexander A. Lisenkov is the author of more than 300 scientific works, including 25 patents for invention, 9 monographs.

Figure 1. Alexander A. Lisenkov (1956–2018).

Since 2001, Alexander A. Lisenkov actively participated in the educational process of the Department of electronic instruments and devices, transferring his knowledge and experience to students and graduate students.

Optimism, the desire to help colleagues in any situation and high professionalism that distinguished Alexander A. Lisenkov, will always be an example of a teacher and a researcher who devoted his life to the University.

In his welcoming speech to the participants of the plenary session, the Vice-rector for research of the Saint Petersburg Electrotechnical University "LETI" Dmitrii V. Gaivoronskii stressed that from the very beginning in the Saint Petersburg Electrotechnical University "LETI" direction associated with "hollow devices", as at that time was called the vacuum technology, was one of the priority and key in the activities of the University. He also noted the leading position of the faculty of electronics of the University in Russia and World in the development of materials and technologies related to vacuum devices and technologies for creating devices based on them.

"For 25 years the Conference has established itself as a meeting of professionals of extra-class, who have the knowledge and competence to look into tomorrow. It is very nice to see young people in the audience, which emphasizes the link between generations and the transfer of experience" – Vice-rector for research of the Saint Petersburg Electrotechnical University "LETI" Dmitrii V. Gaivoronskii.

President of the Russian Scientific and Technical Vacuum Society n. a. Academician S. A. Vekshinskii Sergei B. Nesterov told about the main events that took place in the field of vacuum technology over the last year, including the results of the 14th International Exhibition of Vacuum Equipment "VacuumTechExpo", which was held in Moscow from 24 to 26 April, and the 24th Scientific and Technical Conference "Vacuum Science and Technique" (Crimea), which in September 2018 will celebrate the 25th anniversary.

Plenary reports of the Conference were devoted to theoretical and practical aspects of vacuum technology (figure 2).

IOP Conf. Series: Materials Science and Engineering **387** (2018) 011001 doi:10.1088/1757-899X/387/1/011001

Figure 2. Discussion of the reports at the 25th International Conference on Vacuum Technique and Technology.

IOP Conf. Series: Materials Science and Engineering **387** (2018) 011001 doi:10.1088/1757-899X/387/1/011001

In the report of Vladimir V. Sleptsov, D. Sc., associate professor, head of the Department "Radioelectronics, telecommunications and nanotechnology" of the Moscow Aviation Institute (National Research University) were presented promising developments of the chemical sources of current (CSC) and ultra-high-volume capacitor structures (UCS) for energy storage.

"Vacuum technology is a very serious prospect. Vacuum technology has created electronics. Now, apparently, a period has come when they will actively develop in medicine, energy and other areas that will determine the next level of economic development" – Head of the Department "Radioelectronics, telecommunications and nanotechnology" of the Moscow Aviation Institute (National Research University) Vladimir V. Sleptsov.

Foreign colleagues – chairperson of the TC 1.6 COOMET Institute from Ukraine (Kharkiv) Irina A. Kolozinska and the head of the vacuum laboratory of the TÜBİTAK UME University from Turkey (Kocaeli), Dr. Rifat Kangi highlighted in the reports the development of vacuum technique and technology in their countries.

"There is a motto: vacuum is nothing, but in principle, vacuum is everything. Without vacuum it would be impossible to get into space, nanotechnology is impossible – clean environment is needed everywhere" – Head of the vacuum laboratory of the TÜBİTAK UME University Rifat Kangi.

On June 7, the Conference continued at the All-Russian Research Institute of Metrology n. a. D. I. Mendeleev. In the format of the round table the participants of the Conference discussed the aspects of the development of vacuum equipment and technologies in Russia.

In just three days of the Conference, more than 100 oral and poster presentations were presented on topical issues of vacuum generation, vacuum equipment creation and development of new technological processes. The participants of the Conference paid special attention to solving problems of vacuum technology in the formation of films and coatings by plasma and related methods, study of the properties of coatings and methods of their research, new coating materials, including nanomaterials, new areas of their use, development of modern equipment and technological processes.

Peer review statement

All papers published in this volume of *IOP Conference Series: Materials Science and Engineering* have been peer reviewed through processes administered by the proceedings Editors. Reviews were conducted by expert referees to the professional and scientific standards expected of a proceedings journal published by IOP Publishing.

IOP Conf. Series: Materials Science and Engineering **387** (2018) 012001 doi:10.1088/1757-899X/387/1/012001

The porosity determination of dispersion medium by the method of expansion of pore air into vacuum (on the example of the main soil types in the Chuvash republic)

V V Alekseev[1], S A Vasiliev[2] and M Y Ivanov[3]

[1] Department of information technologies and mathematics, Cheboksary cooperative institute, 428025, Cheboksary, Russia
[2] Department of transport-technological machines and complexes, Chuvash State agricultural academy, 428003, Cheboksary, Russia
[3] Department of «Transport, technological machines and ground transport-technological means», Volzhsky branch Moscow automobile and road construction state technical university, 428028, Cheboksary, Russia

E-mail: av77@list.ru, vsa_21@mail.ru

Abstract. Recently, for the study of the hydrophysical properties of finely dispersed systems, such as soils in Russia and abroad, tomographic studies are beginning to be widely used. The main lack of their use is a significant cost, the lack of the ability to conduct measurements directly in the field. Therefore, we have developed a fairly simple and accessible method for calculating the porosity, based on measuring the state of air both when the pore air expands into the region with a reduced pressure, and when air is pumped from the region of increased pressure. The method makes it possible to take into account the presence of pores covered by menisci and dead-end pores. Statistical processing of experimental results shows the identity of porosity values, with a confidence probability of more than 95 %. However, the relative measurement error is approximately half of that for the case of air injection. On the cultivated lands, after the passage of machine-tractor units, the appearance of a compacted soil layer often occurs. The study of the dynamics of deformations accumulation in the soil is conveniently investigated using maps with isolines of the soil porosity values. They can be used for economically justified planning of agricultural activities. The developed express method for determination of porosity, due to the speed and simplicity of measurements, makes it possible to obtain a large amount of the data in a short time.

1. Introduction

When designing technologies aimed at softening of ground, melioration, and cultivation of soils for a particular field, spatial variability is convenient to represent in the form of maps with isolines of soil porosity values. Many hydrophysical properties of soils are functions of its porosity changing when subjected to mechanical action on the soil. Since the operation of the units involves the use of different modes, it becomes possible to choose the most economically efficient mode. An important place is the assessment of the soil porosity dynamics and the overall structure of the pores around seeds and roots during vegetation. Recently, tomographic studies have begun to be widely used to study the hydrophysical properties of soils in Russia and abroad. Using modern tomographs (with a resolution of up to 16 μm), porosity, bulk density and soil moisture can be measured [1]. Analysis of 3D-images makes it possible to determine the relationship between the sealing effect of the technique and the variation of the network of prevailing water-air canals in soils. The main disadvantage of using tomographs is a significant cost and, consequently, low prevalence in scientific laboratories. In addition, it is not possible to carry out measurements directly in the field. Therefore, we have to develop and use methods that are more accessible. One of these methods is the express method for determining porosity, which is based on taking into account pressures in the pore air before and after the rarefaction/injection. This method, thanks to the speed and simplicity of measurements, allows obtaining a lot of data in a short time, which is extremely important for obtaining isolines and mapping.

From a rather large number of methods for measuring the porosity and structure of the porous space of soils, a number of researchers can calculate the density of the solid phase; micromorphometric (measuring the pore size under the microscope); adiabatic expansion of soil air into vacuum; based on gas diffusion through a soil sample, etc.

One of the rapid methods for determining porosity is the method of adiabatic expansion of soil air into an area of low pressure. For an adiabatic process, the Poisson equation is valid, which has the form [2]

$$pV^{\gamma} = \text{const}, \tag{1}$$

where p is the pressure; V – volume and γ – adiabatic exponent.

Equation (1) makes it possible to calculate the porosity from the pressures before and after the expansion of soil air into the area of a known volume with a low pressure. The method makes it possible to measure the porosity of the same soil sample in multiple replicates (with a fivefold average relative error of less than 7–10 %). For different samples taken from the same depths (0.5 m apart), the average relative error does not exceed 10–12 %.

The most common method is the determination of the porosity in the close association with the porosity and the bulk mass, and density of the solid phase of the soil. The main difficulty of its use is that the solid phase of the soil is multicomponent and has both a mineral and organic components. Therefore, the measurement of the bulk mass of the solid phase is difficult [5, 6].

2. Materials and methods

Schematic diagram of the device for implementing the method of expanding the pore air to the area with low pressure (or injecting air from the area with high pressure) is shown in figure 1, the general view of the device and its use in field measurements – in figures 2 and 3.

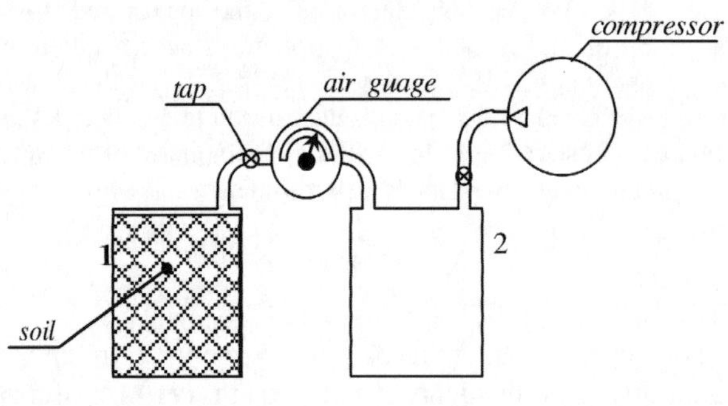

Figure 1. Scheme of the device for measuring soil porosity.

A soil sample of the volume V_{sample} with a pore volume of V_{por} is placed in a vessel 1, which is connected through a tap with a manometer, which in turn is connected to a vessel 2 of the volume $V_{vessel} = V_{sample}$. The vessel 2 is connected to a compressor or a Kamovski pump through a tap. Vessels are tight. If, with an empty vessel 2, raise the pressure to p_{ad} by the compressor and disconnect it with a tap from the compressor, connect it to vessel 1, then part of the air from vessel 2 will pass under pressure into vessel 1 and the total pressure for the system p_{est} will be established.

Figure 2. General view of the device.

Figure 3. Field measurements.

We will write down the Cliperon–Mendeleev law for gas in the vessels before and after the tap opening:

for a vessel with soil,

$$p_{atm}V_{por} = \frac{m_0}{\mu}RT\,, \tag{2}$$

for an empty vessel,

$$p_{ad}V_{vessel} = \frac{m_1}{\mu}RT\,, \tag{3}$$

for the whole system,

$$p_{est}\left(V_{por} + V_{vessel}\right) = \frac{m_0 + m_1}{\mu}RT\,, \tag{4}$$

where m_0 is the mass of the pore gas in the vessel with the sample; m_1 is the gas mass in the vessel 2; p_{atm} is atmospheric pressure; p_{ad} is injected pressure / rarefaction pressure; p_{est} is the pressure established in the system after the junction of the vessels; R is the universal gas constant; T_1, T_2, T are temperatures in the first and second vessels and in the system as a whole; μ is the molar mass of air.

In respect that the temperature after the expansion is equalized to the original value, i.e.

$$T_1 = T_2 = T, \tag{5}$$

from equations (2–4), taking into account the conservation of the gas mass, we can write

$$p_{est} = (p_{atm}V_{por} + p_{ad}V_{vessel})/(V_{por} + V_{vessel}). \tag{6}$$

The porosity $\Pi = V_{por} / V_{sample}$ is calculated by the equation (6):

$$\Pi = \frac{V_{por}}{V_{vessel}} = \frac{p_{ad} - p_{est}}{p_{est} - p_{atm}}\,, \tag{7}$$

To reduce measurement errors, one should take into account the volume of the connecting tubes and so on. For this, a preliminary experiment with a non-porous sample is performed.

With a parallel moisture measurement, the moisture content V_{water} contained in the pores can be calculated. Thus, the total porosity of the sample

$$\Pi = (V_{por} + V_{water}) / V_{vessel}\ . \tag{8}$$

The method is simple, but some nuances should be taken into account when implementing it. The pores in the soil are randomly distributed and therefore the presence of pores overlapped by menisci and dead-end pores is quite likely. The volume of air in such pores changes if the differences in the pressure Δp applied to them are greater than the pressures of the meniscus created by the curved surfaces Δp_m:

$$\Delta p > \Delta p_m. \tag{9}$$

To reduce the effects associated with the displacement of water cuffs, when the pressure changes step-by-step, the experiment is carried out at least five times.

In addition, the velocity of fluid movement in the pores is finite and depends on their radius in accordance with the Poiseuille law

$$v = (\Delta p / 4\eta l)R^2\ , \tag{10}$$

where η is the viscosity of the liquid; l is the length of the pore filled with water; R is the radius of the pore. This circumstance in the experiment requires withstanding some time before recording the steady-state pressure. To ensure that the temperature in the system is sufficiently equalized with the ambient temperature, the vessels must have good thermal conductivity, therefore, metallic ones are more effective.

The active porosity is calculated from the values of the pressure, which is measured immediately after combining the vessels, and the pressure corresponding to the total porosity is measured after the pressure is stabilized, i.e. in 3–5 minutes.

3. Results and discussion

Field experiments and land surveys were conducted in accordance with known, most common methods. Samples of soil for carrying out measurements of hydrophysical parameters were selected with the help of a special drill, which minimally deforms the soil. The soil samples were taken in accordance with GOST 17.4.3.01-83, GOST 28168-89, GOST 17.4.4.02-84, OST 56 81-84. The pycnometric method (GOST 27395-87) was used to determine the density of the solid phase (skeleton) of the soil. To measure porosity, an improved device developed in the laboratory of "Hydrophysics and soil erosion" of the Chuvash State agricultural academy was used [3, 4]. In the absence of electricity, the device can also work with the Kamovsky pump.

To increase the reliability of the data obtained under conditions of strong spatial variation of soil properties, the soil sample is used throughout the cycle of determining the hydrophysical characteristics. The soil moisture was measured by a modified weighing method in a drying cabinet and in a number of cases using a dielectric moisture meter (GOST 20915-75, GOST 28268-89). Data processing was carried out using mathematical statistics.

In the experiment, a sample taken by the drill is used. In our case, soil samples with a volume of about 500 cm^3 were chosen, and the pressure was measured with an exemplary manometer with a fission value of 0.05 atm., which made it possible to measure porosity for the same sample at a fivefold replicate with an average relative error of about 2–6 %.

This method of measuring the porosity and density of the solid phase of the soil can be applied both in field and laboratory conditions.

The porosity of the sample is determined as follows:

1. Place the sample with the cassette in the vessel 2, seal, inject air and stand for 5 minutes, to partially eliminate the liquid "cuffs" that lock the air in the dead-end pores.
2. Transfer the sample to the vessel 1 and seal.
3. Inject air into the vessel 2. Allow to stand for about 2 minutes.
4. Connect with the taps vessels 1 and 2, noting the value of the pressure p_{ad} until the moment of connection and the pressure p_{est} after calming the oscillations of the manometer needle.
5. Calculate the pore volume.
6. Open chamber 1, replace the soil sample with a non-porous body and repeat steps 3, 4, and 5 to

determine the parasitic volume Π'.

According to the volumetric moisture determination w, the total porosity is calculated with the formula:

$$\Pi_0 = \Pi - \Pi' + w. \quad (11)$$

The porosities of the natural state characteristic of the most widespread soils in the Chuvash Republic are given in table 1.

Table 1. Soil porosity in the Chuvash Republic

Soil	Porosity Π_0, m^3/m^3		
	The adiabatic expansion method	Mass accounting method	Calculation of the density of the solid phase
Sod-podzolic light loam	0,48 ± 0,04	0,45 ± 0,02	0,44 ± 0,01
Light gray forest	0,38 ± 0,03	0,36 ± 0,01	0,38 ± 0,01
Dark gray forest	0,47 ± 0,04	0,47 ± 0,02	0,49 ± 0,01
Leached chernozem	0,55 ± 0,04	0,57 ± 0,02	0,58 ± 0,01

Statistical processing of experimental results shows that with a confidence probability of more than 95 %, none of the tested methods is predominant in terms of porosity estimation, but the method based on taking into account the air mass allows us to significantly reduce the relative measurement error.

Several methods for determining porosity using this device have been developed. These are methods based on the isothermal expansion of pore air to vacuum and on adiabatic expansion of pore air to vacuum. However, the air cools with expansion, and the adiabatic process can be considered only approximately. In addition, we have developed a method for calculating the porosity, based on the expansion of pore air to an area of low pressure and taking into account the conservation of the pore air mass. Therefore, the recently received wide distribution and availability of pressure superchargers (for example, an automobile compressor) led to the development of a method for calculating porosity in which soil air is brought into contact with air at higher pressure. Since the main effect depends on the absolute value of the pressure difference, the air injection approach we propose is preferable to the approach in which air expands. When the maximum pressure difference is discharged, the atmospheric pressure is the value, and when injected, this value depends on the compressor's performance characteristics and can exceed the atmospheric pressure several times. The relative error of the measurements decreases by approximately the same factor. The manometers of the compressors are equipped with a well-distinguishable and easy-to-use scale and have high accuracy and minimum error.

4. Conclusions

Anthropogenic compaction on fields allocated for agricultural crops, mostly appears after the passage of machine and tractor units. On the cultivated lands, the emergence of a compacted layer of soil often takes place, which has a significant effect on the growth and development of plants. The study of the dynamics of deformations accumulation in the soil is conveniently investigated using maps with isolines with the soil porosity values. They make it possible to obtain quantitative, easily observable information in general, which was used for economically justified planning of agricultural activities. Therefore, the developed and used express method for determining porosity is described in the paper. Due to the speed and simplicity of the measurements, a lot of data was obtained in a short time, which is extremely important for short-term agricultural field work.

References

[1] Shein E V, Milanovsky E Yu, Khaidapova D D, Dembovetsky A V and Tugai Z N 2014 *Bulletin of the Altai State Agrarian University* **5** 44–8

[2] Alekseev V V 2015 *Hydrophysics of soils in agricultural mechanics* (Cheboksary: New time) 120
[3] Maksimov I I and Alekseev V V 2017 *Hydrophysics of soils in land reclamation* (Cheboksary: New time) 280
[4] Maksimov I I and Maksimov V I 2006 *Energy concept of erosion resistance of anthropogenic agrolandscapes* (Cheboksary: Chuvash State Agricultural Academy Press) 304
[5] Bruce B V and Meunier A 2008 *The Origin of Clay Minerals in Soils and Weathered Rocks* (Springer) 426
[6] Glinski J, Horabik J and Lipiec J 2011 *Encyclopedia of Agrophysics* (Springer) 1075

IOP Conf. Series: Materials Science and Engineering **387** (2018) 012002 doi:10.1088/1757-899X/387/1/012002

Membrane-capacitive transducer with compensation for the primary standard of pressure

I V Andronov[1], V N Gorobey[2], E K Izrailov[2], R E Kuvandykov[2], A A Lobashev[2] and A A Chernyshenko[2]

[1] St.-Petersburg State University, Ulianovskaya 1/1, Petrodvorec, 198504, Russia
[2] D.I. Mendeleev Research Institute for Metrology, Moskovsky 19, 190005, St.Petersburg, Russia

E-mail: v.n.gorobey@vniim.ru, i.andronov@spbu.ru, ruskuw@mail.ru, lobashev@vniim.ru

Abstract. The paper discusses improvements for the primary vacuum-meter transducer for measuring absolute pressure in a wide range from 0.1 to 10 Pa.

1. Introduction

The measurements of low absolute pressures (vacuum) are the basic measurements in a wide area of vacuum technologies. The value of these measurements is high for all branches of industry, related to the use of vacuum in technological processes in view of increasing demands of quality of products. Particularly significant are the vacuum measurements in atomic, aerospace and electronic industry, in metallurgy and other high-technology industries. Due to this, a number of instrumentations for low absolute pressure measurements is developing. So in recent years, a number of different types of precision instruments of a new generation for measuring low absolute pressures, both domestic and foreign, have appeared on in Russian Federation. The metrological support of these instruments requires the improvement of the reference base and the primary standard itself. Therefore, its modernization is an urgent task.

In the D I Mendeleyev FSUE VNIIM, the unit of pressure in this range is carried out on the standard GET 49-2016 [1], which is improved and recertified in 2016. The standard of pressure GET 49-2016 is based on a membrane-capacitive transducer with compensation. The use of a membrane-capacitive transducer with compensation (MEPK) as a primary pressure standard was proposed in the thesis of Ryzhov V A in VNIIM in 1964 [2]. He proposed MEPK with the membrane and the electrode of equal diameters and numerous experimental studies of the prototype have been carried out.

2. Mathematical model of MEPK

The operation of the transducer MEPK is based on the effect of variation of the capacity of the capacitor formed with the membrane and the electrode (see figure 1) under the action on the membrane of the measured pressure P and the electric force F_e caused by applying the compensating voltage U to the electrode. The action of this electric force can be considered as some "electric" pressure P_e . The electric field strength E can be approximated by $U/_d$, where d is the distance between the membrane and the electrode. Then P_e is given by the formula

$$P_e = K_e\,U^2\,, \qquad K_e = \frac{\varepsilon_0}{2\,d^2}\,, \qquad (1)$$

where ε_0 is the dielectric constant of vacuum.

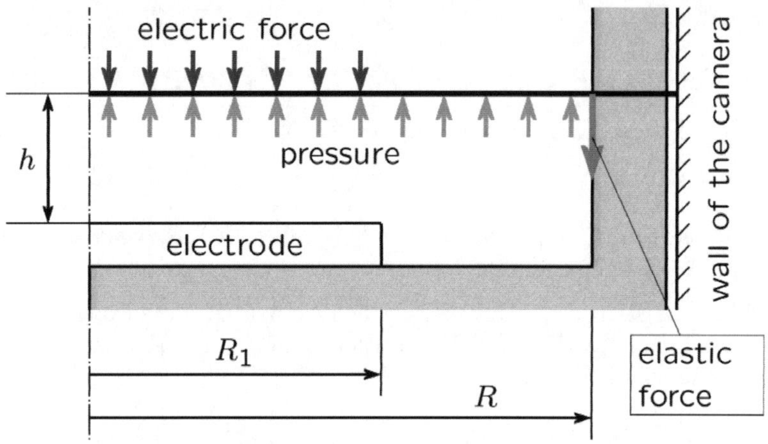

Figure 1. Construction of MEPK and the forces.

The membrane is bended under the action of pressures P and P_e. Let $z(r)$ be its transverse displacement, which we assume small compared to distance d. Below the dimensionless variable $w = {}^{z}/_{d}$ is used. The membrane of MEPK is supposed to have some initial tension displacements are described by the mathematical membrane equation having some initial tension T and its displacements are subject to the membrane equation [3]:

$$\Delta\, w(r,\varphi) = -\frac{q}{Td}\,, \qquad (2)$$

where q is the total load per unit square of the membrane, and the boundary conditions $w(R,\varphi) = 0$ describing the fastening of the membrane on its boundary of radius R. The load q is caused by pressure P and attractive electric force F_e, represented by pressure P_e. Dependence on the angle φ is excluded in view of cylindrical symmetry of the problem and equation (2) is reduced to an ordinary differential equation. As the first order approximation let the force applied to the membrane be piece-wise constant and $F_e = 0$ beyond the electrode, at $r > R_e$. Then equation (2) can be rewritten as two equations:

$$w''(r) + \frac{1}{r}w'(r) = -\frac{P - P_e}{Td}\,, \qquad r < R_e$$

$$(3)$$

$$w''(r) + \frac{1}{r}w'(r) = -\frac{P}{Td}\,, \qquad r > R_e$$

which should be supplied with the conditions of continuity $w(r)$ and $w'(r)$ at $r = R_e$.

When the membrane is bent, the distance between the electrode and the membrane changes. Accordingly, the electric field strength also changes. The operation of the MEPK is based on the fact that the feedback system maintains constant capacity of the capacitor C formed by the membrane and the electrode to which the compensating voltage U is applied. The formula for the capacitance is obtained by summing the elementary capacitances and has the form:

$$C = \frac{\varepsilon_0}{d}2\pi\int_0^{R_e}\frac{r\,dr}{1 + w(r)}\,. \qquad (4)$$

Thus the operation of MEPK is described by the system of equations in which the membrane equation (3) is complemented with the requirement of the capacitance (4) constancy, i.e. $C[w(r)] = const$. Solving this system of equations [4] yields the relation, which expresses pressure P via the voltage U of the compensating electric field:

$$P = K(U^2 - U_0^2)\,, \qquad K = K_e K_1\,,$$
$$K_1 = \frac{4\ln\left({}^{R}/_{R_e}\right) + 1}{2\left({}^{R^2}/_{R_e^2}\right) - 1}\,. \qquad (5)$$

Besides the force of pressure F and the force of electric field F_e the developed mathematical theory of MEPK [4] takes into account the ``elastic'' force F_m that acts on the edge of the membrane due its fastening to the wall of the camera. Thus the force applied to the edge of the membrane is added to the electric force to provide the balance of forces acting on the membrane: $F + F_e + F_m = 0$.

3. The comparison of MEPK and baratron

In the FSUE D I Mendeleev VNIIM, in 2016, an experimental work was performed to comparing the MEPK readings with the baratron readings, the calibration of which was carried out at the manufacturer. Both the MEPK and the baratron simultaneously measured the pressure in a volume filled with air at low pressure.

The membrane and the electrode radii are $R_e = 20.73$ mm, R = 23.02 mm and the distance between the electrode and the membrane $d = 0.097$ mm are conditioned by the requirements of the construction of the of the transducer MEPK, constructed in 1980. The computed transformation coefficient K (5) for these parameters of MEPK is equal to

$$K = 4.553 \cdot 10^{-4}\ [Pa/V^2] \qquad \text{while} \qquad K_1 = 0.9677\,.$$

The relative deviation w of the pressure readings P_{MEPK} of MEPK with respect to the pressure P_b measured by baratron was determined in the three series of experiments:

$$w = \frac{P_{MEPK} - P_b}{P_{MEPK}} 100\,. \qquad (6)$$

The results are shown in figure 2.

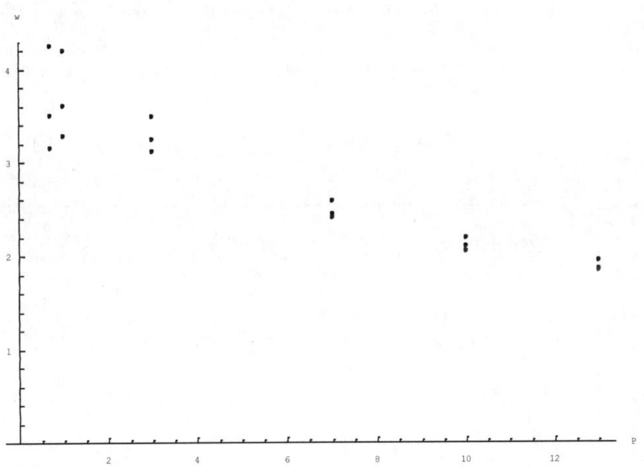

Figure 2. Relative deviation w in percent (6) of the pressure readings P_{MEPK} of MEPK transducer and pressure readings $P_б$ of the baratron dependently on the pressure in Pascals.

4. Results and discussion

Figure 2 shows the difference of readings of the two devices up to 4 % at low pressures which decreases at higher pressures, which is caused by temperature transpiration. The effect of temperature transpiration is observed in the case of temperature difference in different parts of the vacuum system and can be described by the formula [5]:

$$\frac{P_{MEPK}}{P_b} = \left(\frac{T_{MEPK}}{T_b}\right)^{\frac{2L}{2\,(D+2L)}}\,, \qquad (7)$$

where T_{MEPK} is the temperature readings of MEPK, equal to 20 °C, T_b is the temperature of baratron which measurement camera is heated to 45 °C, D is the effective diameter of the connecting pipe

$D = 5$ mm, L is the mean free path of gas molecules at mean pressure. For the air at the temperature of $T = 273$ K it is $L_1 = 6.7 \cdot 10^{-3}$ m at the pressure of 1 Pa and at a different pressure P it is: $L = {}^{L_1}/_{P}$.

Theoretically computed curve of temperature transpiration practically coincides in its shape with the experimental data, which is shown in figure 3.

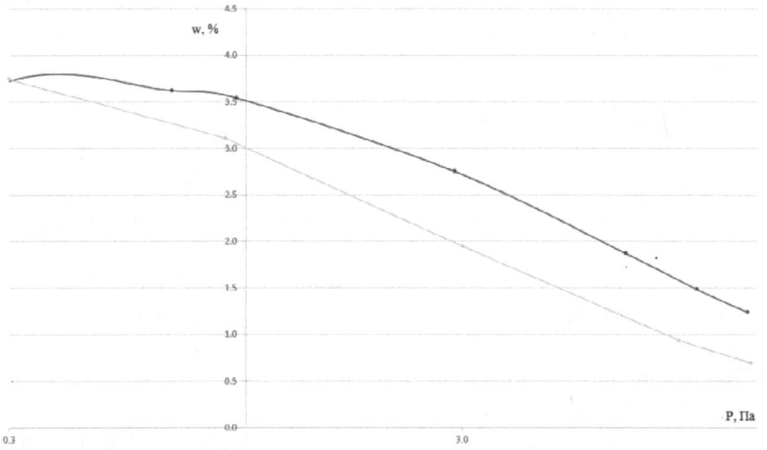

Figure 3. Comparison of the theoretical and experimental transpiration curves.

From the obtained experimental results it can be concluded that the accuracy of pressure measurements is on the order of 1 %. This estimate is based bath on the magnitude of dispersion of the experimental data and on the magnitude of the deviation of the theoretical and experimental transpiration curves.

It is possible to consider the model of MEPK whose membrane has both the tension and the stiffness [4]. In this case coefficient K_1 does not depend on the mechanical characteristics of the membrane only for equal radii of the membrane R and the electrode R_e and is exactly equal to unity (see figure 4).

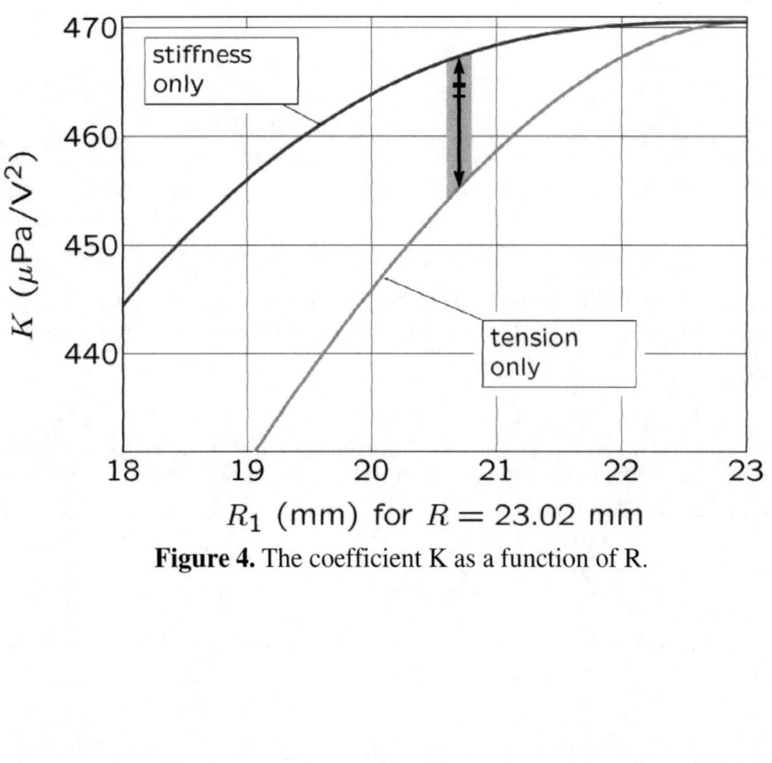

Figure 4. The coefficient K as a function of R.

IOP Conf. Series: Materials Science and Engineering **387** (2018) 012002 doi:10.1088/1757-899X/387/1/012002

To determine the parameters of MEPK it is necessary to perform only geometrical and electric measurements which can be presently performed with very high accuracy. Respectively, the standard of pressure is reduced to only length measurement and voltage measurement. This is especially important in the context of the transition to a new SI system [6], based on quantum effects, and in which electrical quantities become in some sense more primary.

Presently we work on further improvements of GET 49-2016. After this modernization the primary standard of pressure based on MEPK transducer will allow the unit of pressure to be reproduced with the accuracy of about 0.35 %.

References

[1] Gorobey V N 2015 *State primary special reference pressure unit for the absolute pressure range of GET 49-80. Metrological encyclopedia* **1** (St.Petersburg, Liki Rossii publ.) 319–22

[2] Ryzhov V A 1964 *Study of the model of absolute manometer with electrostatic compensation PhD* (Leningrad)

[3] Smirnov V I 1952 *Course of higher mathematics* **2** (Moscow) 627

[4] Andronov I V 2016 *Calculation of the equilibrium state of the membrane of the MEPK transducer of standard GET 49-80 under the influence of electric force and pressure. Reports under the contract of work № 26-T*

[5] Rozanov L N 2012 *Vacuum technological equipment* (St.Petersburg, Polytechnic Univ. publ.) 434

[6] *On the revision of the International System of Units (SI) Draft Resolution A – 26th meeting of the General Conference on Weights and Measures (13-16 November 2018)* 1–3

Temperature characteristics of the photosensitive structures metal-AlGaN developed by vacuum thermal deposition

A A Anikina, S A Tarasov, I A Lamkin and G I Kontsevik

Saint Petersburg Electrotechnical University “LETI”, 197376, Saint Petersburg, Russia

E-mail: ialamkin@mail.ru

Abstract. Broadband ultraviolet photodiodes are developed on the basis of Au-AlGaN Schottky barrier and also the ultraselection ultraviolet photodiodes with a photosensitivity range width on a half-height of 5 nanometers. We studied temperature effect on spectral characteristics of photosensitive structures of Au-AlGaN.

1. Introduction

At present photodiodes are widely used in various fields of science and technology. The use of wide-area semiconductor materials allows developing photodiodes for the ultraviolet range of a spectrum. The most preferable way during the development of photodiodes of an ultraviolet spectral range is solid solutions of AlGaN [1]. The increase in a share of AlN in a solid solution of AlGaN displaces long-wave limit of photosensitivity to the area of short wavelengths [2, 3]. One of the most promising avenues is a development of the selective ultraviolet photodiodes [4]. On the basis of a Schottky barrier of Ag-AlGaN possible to develop the high-performance selective photodiode on a wavelength of 322 nm [5, 6]. In addition, possible to develop the selective photodiodes with the inverse illumination at other wavelengths [7]. The use of the effect of tunnel transition of charge carriers and the filtering semiconductor layers in structure allows implementing the selective photodiode with a photosensitivity range width on a half-height (FWHM) of 5–6 nm [8]. At such small FWHM change of temperature of the photodiode can significantly change photosensitivity range. Therefore the research of temperature characteristics of such ultraselective photodiodes is important.

2. Samples and technique of experiments

Semiconductor structures of AlGaN by method chloride-hydride epitaxy on substrates of sapphire [9]. The ohmic and straightening contacts were put with vacuum thermal deposition. Ohmic Ti/Al contact was double-layer. Ti was an underlayer for Al. Thickness of a layer of Ti was 15 nanometers. Thickness of a layer of Al was 35 nanometers. Because of an interaction of Al and Ti with refractory metals they were sprayed from tungsten boats. During dusting of metals the structure was warmed up to the temperature of 300 °C. Annealing was made for decrease of resistance of ohmic contacts that was described [10] earlier. Au contact forming a Schottky barrier with AlGaN was developed 20 nanometers thick. Au was sprayed from molybdenum boats. During dusting of Au the structure was warmed up to the temperature of 120 °C. The general appearance of structure is shown in the figure 1. The research of ranges of a photosensitivity was conducted in two ways: when irradiating through semidiaphanous contact of Au and when irradiating from sapphire.

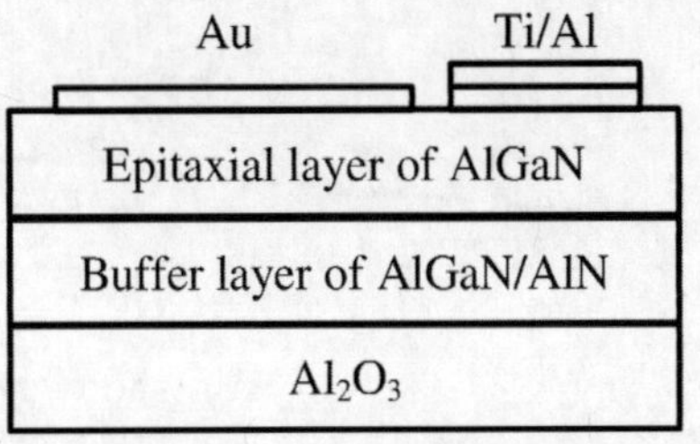

Figure 1. Sample structure of AlGaN.

Monochromator was used for measurement ranges of the diffraction photosensitivity. The xenon lamp was used as a source of radiation. The samples were established on an express holder with a thermoelement. The thermal element allowed to change samples temperature with an accuracy of 1 %.

3. The experiment results and discussion

On figure 2 shown photosensitivity ranges at the frontal flare of a sample from contact Au for different temperatures. At the frontal flare of a sample in a short-wave part of a range of a photosensitivity interband transitions are shown, a decrease of energy of photons the photosensitivity determined by an emission of electrons from contact Au is observed. The photosensitivity arises at the energy of incident photons larger, than barrier height of metal-semiconductor. The research of ranges of sensitivity at different temperatures showed that the different dependence of short-wave and long-wave edge of a photosensitivity on a temperature of a structure is observed. It is apparent that temperature effect on the position of the long-wave edge of a photosensitivity is bound to change of height of a barrier metal-semiconductor from temperature. At the same time, first of all, the short-wave edge is defined by the surface states on the interface metal-semiconductor.

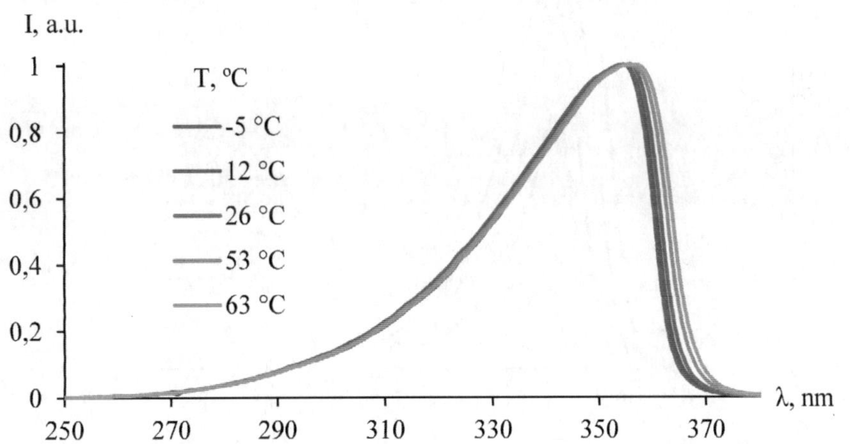

Figure 2. Photosensitivity range with irradiating from contact Au for different temperatures.

The dependence of provision of a wavelength on the temperature at the level of a half of intensity of a photosensitivity for the left-hand and right borders of a range is represented in figure 3. Obviously, temperature more affects on a position of the long-wave edge of a photosensitivity, than short-wave. The range of a photosensitivity of structure moves to the long-wave area and broadens. Photosensitivity range width on a half-height increases from 31 nanometers at a temperature of –50 °C to 38 nanometers at a temperature of 150 °C. Such change of a range of a photosensitivity though has fundamental character, but it is necessary to recognize its influence on performance characteristics of broadband photodiodes slight.

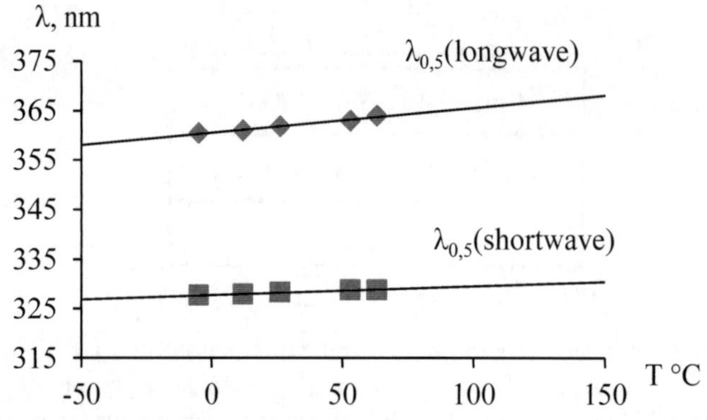

Figure 3. The dependence of the left-hand and right limits of photosensitivity on temperature of broadband photodiodes.

Temperature effect on ranges of a photosensitivity of the selective photodiode is more appealing. In the figure 4 are presented photosensitivity ranges at an illumination of an exemplar from the transparent sapphire substrate for different temperatures. In this case, the photosensitivity is caused only by an emission of electrons from metal. The position of a short-wave edge of a range is defined by the size of a width of the forbidden region of the semiconductor acting as an optical filter in structure. The position of a long-wave edge of a range is defined by a value of the height of a barrier metal-semiconductor.

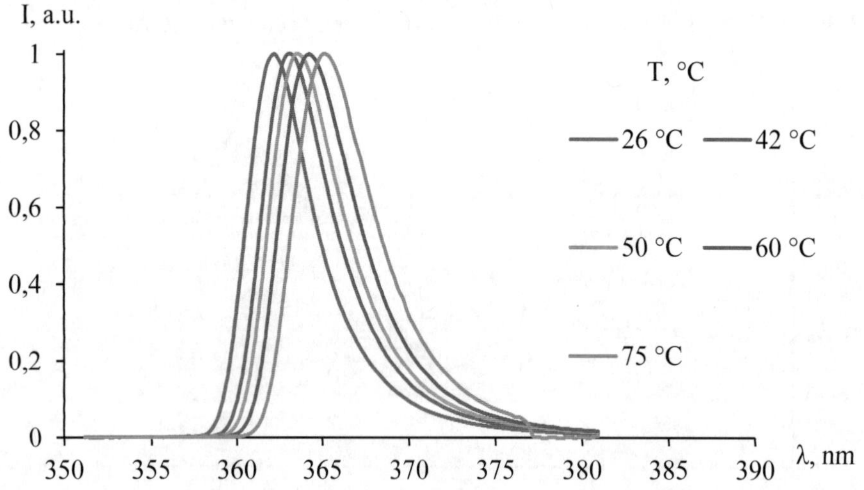

Figure 4. A photosensitivity range with irradiating from a sapphire substrate for different temperatures.

Obviously, a change of temperature can significantly shift a range of a photosensitivity of the selective photodiode. Dependence of provision of a wavelength on a temperature in a photosensitivity maximum, and at the level of 0,1 from a photosensitivity maximum for a long-wave and short-wave spectral range it is shown in figure 5. It was established that a change of temperature can shift a photosensitivity maximum from 356 nanometers at –50 °C on a wavelength of 370 nanometers at a temperature of 150 °C.

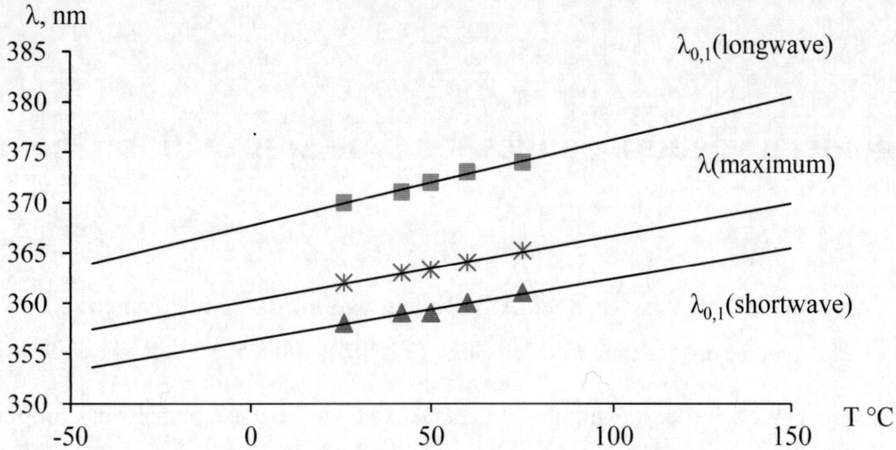

Figure 5. The dependence of the left-hand and right limits of photosensitivity on temperature with irradiating from a sapphire substrate for different temperatures.

Such change of provision of a maximum of a photosensitivity with FWHM of 5 nm is important for the selective photodetector since the spectral area of the photoresponse significantly changes. For example, in our case, the range of a photosensitivity can be both in the visible spectral region, and in ultraviolet.

4. Conclusion

Broadband ultraviolet photodiodes are developed on the basis of Au-AlGaN Schottky barrier with a long-wave limit of photosensitivity about 365 nanometers were created. The selective photodiodes with the inverse flare with FWHM of 5 nanometers with a photosensitivity maximum wavelength of 362 nanometers at a temperature of 26 °C.

Temperature effect on ranges of a photosensitivity of the developed broadband and selective photodiodes is investigated. It is established that for the broadband photodiode change of temperature has a slight influence. The temperature of the selective photodiode is important as significant changes in an area of its photosensitivity.

References

[1] Hahna L, Fuchsb F, Kirste L, Driad R, Rutz F et al. 2018 *Applied Physics Letters* **112** 151102
[2] Lamkin I and Tarasov S 2013 *Journal of Physics: Conference Series* **461** 012025
[3] Liwen Sang, Meiyong Liao and Masatomo Sumiya 2013 *Sensors* **13** 10482–518
[4] Chang-Ju Lee, Chul-Ho Won, Jung-Hee Lee et al. 2017 *Sensors* **17** 1684
[5] Tarasov S A, Lamkin I A, Mikhailov I I, Evseenkov A S and Solomonov A V 2017 *Journal of Communications Technology and Electronics* **62** 1074–7
[6] Lamkin I A, Tarasov S A, Solomonov A V, Andreev M Y and Kurin S Yu 2015 *Journal of Physics: Conference Series* **661** 012040
[7] Averin S V, Kuznetsov P I, Zhitov V A, Alkeev N V, Kotov V M, Dorofeev A A and Gladysheva N B 2011 *Technical Physics* **56** 295–7
[8] Altimime S M, Evseenkov A S, Mikhailov I I, Tarasov S A, Lamkin I A, Andreev M Y, Solomonov A V and Kurin S Yu 2016 *Journal of Physics: Conference Series* **769** 012054
[9] Lamkin I A, Tarasov S A, Petrov A A, Menkovich E A, Solomonov A V and Kurin S Yu 2014 *Journal of Physics: Conference Series* **572** 012063
[10] Lamkin I A, Tarasov S A, Petrov A A, Menkovich E A, Solomonov A V and Kurin S Yu 2014 *Journal of Physics: Conference Series* **572** 012062

Tests of modern vacuum gauges for type approval for the last 10 years

A I Antsukova, V N Gorobei, A A Pimenova and A A Chernyshenko

D. I. Mendeleyev Institute for Metrology (VNIIM), 190005, St. Petersburg, Russia

E-mail: V.N.Gorobey@vniim.ru, a.i.antsukova@vniim.ru, a.a.pimenova@vniim.ru

Abstract. The article presents data on modern vacuum gauges designed to measure absolute pressure. The gauges have been tested for type approval after 2008. The principle of operation and the main metrological characteristics of the following types of vacuum meters are analyzed: membrane-capacitive, capacitive, wide-range, inverse-magnetron, deformation, electronic with integrated display, thermal, ionization and magnetic discharge. Information on these gauges is included in the Federal Information Fund for ensuring the uniformity of measurements and possibility to be applied for state regulation of measurements.

1. Introduction

According to the Federal Law No. 102-FZ " On ensuring the uniformity of measurements" (Chapter 3, Article 11) the approval of the type of Measuring Instruments (MI) is one of the forms of state regulation for ensuring the measurements uniformity.

The decision about approval the MI type is made by the Federal Authorization organization (Rosstandart) on the basis of the positive results of the MI tests for the purpose of the type approval (Chapter 3, Article 12, paragraph 2.). The information about approved MI is entered in the Federal Information Fund for Ensuring Measurement Uniform.

The approval of the type of MI is certified by a certificate of type approval issued by the Federal Executive Authority. The validity of the certificate is usually 5 years.

The data on modern vacuum gauges from 2008 are shown below. The gauges have been tested for approving the type. The information about these gauges is included in Federal Information Fund and can be applied for state regulation of measurements.

2. Thermal vacuum gauges

The principle of operation of such vacuum gauges is based on the dependence of the thermal conductivity of the discharged gas on pressure [1].

During the pumping the gas in working volume of the measuring sensor of the vacuum gauge is discharged, which leads to the change of thermal conductivity of the gas and, as a consequence, to the change in the temperature of the thin heated wire (the filament of the sensor). The wire is the shoulder of the measuring bridge circuit. The change in the resistance of this arm, caused by the change in its temperature, leads to unbalance of the bridge. An electrical voltage signal which is proportional to the pressure arises. The metrological characteristics of thermal vacuum gauges are given in table 1.

3. Ionization vacuum gauges

The principle of operation of vacuum gauges is based on the dependence of the current of positive ions, formed as a result of ionization of rarefied gas molecules on the measured pressure. Metrological characteristics of ionization vacuum gauges are indicated in table 2.

Table 1. Metrological characteristics of thermal vacuum gauges.

Type of gauges	Range of reading, Pa	Measurement range, Pa	Relative error, %	Manufacturer	Registration number №
VTT-18-2	$1{,}33\cdot10^{-2}...66$	$1\cdot10^{-1}...66$	± 25	"S.A. Vekshinskii Institute of vacuum technique", Russia	37033-08
VTS-18-2 (resistance)	$6{,}6\cdot10^{-2}...1\cdot10^{5}$	$1{,}3...3{,}9\cdot10^{3}$	± (60..40)	"S.A. Vekshinskii Institute of vacuum technique", Russia	41803-09
VD		$5\cdot10^{-2}...10^{4}$	± (100..15)	«Thyracont Vacuum Instruments GmbH», Germany	43629-10
VSP	$10^{-2}...10^{5}$	$10^{-2}...10^{4}$	± (100..15)	«Thyracont Vacuum Instruments GmbH», Germany	44060-10
APG (resistance)	$10^{-2}...10^{5}$	$10^{-2}...10^{3}$	± 20	«Edwards Limited», UK	44391-10
CVG101, CVM201, CVM211	$10^{-2}...1{,}33\cdot10^{5}$	$10^{-2}...10^{3}$	± (35..15)	«InstruTech, Inc.», USA	51328-12
CVC 3000 VSP 3000, DCP 3000 VSP 3000	-	$10^{-1}...10^{5}$	± (30...15)	«VACUUBRAND+ CO KG», Germany	51918-12
Televac 2A, 4A	$1{,}3\cdot10^{-1}...1{,}3\cdot10^{5}$	$5\cdot10^{-1}...10^{3}$	± 30	«The Fredericks Company», USA	56243-14
TPR, TPG 201, PPT 200	$10^{-2}...10^{5}$	$10^{-2}...10^{3}$	± (50...30)	«Pfeiffer Vacuum GmbH», Germany	63031-16
ERSTEVAK	-	$5\cdot10^{-8}...10^{5}$	± (40...10)	«Thyracont Vacuum Instruments GmbH», Germany	64910-16
THERMOVAC	$5\cdot10^{-3}...1{,}5\cdot10^{5}$	$5\cdot10^{-2}...1{,}5\cdot10^{5}$	± (50...10)	«Leybold GmbH», Germany	68784-17

4. Magnetic electro-discharge vacuum gauges

The principle of operation of magnetic vacuum gauges is based on the dependence of the measured pressure on the ion current of the gas discharge arising in the discharge vacuum gap between cathode and anode of the pressure sensor as a result of the interaction of intersecting electric and magnetic fields. The voltage of several kilovolts is supplied to the sensor anode from the controller, while the cathode is at zero potential. The electrons appear between the electrodes under the action of electric and magnetic fields, whose voltage vectors are mutually perpendicular, begin to move. The ionization of gas molecules occurs during this motion as a result of collisions of electrons with gas molecules. Positively charged ions are directed to the cathode. The ionization current is proportional to the measured pressure. The metrological characteristics of magnetic electro-discharge vacuum gauges are given in table 3.

5. Diaphragm-capacitive vacuum gauges

The principle of operation of this vacuum gauge is based on the use of the relationship between the measured pressure and the elastic deformation of the sensor element. Flat practically gas-tight membrane is applied as the sensitive element. One side of the membrane is the evacuated volume pumped to residual pressure which does not exceed 10^{-7} Pa. the pressure is maintained for a long time by a chemical getter (getter). The measured pressure is applied to the other side of the membrane causing deformation of the membrane, which leads to the change in the electrical capacitance between the membrane and the electrodes located in the vacuum volume. The capacity is included in one of the shoulders of the bridge circuit. This creates an imbalance of the bridge. The electrical signal appears proportional to the measured pressure. The metrological characteristics of membrane-capacitive vacuum gauges are given in table 4.

Table 2. Metrological characteristics of ionization vacuum gauges

Type of gauges	Range of reading, Pa	Measurement range, Pa	Relative error, %	Manufacturer	Registration number №
VI-18		$1{,}3\cdot10^{-5}$ ...$8\cdot10^{-1}$	±35	"S.A. Vekshinskii Institute of vacuum technique", Russia	37034-08
AIGX –		$6{,}6\cdot10^{-8}$... 6,6	± 15	«Edwards Limited», UK	44388-10
IGM400, IGM401		$1{,}3\cdot10^{-7}$... 6,7	± 25	«InstruTech, Inc.», USA	51327-12
CCM		$1{,}3\cdot10^{-7}$... 1,3	± 30	«InstruTech, Inc.», USA	66042-16
IONIVAC	$5\cdot10^{-8}$... $1\cdot10^{5}$	$5\cdot10^{-6}$... $1\cdot10^{5}$	± (25...15)	«Leybold Germany	67723-17

Table 3. Metrological characteristics of magnetic electro-discharge vacuum gauges.

Type of gauges	Range of reading, Pa	Measurement range, Pa	Relative error, %	Manufacturer	Registration number №
Televac 7B2,7E,7F, 7FC,7FCS	от 1,3 10–9 до 1,3	от 1,3 10–7 до 1,3 10–1	±60	«The Fredericks Company», USA	56242-14

Table 4. Metrological characteristics of membrane-capacitive vacuum gauges.

Type of gauges	Range of reading, Pa	Measurement range, Pa	Relative error, %	Manufacturer	Registration number №
CMR, CCR	10^{-3}...$1{,}33\cdot10^{5}$	10^{-3} ... $1{,}33\cdot10^{5}$	± (30...2)	«Pfeiffer Vacuum GmbH», Germany	63021-16

6. Capacitive vacuum gauges

The principle of operation of capacitance vacuum gauges is based on the relationship between the measured pressure and the change in the electrical capacitance caused by the elastic deformation of the sensitive element. The alumina ceramic diaphragm is used as sensitive element of capacitive vacuum gauges. The metrological characteristics of capacitive vacuum gauges are given in table 5.

Table 5. Metrological characteristics of capacitive vacuum gauges.

Type of gauges	Range of reading, Pa	Measurement range, Pa	Relative error, %	Manufacturer	Registration number №
DVR 2, DVR3, CVC 3000, DCP 3000 VSK 3000, CVC 3000 VSK 3000	-	0,1-1080	±1	«VACUUBRAND +CO KG», Germany	51917-12

7. Wide-range vacuum gauges

Wide-range vacuum gauge is a combined device in which two sensors are combined in one canopy: Pirani thermal sensor and a magnetic electric discharge sensor (cold cathode) or Pirani thermal sensor and an ionization transducer (hot cathode). The metrological characteristics of the wide-range vacuum gauges are given in table 6.

Table 6. Metrological characteristics of wide-range vacuum gauges.

Type of gauges	Range of reading, Pa	Measurement range, Pa	Relative error, %	Manufacturer	Registration number №
Meradat -VIT	$1{,}33\cdot10^{-5}\ldots1\cdot10^{5}$	$1{,}33\cdot10^{-5}\ldots1\cdot10^{5}$	± (50…30)	JSC «Meradat», Russia	41616-09
IGM402	$1{,}3\cdot10^{-7}\ldots1{,}33\cdot10^{5}$	$1{,}3\cdot10^{-7}\ldots10^{3}$	± (25...15)	«InstruTech, Inc.», USA	51329-12
PKR, PBR, IMR, HPT MPT	$5\cdot10^{-8}\ldots10^{5}$	$10^{-7}\ldots10^{3}$	± (100…30)	«Pfeiffer Vacuum GmbH», Germany	62158-15

8. Deformation vacuum gauges

The principle of operation of these vacuum gauges is based on the use of the relationship between the measured pressure and elastic deformation of the piezoresistive sensing element. The metrological characteristics of deformation vacuum gauges are given in table 7.

Table 7. Metrological characteristics of deformation vacuum gauges.

Type of gauges	Range of reading, Pa	Measurement range, Pa	Relative error, %	Manufacturer	Registration number №
VDTO-3		$1{,}33\cdot10^{-3}\ldots1{,}06\cdot10^{5}$	± 10 (relative)	"Alex S&E" SRL, Moldova	40410-09
APR, CPT 200	$10\ldots5{,}5\cdot10^{6}$	$10\ldots5\cdot10^{6}$	± 2 (given)	«Pfeiffer Vacuum GmbH», Germany	63314-16

9. Electronic vacuum gauges with integrated display

Electronic vacuum gauges with integrated display are made in one housing, where are located two measuring sensors: the crystal quartz sensor and the double inverse magnetron sensor with cold cathode. Metrological characteristics of electronic gauges with an integrated display are given in table 8.

Table 8. Metrological characteristics of electronic gauges.

Type of gauges	Range of reading, Pa	Measurement range, Pa	Relative error, %	Manufacturer	Registration number №
ERSTEVAK		$5\cdot10^{-8}\ldots1\cdot10^{5}$	± (40...10)	"Thyracont Vacuum Instruments GmbH", Germany	64910-16
CC-10		$1{,}33\cdot10^{-7}\ldots10^{5}$	± (70...35)	«The Fredericks Company», USA	65023-16
MP2AR		$1{,}33\cdot10^{-1}\ldots2{,}67\cdot10^{3}$	± (100…20)	«The Fredericks Company», USA	65198-16
MP4AR,MX4A		$1{,}33\cdot10^{-1}\ldots10^{5}$	± (100...15)		
MP7ER		$1{,}33\cdot10^{-6}\ldots1$	± (50…35)	"Televac A Division of The Fredericks Company", USA	66274-16
MP7FR		$1{,}33\cdot10^{-8}\ldots1$	± (70…35)		

10. Inverse-magnetron vacuum gauges

The inverse-magnetron vacuum gauges are made in a single canopy, in which there are two sensors: the inverse magnetron and the Pirani sensor. The inverse-magnetron sensor is a discharge tube, inside

IOP Conf. Series: Materials Science and Engineering **387** (2018) 012004 doi:10.1088/1757-899X/387/1/012004

of which the cathode plate on the grounded potential and the anode with high positive charge are placed parallel to each other. Two permanent magnets are outside of the discharge tube. The metrological characteristics of the inverse magnetron vacuum gauge are given in table 9.

Table 9. Metrological characteristics of inverse-magnetron vacuum gauge.

Type of gauges	Range of reading, Pa	Measurement range, Pa	Relative error, %	Manufacturer	Registration number №
MPG400		$5 \cdot 10^{-9} \ldots 1 \cdot 10^{3}$	± 30	«Inficon», Lichtenstein	63642-16

11. Viscous vacuum gauges

The principle of operation of vacuum gauges is based on the dependence of the viscosity (internal friction) of the rarefied gas on the measured pressure. The metrological characteristics of viscous vacuum gauges are given in table 10.

Table 10. Metrological characteristics of viscous vacuum gauges.

Type of gauges	Range of reading, Pa	Measurement range, Pa	Relative error, %	Manufacturer	Registration number №
SRG		$5 \cdot 10^{-5} \ldots 1 \cdot 10^{2}$	± (10...1,3)	"MKS Instruments", Germany	40361-09

12. Conclusion

The mentioned vacuum gauges can be used for state regulation of measurements along with gauges that were manufactured before 2008 and had valid certificates of the type approval for that period. It should also be noted that in general since 2008 imported vacuum gauges have been tested unfortunately.

Now D. I. Mendeleev Institute of Metrology develops domestic modern vacuum gauges with a sensor based on MEMS technology.

References

[1] Rozanov L N 1990 *Vacuum Technique* (Moscow: Higher School) 320

Peculiarities of using nanomaterials based on SiO_2-SnO_2-In_2O_3 as sensitive elements of vacuum sensors

I A Averin[1], A A Karmanov[1], S E Igoshina[1], I A Pronin[1], N D Yakushova[1] and V A Moshnikov[1,2]

[1] Department of Nano- and Microelectronics, Penza State University, 440026, Penza, Russia
[2] Department of Micro- and Nanoelectronics, Saint-Petersburg Electrotechnical University, 197376, Saint-Petersburg, Russia

E-mail: nano-micro@mail.ru

Abstract. Experimental data on the use of nanomaterials based on SiO_2-SnO_2-In_2O_3 as sensitive elements of vacuum sensors are presented. It is shown that the sensory response of a three-component system under consideration is determined by the mass fraction of indium oxide. It was determined that depending on the content of In_2O_3 the nanomaterials under study can be characterized by a decrease and an increase in resistance when the pressure is reduced lower than atmospheric one. The mechanisms that potentially correspond to the observed phenomenon are considered.

Recently, multicomponent systems based on combinations of various wide band gap oxides, such as SnO_2, In_2O_3, ZnO, TiO_2, VO_2, etc., have been of considerable interest for researchers and developers in the field of nanoelectronics [1]. In general, there is a common tendency, according to which the transition from the use of individual oxides to the use of their compound compositions is fixed. It is due to the need for nanomaterials and devices based thereof, having a complex of properties, which is not possible to ensure with the use of the only oxide [2].

For example, for SnO_2 being widely used in the field of gas sensors, the selective detection of reducing gases or oxidizers, which ensure the preservation of a high sensory response, is still an unsolved problem. At the same time, the known approaches, based on variations of surface morphology and operating temperature of the gas sensitive layer, catalytic additions and creation of conditions for the spillover effect, have a significantly limited range of fundamental and applied applications [3]. In particular, a significant difficulty is the separation of the signal due to the presence of atmospheric humidity and the concentration variation of the controlled reducing gas (for example, ethanol, ammonia vapors, etc.) [4].

There are also significant difficulties in the field of photocatalytic applications that cannot be overcome with the use of individual oxides. For example, TiO_2 and ZnO, being promising materials for photocatalysts, are only effectively activated by UV radiation and have very weak activity in the visible range [5], having a significant fraction of the solar spectrum. In this case, the known ways for expanding the operating wavelength range, based, for example, on the formation of defects in the crystal lattice of titanium dioxide, are inevitably accompanied by the appearance of synergistically interrelated effects, for example, by the rapid degradation of photocatalytic activity.

Similar difficulties arise in a number of other fields, where the use of wide band gap semiconductor oxides is already classical. For example, the creation of high-quality transparent conducting coatings having a high transmittance and a low resistivity is a complex task using only a single-component system. However, it can be easily solved using oxide compositions, as it has been shown, for example, for ITO [6].

The use of multicomponent systems based on wide band gap oxides allows solving the indicated problems to some extent. For example, in the field of gas sensors, the use of two- and three-component oxide systems allows to significantly improve the selectivity, and to reduce the influence of air humidity [7]. In the field of photocatalytic applications, the use of oxide compositions provides not only an expansion of the operating spectral range, but also increases the activity in the field of UV radiation in a number of cases [8].

As for the issue on measuring the pressure below atmospheric, nanomaterials based on individual ZnO [9] and VO_2 oxides [10], a two-component system of SiO_2-SnO_2 [11] and a three-component system of SiO_2-SnO_2-ZnO [12] have proven to be useful. However, this field of research and development has a number of difficulties, and in general, the peculiarities of using nanomaterials based on wide band gap semiconductor oxides as sensitive elements of vacuum sensors have not been fully studied. In particular, the mechanism of the sensory response of such nanomaterials to variations of the vacuum level is not clear until now, wherein in the majority of works the dominating role of adsorption/desorption of oxygen processes is assumed [13].

The purpose of this work is to expand the spectrum of materials potentially suitable for measuring the pressure below atmospheric, and to study the three-component system SiO_2-SnO_2-In_2O_3 within the framework of this aspect. Synthesis of the analyzed nanomaterials was carried out in the framework of the sol-gel technology methods using the following precursors: tetraethoxysilane ($Si(OC_2H_5)_4$), tin (II) chloride dehydrate ($SnO_2 \cdot 2H_2O$), 4.5-aqueous indium nitrate ($In(NO_3)_3 \cdot 4,5\ H_2O$), wherein hydrochloric acid being used as a catalyst, and ethanol as a solvent. The mass fraction of tin dioxide for all experimental samples was 50 wt. %, and the content of indium oxide varied from 5 to 15 wt. %. Nanomaterials were formed on substrates made of monocrystalline oxidized silicon of 5x5 mm^2; annealing was carried out at a temperature of 600 °C in the air. Contact pads were made using silver conductive adhesive.

The choice of the three-component system SiO_2-SnO_2-In_2O_3 as an object for investigation is due to the presence of primary experimental data demonstrating the significant contribution of oxygen to the change of its resistance. Figure 1 shows the temperature dependence of the conductivity of the studied nanomaterials, obtained in dry air. The studied samples were placed in a muffle furnace together with silica gel to reduce the effect of water adsorption on their surface; the sample was subjected to temperature control for 10 minutes before each subsequent measurement.

The analysis of the dependence shown in figure 1 shows the presence of a maximum in the low-temperature region of 300–400 K, a deep minimum in the range of 550–650 K, and a sharp increase of conductivity in the high-temperature region of 700–900 K. Such temperature dependence of the conductivity is apparently determined by various forms of oxygen chemisorbed on the surface and in the volume of the three-component system SiO_2-SnO_2-In_2O_3. In the low-temperature region, oxygen has dominant influence in the form of O^2-, and in the high-temperature region in the form of O^-. Simultaneous existence of O^2- and O^- corresponds to the minimum of specific conductivity, the contribution from which is summed, and the nanomaterial is maximally depleted by the main charge carriers. The role of indium oxide in the three-component system under consideration is most likely reduced to an increase of the grouping stability of the Lewis type center (a metal ion with chemisorbed oxygen). The introduction of this oxide into a two-component system of SiO_2-SnO_2 in low concentrations leads to the formation of a solid solution, and the limiting value of the solubility essentially depends on the defectiveness of the initial tin dioxide. There is a segregation of In_2O_3 in the form of an island structure (figure 2) at its high concentration.

IOP Conf. Series: Materials Science and Engineering **387** (2018) 012005 doi:10.1088/1757-899X/387/1/012005

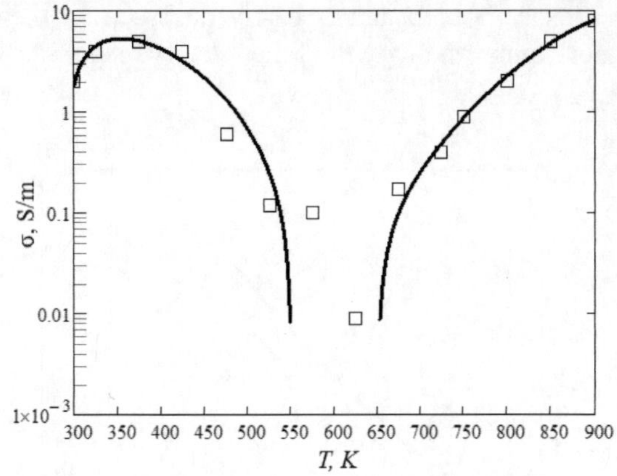

Figure 1. Temperature dependence of conductivity of nanomaterials based on SiO_2-SnO_2-In_2O_3 in dry air.

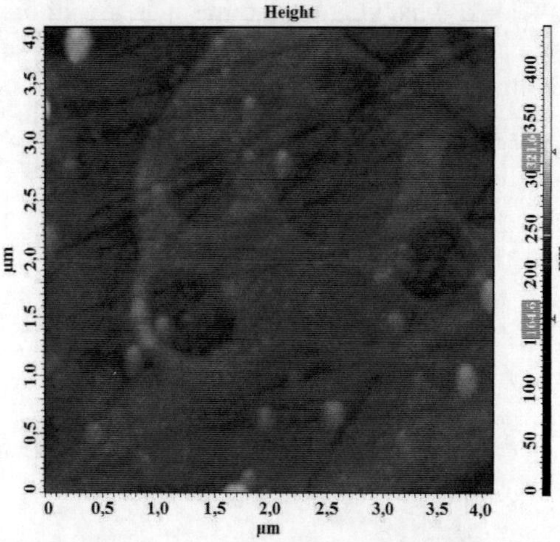

Figure 2. An AFM image of the surface morphology of nanomaterials based on SiO_2-SnO_2-In_2O_3 with a high concentration of indium oxide.

Such a character of the interaction of the investigated nanomaterials with atmospheric oxygen and the concentration features of indium oxide additions determine the sensory response of the system under consideration to a decrease of pressure below atmospheric. Figure 3 shows a relative change in resistance (R/R_0) of sensitive elements of vacuum sensors based on nanomaterials of SiO_2-SnO_2-In_2O_3 composition as a function of pressure. The measurement was carried out using a multimeter Mastech MS8229 when applying constant voltage in 1 B to the sample. The pressure inside the vacuum chamber was fixed by a calibrated gauge thermocouple head. An analysis of the obtained experimental data shows that, depending on In_2O_3 content, the decrease (curves 2 and 3), and the increase of resistance with decreasing pressure below atmospheric (curves 4 and 5) can be characteristic for the system under study. In this case, the mass fraction of indium oxide determines not only the form of the presented dependences, but also the value of the sensory response, which can be calculated from the formula: $S = |1 - R/R_0| \cdot 100\%$, where R_0 is the initial resistance at the pressure chosen as the reference point; R is the resistance of the sample for a given value of the measured pressure.

Figure 4 shows the dependence of the sensory response of the sensitive elements of the vacuum sensors based on the three-component oxide system SiO_2-SnO_2-In_2O_3 on the mass fraction of In_2O_3 (R_0 was estimated at a pressure of 0.014 mm Hg, and the initial samples were not subjected to additional heating). The analysis of the presented dependencies shows that the introduction of a small amount of indium oxide into a two-component system based on SiO_2-SnO_2 leads to a significant increase of the sensory response of the nanomaterials under study. At the same time, samples containing 15 wt. % of In_2O_3 have the maximum sensitivity to the change of pressure. A further increase in the mass fraction of indium oxide leads to a sharp decrease of the sensory response (curve 5 in figure 3). This circumstance has several possible explanations, however, the fundamental role, apparently, is assigned to the structural and morphological changes of the nanomaterials under consideration.

The sensory response of the three-component oxide system SiO_2-SnO_2-In_2O_3 at low In_2O_3 concentrations is explained by the dominant contribution of oxygen in the form of O^{2}-, the existence of which on highly defective surfaces is possible over a wide range of temperatures. Decrease of pressure below atmospheric leads to desorption of oxygen from the surface and from the volume of nanomaterials, as a result of which the electrons return to the conduction band, providing a decrease in resistance (curves 2 and 3 in figure 3). A large S value characteristic for the samples of SiO_2-SnO_2-In_2O_3 compared to the two-component system based on silicon dioxide-tin dioxide is due to the growth of the stability of the grouping type of the Lewis center. In^{3+} cations are stronger Lewis centers than

Sn^{4+} ions, which determine the increase in the concentration of all forms of the adsorbed oxygen. In this case, the low-temperature form of O^{2-} becomes loosely bound, and is easily desorbed as the pressure decreases below atmospheric pressure, which determines the growth of the sensory response.

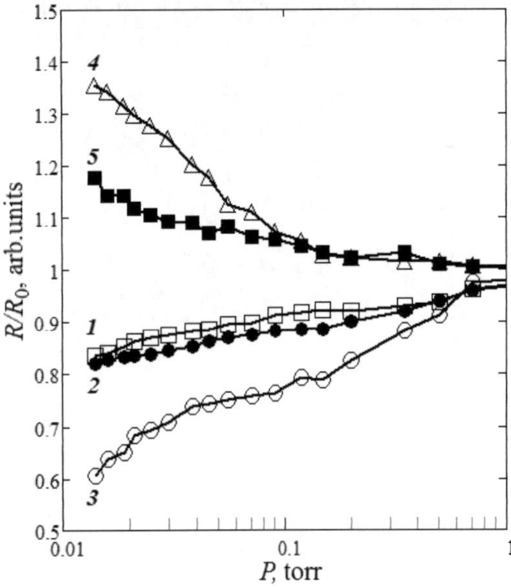

Figure 3. Relative change in resistance of sensitive elements of vacuum sensors based on three-component oxide system SiO_2-SnO_2-In_2O_3 with different mass fraction of indium oxide: 1 – the sample does not contain In_2O_3; 2 – 5 wt. %; 3 –10 wt. % 3; 4 – 15 wt. %, 5 – 20 wt. % of In_2O_3.

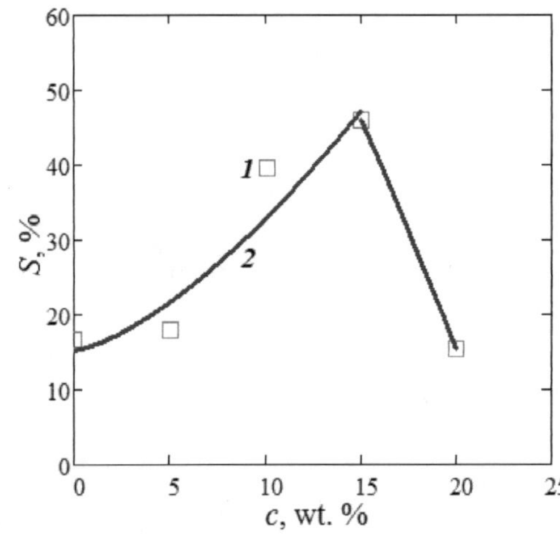

Figure 4. Dependence of sensory response of sensitive elements of vacuum sensors based on three-component oxide system SiO_2-SnO_2-In_2O_3 on the mass fraction of indium oxide: 1 – experimental data; 2 – approximation.

An atypical action of wide band gap semiconductor oxides is characteristic for high concentrations of indium oxide in a three-component system based on SiO_2-SnO_2-In_2O_3 consisting in the growth of resistance with decreasing ambient pressure. It cannot be explained explicitly from the position of adsorption/desorption of oxygen, and requires consideration of additional mechanisms responsible for the sensory response, for example, the interaction of nanomaterials with atmospheric moisture. An explanation of such mechanisms is the subject of a separate study. It can be assumed that the formation of small *p-n* transitions at the SnO_2/In_2O_3 boundary, which arise as a result of the segregation of indium oxide in the form of an island structure, will play a significant role in the observed phenomenon.

Thus, the peculiarities of using nanomaterials based on SiO_2-SnO_2-In_2O_3 as sensitive elements of vacuum sensors are considered. It is shown that depending on the mass fraction of indium oxide the nanomaterials under study can be characterized both by a decrease and an increase in resistance with decreasing pressure below atmospheric. Proceeding from the formation of a stable grouping of the Lewis type center, the growth of the sensory response of a three-component system with a low content of In_2O_3 is explained.

Acknowledgment

The work has been financially supported by the Ministry of Education and Science of the Russian Federation within the framework of the project part of the state assignment for Penza State University No. 16.897.2017/4.6, as well as the Scholarship of the President of the Russian Federation No. SP-3800.2018.1.

References

[1] Faber H, Das Sa, Lin Y-H, Pliatsikas N, Zhao K, Kehagias T, Dimitrakopulos G, Amassian A, Patsalas P A and Anthopoulos T D 2017 *Science Advances* **3** e1602640

[2] Wang L, Li J, Wang Y, Yu K, Tang X, Zhang Y, Wang S and Wei C 2016 *Scientific Reports* **6** 35079

[3] Korotcenkov G, Han S-D, Cho B K and Brinzari V 2009 *Critical Reviews in Solid State and Materials Sciences* **34** 1–17

[4] Dimitrov D T, Nikolaev N K, Papazova K I , Krasteva L K , Bojinova A S, Peshkova T V, Kaneva N V, Pronin I A, Averin I A, Yakushova N D, Karmanov A A, Georgieva A T and Moshnikov V A 2017 *Applied Surface Science* **392** 95–108

[5] Raza W, Faisa S M, Owais M, Bahnemannc D and Muneer M 2016 *RSC Adv.* **6** 78335-50

[6] Triambulo R E, Kim J, Na M, Chang H and Park J 2013 *Applied Physics Letters* **24** 241913

[7] Pavelko R G, Vasiliev A A, Llobet E, Vilanova X, Sevastyanov V G, Kuznetsov N T 2010 *Procedia Engineering* **5** 111–4

[8] Mohamed W S, Abu-Dief A M 2018 *Journal of Physics and Chemistry of Solids* **116** 375–85

[9] Wu L, Song F, Fang X., Guo Z-X and Liang S 2010 *Nanotechnology* **21** 475–502

[10] Bu Y, Zou J, Liu Y, Zhu Z, Deng W, Peng X and Tang B 2017 *Thin Solid Films* **638** 420–5

[11] Averin I A, Igoshina S E, Moshnikov V A, Karmanov A A, Pronin I A and Terukov E I 2015 *Technical Physics. The Russian Journal of Applied Physics* **6** 928–32

[12] Karmanov A A, Averin I A, Pronin I A, Yakushova N D, Igoshina S E, Moshnikov V A and Vishnevskaya G V 2017 *Journal of Physics: Conf. Series* **872** 012005

[13] Zheng X J, Cao X C, Sun J, Yuan B, Li Q H, Zhu Z and Zhang Y 2011 *Nanotechnology* **43** 435501

IOP Conf. Series: Materials Science and Engineering **387** (2018) 012006 doi:10.1088/1757-899X/387/1/012006

Estimation of the position error of the magnetorheological elastomers active vibration control platform for precision vacuum equipment

A M Bazinenkov, I V Makeev, A P Rotari and D A Ivanova

Bauman Moscow State Technical University (National research university), 105005, Moscow, Russia

E-mail: ambazinenkov@bmstu.ru

Abstract. Most of the modern vacuum nanotechnological and research equipment is associated with the vibration control systems. Recently, systems based on composite materials, such as magnetorheological elastomers, are becoming more widely used. Such materials make possible to combine the functions of active, semi-active and passive vibration control in one device. This paper presents a research on the position error of the vibration control platform based on magnetorheological elastomers. This error occurs due to the inevitable unevenness of the top panel of the platform caused by the special properties of the elastomeric actuator. It leads to deviation of the platform from the horizontal plane position. The research on the measurement error was made by the contour estimations method and the correlation coefficients of the actuators movements were used.

1. Introduction

For ultra-precision technological processes (nanolithography, scanning probe microscopy, precision metalworking, alignment of fiber optic systems) it is necessary to provide vibration control at disturbance frequencies in a wide frequency range from 0.5 to 200 Hz with vibration displacement amplitudes up to 200 μm. The most effective vibration protection method is vibration isolation which can be divided into passive, semi-active and active isolation.

Magnetorheological (MR) and electrorheological (ER) fluids and elastomers form a type of smart materials that can change its' properties under the influence of external magnetic and electric fields, respectively. These materials used in precise positioning and vibration control systems make it possible to decrease the positioning error and response time of such positioning systems and the vibration control transfer coefficient of vibration control systems [1, 3].

2. Research object, platform for active vibration control

The object of this research is the vibration control platform based on MR elastomers (figure 1). Magnetorheological elastomers (MRE) are composite materials, compounds based on a non-magnetic matrix of elastomer (mixture of rubbers, silicone) with soft magnetic fillers, particles of micron size., The particles tend to align themselves into a chain structure so the MRE changes its' elastic-viscosity-plasticity characteristics, when an external magnetic field is applied. The MRE deformation will depend on the magnitude of the external magnetic field. The nonmagnetic matrix returns the particles to their nominal position due to its' elastic properties, when the external magnetic field disappears.

IOP Conf. Series: Materials Science and Engineering **387** (2018) 012006 doi:10.1088/1757-899X/387/1/012006

The purpose of the investigated platform is to reduce the oscillations of the object caused by external vibration and shock disturbances [3]. It can work in all three modes of vibration control: active, semi-active and passive.

Passive vibration control is provided by the quasi-zero stiffness suspension system with (QZSS) [2]. QZSS is a combination of loads, levers and spring elements designed to reduce the load of platform and its' actuators (MR dampers) created by the mass of the isolated object. Elements of the system support the movable platform plate at four points symmetrically from its' center.

Active and semi-active vibration control is provided by four MR dampers located symmetrically from the platform center. An electromagnetic coil controls the damper actuation [3].

Figure 1. Active vibration control platform based on MRE.

The four MR actuators are fixed on the platform base. Each MR damper has a MRE disk membrane and one electromagnetic coil in the armored core [3]. Each actuator can be controlled independently by an automated control system which creates a signal transformed into the current in the coil.

3. Experimental research

The MREs mounted on MR dampers can have different characteristics even within the one technological lot. It happens because the distribution of metal particles in the elastic matrix is difficult to control during the manufacturing process. This leads to deviation of the platform movable plate from the horizontal plane. Thus, the purpose of the study in this work is to research the positioning error of the platform caused by these differences.

The measurements were carried out on a four-channel displacement measuring system capaNCDT 6200 (Micro Epsilon, Germany). The system has four capacitive position sensors and one controller. The system can measure platform displacement with a resolution of up to 10 nm and a data acquisition rate of 3.9 kHz. The sensors are installed onto supports above the connection points of the MR dampers and the platform moving plate with a distance of about 0.2 mm to the platform movable plane (figure 2).

Figure 2. Experimental bench.

As a result of the research, the maximum deviation of the platform moving plate from the horizontal plane is 110 μm, which is unacceptable for the nanotechnological purpose. The control signal needs to be adjusted for each damper according to the experimental curves (figure 3).

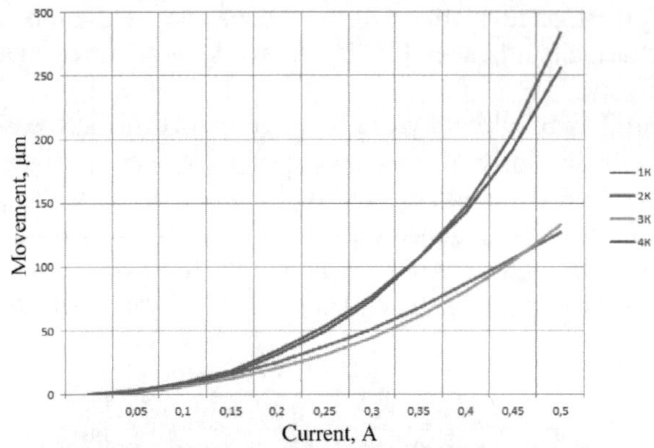

Figure 3. Dependence of displacement of platform MR dampers on the different control current.

4. Experimental results processing

The estimation of measurement error was carried out by the contour estimations method [4]. The required number of parallel experiments of the position of the actuator in each experiment was determined, when experimental measurement results were processed. A number of parallel experiments was 30 with a measurement results standard deviation of 0.1 µm.

Verification of the distribution of the experimental results by the Pearsons' test showed its' compliance with the normal distribution law. The contour estimations method showed that the maximum random measurement error does not exceed 0.5 µm. This means the experimental results have the sufficiently high accuracy of the measurements and show the reliability of the obtained results of active vibration control platform actuators displacements.

The uniform distribution law was chosen as the worst possible form of experimental displacement results distribution.

5. Conclusion

The experimental measurements of the MRE platform displacement as a function of the control electric current showed that the deviation of the platform moving plate movements from the nominal value (0 A current) was about 110 µm with a maximum control current 0.5 A.

The reasons of deviation are the different MRE properties of each actuator due to its' manufacturing technology.

The systematic measurement error is up to 0.4 µm and random errors is up to 2 µm. The errors were detected and significantly reduced. Thus, the experimental results allowed to achieve high accuracy of the measurements. It confirms the high reliability of the results.

Acknowledgment

This work is supported by the Ministry of education and science of the Russian Federation under the project part of the basic part State job No. 9.8503.2017/BP in the sphere of scientific activity.

References

[1] Mikhailov V P, Bazinenkov A M and Akimov I Yu 2007 *High technologies in Russian industry Materials of XIII International science conference (Moscow)* pp 150–8

[2] Alabuzhev P M, Gritchin A A and Kim L I 1986 *Vibrozaschitnie sistemy s kvazinulevoi zhestkostiu (Vibration protection systems with quasi-zero stiffness)* ed K M Ragulskis (Leningrad, Machinery) 96

[3] Mikhailov V P, Bazinenkov A M 2016 *Instruments and Experimental Techniques* **1** 131–5

[4] Levin S F *Ob osnovaniyakh teorii izmeritelnykh zadach (About fundamentals of measurement tasks theory)* URL: http://pribory-si.ru/publication/index.php?ELEMENT_ID=5119

IOP Conf. Series: Materials Science and Engineering **387** (2018) 012007 doi:10.1088/1757-899X/387/1/012007

AFM investigation of MoS_2 thin films

A I Belikov, Do Thi Nhan and V N Kalinin

Department of electronic technologies in mechanical engineering, Bauman Moscow State Technical University (BMSTU), Moscow 105005, Russia

E-mail: belikov@bmstu.ru

Abstract. Molybdenum disulfide thin films deposited on silicon substrates by the magnetron sputtering structural features investigation by atomic force microscope results are presented.

1. Introduction

At the present time electronic properties and technologies for two-dimensional (2D) materials formation for the perspective hardware components that may be used in nanoelectronic devices, for ensuring functional and operational characteristics of electronic systems further increasing are widely and actively studied. Molybdenum disulfide (MoS_2) with one or more molecular layers thickness is most widely studied. MoS_2 monomolecular layers is estimated 100–480 cm^2 $(V\ s)^{-1}$) charges mobility. As distinct from graphene, the monomolecular layer of molybdenum disulfide possess 1.8 eV direct bandgap and it is the direct-band-gap semiconductor, more thicker MoS_2 films demonstrate the indirect semiconductor properties with a bandgap from 1.3 to 1.8 eV, depending on a thickness [1]. A much more number of the modern works performed by different authors, where engineering development samples of electronic devices based on MoS_2 films were prepared, demonstrate the prospects of its application in field effect transistors (FET), light-activated transistors, organic electronics and other devices. Engineering development devices are created in a laboratory environment, generally using molecular layers mechanical splitting and separation from bulk crystal methods.

The method of mechanical splitting MoS_2 layers is suitable for monolayer 2D materials laboratory samples creation, which is widely known as «Scotch method», is inapplicable in industrial production. Therefore, today a lot of researchers solve the concerning technology problem of MoS_2 films' formation on large area substrates renewable at industrial scale. For these purposes the following possible methods are investigated: chemical vapor deposition (CVD), atomic layer deposition (ALD), the technology based on the mechanical printability by transfer elements from a stamp on a substrate surface [2], ion sputtering target methods in vacuum and others. In this work the research results of MoS_2 thin films forming by magnetron sputtering cathode target in the argon environment on a silicon substrates are presented.

MoS_2 films surface researches were conducted by atomic force microscopy (AFM) methodic. Films' structural features analysis consisted in a surface texture scanning by the AFM method and processing received images by IA-P9 program reference techniques for the identification films formation technological modes influence on a films surface features.

2. Experimental

MoS_2 thin films were deposited at vacuum equipment fitted with a balanced magnetron sputtered systems. Before films deposition liquids and ions cleaning substrates surface treatment were produced.

For the magnetron sputtering parameters influence on a films structural features researching, ten samples with a MoS_2 thin films were deposited on silicon substrates at the different conditions. Table 1 shows the main technological parameters.

Table 1. MoS_2 thin films deposition, magnetron sputtering technological parameters.

Sample number	Argon pressure, Pa	Discharge power, W	Substrate bias, Volt	Substrate heating temperature, °C	Deposition time, min
1	$7.50x10^{-1}$	52	20	200	65
2	$7.30x10^{-1}$	45	30	200	60
3	$4.50x10^{-1}$	30	20	200	30
4	$7.50x10^{-1}$	45	27	200	75
5	$4.40x10^{-1}$	45	22	250	32
6	1.00	30	–	250	30
7	1.00	30	–	250	45
8	$6.00x10^{-1}$	30	–	250	110
9	$1.00x10^{-2}$	10	–	250	30
10	$6.10x10^{-1}$	30	–	300	60

Three-dimensional surface topography AFM-images gives nano-dimensional data about films structure, including a roughness, imperfections, amorphous and crystalline phases, nucleation specifics and a mode of growth [3, 4].

Films surfaces AFM-scanning at contact mode was carried out at atomic-force microscope "Solver-NEXT" designed by "NT-MDT" company. Samples have been attached to a metal holder, and then on a surface of each sample the area in the size 30×30 μm has been in a random way selected and scanned. The real thickness of the received films was defined throw the step prepared by masking part of a surface during deposition, and for samples with the maximum deposition time measured up to 2 microns.

Surface texture scanning results visualization was received as topographic maps and 3D images. On topographic maps sections along which surface profiles were under construction were selected. With IA-P9 software using the roughness statistical parameters of the chosen profiles was received.

Table 2. Surface profile parameters for samples number 6, 7, 8.

Sample number	**№ 6**	**№ 7**	**№ 8**
Height of the surface lowermost points h_{min}, nm	-2,444	-8,237	-16,326
Height of the surface uppermost points h_{max}, nm	2,453	6,214	21,552
The maximum overall of heights R_y, nm	4,897	14,451	37,878
Average square roughness R_q, nm	0,868	2,635	7,822
Average arithmetical roughness R_a, nm	0,691	2,092	6,417
Quantity of peaks	19	24	28
Distance between peaks, μm	1.58	1.25	1.07

For the samples number 6, 7 and 8, prepared at the identical target sputtering conditions (power discharge 30 W, substrate temperature 250 °C) but various deposition times (30, 45, 110 minutes), has been realized surface profiles additional handling at the most typical for all scans regions. The prepared profiles were used for average distance between peaks on a surface calculation for the purpose of its planarity estimation. Roughness parameters and average distances between profile peaks calculated values are represented in table 2.

3. Results and discussion

As the received results displayed, with the deposition time increasing and the MoS_2 films thickness growth the surface roughness rise appreciably (essentially several times) that can testify about three-dimensional structure growth character tendencies reinforcement. For small deposition time, at small film thickness higher surface planarity is observed for the sample number 6, and the average distance between peaks of its profile essentially higher, than at samples numbers 7, 8.

The space of roughness R_a values for various technological conditions depending on MoS_2 films deposition time is presented at Figure 1. It is possible to mark the most essential influence of a preheating substrate temperature on a deposited films roughness magnitude, and with a temperature reducing roughness essential growth is observed. With the increasing deposition time, the steady tendency to roughness growth, for all another technological conditions, is observed too.

For the samples received at temperature 250 °C and deposition time 30 min, changing argon pressure (from 0.01 to 1 Pa) and discharge power (10 W, and 30 W) had no such significant effect on a roughness, as in a case with such influencing parameters, as temperature and deposition time.

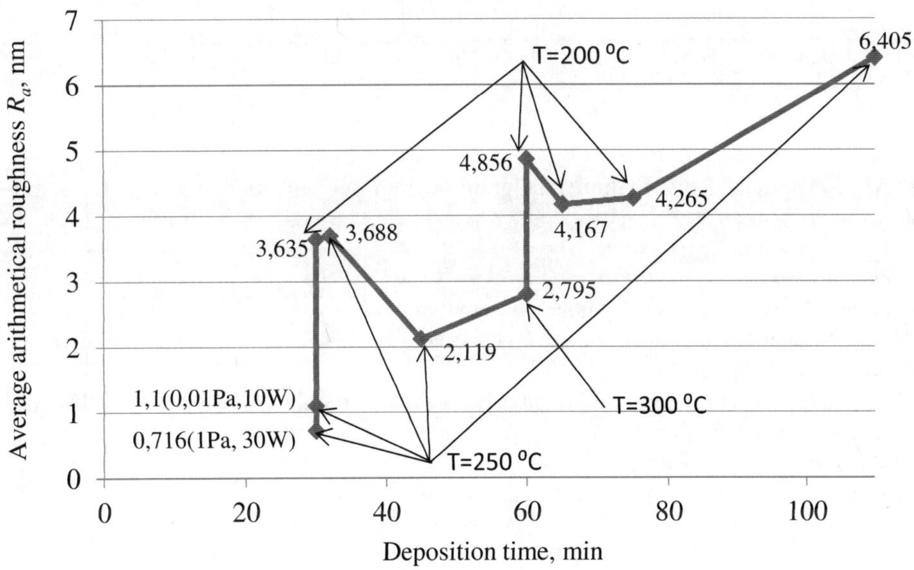

Figure 1. Deposition time influence on a roughness at a different technological condition.

Figure 2 shows films numbers 5 and 6 AFM-images and also surface heights distribution histograms respectively for these films, calculated for all scanned area 30 μm × 30 μm. It is revealed that surface heights distribution for these samples differ substantively.

So, for a sample number 5 that was formed at higher discharge power and smaller pressure in comparison with a sample number 6, the considerable positive asymmetry of distribution is observed (figure 2(a)). At the same time the sample number 6 has smoother surface (figure 2(b)) and there is a slight negative asymmetry at almost symmetrical distribution. It allows making a conclusion that discharge power takes effect on a roughness and structural features of the formed MoS_2 films.

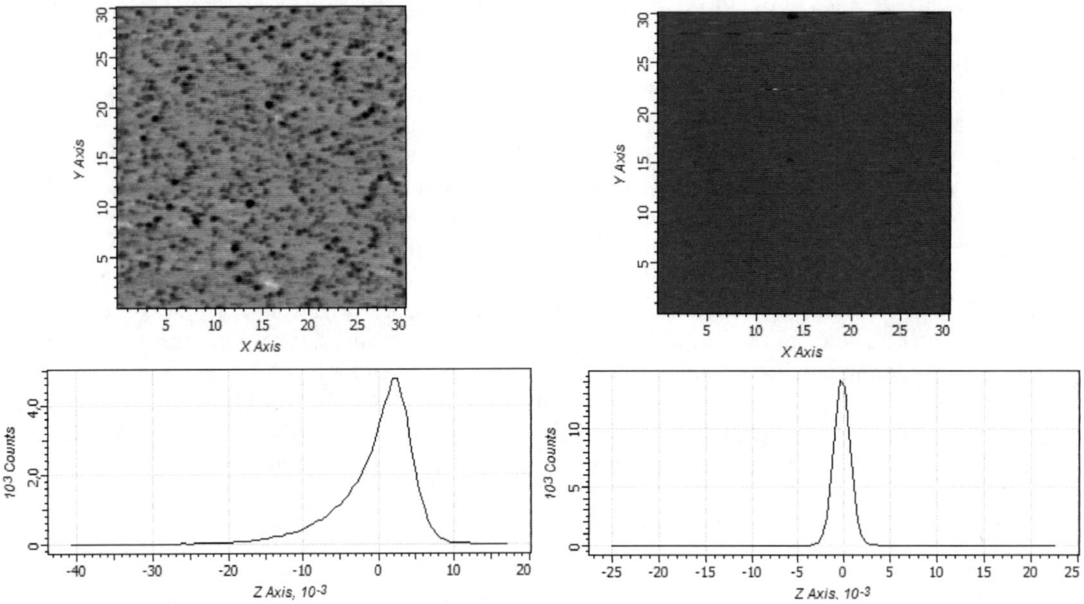

a) b)

Figure 2. AFM-images and distribution histograms of a films surface heights: (**a**) – sample 5, (**b**) – sample 6.

Conclusions

This work received results allows making some outputs:

1. Substrate temperature, and also discharge power and argon pressure has significant effect on a structural features MoS_2 thin films.
2. With increase deposition time the increase films surface roughness that can be linked with a growth features changes at increase film thickness and surface influence at substructure lowering, characteristic growth for an initial stage is observed.
3. AFM investigation method, taking into account with the known features of MoS_2 anisotropic structure can be used for the analysis of molybdenum disulfide thin films.

References

[1] K F Mak et al. 2010 *Phys Rev Lett* **105** 136805
[2] Nam, H. et al. 2013 *ACS Nano* **7** 5870–81
[3] Bykov V A et al. 1997 *Electronics: Science, Technology, Business* **5** 7–14
[4] Arutyunov P A et al. 1999 *Zavodskaya laboratory* **9** 31–41

IOP Conf. Series: Materials Science and Engineering **387** (2018) 012008 doi:10.1088/1757-899X/387/1/012008

MoS_2 thin films spectrophotometry

A I Belikov and Kyaw Zin Phyo

Department of electronic technologies in mechanical engineering, Bauman Moscow State Technical University (BMSTU), Moscow 105005, Russia

E-mail: belikov@bmstu.ru

Abstract. Molybdenum disulphide (MoS_2) thin films deposited on silicon substrates by magnetron sputtering spectrophotometry study results are presented. Graphical-calculated method for MoS_2 thin film samples bandgap values determination was used.

1. Introduction

In recent years two-dimensional materials such as MoS_2, for example, attract increasing attention as promising materials group for using in a new generation of nanoelectronic devices because of its unique properties as high electron mobility, excellent current ON/OFF ratio and etc. [1]. MoS_2 thin films as semiconductor has tunable bandgap, that varies with a thickness. As known, bulk MoS_2 has an indirect band-gap, however, with reducing their thickness to monomolecular layer, formed by two sulfur atoms and one molybdenum atom, electronic structure of it transformed to the direct bandgap [2]. The effectiveness of the molybdenum disulphide using at the digital electronics area stayed actual due to the presence of a bandgap, whose width reaches 1.8 eV value for a monolayer film [2]. And the using such films in the field-effect transistors prototypes provides a high ratio of currents through the channel at the open / closed states (ON/OFF), which are 6-8 orders of magnitude.

To study and measure the spectral characteristics and properties of promising two-dimensional materials are widely used optical methods, including, based on the reflectance spectrum detection, implemented at simple equipment. The reflection spectroscopy method can be used to study thin films with the thicknesses in the range from 1 nm to 1 mm [3]. Transmittance T and reflectance R optical coefficients values using allow with a much more accuracy determinate optical absorption coefficient (α) at the region around the optical gap, and rest upon the equations of α as a wavelength λ function can be implemented bandgap energy. Group of Miika Mattinen et. al 2017 [4] produced by ALD method MoS_2 thin films at borosilicate glass substrates and determine bandgap by applying Tauc graphical method, using absorption spectrum recalculated from the transmission spectrum. At report [6] also shown that the bandgap of thin films TiO_2 can be obtained by using Tauc method from the absorption coefficient, derived from the diffuse reflectance.

MoS_2 thin films optical properties at the different thickness measuring by spectrophotometer «Epsilon» and its bandgap estimation by using Tauc method results is presented at this work.

2. Experimental

MoS_2 thin films at silicon substrates were prepared by magnetron sputtering MoS_2 target at 10^{-1} Pa argon pressure. Substrate surfaces before the deposition passed preliminary preparation with two stages: cleaning at ultrasonic bath in alkaline solution and ethanol consistently, and surface treating with argon ions bombarding from an independent ion source in a single vacuum, immediately before the material target sputtering. The substrate temperature for different process parameters was set at the range from 200 to 300 °C. The thicknesses of the samples were determined by masking part of the substrate surface during deposition in order to have a sharp step between coated and uncoated zone, and then measuring the height profile.

MoS_2 thin films measurements were carried out at atomic force microscopy (AFM) "Solver-NEXT" by "NT-MDT". And the "Epsilon" spectrophotometer with the wavelength range 380-1100 nm was used for the diffuse reflectance spectral characteristics at prepared samples with coatings optical measurements.

The "Epsilon" spectrophotometer that used at this work provides spectral reflectance *R* for flat optical components and coatings on the substrate measurements. An optical scheme equipped with a white light source collected the reflected light signal from the sample. An optical fiber used to selectively couple part of the reflected light signal into a spectrophotometer. At the device optical system the spectral reflector uses fixed angle (90°) to a surface and carry out spectroscopic measurement that scans the frequency of the probe light. Conditionally, the elementary diagram for MoS_2 thin films on Si substrate reflectance spectrum measurements is shown in Figure 1. At transmittance spectrums registration except components of light which as it is shown in Figure 1, has passed through a substrate and researched layer, it is possible to evolve some component which has passed through film layer, but it was reflected from the film-substrate border. This component spectrum in the general stream of the light reflected from the researched sample represents resulting reflectance-absorption spectrum, taking into account absorbed light.

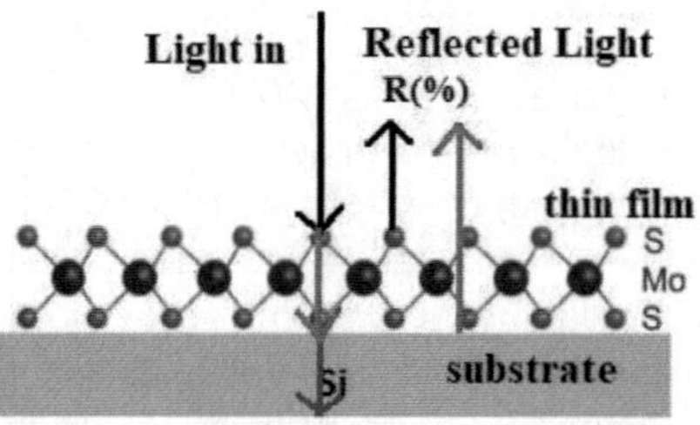

Figure 1. Scheme for the spectral reflection measuring of MoS_2 thin films on Si substrate.

The most widespread approach to the light behavior theoretical analysis in the dispersing medium consists in the decision first-order differential equations system that describes an intensity decreasing in a sample own due to the arising phenomena from absorption and scattering. On this approach based P. Kubelka and F. Munch, and also M. Gurevich have positioned interrelation between absorption coefficients α, medium dispersion *S*, and diffusive reflectance quotient (*R*) which can be received experimentally:

$$F(R) = \frac{(1-R)^2}{2R} = \frac{\alpha}{s} \quad (1)$$

Function *F(R)* (1), is the absorption function which is in proportional dependence with a layer material absorption coefficient, therefore, having executed spectrum registration at a diffusive scattered radiation and plotted absorption function from the light wave length dependence it is possible to receive actually investigated material absorption spectrum. Its difference from a real absorption spectrum will consist only at a difference value of a light dissipation factor *S* at a sample surface.

To estimate materials optical bandgap modified Kubelka–Munk function can be applied by receiving function *F(R)* on photon energy *hv* multiplication, using the corresponding coefficient *(n)* that associated with an electronic transition, as follows:

$$(F(R) \times hv)^n \quad (2)$$

By extrapolating equation (2) as an energy function in eV, the semiconductor films bandgap can be obtained. It is known that the monomolecular MoS_2 layer has a direct bandgap, and two or more MoS_2

layers have an indirect gap, so for the calculations performing the corresponding value of the semiconductors indirect gap coefficient n = ½ is using.

To determine the optical bandgap in accordance with the Tauc method [5], that based on relation (3), $(F(R)\times h\nu)^{1/2}$ from $h\nu$ dependences were plotted and then extrapolated to the $h\nu$ axis (Figure 3b) with photon energy value that corresponding to the bandgap (E_g) value.

$$(F(R)h\nu)^{1/2} \approx (h\nu - Eg) \quad (3)$$

3. Results and discussion

In experiments series five samples MoS_2 thin films on Si substrates with a difference in the thickness and surface roughness have been prepared. The measured values of a thickness and average square roughness R_a are resulted in table 1. Figure 2 shows the reflection spectra of MoS_2 films for samples A, B, C, D, E, and also for an initial silicon substrate.

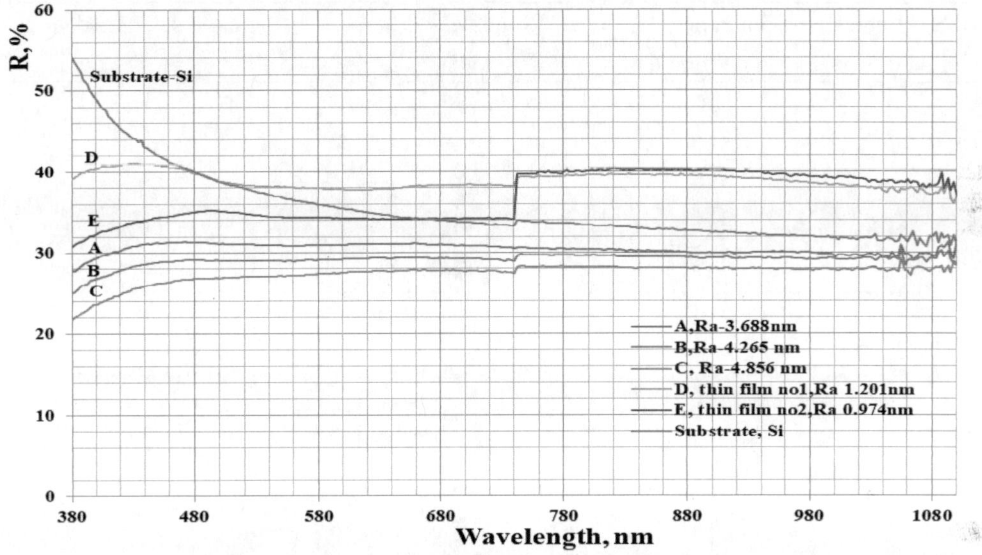

Figure 2. MoS_2 films samples on Si substrate spectra's.

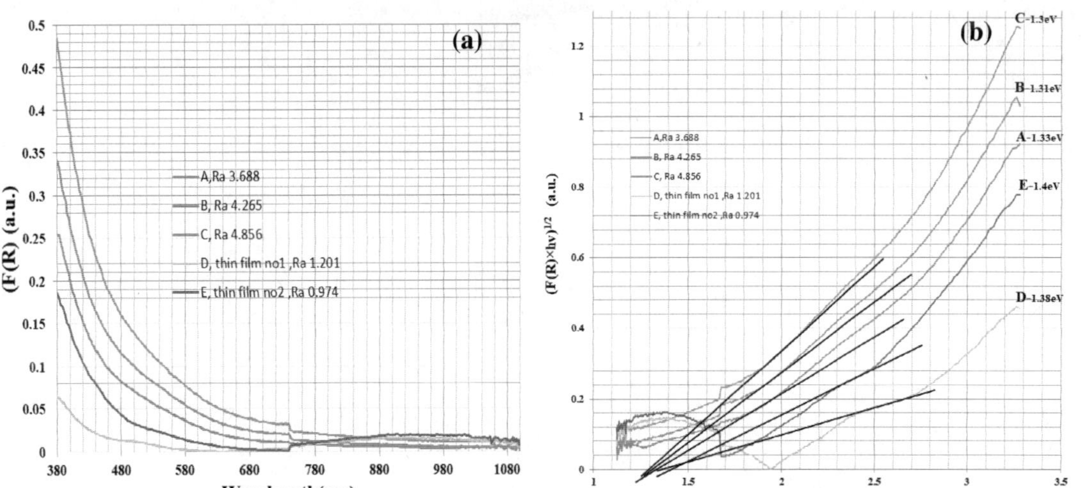

Figure 3. Graphical representation of *F(R)* spectra – absorbance coefficient from the light wavelength dependences (**a**); Graphical representation of $(F(R)h\nu)^{1/2}$ from E_g dependences for MoS_2 samples (**b**).

Based on the dependence (1) and using the obtained reflection coefficient values for different wavelengths, the values for the absorption function *F(R)* were determined.

The calculated absorption coefficient values were used to plot the dependence $(F(R)h\nu)^{1/2}$ from $h\nu$, whose fragment for the energy range of wavelengths from 1.2 eV to 3.3 eV is shown in figure 3(b). Subsequently, by extrapolation at the linear sections of the plots, the bandgap values for each MoS_2 sample were determined.

Using the Tauc method were obtained bandgap values for MoS_2 films samples at different thickness and roughness results are summarized in table 1. For D and E samples with the same thickness but deposited at different parameters, were observed a difference in E_g values that can be caused by differences in a films structure, which reflected on a surface roughness.

Table 1. MoS_2 films on Si substrates geometric characteristics and bandgap E_g experimental values.

MoS_2 samples	Thickness, nm	Roughness R_a, nm	Bandgap calculated value E_g, eV
A	500	3.688	1,33
B	530	4.265	1,31
C	680	4.856	1,3
D	20	1.201	1,38
E	20	0.974	1,4

Conclusions

At the present results shows that for MoS_2 films with a more than 600 nm thickness, the observed bandgap values had compatibility with the MoS_2 bulk crystals characteristic literature data's (1.3 eV). With the films thickness decreasing to 530 nm or less, bandgap values increasing and reaching 1.4 eV at a 20 nm film thickness is observed.

References

[1] Radisavljevic B et al. 2011 *Nature Nanotech* **6** 147–50
[2] Mak K F et al. 2010 *Phys Rev Lett* **105** 136805
[3] *Filmetrics* https://www.filmetrics.com/ellipsometry
[4] Mattinen Miika 2017 *Adv Mater Interfaces* **4** 1700123
[5] Tauc J 1972 *Journal of non-crystalline solids* **10** 569–85
[6] Murphy A B 2007 *Solar Energy Materials & Solar Cells* **91** 1326–37

Experimental leak tests of gas-filled instruments using automatic leak detection

S A Bushin and S S Galkin

Dukhov Automatics Research Institute (VNIIA), Federal State Unitary Enterprise, 22 Suschevskaya ul, Moscow 127055, Russia

E-mail: vniia4@vniia.ru

Abstract. The experimental leak tests of several gas-filled instruments (dischargers) are described. The tests are based on accumulation mass-spectrometry. The technique is implemented by means of VNIIA-developed and produced prototype for automated final leak test facility (FLTF). The tests show that a part of tested instruments have leakage within the range of filling gas (neon) equal to $1.5 \cdot 10^{-12} - 10^{-7}$ Pa·m^3/s.

1. Introduction

In order to maintain a gas composition either inside a vacuum instrument or in gas-filled instrument, their enclosure has to be leak-tight and it depends on quality and mechanical integrity of leak-tight welded joints of parts and materials.

An automated final leak test facility (FLTF) was used for series of tests to check the key requirements of leak tightness testing technique developed to control the filling (control) gas leaks in small-sized gas-filled instruments [1–4].

An object to be tested is a process prototype for ceramic and aluminum alloy gas-filled instruments with a tube and basing (at the end of the tube there is a compound bead) fabricated in different time.

The following equipment was used for tests: 1) FLTF prototype [5] with neon detection limit as much as $5 \cdot 10^{-14}$ Pa·m^3/s and 2) *Adixen ASM*142 *Graph D* leak detector [6] with detection limit as much as $5 \cdot 10^{-13}$ Pa·m^3/s with accessories.

2. Testing technique

Leak tightness of gas-filled instruments was tested using an automatic final leak tightness testing technique [7, 8]. It is a leak tightness detection based on accumulation mass-spectrometry. When applying the detection like this the tested instrument is placed in an accumulation chamber. A filling gas (neon) which exhausts from the instrument accumulates in the chamber. When a gas portion is accumulated and transported to the ionization area of gas-detector, mass-spectrometer registers gas pressure. A gas leak is detected by mass-spectrometer signal, which is a function of time, provided that the tested instrument and gas-detector are isolated from the exhausted filling gas.

The instrument leak rate is calculated by expression (1):

$$Q = \frac{\Delta I \cdot V_{\Sigma}}{S \cdot \tau}, \tag{1}$$

where: Q is a leak rate, Pa·m^3/s; ΔI is a mass-spectrometer signal increment when the accumulation chamber is re-filled with gas (Figure 1), A; V_Σ is a total volume of free space within the measurement system and the accumulation chamber, m^3; S is a sensitivity factor of measurement system, A/Pa; τ is an accumulation time, s.

Four groups of instruments were tested in four stages:

Group I – The instruments were tested for leaks with FLFT;

Group II – The instruments were re-tested after being exposed to (200 ± 10) ^{0}C for 3.5 hours and exhausted at the same time;

Group III – The testing technique [7] was applied to the instruments having been pressurized in neon up to pressure (≈ 1 atm.), which was directly set while the instrument is in the chamber without its removal and exposure to the ambient air for set exposure time. All instruments including test ones were pressurized (see Note);

Note:

1) Pressurization is a technique applied to leaky gas-filled instruments being stored. The procedure is used to restore an initial pressure of filling gas within the instruments to be unsoldered and removed from the pack rack in order to improve measurements of leak rate.

2) Two based instruments were used as test ones at different times. The first one was applied to the first group of instruments to be tested and the second – instead of the first – for tests of all other instruments.

Group IV – The instruments with neon leak more than 1·10^{-8} Pa·m^3/s were additionally pressurized in helium and then tested for leak tightness using helium mass-spectrometer leak detector.

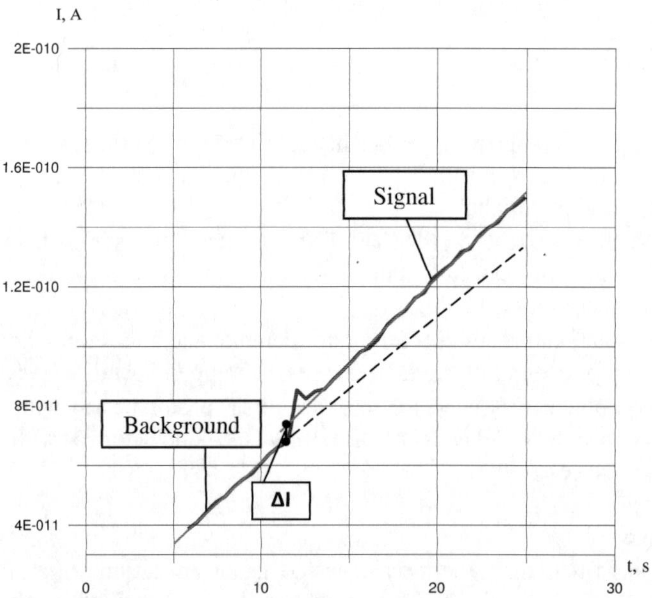

Figure 1. Typical frame of *QMG* 220 mass-spectrometer signal at mass 20 for filling gas leakage.

Iocurrent increment per mass 20 for each instruments, $\Delta I'$, was calculated as difference of existing peak amplitude I_i with respect to the system background and ion current increment, ΔI_1. The last parameter is described by mass-spectrometer response to expansion of gas portion accumulated within the inspection equipment volume. The mass-spectrometer response is also overlapped by gas-detector response to relatively intensive gas escape from the surface of the tube with compound bead at the end. ΔI' as calculated by expression (2):

$$\Delta I' = I_i - \Delta I_1. \quad (2)$$

For comparative analysis of test results obtained using accumulation mass-spectrometry (FLTF) and a sniffer probe of *ASM142D Graph* leak detector, a neon leak rate (Q_{Ne}) was converted into helium leak rate (Q_{He4}) by expression (3):

$$Q = Q_{Ne} \cdot \sqrt{\frac{M_{Ne}}{M_{He}}}, \quad (3)$$

where: Q_{Ne} is a *neon* leak rate, Pa·m^3/s; M_{Ne}, M_{He4} are *Ne* and He_4 molecular masses.

3. Test results

The tests of group I (4 instruments) showed a leak rate of $1.2\cdot10^{-12}$ Pa·m^3/s (allowable leak rate).

Group I instruments were vacuum-heated. When the instruments were cooled, the neon portions were inlet in each of accumulation volumes up to pressure 1 atm. Figure 2 (a) and (b) shows curves demonstrating the test results for two instruments. Neon leak rates were $7.12\cdot10^{-12}$ Pa·m^3/s and $1.99\cdot10^{-8}$ Pa·m^3/s (large leak); a leak rate for other instruments in the group does not exceed $1.2\cdot10^{-12}$ Pa·m^3/s.

In order to find a leak location in the instrument with large leak, the instrument after being removed from FLTF accumulation chamber was placed in special in-process storage box (V ≈ 1.5 l) and then exhausted and pressurized in He_4 at pressure ≈ 2.5...3 atm. for 35 days and tested by sniffer probe of helium leak detector. As a result, the location of compromised integrity on the instrument enclosure was found. It was a surface section of disk-shaped electrode with tube welded at the base. A leak rate registered by leak detector was (2.2...2.6)·10^{-7} Pa·m^3/s.

The test of group II (6 instruments) did not reveal leaks. After heating at (190...195) ^{0}C the leaks were also not detected.

The test of group II of instruments being neon-pressurized at 1.1 atm. for 42 hours 50 minutes revealed a leak rate (after approximation and extrapolation while the instrument was unsoldered) (see Figure 3) equal to Q=$9.95\cdot10^{-8}$ Pa·m^3/s for *Ne* and Q=$2.22\cdot10^{-7}$ Pa·m^3/s for He_4, respectively.

The instrument when removed from the accumulation chamber was placed in special in-process storage box and pressurized in helium at pressure ≈ 1 atm. for 30 days. A sniffer probe of leak detector was then used to find a leak location, which was a ceramic and aluminum alloy joint directed towards the tube and a leak rate was (1...2)·10^{-7} Pa·m^3/s.

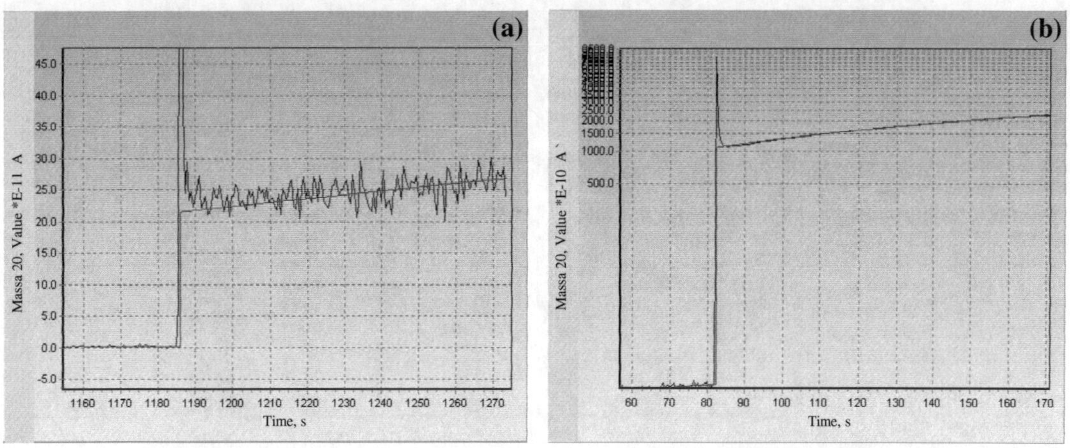

ΔIo=9.649E-12 A, ΔIi=2.142E-10 A
Q=7.115E-12 Pa·m^3/s
Mass 28 (p/b)=1.0; Mass 32 (p/b)=0.9; Mass 40 (p/b)=0.7
I10/I20=0.027% (1.4725%-1.6275%)
I21/I20=0.125% (0.306%-0.374%), I22/I20=8.595% (9.894%-10.506%)
Conclusion: Object is NON-TIGHT!

ΔIo=9.649E-12 A, ΔIi=1.719E-6 A
Q=1.992E-8 Pa·m^3/s
Mass 28 (p/b)=3.0; Mass 32 (p/b)=9.1; Mass 40 (p/b)=6.8
I10/I20=0.027% (1.4725%-1.6275%)
I21/I20=0.260% (0.306%-0.374%), I22/I20=8.479% (9.894%-10.506%)
Conclusion: Object is NON-TIGHT!

Figure 2. Frames from single-peace leak tightness test record for instrument 3 (**a**) and 4 (**b**).

IOP Conf. Series: Materials Science and Engineering **387** (2018) 012009 doi:10.1088/1757-899X/387/1/012009

The test of group III (6 instruments) showed that only two of them after being pressurized in neon (at 1 atm. for 20h 41 min) had the leak traces. A total leak rate was $6.03 \cdot 10^{-12}$ Pa·m^3/s ($1.34 \cdot 10^{-11}$ Pa·m^3/s for He_4) and $2.39 \cdot 10^{-11}$ ($5.34 \cdot 10^{-11}$ Pa·m^3/s for He_4), respectively.

The test of instruments (5 pcs) from group IV after first and second stage showed that a leak rate did not exceed an allowable value. Three instruments revealed a higher value only after neon-pressurization at 0.995 atm. for 30 h 47 min. Their leak rate for neon was $5.57 \cdot 10^{-11}$ Pa·m^3/s, $6.96 \cdot 10^{-11}$ Pa·m^3/s and $2.04 \cdot 10^{-11}$ Pa·m^3/s, respectively (see figure 4).

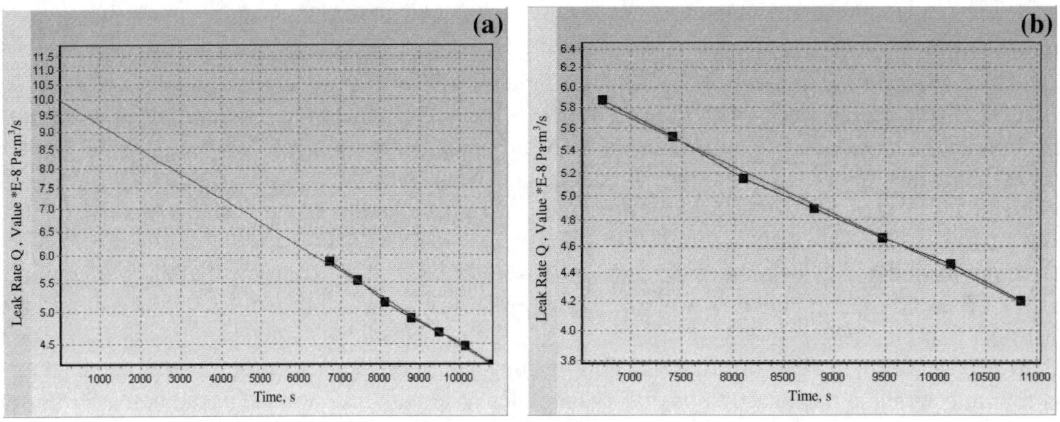

Figure 3. Approximation (**b**) and extrapolation (**a**) curves for variable leak rate during leak test of neon-pressurized instruments of group II with large leak in mode Large Leak Rates

The instruments of group IV during leak tests revealed a leak rate higher than allowable one. A leak rate for three instruments subject to single-piece leak tightness tests was$1.32 \cdot 10^{-10}$ Pa·m^3/s, $9.63 \cdot 10^{-11}$ Pa·m^3/s and $1.83 \cdot 10^{-10}$ Pa·m^3/s (see figure 5(a)).

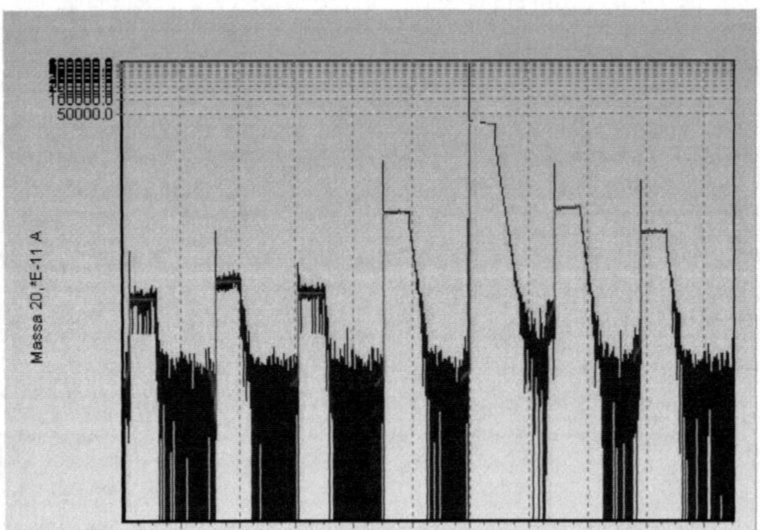

Figure 4. Amplitude behavior at mass 20 for six neon-pressurized instruments including a reference one (ref.) and an instrument X [with leak rate equal to $4.5 \cdot 10^{-9}$ Pa·m^3/s for comparison] during single-piece leak tightness tests.

Note that the leak rates in figure 5(a) are much different from those in figure 5(b) (the processed data of large leak rates). The leak rate history in figure 5(b) (instrument *X*) is almost unchanged and a leak rate is ≈ $4.5 \cdot 10^{-9}$ Pa·m³/s. It indicates an actual leakage and is proved by relevant calculations and observations over the instruments for relatively long time (since August 2013) unlike the leak test results for instrument A, B, C in figure 5(a).

It should be noted that the instrument external surface cleaned with ultrasound when manufactured and made of ceramics (70% of surface) has a grained physical structure contrary to its geometric area. There can be microvoids and holes which can quite intensively inlet the filling gas in the vacuum volume after pressurization. Moreover, the micro-bubbles on the compound bead spread over plasticizer and partially extricated and filled during pressurization can greatly distort the results. Point-by-point measured leak rate in 10 – 15 minutes after pressurization reaches about $(1...3) \cdot 10^{-9}$ Pa·m³/s which is a level of integral gas leak rate from the surface. In 5–8 hours after high vacuum evacuation a level of leak rate drops to about $(0.5...2) \cdot 10^{-10}$ Pa·m³/s. The initial background (before pressurization) is resumed in 1.5–3 days with total evacuation time per a shift equal to 3–5 hours.

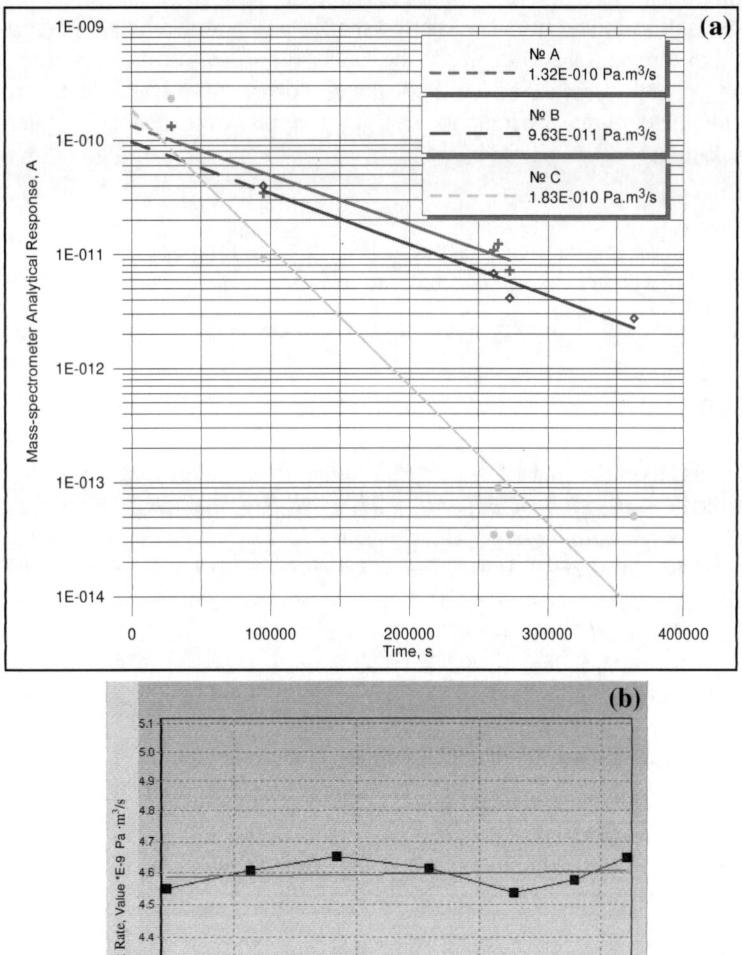

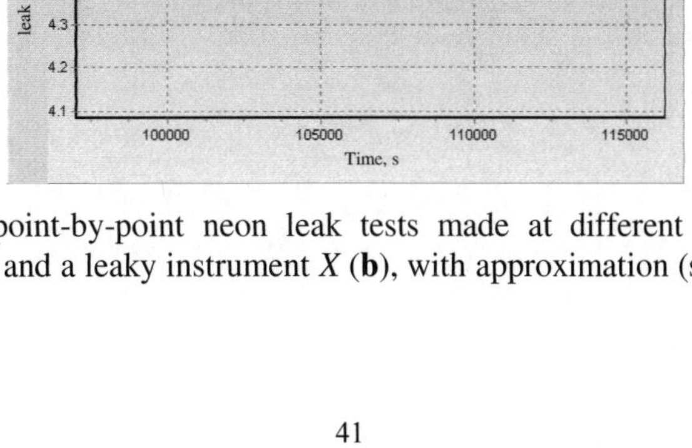

Figure 5. Curves of point-by-point neon leak tests made at different time on three pressurized instruments A, B, C (**a**) and a leaky instrument *X* (**b**), with approximation (solid line) and extrapolation (dash line).

Upon the above results there are no reasons to say that the instruments A, B, C are leaky.

4. Summary

The leak tightness tests of instruments carried out with FLTF had the following results:

– Two instruments showed a neon leak rate as much as $1.99{\cdot}10^{-8}$ Pa·m^3/s, $9.95{\cdot}10^{-8}$ Pa·m^3/s, respectively;

– Eight instruments demonstrated a neon leak rate (after pressurization) as much as $1.5{\cdot}10^{-12}$ to 10^{-11} Pa·m^3/s.

Other instruments had a leak rate within an allowable value.

5. Result analysis

The tests performed showed that detection of the filling gas leak rate using FLTF enabled to find both relatively large (10^{-7}–10^{-9} Pa·m^3/s) and small (less 10^{-11} Pa·m^3/s) leakages. But the leak test results will be reliable if a number of requirements are followed: the leak tests should be conducted right after pressurization of instruments and before coating the external surfaces of enclosure. If the requirements are ignored, prior to leak tests additional measures should be undertaken: the instruments have to be vacuum-heated at high temperature to degasify the external surfaces and exposed to high pressure in filling gas medium to restore pressure inside the leaky enclosure. So the leak tests prove that only two instruments are leaky and eight instruments can have a probable leakage. Other ten pieces are considered to be leak-tight.

6. Conclusion

The key requirements of filling gas leak tightness testing technique applied to small-sized gas-filled instruments using FLTF were checked. The technique is showed to enable highly sensitive tests of small-sized gas-filled instruments.

References

[1] Bushin S A 2011 *Materials of 18-th Research Conference "Vacuum Science and Technique"* (Sudak)

[2] Bushin S A and Galkin S S 2014 *Vacuum Equipment and Technology* **1** 39–41

[3] Bushin S A 2017 *Collection of papers of 21-st All-Russian Conference for Non-destructive Tests and Technical Diagnostics* (Moscow) pp 275–88

[4] Bushin S A and Kozlovskaya T I 2017 *Proc. 15th Asia-Pacific Conference on Non-Destructive Testing (APCNDT2017)* (Singapore) URL: http://www.ndt.net/events/ APCNDT2017/app/content/Paper/31_Bushin_Rev4.pdf.

[5] Smirnov G A, Khapov A S, Kozlovskaya T I, Bushin S A, Bakanov A P, Chernykh E V, Galkin S S *RF Patent No. 101072 for Prototype* G 01 M 3/02, 3/26. *Automated Final Leak Tightness Facility* 28.11.16

[6] *Adixen Vacuum Products by Pfeiffer Vacuum, France*

[7] *Neon Leak Tightness Tests Instruments: Technique No. PM T401/22-2012* (Moscow: VNIIA) 2012

[8] Bushin S A, Kamenev V G and Galkin S S 2015 *Vacuum Equipment and Technology* **2** 77–82

Determination of heat fluxes and surface temperatures during magnetron sputtering with a cold and hot target

S P Bychkov

Department Electronic Technologies in Mechanical Engineering N.E. Bauman Moscow State Technical University, 2nd Baumanskaya ul., 5, 105005 Moscow, Russian Federation

E-mail: bychkov@bmstu.ru

Abstract. The method of calculation of heat fluxes and surface temperatures during magnetron sputtering of cold and hot targets developed from using the algebraic method for calculating radiant fluxes is presented, which allows to predict the required power characteristics for obtaining the desired temperature on the target and the substrate. Comparison of the experimental data and the demonstrated results of the calculations testify to the possibility of applying this technique. The temperature on the substrate can be controlled by changing the distance between the substrate and the target within certain limits, and the value of the distance can be estimated using the suggested calculation model.

The processes of obtaining films by the method of magnetron sputtering are accompanied by heating of the substrate [1–3]. When forced heating is used, the actual substrate temperature can be significantly higher than the assumed value due to the thermal processes during sputtering [4, 5]. The determination of the substrate temperature is essential for various processes [6, 7]. To solve it calculation of the heat fluxes in the process under consideration is required. Such calculation makes it possible to compare the thermal effect of the process with the geometric parameters of the system of bodies under consideration for the magnetron sputtering flowsheet specified, as well as with the target size and mass, making it possible to estimate the temperature and time of substrate heating depending upon the shape, size and material of the elements in the working chamber of the process unit. This approach allows calculations for processes with both cold and hot targets. The thermal scheme of the process in hand is shown in figure 1.

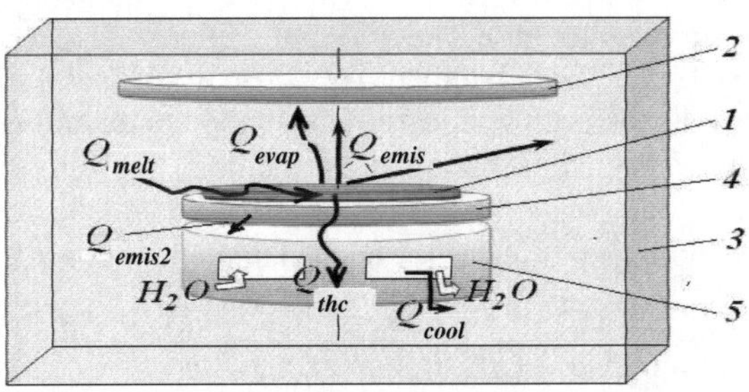

Figure 1. The thermal scheme of the process in hand:
1 – target, 2 – substrate, 3 – working chamber, 4 – crucible, 5 – magnetron.

The thermal flow P is the power released in a magnetron sputtering unit which generates a stream of ions bombarding the target 1. The part of the full power which provides heating and melting the target to the liquid state, when needed, will be referred to as the net power P_{net} within the framework of the task. After warming up the target to the preset temperature the net power is spent on the compensating of heat loss fluxes, of which the following will be considered significant: the heat flow Q_{emis} of the heated target emission onto the substrate *2* and the walls of the working chamber *3*; the heat flow Q_{thc} of the thermal conductivity from the target to the magnetron *5* through the crucible (target holder) *4* or the heat flux Q_{emis2} of crucible emission onto the surface of the water-cooled magnetron; the heat flux Q_{evap} transported onto the substrate and the walls of the working chamber by the target material evaporated from the target; the heat flux Q_{melt} spent on the target melting.

The heat loss power P_{loss} is a part of the full power spent on heating the magnetron itself, which equals the value of the heat flux Q_{cool} given to the cooling liquid in case of magnetron cooling. It should be borne in mind that the full value of the heat flux taken from the inner surface of the magnetron during its liquid cooling equals the sum $Q_{cool} + Q_{thc}$ or $Q_{cool} + Q_{emis2}$.

Let's compose the heat balance equation on the assumption that the system goes to steady-state condition during the process under consideration. In this case, the released thermal energy has a useful component, distributed in accordance with the above mechanisms, and the heat loss power

$$P = P_{net} + P_{loss},$$

$$P = Q_{emis} + Q_{thc}(\text{or } Q_{emis2}) + Q_{evap} + Q_{melt} + Q_{cool}.$$

The heat flux during the target material melting Q_{melt}, W, in a time τ, sec

$$Q_{melt} = \frac{Q_{\tau_{melt}}}{\tau}.$$

Where the heat of the target material melting $Q_{\tau_{melt}}$

$$Q_{\tau_{melt}} = c_{melt} m.$$

The heat flux Q_{evap} of the evaporated target material, W

$$Q_{evap} = 5{,}83 \cdot 10^{-5} \cdot kN_{\text{A}}\, p_{sat} \left(\frac{T_{evap}}{M}\right)^{\frac{3}{2}} F_{evap} \ \ [\text{W}],$$

where T_{evap} – the evaporation temperature, K; k – Boltzmann constant, J/K; N_{A} – Avogadro constant, atom/mol; M – molar mass of the evaporant, kg/mol; p_{sat} – saturated vapour pressure of target material, Pa; F_{evap} – evaporation surface area, m^2.

In case of a gap, the heat flux from the target to the magnetron is defined by the value of the emission heat flux from the crucible to the magnetron

$$Q_{emis2} = \sigma_0 \frac{T_{cr}{}^4 - T_{mag}{}^4}{\dfrac{1}{\varepsilon_{cr}} + \dfrac{1}{\varepsilon_{mag}} - 1} F_{cr}.$$

The greatest difficulty is the calculation of the emission heat flux Q_{emis} from the target surface onto the substrate and the walls of the working chamber. The difficulty is that in this case it is the matter of the radiant interaction of several surfaces with different temperatures and emission coefficients. It is expedient to solve similar problems using the algebraic method. Let's use the computational scheme shown in figure 2.

The algebraic system of radiant heat transfer for the scheme under consideration is as follows:

$$Q_{res1} = \varphi_{21}(E_{02}F_2 + R_2 Q_{res2}) + \varphi_{31}(E_{03}F_3 + R_3 Q_{res3}),$$
$$-(E_{01}F_1 + R_1 Q_{res1}),$$
$$Q_{res2} = \varphi_{12}(E_{01}F_1 + R_1 Q_{res1}) + \varphi_{32}(E_{03}F_3 + R_3 Q_{res3}),$$
$$-(E_{02}F_2 + R_2 Q_{res2}),$$

$$Q_{res3} = \varphi_{13}(E_{01}F_1 + R_1 Q_{res1}) + \varphi_{23}(E_{02}F_2 + R_2 Q_{res2}),$$
$$-(1 - \varphi_{33})(E_{03}F_3 + R_3 Q_{res3}).$$

We'll make the calculations for the case that the substrate has a thermal decoupling with the walls of the working chamber. T1 and T3 are known, T2 on the substrate holder is unknown, the heat equilibrium is (Q_{res2}=0). It is possible to determine the power emitted by the target using the condition $Q_{emis}=Q_{res}=Q_{res1}=-Q_{res3}$, as the equation

$$Q_{res1} + Q_{res2} + Q_{res3} = 0$$

is valid for the entire system in the steady-state condition.

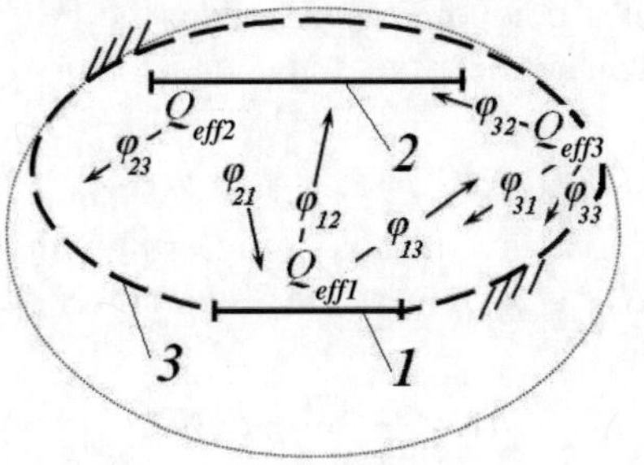

Figure 2. The computational scheme for determining the emission heat flux from the substrate surface: 1 – target, 2 – substrate, 3 – walls.

After calculating the corresponding determinants, taking into account the values of the angle coefficients φ_{ij}, the emission coefficients ε_i or the fractional reflectivities R_i, the known temperatures and areas T_1, T_3, F_1, F_2, F_3, we determine the unknown values of the resultant flux Q_{res} and the intrinsic emission of the surface *2* as an absolute black body

$$Q_{res} = \frac{\Delta_{Q_{res}}}{\Delta},$$

$$E_{02} = \frac{\Delta_{E_{02}}}{\Delta}\frac{1}{F_2}.$$

When determining the heat loss flux of the magnetron sputtering system, namely, the heat flux vented by the cooling liquid, it is problematic to distinguish the component Q_{cool}, which is determined by the heat liberation in the magnetron itself. It is more correct to talk about the heat dissipation of the total heat flux $Q_{cool} + Q_{thc}$ or $Q_{cool} + Q_{emis2}$. This value can be most easily determined for any time point during the system operation by registering the coolant temperature upstream of the cooling system t_{cool_in} and downstream of it t_{cool_out}, as well as the flow rate W_{cool} of the coolant with the heating capacity c_{cool} by the formula

$$Q_{cool} + Q_{thc}(or\ Q_{cool} + Q_{emis2}) = c_{cool}W_{cool}(t_{cool\_out} - t_{cool\_in}).$$

In the absence of such data, the power Q_{cool}, released as heat during the interaction between the electromagnetic field and the magnetron, can be estimated as a fraction of the total power

$$Q_{cool} = \eta P.$$

According to the available estimates, the coefficient value is η = 0,6 ... 0,8. Using this approach, it is possible to determine the values of all the components of the heat fluxes when the desired temperature on the target is reached. Having made a similar calculation for various intermediate temperature values in the range of room temperature to the required maximum, one can obtain the temperature dependences

$$P(T) = Q_{emis}(T) + Q_{thc}(T)[or\ Q_{emis2}(T)] + Q_{evap}(T) + Q_{melt}(T) + Q_{cool}(T).$$

If the supplied power is fixed at the level $P_{roundup}(T_{max})$ which is somewhat larger than the total loss power generated when the required maximum temperature on the product T_{max} is reached, assuming that the magnetron is almost immediately output to this power upon power-up, it is possible to determine the power Q_{heat} , spent on heating the target and the heat-insulated substrate during a temperature change in time as a function of $Q_{heat}(\Delta T)$

$$Q_{heat}(\Delta T) = P_{roundup}(T_{max}) - P(\Delta T),$$

$$Q_{heat}(\Delta T) = P_{roundup}(T_{max}) - [\, Q_{emis}(\Delta T) + Q_{thc}(\Delta T)[or\ Q_{emis2}(\Delta T)] + Q_{evap}(\Delta T) + Q_{melt}(\Delta T) + Q_{cool}(\Delta T)].$$

where ΔT is the temperature excess between the current temperature of the target and its initial temperature equal to the ambient temperature.

Let's depict the obtained dependence $Q_{heat}(\Delta T)$ graphically and form the corresponding trend line, for example, in the form of a polynomial. Assuming the heat accumulated by the substrate is insignificant, it is possible to calculate the target temperature history from the following fundamental equation

$$Q_{heat}(\Delta T)d\tau = c_v m d(\Delta T),$$

where $d\tau$ is the time interval during which the body of mass m with the specific mass heat capacity c_v is heated by the value $dT = d(\Delta T)$ at a temperature T at a power $Q_{heat}(\Delta T)$, spent on heating. Let's transform the formula

$$d\tau = \frac{c_v m}{Q_{heat}(\Delta T)} d(\Delta T).$$

Taking $c_v = const$, we can integrate and determine the minimum time of the target heating to a given temperature

$$\tau = \int_0^{\tau_{heat}} d\tau = \int_{T_0}^{T_{end}} \frac{c_v m}{Q_{heat}(\Delta T)} d(\Delta T).$$

The time of heating the target to the maximum temperature is much less than the time of heating the substrate to the maximum temperature. To calculate the substrate heating time, let's use the same approach as for calculating the target heating time, using the dependence of the power $Q_{heat\_s}(\Delta T)$ spent on the substrate heating during the magnetron sputtering process at the maximum target temperature instead of $Q_{heat}(\Delta T)$.

Let us consider the results of the calculations performed for some experimentally studied processes [8]. The experiments on sputtering copper and titanium targets were carried out by the authors in a vacuum chamber with a volume of $7.8 \cdot 10^{-2}$ m^3, equipped with a flat magnetron with a diameter of 160 mm, operating on a direct current. A target of 1 mm thickness was bolted to a copper water-cooled card with a 1 mm gap. The residual pressure did not exceed 10^{-2} mT. The sputtering was carried out in an argon medium under a pressure of p = 5–10 mT and at a current density of j = (5.5–14.4) mA/cm^2. A thermocouple sensor was used to study the thermal processes on the substrates, its sensitive element being a copper plate 10×10 mm^2 with a thickness of 1 mm, on which a junction of the thermocouple was fixed. In each experiment, the sensitive element temperature change was measured for a given value of p and j during the magnetron operation.

According to the computational model, to heat a copper target in the solid state to a temperature close to the melting point of 1084 °C, the required useful power of the magnetron should be in the range of 465–560 W. At the same time, when the distance between the target and the substrate varies from 40 to 160 mm with the above dimensions of the interacting surfaces, the maximum substrate heating temperature falls from 511 to 203 °C (figure 3, curve 1). For a titanium target in the solid state around the melting point of 1668 °C, the estimated required useful power of the magnetron is in the range of 1220–1500 W, and the maximum substrate heating temperature falls from 712 to 314 °C with a change in the distance from the target in the same range (figure 3, curve 2).

The calculations of the substrate heating time as a function of the target heating temperature and the distance from the target to the substrate were carried out for the distance between the substrate and the target of 120 mm. In this case, for a given system configuration, the substrate heating time for a copper target with a maximum temperature of 800 and 1000 °C is shown in figure 4, for a titanium target – in figure 5.

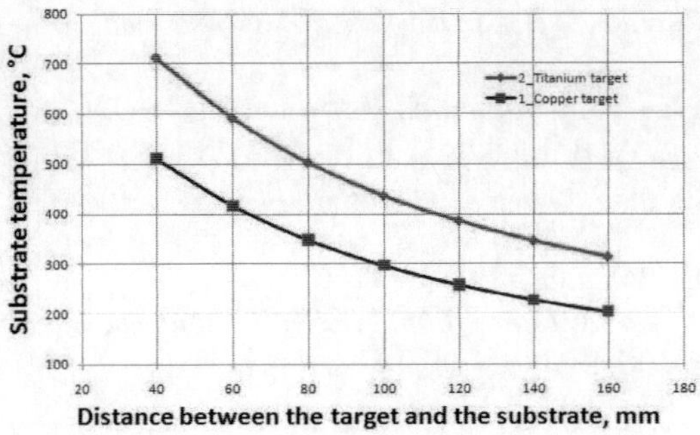

Figure 3. The substrate temperature, °C, change as a function of the change of the distance between the target and the substrate, mm, at the solid target temperature: 1 – copper at 1084 °C, 2 – titanium at 1668 °C.

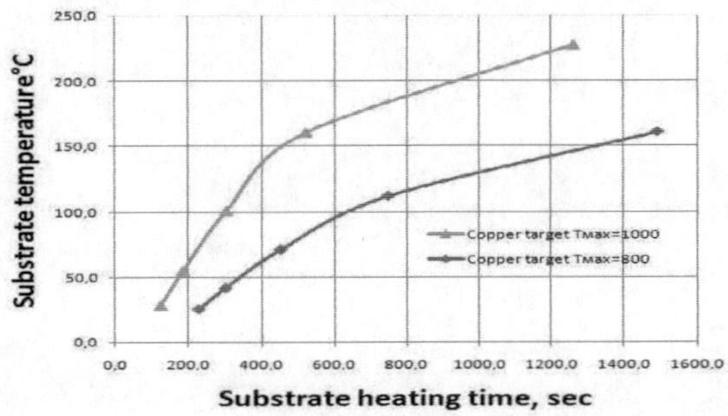

Figure 4. Temperature and time of the substrate heating in the system with a copper target with the maximum temperature of 800 and 1000 °C.

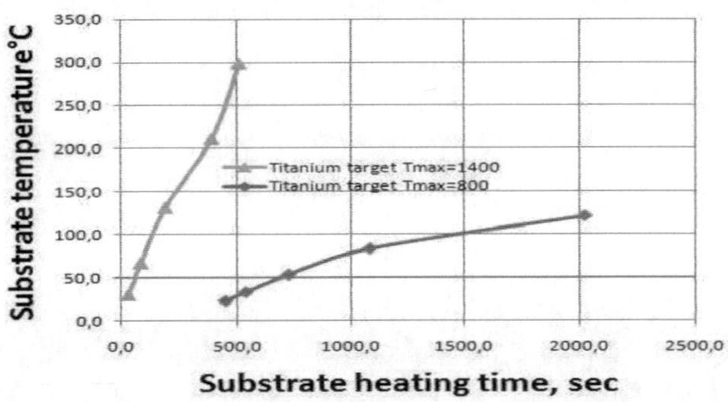

Figure 5. Temperature and time of the substrate heating in the system with a titanium target with the maximum temperature of 800 and 1400 °C.

Hence, a comparison of the experimental data and the presented calculations indicate the possibility of using the presented technique for calculating the heat fluxes and surface temperatures during magnetron sputtering of cold and hot targets, allowing to predict the required power characteristics to obtain the required temperature on the target. The temperature on the substrate can be controlled within certain limits by changing the distance between the substrate and target the value of which can also be estimated through the proposed computational model.

The computational task complexity was associated primarily with the lack of all the necessary data in the source used, for example, concerning the size and relative position of the target and the substrate. Therefore, the values of some important system parameters, which can significantly differ from the real ones, were used in the calculations, what is probably one of the main reasons for the deviation of the calculated results from the experimental ones. The refinement of the system characteristics will allow estimating the accuracy of the model more qualitatively.

References

[1] Shapovalov V I 2010 *Glass Phys. Chem.* **2** 121–57
[2] Poluektov N P, Tsar'gorodsev Y P, Usatov I I and Evstigneev A G 2017 *Thin Solid Films* **640** 60–6
[3] Kravchenko I N, Kolomeychenko A V, Pupavtsev I E, Baranov Y N, Puzryakov A A and Solovev R Y 2017 *Welding International* **10** 809–13

[4] Shapovalov V I, Komlev A E, Bondarenko A S, Baykov P B, Karzin V V 2016 *Phys. Lett. A.* **7–8** 882–5

[5] Loktionov E Yu, Protasov Yu S and Protasov Yu Yu 2014 *Quantum Electronics* **3** 225–32

[6] Ashkihazi E E, Sedov V S, Sovyk D N, Khomich A A, Bolshakov A P, Ryzhkov S G, Khomich A V, Vinogradov D V, Ralchenko V G and Konov V I 2017 *Diamond and Related Materials* **75** 169–75

[7] Bondarenko G G, Fisher M R and Kristya V I 2016 *Vacuum* **129** 188–91

[8] Zavyalov A V, Morozova A A and Shapovalov V I 2016 *International Scientific and Technical Conference INTERMATIC–2016 (Moscow)* pp 106–8

IOP Conf. Series: Materials Science and Engineering **387** (2018) 012011 doi:10.1088/1757-899X/387/1/012011

Gas-discharge laser with controlled output power for medical applications

V V Chernigovskiy, S A Martsinukov, D K Kostrin and A S Kiselev

Department of electronic instruments and devices, Saint Petersburg Electrotechnical University "LETI", 197376, Saint Petersburg, Russia

E-mail: sergm2006@mail.ru

Abstract. In this work the possibility of control of power of radiation of the carbon dioxide laser intended for surgeries is considered. It is shown that the electromagnetic control system allows regulating laser radiation power in the wide range of values that gives the chance of use of the only laser machine in various medical applications. Experimental results of impact of the modulated laser radiation on biological tissues are given.

The use of lasers in surgical practice has a number of advantages due to the specific impact of laser radiation on biological tissues. The possibility of a high concentration of light energy in small volumes makes it possible to selectively affect the biotissue and dose the degree of this effect from coagulation to their evaporation and cut.

Laser radiation allows removing pathological tissue without damaging the surrounding healthy tissue, laser operations are almost bloodless; healing of laser wounds occurs faster and better than using other methods of surgical treatment. Non-contact removal of biological tissues is carried out with minimal trauma and with high accuracy. Good hemostasis in the zone of laser action leads to the fact that there is practically no edema in the wound area, and, consequently, the postoperative period proceeds without pain. Laser radiation has a bactericidal effect, so laser wounds are sterile. Penetrating deep into the tissue, the laser activates the cells, resulting in faster healing of laser wounds. Considering all the advantages of high-energy lasers, it is clear that they are widely used in virtually all areas of surgery.

Very popular are the CO_2-lasers, known as "laser scalpels". For the best use of a CO_2-laser, it is necessary to control the output power of the radiation [1, 2].

Of the main methods used, it should be noted: mechanical, acousto-optical, electro-optical, current. Each of these methods has its own specific drawbacks: the mechanical method does not allow changing the amplitude of the radiation, the acousto-optical modulator has a small radiation stability of the crystal and a large inertia; in the electro-optical one the strong sensitivity of the crystal to damage and the small radiation resistance; with the current control method, oscillatory processes arise establishment of the coherent power of the CO_2-laser, caused by a change in the temperature of the resonator and a change in the composition of the gas mixture [3].

To eliminate these drawbacks, it is advisable to use electromagnetic modulation. In this case, a transverse magnetic field is used to modulate the power of the CO_2-laser. The electromagnetic method of modulation is based on the plasma-optical effect, which is caused by the interaction of the magnetic field with the active medium of the laser. This method allows controlling the output power of a continuous laser, changing its value from maximum to zero according to a linear law with sufficient speed [4].

The principle of operation of the CO_2-laser is as follows (figure 1). When it is turned on in the discharge tube, a glow discharge is formed [5]. The electrons interact with molecules of the gas mixture and excite the upper metastable levels of the CO_2 molecule, thereby providing a population inversion of the active medium. The resulting coherent quanta due to induced transitions lead to additional induced transitions. The energy density of these coherent quanta begins to increase. In accordance with the Einstein relation, the probability of induced transitions increases. The process itself is avalanche-like. As a result, the laser goes into a quasi-stationary mode. The amplification of coherent radiation balances the output of the coherent laser power and the loss of radiation in the active medium. Between the mirrors of the optical resonator of the laser, a standing electromagnetic wave is established. The energy density of this wave in the cross section of the discharge has a Gaussian character, i.e., it is maximal on the discharge axis and drops practically to zero to the walls of the discharge tube. The same distribution has the density of discharge electrons and excited molecules of carbon dioxide.

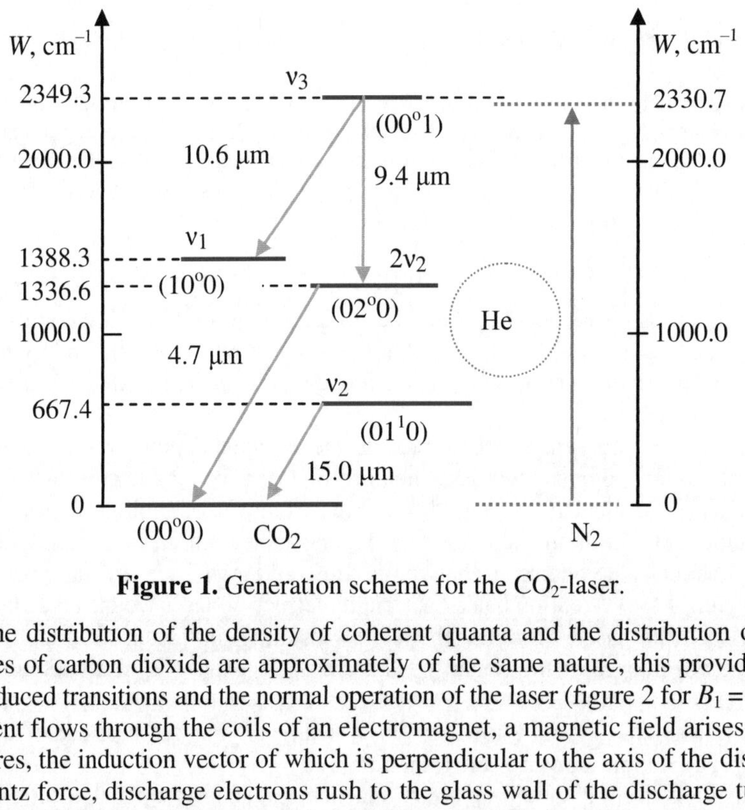

Figure 1. Generation scheme for the CO_2-laser.

Consequently, the distribution of the density of coherent quanta and the distribution of the density of excited molecules of carbon dioxide are approximately of the same nature, this provides the necessary probability of induced transitions and the normal operation of the laser (figure 2 for $B_1 = 0$ and curve *1*).

When a current flows through the coils of an electromagnet, a magnetic field arises in the air gap of the magnetic cores, the induction vector of which is perpendicular to the axis of the discharge tube [2]. Due to the Lorentz force, discharge electrons rush to the glass wall of the discharge tube and create a significant negative potential. Under the action of this potential, the discharge ions also rush to the wall and the entire gas discharge densely presses against one of the walls of the discharge tube. The distribution of electrons and, respectively, excited molecules of carbon dioxide in a plane perpendicular to the axis of the discharge begins to be of an inhomogeneous nature: the concentration of excited molecules of carbon dioxide on the axis decreases practically to zero, and at one of the walls of the discharge tube increases (figure 2 for B_2, B_3 and curves *2, 3*).

In modern lasers, the mirrors of an optical resonator of spherical type are used and the coherent quantum arising near the wall (outside the resonator axis) will be largely lost and do not establish a stationary standing wave in the resonator. In this case, the laser ceases to generate coherent radiation.

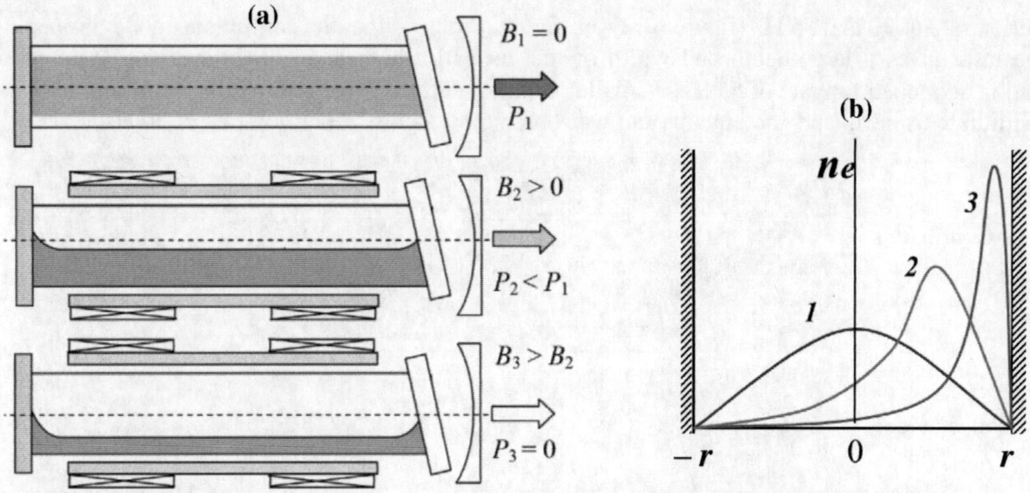

Figure 2. Distribution of charged particles: **(a)** – in the discharge tube of laser, **(b)** – in the discharge gap; *1* – $B_1 = 0$; *2* – $B_2 > 0$; *3* – $B_3 > B_2$.

The electromagnetic modulation system allows adjusting the duration and frequency of the pulses, and also allows adjusting the output power of the laser radiation. This fact makes it possible to use such a laser in many areas of surgery, rather than having many different lasers with different output powers.

Using a CO_2-laser with electromagnetic control of the output power of the radiation, a number of experimental studies were carried out. Investigations were carried out on dead tissues of a pig (skin, kidney and others).

The results of the investigation of the depth of dissection from the duration of pulses show that the dependence has an optimum value in the pulse duration range of 5...10 ms (figure 3). Also, studies have shown that with increasing pulse duration, the efficiency of laser destruction and the level of traumatism increase (figure 4). Therefore, it is necessary to use pulses of short duration.

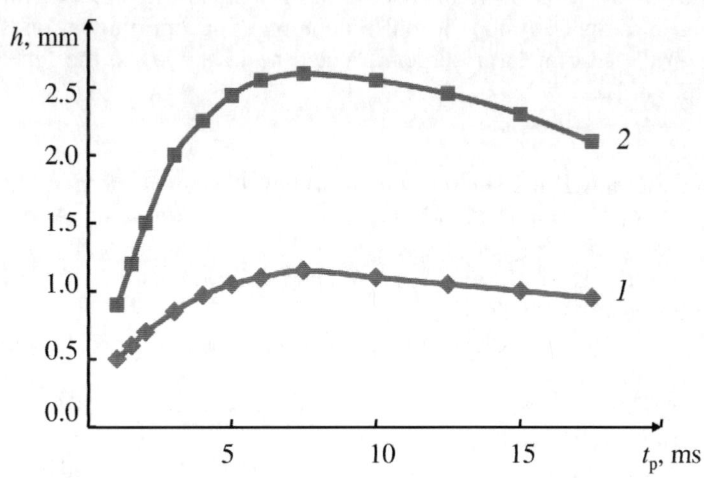

Figure 3. Dependence of depth of opening on pulse duration: *1* – bone, *2* – kidney.

The mode of operation of the laser has a strong effect on the width of the cut, with increasing frequency and duration of laser pulses, the width of the cut and the zone of traumatism increase [6]. The smallest zone of traumatism is observed at a repetition rate of pulses per unit Hz.

Studies were carried out on the quality of the biotissue cut by the laser beam at a relative speed of 3 cm/s in the continuous mode of operation of the laser and pulsed with duration of 5 ms and a repetition rate of 30 and 55 Hz (figure 4). From figure 4 follows that the continuous mode of operation of the laser gives a large depth and width of the cut and a high level of traumatism. With pulsed operation at a repetition rate of 55 Hz, a smaller depth of cut was observed, but the level of traumatism was still high. A clean and good quality cut was obtained at a pulse repetition rate of 30 Hz.

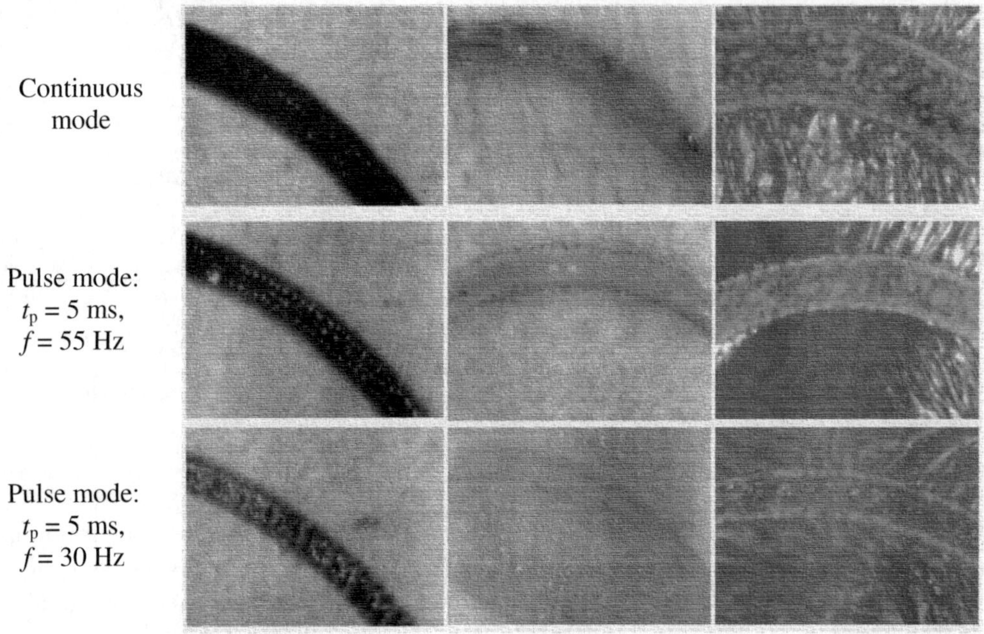

Figure 4. The top view of the cut made by a laser beam with a speed of 3 cm/s and its different modes of operation.

Thus, in order to ensure a low level of traumatism in healthy tissues, it is preferable to use CO_2-laser installations operating in the pulsed mode at pulse duration of several milliseconds and a repetition rate of up to 10 Hz in the case of laser surgery. When the laser moves, the repetition rate can be increased to several tens of Hz.

References

[1] Kiselev A S and Smirnov E A 2017 *Journal of Physics: Conference Series* **929** 012050
[2] Chernigovskiy V V, Kostrin D K, Martsinukov S A and Lisenkov A A 2016 *Vakuum in Forschung und Praxis* **6** 34–7
[3] Kiselev A S, Kostrin D K, Lisenkov A A and Smirnov E A 2017 *Journal of Physics: Conference Series* **789** 012027
[4] Martsinukov S A, Kostrin D K, Chernigovskiy V V and Lisenkov A A 2016 *Journal of Physics: Conference Series* **729** 012023
[5] Ramazanov A N, Kostrin D K, Goncharov V D and Lisenkov A A 2016 *Journal of Physics: Conference Series* **729** 012004
[6] Martsinukov S A, Kostrin D K, Chernigovskiy V V and Lisenkov A A 2016 *Journal of Physics: Conference Series* **735** 012049

Experimental study of the fragmentation of van der Waals gas clusters by electron ionization

S T Chinenov, A E Zarvin, V E Khudozhitkov, V V Kalyada and A S Yaskin
Department of Applied Physics, Novosibirsk State University, 630090, Novosibirsk, Russia

E-mail: s.chinenov@g.nsu.ru

Abstract. The fragmentation of gas clusters of argon, nitrogen and methane during ionization by electron impact have been examined. Clusters were formed when the gas flowed through the supersonic nozzle into the vacuum. A molecular beam was selected from the supersonic jet by the skimmer, then it reached the analyzer of the quadrupole mass spectrometer Hiden Epic 1000. The ionization of the gas was carried out in the molecular beam by the intrinsic ionizer of the detector with an electron energy of 70 eV and in the supersonic jet by an electron beam (10 keV). A comparison of mass spectra obtained by both methods with different parameters of the supersonic jet has been done. The difference of the obtained cluster distribution has been substantiated; the disadvantages of using the classical configuration of quadrupole mass spectrometry have been analyzed. Dependences of the percentage composition of the recorded flows have been obtained. Distinctive "magic" cluster numbers have been found.

Cluster is considered to be a composition of atoms or molecules located in a state between a molecule and a solid [1]. The use of these particles lies in a wide range: from polishing surfaces to creating organic structures. The most common method of creating gas clusters is the use of gas flowed into the vacuum through sound or supersonic nozzles. Such adiabatic expansion of the flow leads to the formation of associates within which the particles are retained by van der Waals forces and have a wide size distribution. Since the study of clusters and their characteristics involve consideration of the composition of jets, one of the best approaches to the study of these objects is the ionization of clusters with subsequent detection of the flow by a mass spectrometer. However, the application of this technique requires special conditions for analysis due to a number of reasons.

Firstly, quadrupole mass spectrometers discriminate particles by mass [2]. This happens because the energies of the ionized particles and the energies of the fragments detached from them can vary greatly. At the same time, focus parameters are calculated for the particles with the same energy. Therefore, the difference in the phase volumes of identical particles leads to signal losses. So it can significantly affect the correctness of the different masses signals ratio for cluster flows with a wide size distribution.

Secondly, it is known from the literature [1, 3–5] that one of the features of gas clusters is their intense fragmentation during the electron ionization. Multiple ionization results in the appearance of Coulomb repulsion that in turn leads to the collapse of the cluster into fragments the size of which can vary greatly. On the other hand, the electron energy used in ionizers of quadrupole mass spectrometers usually lies in the region of 40–70 eV that corresponds to the maximum ionization cross section of most molecules and considerably exceeds the binding energy of van der Waals associates. So, if the task is to analyze the composition of cluster flows, it is better to increase electron energies in order to reduce the fragmentation effect by decreasing the time of collisions.

In addition, it is worth mentioning the distortion possibility of the recorded flow due to the effect of jet selection [6]. This feature is connected to the fact that the velocity ratio is proportional to the mass of the particle. This results in the dependence of the particle masses on the distance to the axis of the jet. Therefore, the selection of the axile part by the narrow skimmer for the purpose of forming a molecular beam, that is used in the most configurations for flow analysis, facilitates the selection of large clusters. Thus, it is necessary to take into account these features when analyzing gas clusters.

The study was performed on an experimental gas-dynamical assembly LEMPUS-2 at Novosibirsk State University [7]. The scheme of experiments is shown in figure 1.

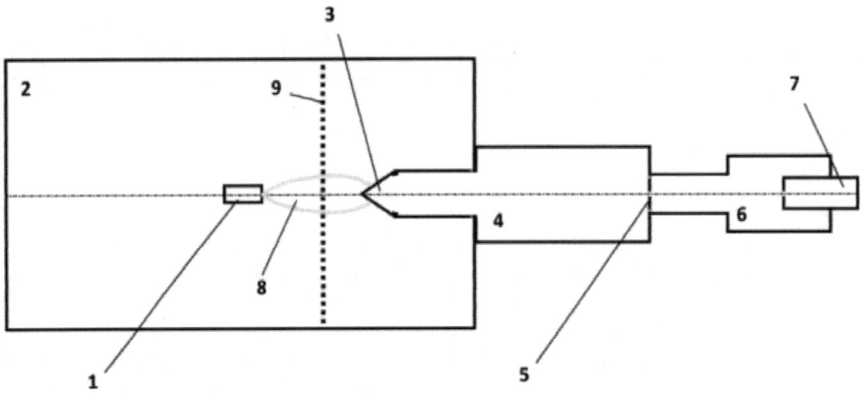

Figure 1. The scheme of experiments: 1 – nozzle supply; 2, 4, 6 – extension section, section after the skimmer and detector section, respectively; 3 – skimmer; 5 – collimating diaphragm; 7 – mass spectrometer; 8 – flowing jet; 9 – electron beam.

Clusters were formed when the gas flowed from the nozzle supply (1) through the sound or supersonic nozzle into the extension chamber (2). A molecular beam was selected by the skimmer (3) with the diameter of 0.44 mm and the angles of 40–45° and then through the section after the skimmer (4), the collimating diaphragm 3 mm in diameter (5) and the detector section (6) reached the input aperture of the mass spectrometer (7) with the diameter of 4 mm. The gas flow (8) was ionized in the expansion chamber by the high voltage (10 keV) electron beam (9) the source of which was an electron gun with a hollow cathode or the intrinsic ionizer of the mass spectrometer with the electron energy of 70 eV. In the first case the skimmer and the collimator were used as electrostatic lenses in order to minimize ion losses during the transport of charged clusters along the assembly to the detector, a series of experiments is provided to optimize the delivery of ions. The quadrupole mass spectrometer Hiden Analytical EPIC 1000 which can make records in the range from 1 to 1000 a.m.u., was used in this work. According to the received mass spectra, the sensitivity of the instrument is 7 orders of magnitude. The pressure in the nozzle supply P_0 ranged from 21 to 500 kPa.

Experiments with the sound nozzle 0.55 mm in diameter on argon and with the supersonic nozzle of 0.17-8.2-2.2 and 0.215-17-3.8 (the diameter of the critical section – the length of the diffuser - the diameter of the outlet section) on methane and nitrogen were carried out. The gas temperature in the pre-chamber in the experiments was 298 ± 1 K. The formation of the molecular beam was in the region of absence of both skimmer effect and the influence of the Mach disk to avoid the gas dynamics distortion of the flow. The choice of the sound nozzle for the analysis of argon clusters is caused by the large average sizes <N> of neutral clusters (calculated by the Hagena's formulae [8]) in comparison with nitrogen and methane. In order to increase the amount of condensate in the jets of methane, helium was added in various proportions.

The comparison of the mass spectra of argon at P_0 = 100 kPa (average cluster size <N> = 25) is shown in figure 2. The mass spectrum obtained as a result of supersonic jet ionization (blue markers e-b) shows modal distribution behaviour. In this case the amplitudes of the cluster peaks are quantitatively higher than the values obtained during ionization by the intrinsic block. Consequently, the use of the high voltage electron beam gives more informative signals. This result is also confirmed with other values of stagnation pressure, i.e. with other average cluster size.

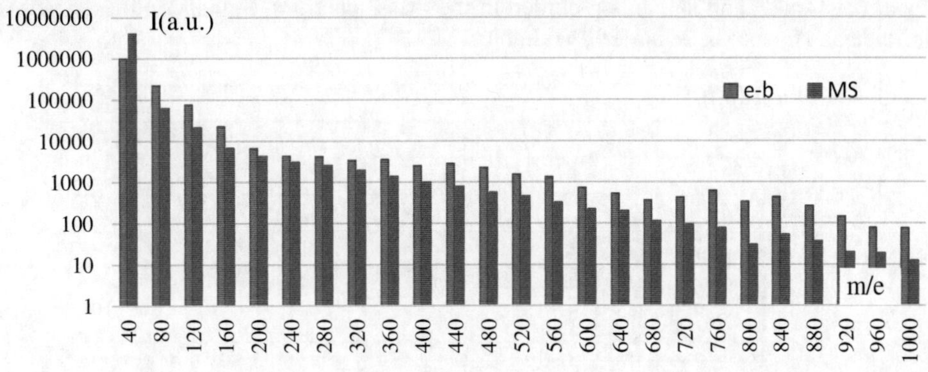

Figure 2. Comparison of the mass spectra of argon clusters during different ionizations: red – ionization in the mass spectrometer; blue – ionization in the expansion section.

Figure 3 shows the percentage ratios of amplitudes of the recorded clusters depending on their size N for three different stagnation pressures, respectively, for the three average cluster sizes calculated by the formula of O. Hagena [8]. At the same time, an increase of <N> resulted in not only to the broadening of the distribution in the direction of large masses but also to the significant reduction of the ratio of the amplitudes between the signals of the monomers and other small clusters; that corresponds to the concept of the distributions behaviour when the pressure changes. Interestingly, the presence of regular singularities for m/e = 11, 14, 19 and 21 that coincides with the data on argon "magic" cluster numbers [9]. However, the monomer signal remained predominant for the selected pressure range.

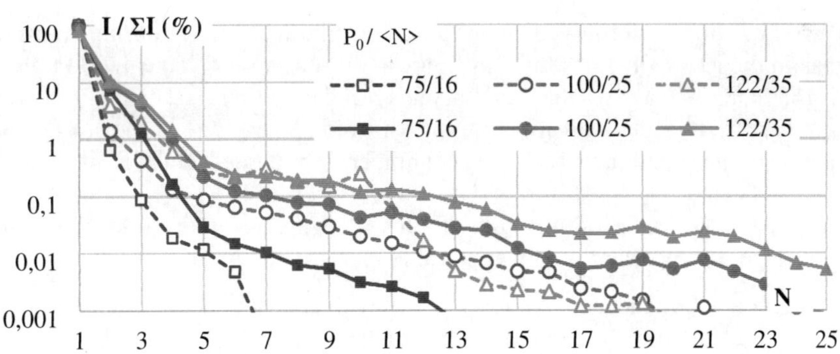

Figure 3. Percentage composition of the recorded mass spectra for different P_0 and, respectively, values of <N>: dashed lines – ionization of the molecular beam; solid lines – ionization of the jet.

Another result is observed when nitrogen jets flow in the conditions of large clusters formation. Figure 4 shows the data obtained at stagnation pressure P_0 = 400 kPa. In this mode the evaluation according to the parameter of O. Hagena gives an average cluster size of <N> = 254. The mass spectrum recorded by the intrinsic ionizer of the mass spectrometer, figure 4(a), consists of a monomer signal (m/e = 28) that is almost three orders of magnitude smaller than a dimer signal (m/e = 56) and is about an order of magnitude smaller in the amplitude of the group of oligomers from 6 to 12 (m/e from 168 to 336).

In the case of ionization by the high voltage electron beam, figure 4(b), a set of cluster peaks is much larger in amplitude and composition. The maximum amplitude of the signal is recorded on hexamers of nitrogen; the amplitude of the monomers is smaller than the one of the clusters, from dimers to hexamers inclusive. This is completely different from the spectrum of the device's intrinsic block. The change of the stagnation pressure results in the displacement of the maximum peak towards larger or smaller masses which is not observed with the standard configuration of the ionizer. Also a group of peaks at N = 24 and above is recorded in the mass spectrum. Signals on these masses have a small amplitude and cannot be regarded as reliable.

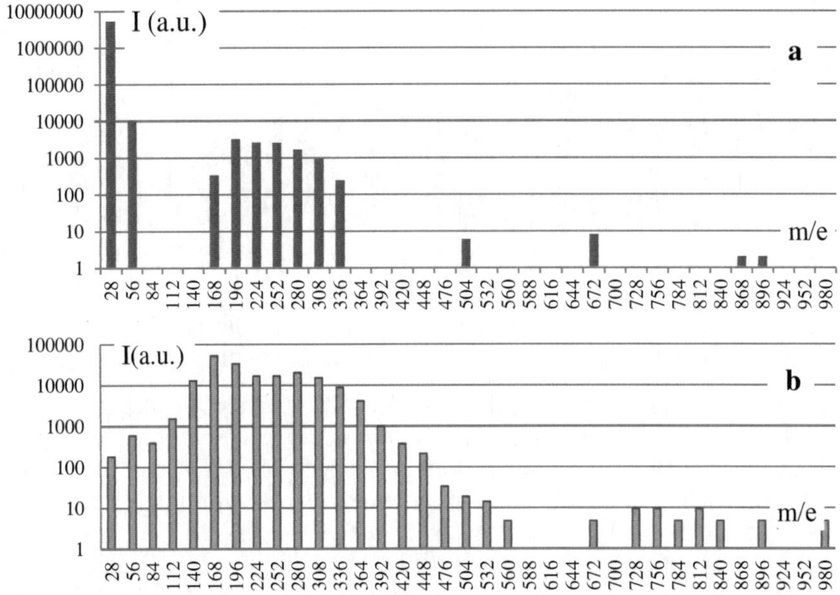

Figure 4. Mass spectra of nitrogen clusters ionized in the detector of the mass spectrometer (**a**) and in the expansion chamber (**b**), P_0 = 400 kPa, <N> = 254.

Several local maxima are also observed in the mass spectrum recorded during ionization by high voltage electrons in the jet. As it is observed in figure 4(b), at a pressure P_0 = 400 kPa the maxima are at m/e = 56, 168 and 280 a.m.u. that corresponds to the cluster size N = 2, 6, and 10. The measurements at variations of the stagnation pressure proved that the size of the clusters with absolute or local maximum of the amplitude shifts. The number of recorded clusters of a larger size also changes.

The dependence of the value of the size of clusters with absolute or local maximum of the amplitude on the stagnation pressure is shown in Table 1.

Table 1. Dependence of cluster size N with absolute (A) or local (L) maximum of the peak amplitude on the stagnation pressure P_0.

	N											
P_0, kPa	1	2	3	4	5	6	7	8	9	10	11	12
100	A											
200		A										
300		L				A						
400		L				A				L		
500		L				L					A	
600		L				L						A

Results that have been obtained in molecular nitrogen during ionization in the jet by a high voltage electron beam on "half" masses have been formed as a result of dissociative ionization of molecular nitrogen and its clusters are also interesting. The obtained mass spectrum at recording of all peaks with nonzero amplitude at stagnation pressure P_0 = 500 kPa is shown in figure 5. It is noticeable that the amplitudes of the peaks $(N_2)_{n-1}N^+$ are less than $(N_2)_n^+$ for all values of n, except n = 2, at the same time, when n grows, the ratio first increases to 40-fold but then decreases to 20-fold.

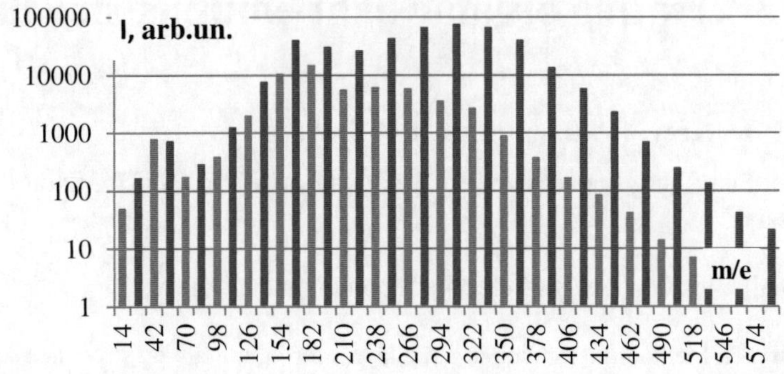

Figure 5. Mass spectra of nitrogen clusters ionized in the expansion chamber. P_0 = 500 kPa. The blue lines - the ion amplitudes $(N_2)_n^+$, the red lines - $(N_2)_{n-1}N^+$, where n is whole numbers.

Undoubtedly, the processes of interaction of ions and neutral particles on the path from the site of ionization to the ion detector have a great impact in such difference of the results using different methods of ionization, that was also noted by other authors [2–3, 10]. Also it can be assumed that the effect obtained in nitrogen jets unlike argon can be caused by the physical features of molecular clusters in comparison with the atomic ones. In addition, it is possible that there is influence of volume charge compensation by negative nitrogen ions during the transportation of ions formed by the high voltage electron beam in the jet to the detector of the mass spectrometer.

The results of this study demonstrate the possibilities and advantages of using external ionization in the molecular beam cluster gas mass spectrometry.

Acknowledgment
The study was carried out using the equipment of the Center for Collective Use "Applied Physics" of NSU with the financial support of grants of the Ministry of Education and Science of the Russian Federation 3.5918.2017/ITR and 3.5920.2017/ITR.

References
[1] Bonhommeau D, Halberstadt N and Viel A 2006 *J. Chem. Phys.* **124** 184–8
[2] Almazov V B, Andreeva A D, Gaal L N and Pomosov T V 2012 *Sci. Instr. Making* **22** 107–12
[3] Buck U and Schutte S 2002 *Int. J. Mass Spect.* **220** 183–92
[4] Van der Burgt P J M and McConkey J W 1995 *J. Chem. Phys.* **102** 8414
[5] Haberland H 1985 *Surf. Sci.* **156** 305–12
[6] Zarvin A E, Kalyada V V and Khudozhitkov V E 2017 *T&A* **24** 691–702
[7] Zarvin A E, Kalyada V V, Madirbaev V Gh, Korobeishchikov N G, Khodakov M D, Yaskin A S, Khudozhitkov V E, Gimelshein S F 2017 *IEEE Trans. Pl. Sci.* **45** 819–27
[8] Hagena O F 1981 *Surf. Sci.* **106** 101–16
[9] Eletskii A V and Smirnov B M 1989 *Sov. Phys. Usp.* **32** 763–82
[10] Khodorkovskii M A, Artamonova T O, Murashov S V, Michael D, Rakcheeva L P, Belyaeva A A, Timofeev N A, Mel'nikov A S, Shakhmin A L and Dement'ev I A 2009 *Tech. Phys.* **1** 1–6

Technology of creating organic light-emitting structures

A E Degterev, I A Lamkin and **S A Tarasov**

Saint Petersburg Electrotechnical University "LETI", 197376, Saint Petersburg, Russia

E-mail: alex.degterev@rambler.ru, ialamkin@mail.ru

Abstract. Methods of creating organic layers for light-emitting structures were investigated in this work. The photo- and electro-luminescence spectra of the made structures were measured. The necessary conditions for the creation of optimal layers were experimentally revealed.

1. Introduction

Organic light-emitting structures (OLED) are semiconductor devices created from a number of thin organic films. Light is emitted, when the electric current passes through the structure. The structure of OLED consists of a transparent substrate, a transparent anode, transport layers with a hole and electronic conductivity, an active layer, and a cathode [1].

Each material must be applied in a specific way. In this paper we used the method of vacuum thermal evaporation and the spin-coating method. The choice of the way to create the structure is a serious task. The properties of each of the layers, including the electrodes, greatly influence the parameters of the manufactured sample.

2. Experiment: change of substrate heating temperature

Organic light-emitting structures FTO/PEDOT:PSS/TPD/Alq3/Al were created. Transport layers such as PEDOT:PSS, TPD and Alq3 were alternately applied on the glass substrate with a layer of FTO. Then were created aluminum contacts.

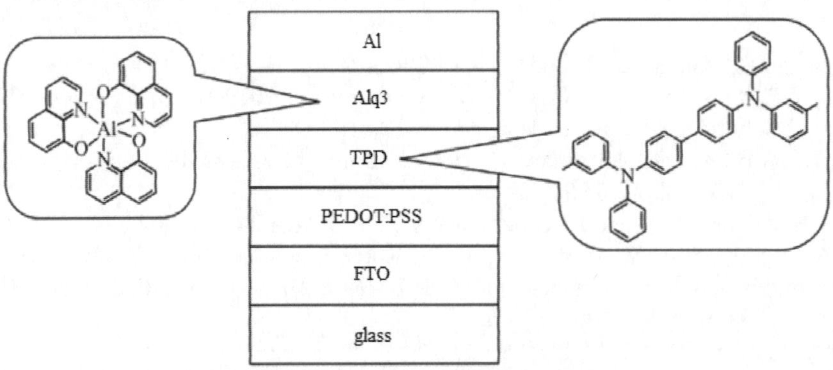

Figure 1. Structure of created OLED [2].

Substance PEDOT: PSS is a polymeric compound. Its properties are p-type conductivity and electron-blocking. The TPD (N, N'-Bis (3-methylphenyl) -N, N'-diphenylbenzidine) is a small molecular compaund. It has a hole conductivity. The material Alq3 (Tris- (8-hydroxyquinoline) aluminum) has an electronic conductivity and it is an active layer in the created structures.

Due to the large molecular weight of the material, the evaporation temperature of the compound exceeds the decomposition temperature, so PEDOT: PSS was applied to the substrate with FTO by the spin-coating method. The centrifugation time was 30 seconds, the speed was 3000 r/s, the number of layers was 1. The TPD and Alq3 materials are small-molecular compounds with low solubility. Because of that, they were applied by the method of vacuum thermal evaporation. The precipitation was carried out from a ceramic melting pot heated by a molybdenum heater. The pressure of the residual gases did not exceed 10^{-5} Torr. It was made to maintain the maximum purity of the evaporated material upon deposition on the substrate.

It was established, the temperature of the glass substrate's heating affects the quality of the obtained layers. The photoluminescence spectra of the TPD and Alq3 layers were investigated to determine the optimum substrate heating temperature.

Investigation of the materials photoluminescence was done to determine the homogeneity of the created layers. It was carried out with a scanning excitation laser beam over the entire area of the sample. The optical power of the emitting, measured in relative units, was calculated from the spectral characteristics. Then the surface of the photoluminescence emission power deviation from the mean value was compiled for different substrate heating temperatures. Figure 2 shows the surface of the deviation of the Alq3 photoluminescence emission power deposited at a substrate temperature 60 °C from the average power value. The data for all the measured layers are summarized in table 1.

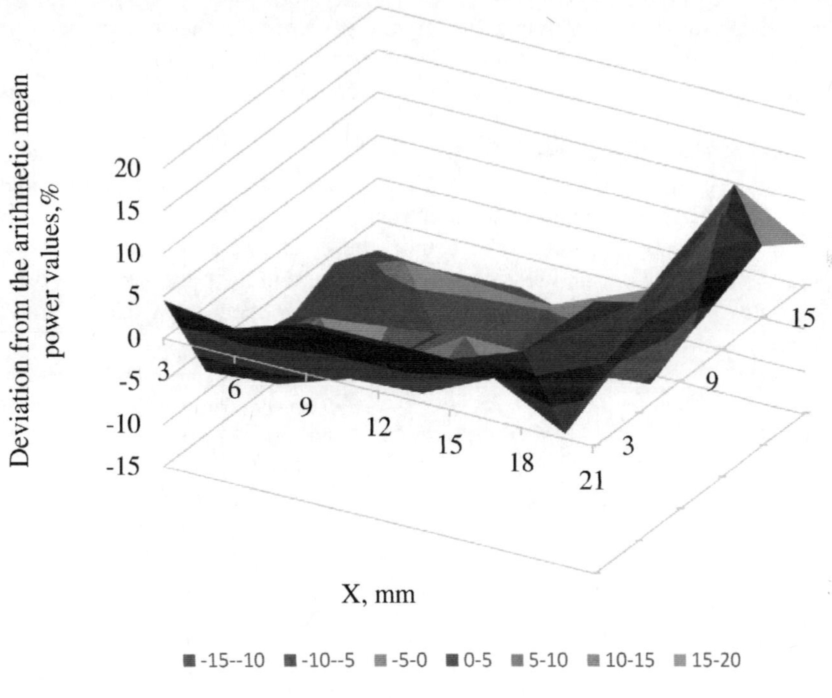

Figure 2. The surface of the deviation of the photoluminescence emission power of Alq3 deposited at a substrate temperature 60 ° C from the average power value.

It turns out that the most homogeneous layer of TPD is obtained, when the substrate is heated to 50 °C, and for Alq3 the temperature should be 60 °C. Thereafter, it is better to keep the above said temperatures when these materials are alternately sprayed. There is no necessity to cool down the substrate to room temperature after the deposition of TPD. It could be immediately heat up to 60 °C and then spray Alq3. This procedure didn't affect on the uniformity of the layers.

Table 1. Investigation of the layers photoluminescence as a function of the substrate heating temperature.

Material	Annealing temperature, °C	The average emitting power over the surface of the structure, rel. units	Deviation from the average power, %
TPD	51	1173	10
TPD	61	967	9,4
TPD	67	320	21
TPD	77	170	17
Alq3	50	690	14
Alq3	60	727	17
Alq3	74	670	16
Alq3	80	477	22
Alq3	88	606	15
TPD + Alq3	51	950	8
TPD + Alq3	TPD at 51 Alq3 at 60	1384	11

It is important to say, there are some increase in the amount of deposited material to the edges of the substrate. The search for a solution to this feature of sputtering is currently underway. However, as recommendations to create better layers, contact shouldn't be created near the edges of the substrate.

3. Experiment: change in residual gas pressure

The next part of the study in the work had become a comparative characteristic of two identical in composition light-emitting structures created at different degrees of evacuation of the vacuum. By the method of vacuum thermal evaporation for one structure, the layers were deposited at a pressure of residual gases no higher than 10^{-5} Torr, and for another at a pressure in the region of 10^{-4} Torr.

The created structures FTO/PEDOT:PSS/TPD/Alq3/Al have their own electroluminescence spectrum shown in figure 3.

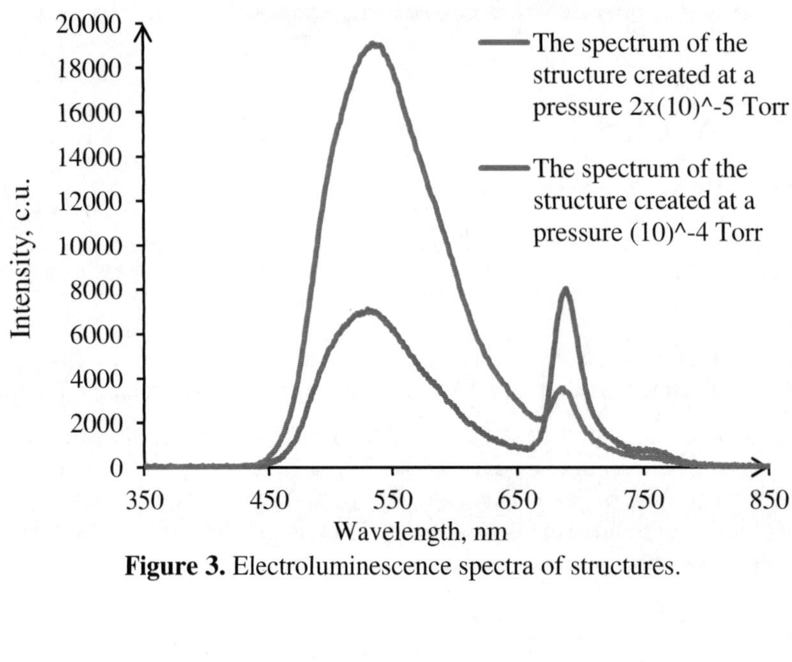

Figure 3. Electroluminescence spectra of structures.

This dependence clearly demonstrates the need to reduce the pressure of residual gases in the creation of organic films. The emitting intensity of the structure is strongly influenced by the uniformity of the layers deposition and the preservation of the material purity upon deposition on the substrate.

The number of particles collisions of the evaporated material increases because of the high content of air molecules and other impurities in the chamber. This leads to hence the length of their free path decreases and that's why amount of pure precipitated material decreases. The uniformity of the deposition of layers is destroyed. More impurities appear in film, that leads to a distortion of the spectral characteristic. The solution to this problem is to fill the chamber with an inert gas, for example argon [2]. Also any contact of structures with the environment should be prevented.

4. Conclusions

Organic light-emitting structures FTO/PEDOT:PSS/TPD/Alq3/Al were created. Optimal temperatures of substrate heating for organic materials such as TPD and Alq3 were found in order to ensure high uniformity of the deposited layers using method of vacuum thermal evaporation.

It also was demonstrated the necessity of the maximum possible reduction of residual gas pressure to create more efficient light-emitting structures.

Acknowledgment

The work was supported by the Ministry of Education of the Russian under project of the state task in the field of scientific activity, project number 16.1750.2017/4.6.

References

[1] Stepanov E M, Mikhailov I I, Tarasov S A and Solomonov A V 2016 *Proceedings of the 2016 IEEE North West Russia Section Young Researchers in Electrical and Electronic Engineering Conference (ElConRusNW)* pp 82–4

[2] Vitukhnovskii A G, Vaschenko A A and Lebedev V S 2013 *Physics and Technology of Semiconductors Mechanism of electron excitation transfer in organic light-emitting devices based on semiconductor quantum dots* **47(7)**

[3] Vyavahare O 2009 *Fabrication and Characterization of Organic Light Emitting Diodes for Display Applications: Master's thesis* (Rochester Institute of Technology, Henrietta, NY, USA)

Development of calculation method of main parameters of vacuum liquid transportation system

K E Demikhov, A A Ochkov and U S Gordeeva

Department of Vacuum and Compressor technology, faculty of Power engineering, Bauman Moscow State Technical University, Moscow 105005, Russia

Abstract. The design of a vacuum pumping system based on the principle of the displacement membrane was proposed. This principle does not imply contact of the working metal surfaces with the transported liquid. This article presents a calculation method of the main parameters of the proposed vacuum fluid transportation system. As a result of the calculation, the values of the pumping speed, the dimensions of the displacement membrane, the required chamber height, the pressure drop in the chamber, required to ensure the deflection of the displacement membrane, were obtained. Based on the proposed method, a program for calculation the main parameters of a liquid transfer system based on the mathematical software MathCAD was developed.

1. Introduction

Systems for dosed transportation of liquids without allowing it to contact with various metallic surfaces of the system are required in certain areas of science and technology [1]. For example, to restore blood circulation in the damaged areas of the heart, in addition to medical treatment, reconstructive surgeries are performed [2]. If the disease is inoperable organ transplantation is performed. Thus, the creation of a system capable of creating and maintaining the required differential pressure to ensure blood circulation is a relevant problem. For determination of the required parameters of a vacuum system that provides specified characteristics, development of a calculation method is required [3, 4]. The developed method is considered in the context of the blood transportation process by heart.

2. Description of the system design for which the calculation method is developed

A schematic diagram of the proposed construction of a vacuum fluid transportation system is shown in figure 1. The principle of the system operation is the pumping of liquid by creating and maintaining a required pressure drop by a vacuum pump. The chamber is divided into two compartments: hydraulic and vacuum, using of an elastic membrane [5]. The upper hydraulic compartment of the chamber has two inputs for supply and outflow of liquid. The lower part of the chamber also has two inputs for connecting the vacuum pump and control valves. One input serves for pumping the chamber down, the other one for gas injection. After starting the vacuum pump and opening the valves in the inlet *VE2*, *VE4*, the chamber is pumped down and the elastic membrane is bent due to the pressure difference between vacuum and hydraulic compartments. At the same time, the control hydraulic valve *VE5* is opened in the upper part of the chamber, the liquid enters the chamber under the effect of pressure difference [6]. Valves *VE, VE3, VE6* are closed at this time. After closing the valves *VE2, VE4* and the valve *VE5*, the outlet valves *VE1, VE3* are opened and the gas from the pump discharge of the vacuum pump enters the chamber and, equalizing the pressure, restores the elastic membrane element into its undeformed shape that simultaneously pushes the liquid located in the upper part of the chamber. At the same time, the second hydraulic valve *VE6* is opened and the liquid is ejected from the chamber. Valves *VE2, VE4, VE5* are closed at the same time. Then the cycle repeats. The pressure in the upper and lower parts of the chamber is controlled by pressure sensors throughout the process.

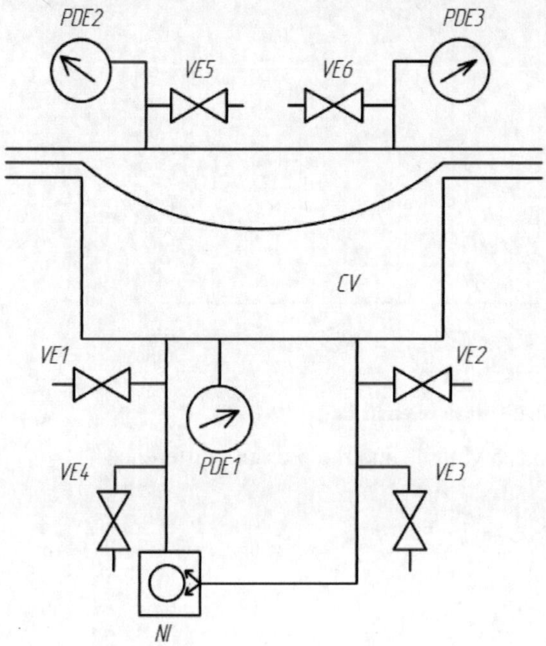

Figure 1. A schematic diagram of a vacuum fluid transportation system, where
CV – vacuum chamber;
NI – mechanical vacuum pump;
PDE1,2,3 – differential pressure sensors;
VE1-6 – solenoid valves.

3. To determine the main parameters of the vacuum system, the following calculation algorithm was proposed

The main parameters of the developed system can be divided into performance characteristics and control parameters [7]. The control parameters include such geometric parameters of the system as:

D_{in} – internal diameter;
t_{membr} – membrane thickness;
Hydraulic compartment parameters such as:
V_{disp} – liquid volume per stroke;
Membrane mechanical properties:
E - Young's modulus of the membrane material;
μ - Poisson ratio;
σ_u – ultimate tensile strength.
The main performance characteristics are:
S_n – pumping speed;
Δp – vacuum compartment pressure difference.

The design process starts with setting the value of D_{in} based on the system's initial conditions and constraints. Finite element is conducted in order to determine the required pressure difference value whereby the displaced volume created due to membrane deflection is equal to the required volume per stroke. After that the chamber height is determined based on the vertical deflection of the membrane's centre when the required pressure difference acts upon it. The relationship between the displaced volume and the pressure difference is presented in figure 2.

It is assumed that the resistance influence of all quick coupling elements of the system reduces the pumping speed by no more than 10 % [8].

The flowchart of the algorithm for calculating the main system's parameters is shown in figure 3.

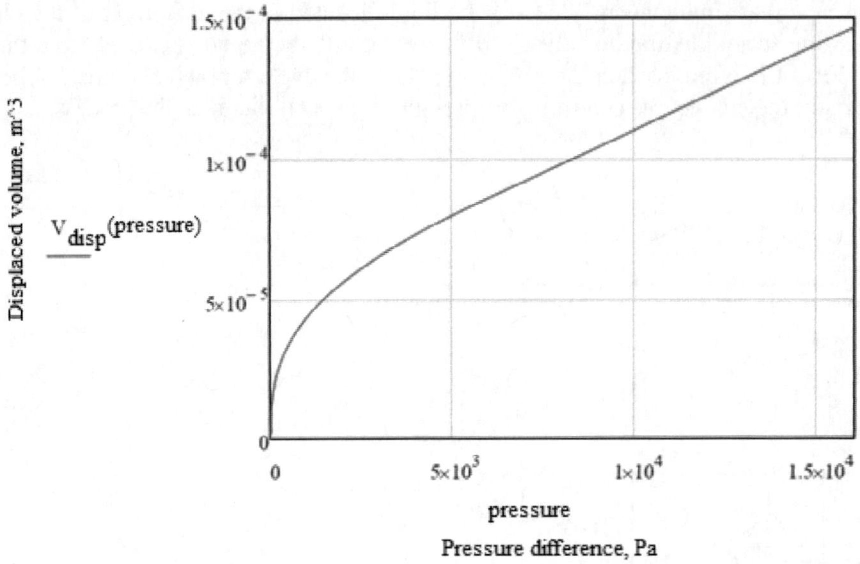

Figure 2. The relationship between the displaced volume and the pressure difference.

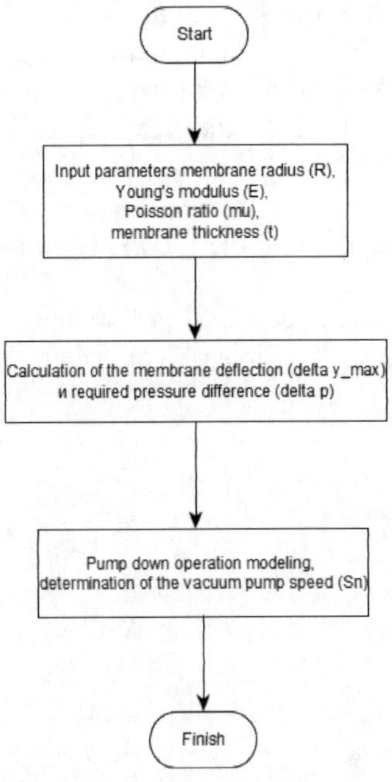

Figure 3. Flowchart of the algorithms of vacuum liquid transportation system main parameters calculation.

The chamber's volume changes caused by deflection of the membrane during pump down necessary to take into account for the pumping speed determination. The problem is solved by a numerical analysis. The whole time interval of modelling is divided into sufficiently small segments, during which it is assumed that $p = const$.

The following iterative algorithm is used:

1. Determination of the gas volume pumped down during a small time interval;
2. Determination of the pressure value at next iteration using equation (1)

$$p_{i-1} \times V_{i-1} = p_i \times (V_{i-1} + \Delta V) \quad (1)$$

3. Determination of the displaced volume caused by pressure difference $p_{hydr} - p_i$;
4. Calculation of the vacuum chamber's volume with the displaced volume taking into account

$$V_i = V_{i-1} - V_{disp} \tag{2}$$

5. Go to the next step.

The relationship between the displaced volume, the pressure difference during the stroke time (0.4 s) and the pumping speed are determined. Using this pumping model, it is possible to determine the required value of the pumping speed S_n.

The application of the proposed calculation method of main parameters of vacuum liquid transportation system based on the example of the heart, yielded the values required vacuum pumping speed: S_n=*0,744* L / min and pressure difference $\Delta p = 3.659 * 10^3$ *Pa.*

The pump down chart for two strokes (1 cycle) is shown in figure 4.

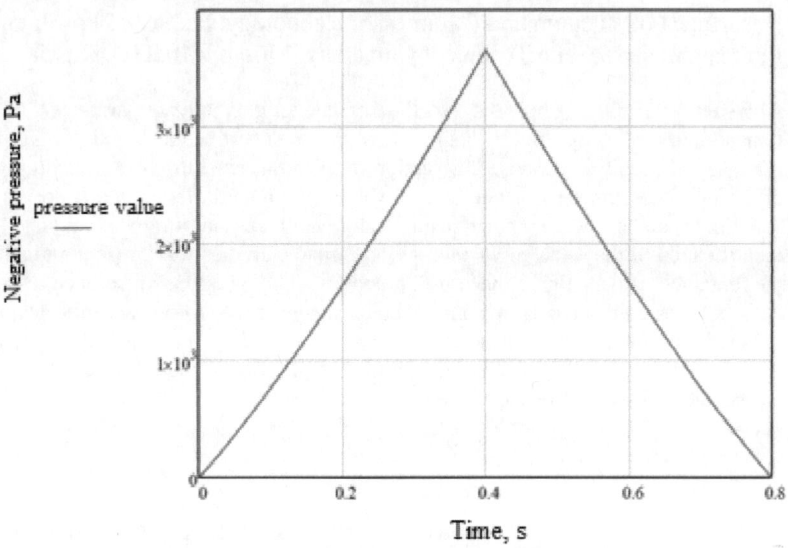

Figure 4. The pump down chart for two strokes.

4. Conclusion

The proposed calculation method allows to determine the main parameters of the vacuum system for obtaining required liquid flow rate. The future work will be devoted to the designing and manufacturing an experimental facility for the verification of the proposed calculation method.

References

[1] Demikhov K E, Panfilov Yu V, Nikulin N K and Avtonomova I V 2009 *Mechanical Engineering* 590
[2] Bokeriya L A 1999 *Lektsii po serdechno-sosudistoy hirurgii* (Moskva, Izd-vo *NTsSSH im. A.N. Bakuleva RAMN*) 311–39
[3] Borisov Yu A and Chernishev A V 2015 *Biomedical Engineering* **4** 246
[4] Gudkov A G et al. 2016 *Biomedical Engineering* **3** 214
[5] Pugachuk A S and Chernishev A V 2015 *Biomedical Engineering* **4** 231–5
[6] Volkov V Yu et al. 2013 *Thermal Engineering* **6** 429–33
[7] Volkov V Yu et al. 2015 *Chemical and Petroleum Engineering* **11–12** 799–805
[8] Kyurdzhiev Yu V, Ochkov A A, Darutkin A A and Avilenko KV 2016 *Mashinostroitel* **4** 61–3
[9] Petrov I B 2009 *Trudyi Moskovskogo fiziko-tehnicheskogo instituta* **1** 5–16

IOP Conf. Series: Materials Science and Engineering **387** (2018) 012015 doi:10.1088/1757-899X/387/1/012015

Experimental study of a vacuum fluid transportation system

K E Demikhov, A A Ochkov and U S Gordeeva

Department of Vacuum and Compressor technology, faculty of Power engineering, Bauman Moscow State Technical University, Moscow 105005, Russia

Abstract. The authors proposed a method of calculation the main parameters of a vacuum fluid transportation system. In this paper a designed experimental stand for verification of the proposed method was proposed. Selection of the required equipment including vacuum pumps, differential pressure sensors and control valves was made. The body and covers of the pumped vacuum chamber were made using additive manufacturing techniques. A subsequent identification of the selected vacuum pumps' characteristics was carried out taking into account the conductivity of the connecting pipelines. The pressure difference established in the chamber was calculated taking into account the relationship between the pumping speed and the pressure obtained during the experiment. As a result of the experiment, the pressure values in the mechanical and hydraulic compartments of the chamber, the volume of liquid pumped by the vacuum system were obtained.

1. Introduction

One of the most relevant problems in the areas of science and technology is the development of a system for transportation various types of fluid without changing its physical and biological characteristics. The authors proposed a method for calculation of the main parameters of a vacuum fluid transportation system. This system allows the movement of the liquid without contact with inorganic surfaces [1]. An experimental stand for the method's verification was designed and manufactured.

2. Description of the design of the vacuum fluid transportation system

The photo of the developed system is shown in figure 1. The system works by creating a pressure difference between the hydraulic and the vacuum compartments separated by the elastic membrane. Two ports equipped with the solenoid valves *VE5, VE6* for liquid input and discharge are located in the hydraulics compartment of the chamber. Similarly, vacuum chamber has two ports *VE1, VE2* for pumping down and injecting gas into the chamber.

Pressure values in the vacuum and the hydraulic compartments of the chamber are measured by differential pressure sensors *PDE2, PDE3.*

During the suction stoke the chamber is pumped down by the vacuum pump and the elastic membrane is bent thus pulling the liquid from the input port.

During the discharge stroke the vacuum pump injects gas into the chamber and increases the pressure in it. The elastic membrane springs back to its undeformed shape because of the eliminated pressure difference between the hydraulic and the vacuum compartments and pushes out the liquid through the discharge port.

The valves *VE1, VE2, VE3, VE4* are used for the connection of the different ports of the vacuum pump during the suction and discharge strokes.

IOP Conf. Series: Materials Science and Engineering **387** (2018) 012015 doi:10.1088/1757-899X/387/1/012015

Figure 1. A schematic diagram of a vacuum fluid transportation system, where CV – vacuum chamber; NI – mechanical vacuum pump; PDE1,2,3 – differential pressure sensors; VE1-6 – solenoid valves.

3. Equipment selection

Some of the system components, such as the pumped vacuum chamber and the cover, were made using the additive technologies, figure 2. In the papers [2, 3] the authors used 3D printing to create system components for middle- and high-vacuum applications. Laminated application of the material may cause the micro gaps forming, this problem was taken into account this problem was dealt with by setting the slicer's fill value to 100%. It makes possible to take the leakage flow value negligible compared with the pressure values in the vacuum compartment. The material of the chamber is PLA plastic.

Figure 2. The vacuum chamber.

The elastic membrane was made of silicone sheet, figure 3. Uniaxial tension tests were carried out and the Young's modulus for the material used was calculated. The results are E = 1.45 MPa. The ultimate tensile strength is equal to 20 MPa. The modulus of elasticity was calculated based on specimen elongation under the applied load. It was not feasible to determine the transverse deformation of the specimen, hence for further calculations the value of the Poison coefficient will be used.

The pressure was measured using differential pressure sensors Freescale MPX5050DP with measuring range 0...50 KPa for vacuum compartment and Freescale MPX5010DP for the hydraulic compartment. The maximum measurement error of this sensors specified in the datasheet does not exceed 2.5 %.

The data collection was carried out using external 16-bit ADC (Analog-to-digital converter) Adafruit ADS1115 and Arduino Uno R3 microcontroller in conjunction with PLX-DAQ software.

The speed of the vacuum pump enabling the system to operate at required flowrate and create required pressure head is equal to $S_n = 0{,}744\ L\,/\,min$. The experimental stand included two vacuum pumps with pumping speed -$S_n = 5\ L\,/\,min$, $S_n = 3\ L\,/\,min$. The determination of the aggregated pumping speed of two parallel-connected pumps, taking into account the connecting pipelines, was carried out according to the following procedure. A chamber with known constant volume was

IOP Conf. Series: Materials Science and Engineering **387** (2018) 012015 doi:10.1088/1757-899X/387/1/012015

pumped down and the pumping curve was obtained. It was hypothesized that the pumping characteristic of the vacuum system "pumps - connecting pipelines" is described by the Gaussian distribution. A numerical method that takes into account the functional dependence of $S_{(p)}$, figure 4, yields parameters of the Gaussian curve whereby the difference between the pump curve obtained experimentally and the theoretical pumping curve was no more than 1%.

Figure 3. The elastic membrane.

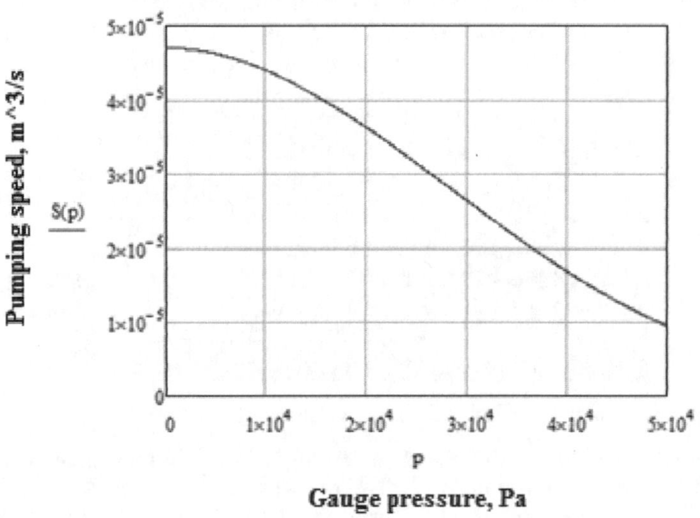

Figure 4. The pumping characteristic curve of the vacuum system "pumps - connecting pipelines".

The connecting diameter of the pipelines and branching fittings was determined by the connecting diameter of the solenoid valves available at the market, the diameter value d is 5,5 mm. The tubes with the internal diameter d_i = 4 mm were used as the connecting pipelines [4, 5]. The material of the tubes is PVC. The quick-disconnecting joint design is shown in figure 5.

The fact of the membrane's buckling during the assembly was established [6, 7]. This is caused by the compressive forces influence in the place where the membrane is fixed. This disadvantage can be eliminated by creating pretension during the assembly to counteract the compressive forces. The relationship between the volume pumped down during the suction stroke depends on the degree to which the membrane is bent. Because of that fact the calculation included the initial membrane deflection caused by buckling [8]. The deflection value is 10,3 mm. The pumping curve obtained using the calculation method taking into account the membrane buckling, and the curve obtained experimentally is shown in figure 6. The chamber's leak rate equal to 0.2 m^3 x Pa / s. was taken into account.

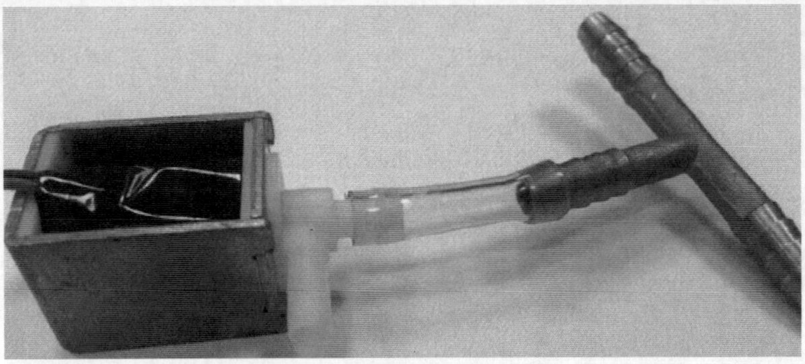

Figure 5. Quick-disconnecting joint “valve-tube-fitting connection”.

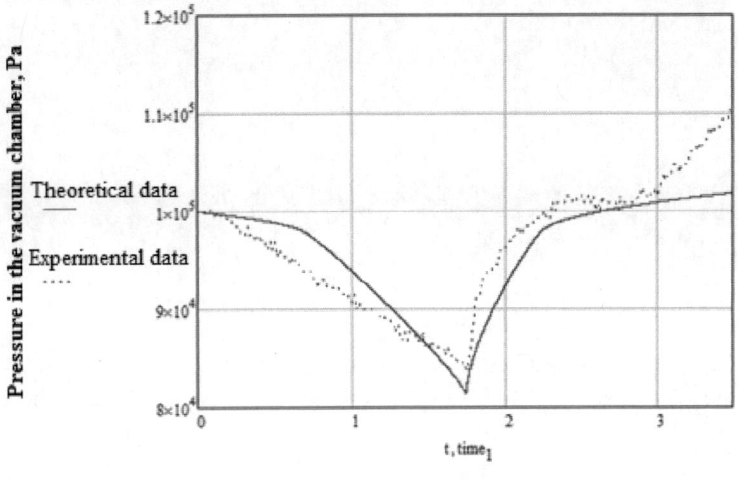

Figure 6. The pumping curve obtained using the calculation and the curve obtained experimentally.

The data obtained theoretically and experimentally show reasonable agreement during the cycle, except for the period (3, 3.4) s. This can be explained by the fact that a membrane that has lost its stability can take a wavy form [9] during the injection process. In the improved model of the installation, it is proposed to use a polymethyl methacrylate vacuum chamber to obtain visual data of the membrane deformation during the system operation.

4. Conclusion

In this paper, an experimental stand was designed and manufactured to verify the developed method for calculation of the main parameters of a vacuum fluid transportation system. The results obtained theoretically and experimentally showed good agreement. A small discrepancy is caused by design deficiencies. To increase the agreement between the theoretical and experimental results mechanical properties of the membrane have to be determined with higher degree of accuracy. To that end we are going to use flexible strain gauges for large strain given that the Young’s modulus is much lower than that of steel. Furthermore fatigue life test of membrane should be carried out in order to determine the pump service life. To eliminate it some proposals have been put forward that will be implemented in the next version of the system.

References

[1] Demikhov K E, Panfilov Yu V, Nikulin N K and Avtonomova I V 2009 *Mechanical Engineering* 590

[2] Povilus Alexander P et al. 2014 *Journal of Vacuum Science & Technology A: Vacuum, Surfaces, and Films* **3** 033001
[3] Chaneliere T 2017 *Vacuum compatibility of ABS plastics 3D-printed objects* (Diss. CNRS, Laboratoire Aimé Cotton)
[4] Gudkov A G et al. 2016 *Biomedical Engineering* **3** 214
[5] Pugachuk A S and Chernishev A V 2015 *Biomedical Engineering* 4 231–5
[6] Volkov V Yu 2013 *Thermal Engineering* **6** 429–33
[7] Volkov V Yu et al. 2015 *Chemical and Petroleum Engineering* **11–12** 799–805
[8] Gavryushin S S and Nikolaeva A S 2016 *Mechanics of Solids* **3** 339–48
[9] Moldovan Dorel and Leonardo Golubovic 1999 *Physical review letters* **14** 2884

IOP Conf. Series: Materials Science and Engineering **387** (2018) 012016 doi:10.1088/1757-899X/387/1/012016

System of vacuum mechanisms failure prediction based on outgassing flows dynamic

E A Deulin, V P Mikhailov and R O Emelianenko

Bauman Moscow State Technical University, 105005, Moscow, Russia

E-mail: deulin@bmstu.ru, mikhailov@bmstu.ru

Abstract. We simulated a vacuum system in order to diagnose vacuum mechanism failure with the goal of estimating dynamic properties of pumps (time to achieving desired pressure in the presence of rapidly varying flows, precision of dynamic pressure maintenance), determining maximum recordable impact frequency of rectangular gas flow pulses when detecting the mechanism state and predicting time to failure. The following parameters are usable for diagnostics: frequency and magnitude of outgassing flow oscillations in the mechanism, residual pressure fluctuations in the vacuum chamber.

The method of diagnosing failure of mechanical components of vacuum equipment that we present [1] is based on frequency analysis of outgassing flow. It entails the following difficulties: detecting the outgassing flows in the working chamber which are released by various mechanism components and decoding the signals recorded. There are several factors contributing to this: outgassing flows are small and only noticeable in ultra-high vacuum; due to evacuation, the outgassing flow being measured decreases in magnitude and alters its shape; properties of equipment performing measurement and signal discretisation affect measurement accuracy; fluctuations of base pressure in the vacuum chamber affect measurement results.

Figure 1 shows the layout of our vacuum system for mechanism failure prediction.

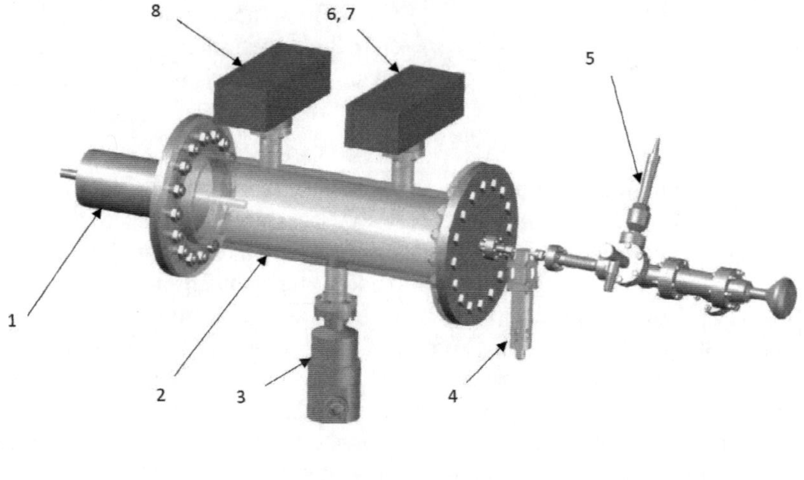

Figure 1. Layout of the vacuum system for mechanism failure prediction.

It comprises the mechanism undergoing diagnostics 1, a vacuum chamber 2, a vacuum pump 3, a pulsating piezoelectric leak valve 4, a thermocouple vacuum gauge 5, ionization 6 and thermocouple 7 vacuum gauges, and an ionization vacuum gauge 8.

Let us consider a bellows-sealed rotary feedthrough as an example of a mechanism undergoing diagnostics (figure 2), comprising an intake eccentric shaft 1, a case 2, bellows 3, the first, second and third vacuum ball bearings 4, 5, 6, a takeoff eccentric shaft 7. Rotation of the takeoff eccentric shaft 7 leads to outgassing flows Q_1, Q_2, Q_3 being released from the friction assemblies of the first, second and third vacuum bearings.

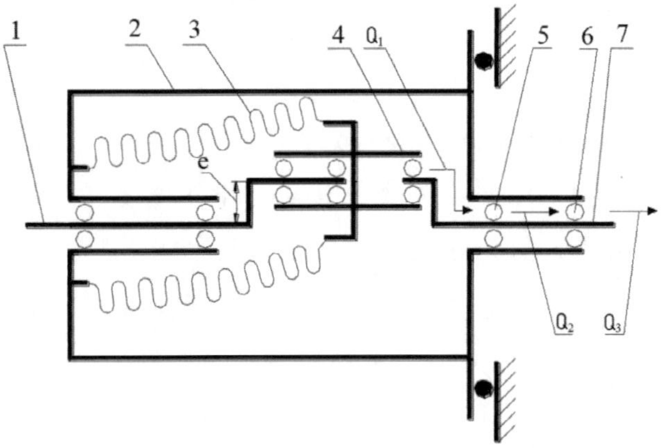

Figure 2. Layout of a bellows-sealed rotary feedthrough.

Figure 3 shows an example of signal summing for the case of outgassing flows released by a mechanism operating in a vacuum and random fluctuations of the total pressure P as a function of time. It clearly demonstrates the difficulty of decoding the operating vacuum chamber pressure signal that responds to the start and stop moments. Linking this signal to the original signal of the bearing rotation frequency requires employing the scientific methods shown below.

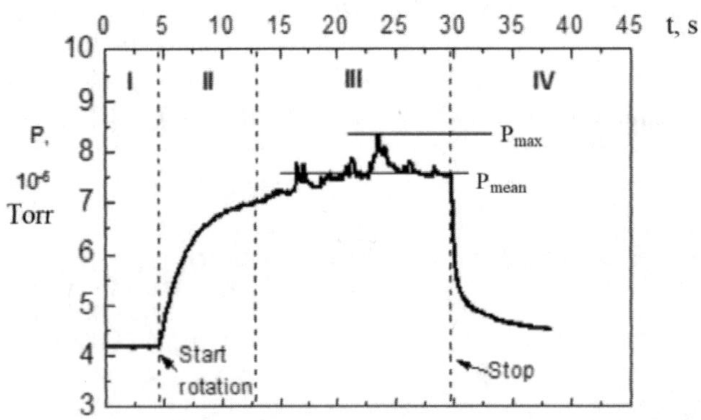

Figure 3. Operating pressure in the vacuum chamber as a function of the start–stop working process of a vacuum mechanism drive equipped with ball bearings.

In practice, the problem of vacuum signal diagnostics involves theoretically reconstructing the magnitude and shape of the initial outgassing signal, knowing which enables us to diagnose the "health" of the mechanism under consideration. A convenient tool for reconstructing the original outgassing flow signal is a pulsating piezoelectric leak valve 4 (figure 1), which has an advantage in the form of being able to create small pulsating gas flows simulating mechanism outgassing pulses with the frequency up to 10^4 Hz and flow magnitude up to 10^2 m^3Pa/s.

Consider an example solution to the diagnostic model problem using a pulsating piezoelectric leak valve 4 (figure 1). We assume that the pulsating piezoelectric leak valve 4 simulates a pulsating outgassing flow released by a vacuum mechanism, for example, a gear, a ball bearing or others. The top plots in figure 4 represent signals originating from the gas flow generated in the vacuum chamber by the piezoelectric leak valve pulses at various frequencies; the plots below show the variations in the vacuum

IOP Conf. Series: Materials Science and Engineering **387** (2018) 012016 doi:10.1088/1757-899X/387/1/012016

chamber pressure. Comparing the pressure variation curves obtained while letting in the gas flows generated by the piezoelectric leak valve shows that it is possible to solve the problem facing the developers of the diagnostic system. To elaborate on the ways of solving this problem, consider the plot in figure 4 that shows input and output signals resulting from varying the input gas flow signal frequency from 0.1 to 1 Hz, for the evacuation rate of 0.05 m^3/s and the vacuum chamber volume of 0.04 m^3.

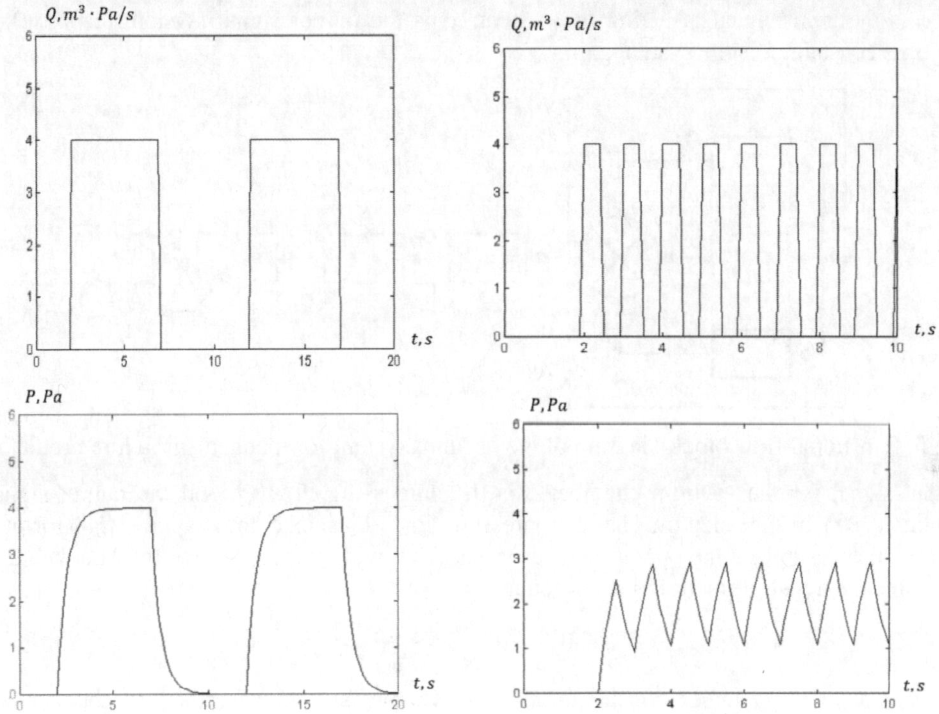

(a) **(b)**

Figure 4. Example curves for the externally introduced flows *Q* and pressure signals *P* under investigation for input signal frequency $F1$ = 0.1 Hz (**a**) and $F2$ = 1 Hz (**b**).

The difficulty in using frequency analysis for decoding the vacuum chamber flow signal stems from the difficulty of separating the spectra of outgassing flows released by different components of the mechanism. This happens because of "signal smearing", which is a reduction in pressure signal magnitude due to the chamber volume and evacuation rate.

Let us now conduct a dynamic simulation and consider the functional flow block diagram of our diagnostic system (figure 5) using failure prediction of the bellows-sealed rotary feedthrough (the layout of which is shown in figure 2) as an example. When the intake shaft rotation is X (figure 5), the outgassing flows Q_1, Q_2, Q_3 released from the friction assemblies in the ball bearings 1, 2, 3 sum and form the total vacuum mechanism 4 flow Q. This flow enters the vacuum chamber 5 that has the volume V, which is being evacuated at the rate S_0. As a result the vacuum chamber pressure P increases. The ionisation pressure gauge 6 records the pressure P by converting it to ionic current I. The ionic current signal I enters the analogue-to-digital converter 7, gets converted to a digital representation of the voltage U and analysed in the control computer 8.

Gas balance in the chamber 5 is described by the equation [2]:

$$Q = PS_0 + V\frac{dP}{dt}, \qquad (1)$$

where Q is the total flow in the vacuum chamber; P is the vacuum chamber pressure; V is the vacuum chamber volume; S_0 is the evacuation rate.

It is possible to transform equation (1) into a lag element equation by dividing both sides by S_0 and obtaining the following factors: $T = V/S_0$, which is the vacuum system time constant, $k = 1/S_0$, or the vacuum system gain:

$$T\frac{dx_2}{dt} + x_2 = kx_1, \tag{2}$$

where x_1 is the input signal affecting the system; x_2 is the output signal (system response); T is the system time constant; k is the system gain.

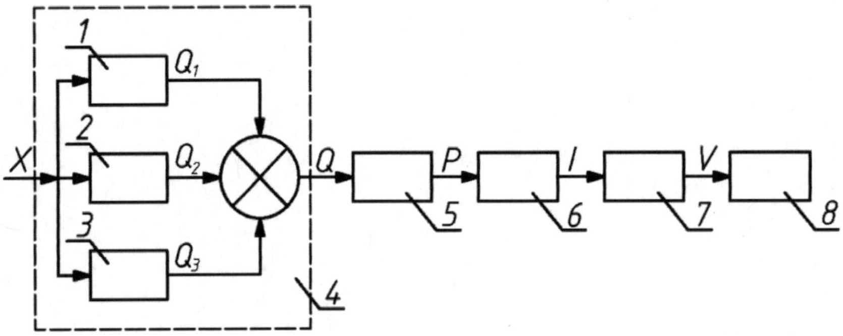

Figure 5. Functional flow block diagram of the vacuum system for mechanism failure prediction.

The input signal for the vacuum chamber 5 is the outgassing flow Q, and the output signal (the system response) is the vacuum chamber pressure P. Let us take the Laplace transform of the equation 2, that is, replace the operation of differentiation with respect to time by a Laplacian S. Let us determine the transfer function of the vacuum system:

$$W_v = \frac{x_2(S)}{x_1(S)} = \frac{k}{TS+1}. \tag{3}$$

The vacuum system under consideration is a dynamic system described by the lag element transfer function [4–8]. The ionisation pressure gauge 6 consists of an ionisation transducer and an electronic unit comprising a microammeter that records the ionic current. Let us determine the transfer function of the ionisation transducer. Writing down the ionisation transducer equation:

$$P = \frac{I}{K}, \tag{4}$$

where I is the ionic current in the transducer; K is the ionisation transducer constant, we arrive at the following ionisation transducer transfer function resulting from the equation 4:

$$W_i = \frac{I(S)}{P(S)} = K. \tag{5}$$

Therefore, the ionisation transducer is an amplifying circuit with the gain K. This transducer is practically inertialess, since the time it takes for the ions to move in response to the voltage from the anode mesh to the ion collector is approximately 0.05 ms. It is these ions that generate the ionic current that assists in determining the vacuum chamber pressure P. A microammeter measures the ionic current over approximately 10 ms. If the signal frequency from the piezoelectric leak valve gas flow or the contact friction zone in the test mode does not exceed 10 Hz, this measurement time may be disregarded as well.

Let us determine the time constant T and gain k of our vacuum system ($V = 4\cdot10^{-2}$ m^3, $S_0 = 5\cdot10^{-2}$ m^3/s): $T = 4\cdot10^{-2}/5\cdot10^{-2} = 0.8$ s, $k = 1/5\cdot10^{-2} = 20$ s/m^3. If the duration of a rectangular gas flow pulse is at least 0.8 s, then the pressure pulse is shaped as an aperiodic signal, as shown in the figure 4*a*. For the duration of a rectangular gas flow pulse that does not exceed 0.8 s, the pressure pulse is shaped as a saw-wave signal with its magnitude reduced (figure 4(b)). Therefore, dynamic properties of a vacuum system constrain the frequency and duration of gas flow pulses generated by a precision piezoelectric leak valve 4 (see figure 1) or friction zones in the vacuum mechanism. It is possible to run vacuum mechanism diagnostics at higher frequencies, but the pressure signal oscillation magnitude will decrease with frequency, which will make diagnostics harder and create an erroneous impression that the mechanism is failing. Calibration of the vacuum system for mechanism failure prediction in the test mode takes place as follows: the piezoelectric leak valve 4 (see figure 1) assigns the frequency and duration to the rectangular gas flow pulses which correspond to the frequency and duration of outgassing in friction assemblies of the vacuum mechanism; the oscillation frequency and the shape of the vacuum chamber pressure 2 signal are recorded using the ionisation transducer 8, and the data captured contributes to decoding the outgassing flow signal for the vacuum mechanism.

Gas exchange theory [3] states that mechanically induced desorption happens when surfaces are no longer in contact, therefore it is important for a diagnostic system to account for the fact that outgassing in friction assemblies occurs in time with vacuum mechanism surfaces coming into contact. It means that a square wave is the signal shape providing the closest match for simulating outgassing flows from kinematic pairs of vacuum mechanisms in operation. The diagnostics model presented may be used for all types of vacuum systems. In this way we arrive at the following conclusions: 1) we show that the dynamics of outgassing flow and pressure variations is affected by the processes happening in zones of contact friction and the vacuum system properties; 2) analysing the dynamic model of a vacuum system enables us to reconstruct the outgassing flow, which represents the "health index" of the mechanism; 3) dynamic properties of a vacuum system constrain the frequency and duration of gas flow pulses generated by a precision piezoelectric leak valve or friction zones in the vacuum mechanism; 4) it is possible to run vacuum mechanism diagnostics at higher frequencies, but the pressure signal oscillation magnitude will decrease with frequency, which will make diagnostics harder.

Acknowledgement
The study was supported by the Ministry of Education and Science of the Russian Federation as part of conducting research and development in the framework of the basic government assignment in the field of science no. 9.8503.2017/8.9. The authors would like to thank A.V. Glushchenkov and E.R. Tyutyukin, students of Bauman Moscow State Technical University, for their contribution to preparing the article for publication.

References
[1] Akhmadiev D R, Gushchin I V, Deulin I A and Usov A B 1993 *Patent SU 1835065 A3 A means of health diagnostics for cyclically loaded components of vacuum equipment*
[2] Rosanov L N 1990 *Vacuum Equipment* (Moscow: Vysshaya Shkola) 320
[3] Deulin E A, Mikhailov V P, Panfilov Y V and Nevshupa R A 2012 *Mechanics and Physics of Precise Vacuum Mechanisms* (Springer Edition) 234
[4] Pugachuk A S, Ten A Yu, Borisov Yu A and Chernyshev A V 2016 *Procedia Engineering* **152** 372–9
[5] Krasinskaja Je M, Ilyina A N and Rukavishnikova A S 2016 *2nd International Conference on Industrial Engineering, Applications and Manufacturing ICIEAM 2016* pp 7910960
[6] Ivanov A S and Fomin M V 2017 *Russian Engineering Research* **12** 1017–22
[7] Tumakova E V 2016 *Measurement Techniques* **6** 600–4
[8] Liu B, Grout V and Nikolaeva A 2018 *IEEE Transactions on Industrial Electronics* **7** 5712–21

The phenomenon model of X-ray beam angular divergence decreasing formed by composite waveguide-resonator

V K Egorov[1] and E V Egorov[1,2]

[1] Institute of Microelectronics Technology RAS, 142432, Chernogolovka, Russia
[2] Peoples Friendship University of Russian, Moscow, Russia

E-mail: egorov@iptm.ru

Abstract. The phenomenon of X-ray fluxes waveguide-resonance propagation is presented briefly. There is described simplest design of the planar X-ray waveguide-resonator – device functioned in frame of the phenomenon. Advantages and shortage of the device are discussed in details. Experimental characteristics of the composite planar X-ray waveguide-resonator allowed to decrease the angular divergence of its emergent beam at the integral intensity conservation are presented. Simplified model of this device functioning is formulated.

1. Introduction

It is well known that the X-ray beam diagnostics of materials is the most widespread and universal means of its properties investigation. In the last time, specific devices for the nanosize objects analysis by X-ray methods were elaborated. There are planar X-ray waveguide-resonance structures [1]. Planar X-ray waveguide-resonator (PXWR) is the simplest device from the structures set. Main feature of these devices is the functioning in frame of the waveguide-resonance radiation propagation phenomenon. The phenomenon discovery was result of the creative rethinking of literature data and our systematic experimental investigations.

First stone into the waveguide-resonance radiation propagation phenomenon was laid down by appearing of X-ray slitless collimators [2]. This specific device was formed by two quartz polished planar plates tightly pressed to each other. The slitless channel between its could not let by optical radiation but allowed to transport X-ray fluxes. The next step in this direction was connected with the interference field of X-ray standing wave registration in conditions of X-ray flux total external reflection on the material interface [3]. The conception of the X-ray standing wave interference field introduced into the interpretation approach by prof. Laue for the Borrmann effect description [4] allowed to find the alternative logical explanation of the slitless collimator functioning. Our systematic investigations of X-ray quasimonochromatic flux propagation through the extended slit clearance in conditions of the slit width variation allowed to conclude that the nanosize slit clearance transports X-ray radiation fluxes by the specific mechanism, which did not mention early [5]. We called the mechanism as the waveguide-resonance radiation propagation one and understood that it is the fundamental phenomenon. Devices functioned in frame of the phenomenon we called planar X-ray waveguide-resonators [6]. It is very interesting that our discovery got the independent confirmation in the work of Japan scientist [7]. They studied the propagation peculiarities variation of MoKα monochromatizating radiation flux through the angular reflecting structure formed by polished silicon plates on the angle in the magnitude range 0–0.5 degrees. Its experiments data showed that the sharp increasing of the structure emergent beam intensity at angles near 0.01 degrees can not be explained by well-known effects and mechanisms.

2. X-ray flux transportation by simplest PXWR

Construction of the planar X-ray waveguide-resonator is shown in figure 1. It device captured of X-ray quasimonochromatic flux by its entrance cut in the angular range, which can not exceed the double critical angle of the radiation total reflection for reflectors material, transports the flux almost without attenuation and forms X-ray emergent beam. It is significant that the emergent beam divergence angle is equal to the radiation capture angle. The conception allowed to build the model of X-ray flux waveguide-resonance propagation was based on the revolutionary hypothesis about the uniform interference field of X-ray standing wave in all space of the planar extended slit clearance. This situation is characterized by the principle difference from the multiple flux total reflection in the planar extended slit clearance. The difference is readily illustrated by the figure 2. X-ray flux of the quasimonochromatic radiation in the wide slit clearance undergoes the multiple external total reflection with set appearing of the local interference fields of X-ray standing wave (figure 2(a)). The longitudinal size of local interference areas "c" is defined by half of the radiation coherent length parameter value L in line with the expression [7]:

$$c = \frac{L}{2} = \frac{\lambda_0^2}{2\Delta\lambda_0} \qquad (1)$$

where λ_0 is the main wavelength of the radiation, $\Delta\lambda_0$ is the radiation monochromatization degree. The transversal size of these areas is defined by the experimental conditions and is approximately equal to the longitudinal dimensions for the total external reflection phenomenon [8]. Because of the interference field is excited not only in the space above material interface but in the reflector material volumes the minimum flux attenuation can be achieved in conditions of the phase-conjugation consecutive reflections, only. In the conjugation violation case, X-ray flux attenuation increases owing to reexcitation the interference field in reflectors volumes.

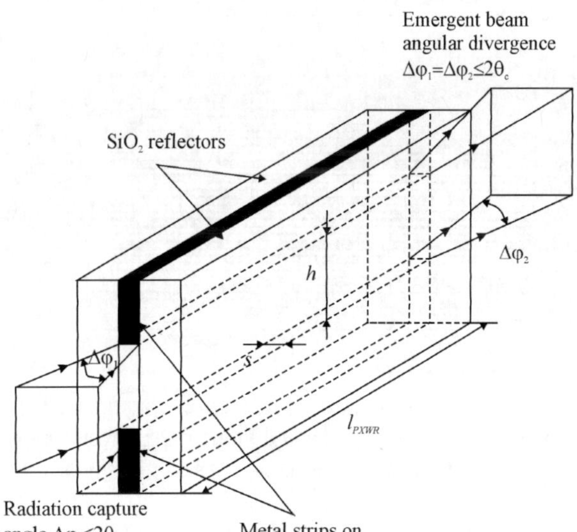

Figure 1. Scheme of X-ray initial flux capture by entrance cut of PXWR simplest design and emergent beam formation by this device.

X-ray flux transportation conditions will change when the distance between reflectors is smaller as half of the radiation coherence length magnitude. This case will be characterized by the local interference field areas confluence and the uniform interference field of X-ray standing wave arising in all space of the planar extended slit clearance (figure 2(b)). The arising of uniform interference field is the necessary and sufficiency condition for the realization of X-ray flux propagation mechanism through the planar extended slit clearance. The mechanism allows to propagate X-ray flux captured under any angles, which are smaller as the total external reflection critical angle transported by the slit clearance almost without attenuation and realized the radiation superfluidity effect.

PXWR of simplest construction demonstrates other specific peculiarities. It is well known that coherence lengths of the characteristic radiation generated by laboratory X-ray sources are characterized by nanosize magnitudes [9]. For example, the coherence length of CuKα radiation is equal to 430 nm. Our systematic investigations of the extended slit clearances emergent beam with distances between reflectors, which were smaller as half of the transported flux radiation showed that beams had nanosize width and heightened radiation density. Its parameter in beams formed by similar waveguides exceeds the radiation density in slit-cut systems emergent beams on 3–4 orders [10]. These facts were very useful for the element material diagnostics improving by TXRF method [11].

In spite of PXWR evident merits this device is characterized by essential shortage. Its emergent beam demonstrates the angular divergence near 0.1 degree. This magnitude is not great but owing it existence the emergent beam losses nanosize width and high value of its radiation density on distance of some centimeters from PXWR outlet. So, the fundamental task of PXWR development is the search of approaches for the angular divergence decreasing.

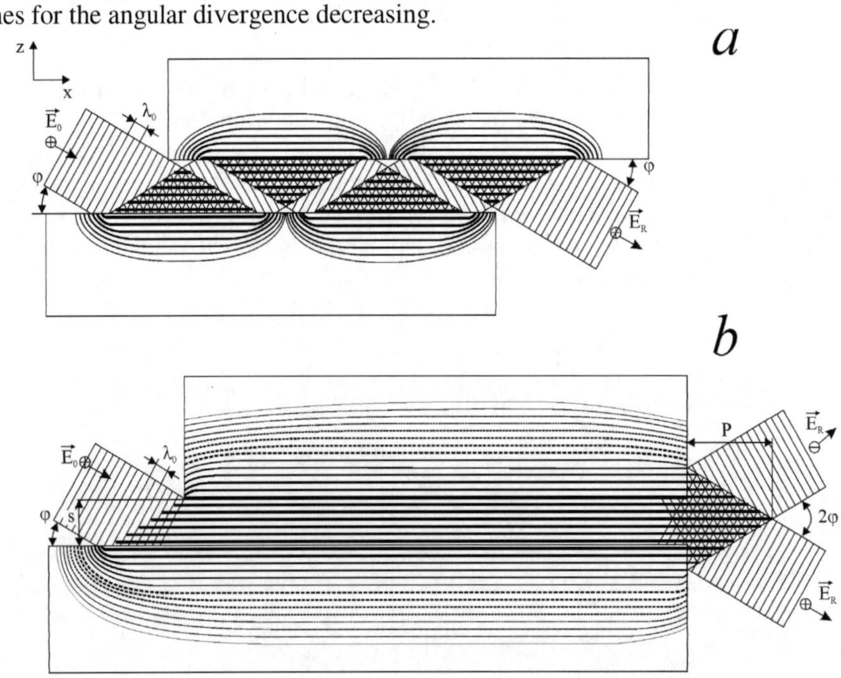

Figure 2. Schemes of X-ray quasimonochromatic flux transportation visualization by the planar extended slit clearance in case of great distance S between reflectors (a) and small magnitude of the distance. P is the protrusion parameter of the uniform interference field.

3. PXWR constructions for angular decreasing of its emergent beams

Experimental search of waveguide-resonance constructions allowed to decrease its emergent beam angular divergence in comparison with the radiation capture angle without reduction of its integral intensity offered to elaboration of modified PXWR designs. Simplest solution of the task was connected with the waveguide-resonance device arrangement by use of nonequivalent length reflectors. In this case, the emergent beam integral intensity does not change but the form of spatial intensity distribution suffers distortion (figure 3(b)). Radiation intensity distribution of emergent beam formed by PXWR with simplest design is characterized by the Gauss outline (Fig. 3a). PXWR construction with nonequivalent reflectors length demonstrate the emergent beam with half-Gauss intensity outline with one half of the outline featured for the waveguide-resonator with simplest construction.

Specific approach allowed to build the original construction completed by two PXWR of simplest design, which were installed one after another on some distance with execution of its mutual alignment (CPXWR) [12]. Experimental data reflected spatial intensity distributions a characteristic for emergent beams formed by the composite PXWR is presented in figure 3(c). CPXWR use for the X-beam formation reduces the spatial distribution divergence magnitude at conservation of the integral intensity and form of the intensity distribution.

Experimental investigations showed that the composite waveguide-resonator demonstrates its unique properties at the specific conditions realization. Conservation of X-ray flux integral intensity is possible if the distance (gap) between PXWRs compiling CPXWR is not exceed the protrusion parameter P (figure 2(b)) defined by the expression [12]:

$$P = \frac{\lambda_0^3}{8\Delta\lambda_0^2} \tag{2}$$

When the gap is higher of this parameter magnitude the construction will produce the small angular divergence. Such construction is not the composite waveguide-resonator. When the gap will be smaller as the interference field protrusion parameter value the interaction possibility of interference fields excited in first and second waveguide-resonators appears. It is precisely, the interference fields interaction leads to X-ray flux integral intensity conservation in the result of partial radiation tunneling in the gap between PXWRs. We displayed that the gap width variation leads to change of the device emergent beam angular divergence. The analytical description of this effect remains to be obtained owing to influence on the effect a form and quality treatment of reflectors butts. At the same time, it is temptingly to propose the model of composite waveguide-resonator function.

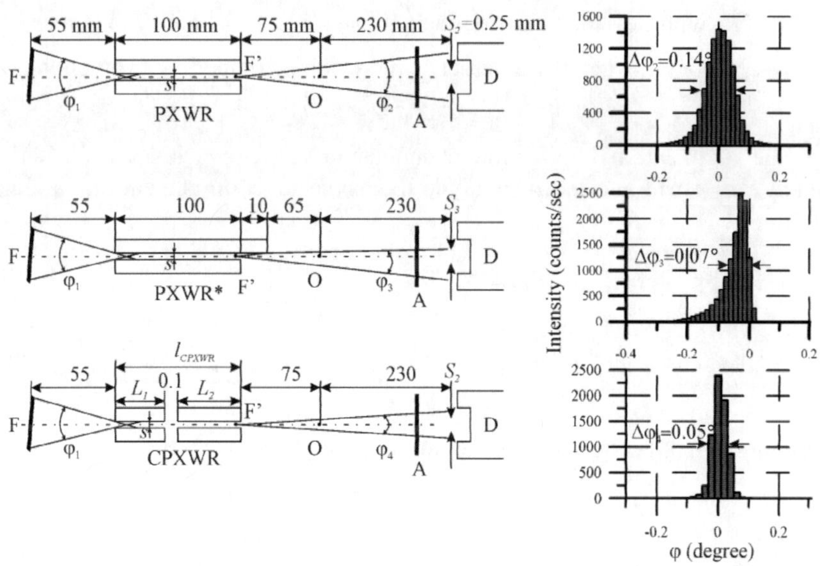

Figure 3. Schemes of X-ray quasimonochromatic beams formation by simplest PXWR, waveguide-resonator built on base (a) of nonequivalent length reflectors (b) and composite waveguide-resonator (c) and intensity spatial distributions in emergent beams of its devices.

X-ray photons flux is the sample of real statistical ensemble. Similar ensemble properties description is subject to the Liouville theorem, whereby the variation of one ensemble parameter must be accompanied by change of its other one. Integral intensity of CPXWR emergent beam does not undergo variation. At the same time, our X-ray precision diffraction measurements showed that its emergent beam is characterized by the degree monochromatization deterioration [13]. In the result, the model must characterize the function connection between X-ray beam angular divergence reduction $\Delta\varphi$ and its degree of radiation monochromatization $\Delta\lambda$.

The energy (E_p) and momentum (P_p) of single photon in the radiation flux generated by laboratory X-ray sources are defined by expressions [14]:

$$E_p = \hbar\omega = \frac{hc}{\lambda_0} = P_p c \tag{3}$$

$$P_p = \frac{E_p}{c} = \frac{h}{\lambda_0} \tag{4}$$

where h=2π, $\hbar$ is the Planck constant, ω is the radiation angular frequency and c is the light velocity in vacuum. Variation of these photon parameters can be obtained in the result of (3) and (4) expressions differentiation:

$$\Delta E_p = -\frac{hc}{\lambda_0^2}\Delta\lambda = \frac{P_p^2 c}{h}\Delta\lambda \tag{5}$$

$$\Delta P_p = -\frac{h}{\lambda_0^2}\Delta\lambda = -\frac{P_p}{\lambda_0}\Delta\lambda = \frac{P_p^2}{h}\Delta\lambda \tag{6}$$

Accordingly to hypothesis about the partial angular tunneling of X-ray flux in the gap between simplest waveguide-resonators one can believe that the tunneled flux part is characterized by $\vec{P}_1$ momentum been equal P_p. Vector of the momentum is directed at φ_1 angle to the composite waveguide-resonator axis. The emergent beam of CPXWR will be characterized by new momentum value $\vec{P}_2 = \vec{P}_p + \delta\vec{P}$ with new direction defined by φ_2 angle to the CPXWR axis. The experiment shows that $\varphi_1 > \varphi_2$. So, the flux angular divergence reduction is defined by expression:

$$\Delta\varphi = \varphi_1 - \varphi_2 \tag{7}$$

Owing to the total external reflection phenomenon of X-ray fluxes of material interfaces is characterized by very small angles it is possible to change function tangent on the function argument in the next form:

$$\Delta\varphi = (\varphi_1 - \varphi_2) \approx \frac{P_{y1}}{P_{x1}} - \frac{P_{y2}}{P_{x2}} \approx \frac{\Delta P_y}{P_p} \tag{8}$$

On the other hand, the scalar momentum value of photon can be presented on the conventional form:

$$P_p^2 = P_x^2 + P_y^2 \tag{9}$$

The momentum variation will be described by the next expression:

$$P_p \frac{\Delta P}{\Delta\varphi} = P_x \frac{\Delta P_x}{\Delta\varphi} + P_y \frac{\Delta P_y}{\Delta\varphi} \tag{10}$$

Owing to the tunneling effect is accompanied by chief variation of ΔP_y ordinate momentum component the expression (8) can be presented in the final form:

$$\Delta P_p = P_y \Delta\varphi = \Delta\varphi P_p \sin\varphi_1 \cong \Delta\varphi P_p \varphi_1 \tag{11}$$

Using the expression (6) we can get connection between X-ray flux angular divergence reduction and deterioration of its radiation monochromatism:

$$\Delta\lambda = \varphi_1 \lambda_0 \Delta\varphi \tag{12}$$

Because of the initial degree monochromatization of X-ray flux is characterized by $\Delta\lambda_0$ the tunneling part of the flux will be described by increasing value of this parameter:

$$\Delta\lambda_2 = \Delta\lambda_0 + \varphi_1 \lambda_0 (\varphi_1 - \varphi_2) \tag{13}$$

In the result, it is clear that the composite waveguide-resonator decreases angular divergence in its emergent beam in comparison with X-ray beam formed by PXWR with simplest construction but it demonstrates worse monochromatism. Similar device can be used with success for X-ray fluorescence analysis but it is not acceptable for X-ray diffractometry.

Acknowledgment
This work produced by Government task #007-0220-18-00 and has been partially financial supported by Russian Foundation of Fundamental Investigations (grant #16-07-00665) and by Ministry of Education and Science of the Russian Federation (the Agreement number 02.a03.0008).

References

[1] Egorov V K and Egorov E V 2017 *Planar X-ray waveguide-resonators. Realization and perspectives* (Lambert Acad. Press: Saarbrucken) 388
[2] Lejkin V et al. 1984 *Instrum. Exp. Tech.* **6** 1333
[3] Bedzyk M et al. 1989 *Phys. Rev. Let.* **69** 1376
[4] Laue M 1960 *Rontgenstrahlinterferenzen* (Acad. Verlag-Ges.: Berlin) 540
[5] Egorov V K and Egorov E V 2003 *Advances in X-ray Analysis* **46** 307
[6] Egorov V K and Egorov E V 2004 *X-ray Spectrometry* **33** 360
[7] Tsuji K et al. *X-ray Spectrometry* **34** 281
[8] Egorov V K and Egorov E V 2007 *X-ray Spectrometry* **36** 381
[9] Mirkin L I *Handbook of X-ray analysis of polycrystal materials* (New York: Consultance Bureau) 920
[10] Egorov V K and Egorov E V 2010 *Trends in Appl. Spectrometry* **8** 67
[11] Egorov V K and Egorov E V 2013 *Advances in X-ray Chem. Anal. Japan* **44** 21
[12] Egorov V K and Egorov E V 2002 *Proceeding of MRS* **716** 189
[13] Egorov V K and Egorov E V 2017 *Uspekhi prikladnoy fiziki* **6** 534
[14] Bohm M 1952 *Quantum theory* (Prentice-Hall: New York) 755

TÜBİTAK UME vacuum laboratory capabilities and activities

Alper Elkatmis and Rifat Kangi

TUBITAK National Metrology Institute (TUBITAK UME)
P.K. 54 41470 Gebze, Kocaeli, TURKEY

E-mail: alper.elkatmis@tubitak.gov.tr

Abstract. TÜBİTAK National Metrology Institute (TÜBİTAK UME) is a research and development institute that operates under the Scientific and Technological Research Council of Turkey (TÜBİTAK). TÜBİTAK UME Vacuum laboratory takes part in under the mechanics group among the groups of physics, mechanics and chemistry group laboratories. TUBITAK UME Vacuum Laboratory aims to develop and maintain national measurement standards, to establish vacuum pressure traceability to SI units and to disseminate this to industry, science and research in Turkey. The laboratory was founded in 1996 and started to provide secondary level calibration services with Vacuum Gauge Calibration System (VGCS-200). Primary vacuum standard called Multi-Stage Static Expansion System (MSSE1) was established in 2000. Commercially available secondary vacuum gauge calibration system was improved with the addition of automation under the name Vacuum Gauge Metrology System (VGMS) in 2010. The main activities of the Vacuum Laboratory are conducted in the following working fields: Primary and secondary vacuum measurements and calibrations, secondary calibrations of leak standards, primary constant pressure flowmeter (CPFM) standard establishment works, renewal works of existing primary static expansion system, European Metrology Research and Innovation Projects, Quadrupole Mass Spectrometer (QMS) characterization measurements on Dynamic Vacuum System (DVS), training and consultancy services in the field of vacuum metrology. In this paper, the Vacuum Laboratory capabilities and activities will be briefly introduced.

1. Introduction

TÜBİTAK National Metrology Institute (TÜBİTAK UME) is a research and development institute that operates under the Scientific and Technological Research Council of Turkey (TÜBİTAK) [1]. The institute aims to ensure the reliability of all measurements conducted in Turkey, to integrate of these measurements into the international system and to develop existing and new measurement technologies by establishing national measurement system. In accordance with this purpose, TUBITAK UME Vacuum Laboratory aims to develop and maintain national measurement standards, to establish vacuum pressure traceability to SI units and to disseminate this to industry, science and research in Turkey.

The laboratory takes part with two main measurement quantities in the BIPM (Bureau International des Poids and Measures)/CMC (Calibration and Measurement Capability) tables of mass and related quantities about the vacuum pressure and it has been accredited in 2005 by TÜRKAK (Turkish Accreditation Agency) in the following scope, given in table 1. The calibration and measurement capabilities are capacitance manometers (CDG), pirani gauges, thermal conductivity vacuum gauges, hot and cold cathode ionization gauges and spinning rotor gauges (SRG).

Table 1. The laboratory scope by Turkish Accreditation Agency (TÜRKAK).

Calibration scope and calibrated gauges	Measurement Range, in Pascal	Best Measurement Capability (Expressed as an expanded uncertainty) (k=2), p in Pascal
Absolute Pressure / Pa / Pirani and Thermocouple vacuum gauges	$1\times10^{-1} \leq p \leq 1\times10^{5}$	4.1×10^{-2} x p to 2.0×10^{-2} x p
Absolute Pressure / Pa / Capacitance Diaphragm Manometers (CDG)	$1\times10^{-1} \leq p \leq 1\times10^{5}$	3.5×10^{-2} x p to 7.9×10^{-3} x p
Absolute Pressure / Pa / Hot and Cold Cathode Ionisation vacuum gauges	$1\times10^{-4} \leq p \leq 1\times10^{-1}$	1.1×10^{-1} x p to 4.0×10^{-2} x p
Absolute Pressure / Pa / Spinning Rotor Gauges (SRG)	$1\times10^{-4} \leq p \leq 1\times10^{-1}$	9.0×10^{-2} x p to 3.5×10^{-2} x p

The laboratory participated in 3-year-joint research project, ended in 2014, EMRP IND12 "Vacuum metrology for production environments" funded by the European Metrology Research Program, in the areas such as stability measurements of leak detectors, characterization of quadrupole mass spectrometers and investigation of the effects of various parameters on the sensitivity of the QMS [2]. Goal of this project is to open new measurement capabilities for vacuum and to help industry to characterize vacuum in industrial environments. In the studies of the characterization of QMS study, the laboratory has taken an active role as a task leader [3]. Further, the laboratory continues his contributions to the European Metrology Programme for Research and Innovation project named EMPIR 14IND06 "Industrial standards in the intermediate pressure-to-vacuum range", which will end in 2018 [4]. The laboratory gives consultancy services and trainings to his customers from industry and military regarding vacuum metrology, calibration methods, and measurement techniques. The primary standard has been involved in key comparisons that were organized in the frame of EUROMET.M.P-K1 (442 a/b) and other international comparisons are still ongoing EURAMET.M.P-K15.1, COOMET.M.P-K15 and COOMET.M.P-S2.

2. The capabilities and activities of the laboratory

Vacuum gauge calibrations are performed on VGMS between 1×10^{-4} Pa to 1×10^{5} Pa with the relative uncertainty of $8.3\text{x}10^{-2}$ to $7.9\text{x}10^{-3}$. Figure 1 shows a simplified view of the test chamber with test ports at the top and test gas entering on the side and being ported to the top of the test chamber that is designed in accordance with ISO/DIS 3567. The main and rough pumps are at the bottom of the chamber. The comparison method is used for the calibration of vacuum gauges, where the device under calibration is compared to a calibrated reference standard. The readings of the two vacuum gauges can be compared only if both are exposed to the same pressure. The main part of the system is vacuum chamber with eight ports to which reference gauges and test gauges to be calibrated are connected. The vacuum chamber and the position of the test flanges have been designed to fulfil this condition. The system provides automated control of vacuum levels from nominally $1.3\text{x}10^{-4}$ Pa to $1.3\text{x}10^{-1}$ Pa dynamic pumping and up to 130 kPa static fill. The system uses a turbo pump with integrated electronic drive unit. A diaphragm pump backs the turbo and a dry scroll provides rough vacuum for the test chamber and gas feed tank on the system, with which ultimate vacuum values of 10^{-6} Pa are obtained. A nude ion gauge is used as feedback transducer to monitor the chambers high vacuum inlet pressure. Traceable pressure measurements are made by either a SRG or one of three high-accuracy CGSs, depending on the pressure range. In the low vacuum mode, one of the capacitance manometers with the ranges of 130 Pa, 1.3 kPa and 130 kPa is used as reference standard; whereas in the high vacuum mode spinning rotor gauge with the range of 10^{-4} Pa to 1 Pa is used. The system is controlled either by manually or by computer in combination of controllers of transfer standards. Feedback transducer is ion gauge or capacitance manometer in low pressure mode and in the high pressure mode, respectively.

Figure 1. Vacuum Gauge Metrology System (VGMS).

Figure 2. TÜBİTAK UME Primary Vacuum System (MSSE1).

The reference standards of VGMS are traceable to the primary multi-stage static expansion system, MSSE1, shown in figure 2. The pressure range of the primary system is between 0.9 mPa to 1000 Pa with the relative uncertainty between 2.1×10^{-3} to 9.5×10^{-4} [5]. The expansion method of generating known low pressures was proposed by Knudsen (1910) and has subsequently been developed by a number of workers as a wide-range and accurate method of vacuum gauge calibration. Two methods have been developed of applying this principle to the generation of the known low pressures for calibration purposes. The principle of the method is to expand a small volume of suitable gas at a measured (relatively high) pressure into a pre-evacuated, larger volume, thereby effecting a pressure reduction which is calculable from the known ratio of the systems' volumes. The generated pressure is directly calculable from the initial pressure and the volume ratio. To generate lowest pressures, a multi-stage system in which the large overall volume ratio is effectively achieved by successive expansions through a series of volume pairs, each of modest volume ratios. MSSE1 static expansion system has three stages each of volume ratio about 10^2. This is known as the 'Series Expansion' method. The lower range of the system may be extended by one decade by using cascaded expansion method. The cascaded expansion method is basically based on the reduction of the pressure value by pumping the expansion volume after the first step with the adjacent transfer volume isolated and then once again expanding the gas from the transfer volume to the calibration volume. Work is continuing with measurements regarding this method. The traceability chain is schematically given in figure 3. Quadruple Mass Spectrometer measurements such as determination of gas impurities and monitoring the vacuum conditions in vacuum chambers in the mass range of 1 to 200 amu can be performed in the laboratory. Needed measurements with different gasses such as N_2, Ar, He and CO_2 were performed at the laboratory in the scope of EMRP IND12 project.

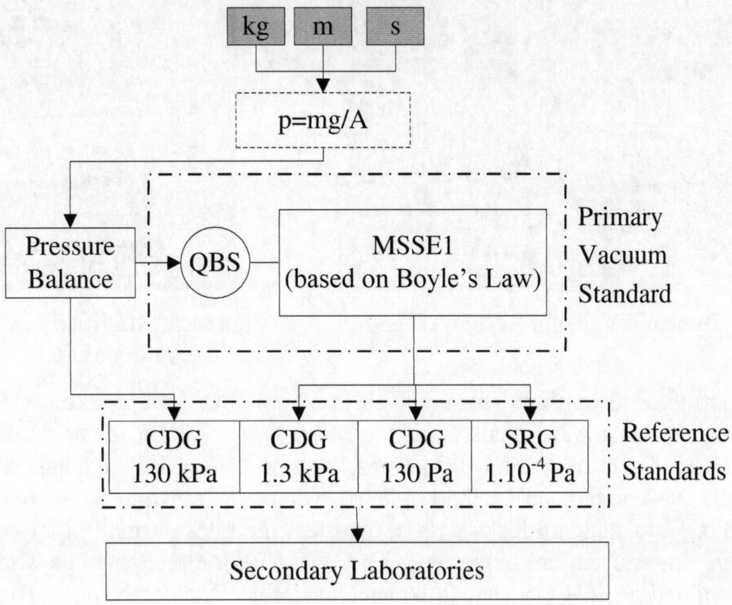

Figure 3. TÜBİTAK UME Vacuum Laboratory SI Traceability Chain; QBS: Quartz Bourdon Tube Sensor.

Under the scope of the project, the laboratory initiated to work in the fields such as leak standard measurements, stability measurements of QMS and Mass Spectrometer Leak Detector (MSLD), and the investigation of the effects of various parameters on the sensitivity of the spectrometer. These kinds of measurements have being conducted on the "Dynamic Vacuum System (DVS)" established in 2012, shown in figure 4. A simple and reliable dynamic vacuum system for in situ calibration of extractor gauges (EXG) and QMS's was realized at TÜBİTAK UME in collaboration with colleagues from national metrology institute of Korea, KRISS. The system can also be used for QMS tests and characterization of dry vacuum pumps. The system consists of two chamber units. In the single chamber unit, throughput method is used with the throughput range higher than 10^{-3} mbar·l/s. In the double chamber unit, orifice method is used with the throughput range higher than 10^{-10} mbar·l/s. A simple way is used for pressure generation and control in this case. The gas admission from the gas regulator into the upper chamber is maintained through the UHV leak valve and through the 10 mm-orifice for the lower chamber. The double chamber unit can be operated by means of 4 replaceable orifices with different sizes (from 10 to 30 mm). The base pressure in the lower chamber is approximately less than $5 \cdot 10^{-7}$ Pa and can be reached within 12 hours by continuous evacuation of the system.

The system was modified with the integration of SCE (Standard Conductance Element) manufactured by National Metrology Institute of Japan, National Institute of Advanced Industrial Science and Technology (NMIJ/AIST) in 2014, figure 5. SCE serves to introduce a test gas with known flow rate to a vacuum chamber, which is evacuated at the pressure of less than 1 mPa. The SCE is made of a stainless-steel sintered filter with the pore size of less than 1 μm. Since the gas flow through the SCE satisfies the molecular flow condition even at the pressure up to 10 kPa, some useful characteristics of molecular flow are available. Molecular conductance is between 1×10^{-10} m^3/s to 2×10^{-9} m^3/s of the SCE. Calibrations for various gas species are available only with this single leak element because the conductance is easily compensated by the molecular mass of gas species.

Figure 4. Dynamic Vacuum System (DVS).

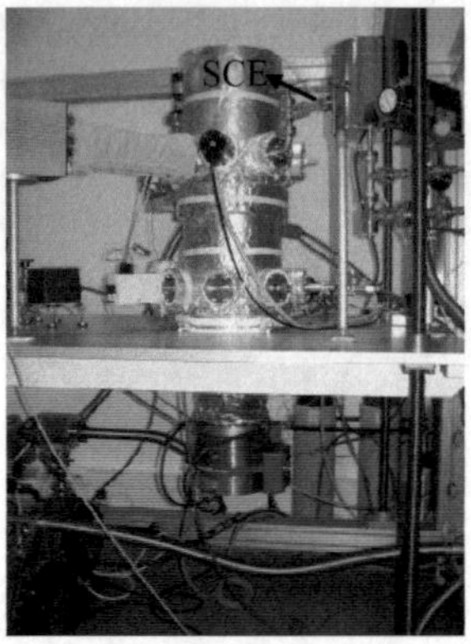

Figure 5. Modified DVS with the integration of SCE.

The laboratory continues its activity in the field of gas flow rate measurements realizing a measurement station to meet the national requirements on the leak standard calibration in the range from 2×10^{-8} mbar·l/s to 2×10^{-5} mbar·l/s with permeation type helium leak standards having different leak rates, figure 6. The comparison method is used, where unknown leak is compared with two known standard leaks. Two standard leaks with known leakage rates normally lying on either side of the unknown leakage rate is used for comparison. The UME calibration apparatus works starting from reference leaks calibrated against primary flowmeter available in other NMIs. The simplicity and efficiency of comparison calibration method make it suitable for the calibration of leaks in industrial level. For this reason it can be considered as a secondary standard.

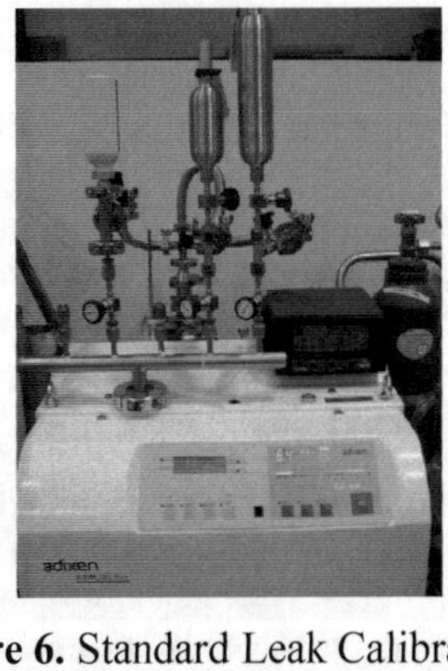

Figure 6. Standard Leak Calibrations.

The temperature and moisture parameters are saved by PC automatically in the laboratory. The values of and their tolerances and calibration/test list are given in tables 2 and 3, respectively.

The use of primary flowmeter both to generate and measure gas flows is essential in continuous expansion systems used as primary vacuum standards and for standard leak calibration, and many metrological laboratories carry out gas flow determination studies [6–9]. Therefore, primary constant pressure flowmeter (CPFM) standard establishment project is continuing under the scope of TÜBİTAK UME Research Infrastructure Renewal and Development Projects funded by Republic of Turkey Ministry of Development, 2017-2019. The other project is also in progress: Renewal of existing primary static expansion system. The static expansion system has been used for many years and needs to be renewed. Targeted gas flow operating range regarding the CPFM is between 4×10^{-12} mol/s to 4×10^{-7} mol/s. The system design, procurement and manufacturing processes are continuing. With the primary gas flow meter installation and integration to the dynamic vacuum system, it will be the primary level standard named continuous expansion system with the estimated range between 10^{-7} Pa to 10^{-3} Pa. In addition, traceability of gas flow rate measurements and standard leak calibrations will also be provided in this way. This works will extend the pressure measurement range at the primary level down to 10^{-7} Pa.

Table 2. The ambient conditions values.

Ambient Parameters	Required Values	Tolerance
Temperature	20 ºC (23 °C optional)	± 0.5 ºC
Moisture	% 45	± 15

Table 3. Vacuum Laboratory Calibration/Test List.

Service code	**Name**
G2VK-1100	Pirani, Thermal Conductivity, Thermocouple Type Vacuum Gauges
G2VK-1200	Capacitance Diaphragm Type Vacuum Gauges (CDGs)
G2VK-2100	Cold and Hot Cathode Type Vacuum Gauges
G2VK-2200	Spinning Rotor Type Vacuum Gauges (SRGs)
G2VK-3100	Full Range Vacuum Gauge Calibration
G2VK-5100	Helium leak rate detection in vacuum ($2x10^{-5}$ mbar·l/s – $2x10^{-6}$ mbar·l/s)
G2VK-5200	Helium leak rate detection in vacuum ($2x10^{-6}$ mbar·l/s – $2x10^{-8}$ mbar·l/s)
G2VK-9900	Special Calibration / Test

3. Conclusion

TÜBİTAK UME Vacuum Laboratory is responsible for supply of reference standards in the area of vacuum metrology in Turkey. The main aim is to provide traceability to National Standards based on SI units, Pa, for industry and government departments. The laboratory continues to keep its place in the national and international metrology network through both with the standards it has and its activities. The laboratory served under the Pressure Laboratory of TÜBİTAK UME until 2015 and it was separated from the pressure laboratory at the beginning of 2015. The studies mentioned above are conducted with three researchers who all have over 20 years of metrology experience.

References

[1] http://www.ume.tubitak.gov.tr/en
[2] https://www.ptb.de/emrp/vacuum.html
[3] Elkatmis A et al. *Measurement* (submitted)
[4] http://pres2vac.eu
[5] Kangi R, Ongun B and Elkatmis A 2004 *Metrologia* **41** 251–6
[6] Peggs G N 1976 *Vacuum* **8** 321–8
[7] Jousten K et al. 1993 *Vacuum* **2** 135–41
[8] Jousten K et al. 1999 *Metrologia* **36** 493–7
[9] Calcatelli A et al. 2003 *Measurement* **34** 121–32

IOP Conf. Series: Materials Science and Engineering **387** (2018) 012019 doi:10.1088/1757-899X/387/1/012019

To the calculation of the equation of measurements of a MEMS vacuum transducer

A Garshin, V Gorobey and R Kuvandykov

FGUP VNIIM n. a. D. I. Mendeleev, 190005, Saint-Petersburg, Russia

E-mail: ruskuw@mail.ru

Abstract. The article is devoted to the numerical method of calculating the vacuum transducer, created by MEMS technology and designed to measure the absolute pressure in a wide range from 10 to 10,000 Pa.

Measurements of low absolute pressures are the most important measurements in vacuum technology. In this regard, it is important to develop low absolute pressure measuring instruments on new principles. At VNIIM n. a. D. I. Mendeleev we works on improvement of the vacuum gauge of resonance type, created since 2015 by MEMS technology [1]. Refinement of the theoretical model using modern numerical computer technology is an significant task.

Below is a calculation of the conversion factor of MEMS vacuum gauge with a given geometric dimensions. Metrologically significant dimensions are shown in figure 1.

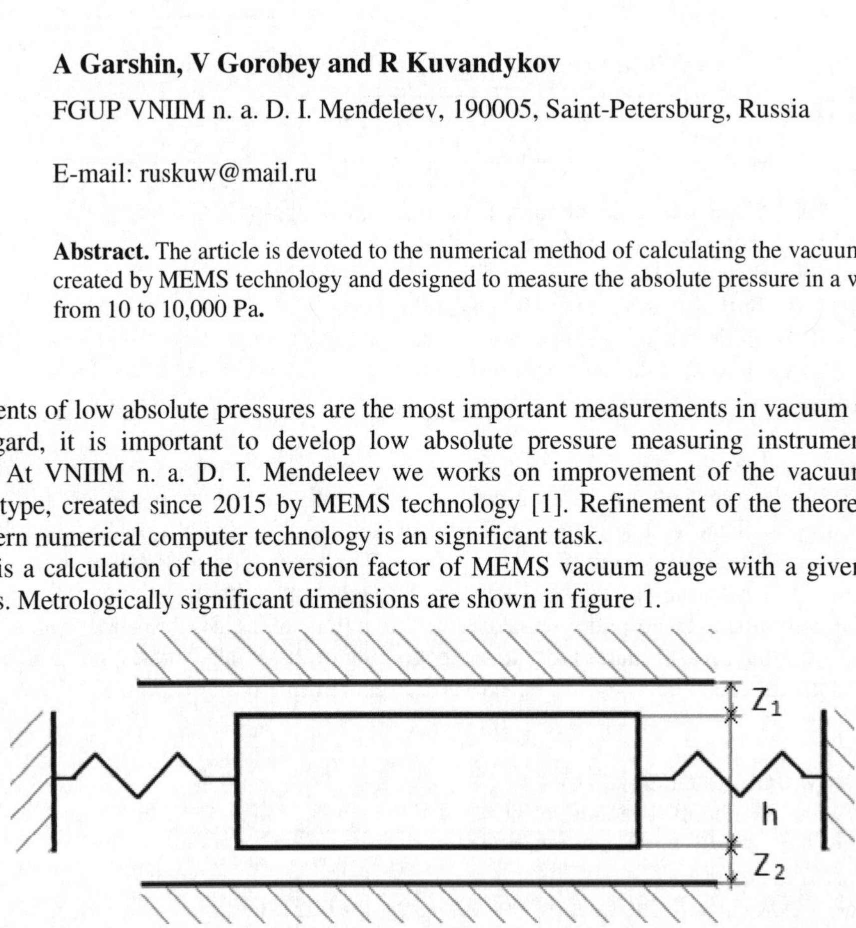

Figure 1. Metrologically significant dimensions: h – the thickness of the movable plate; Z_1 - the thickness of the gap №1; Z_2 - the thickness of the gap №2.

The following is a summary table of the numerical values of the parameters.

Table 1. Metrologically significant dimensions.

Mark	Parameter	Value
h	Plate thickness	100 μm
Z_1	Gap thickness	3 μm
Z_2	Gap thickness	3 μm
L_1	Plate length	3 mm
L_2	Plate width	3 mm

We perform a preliminary calculation of gas leaks through the edges of air gaps at the time of changing the position of the plate during vibrations.

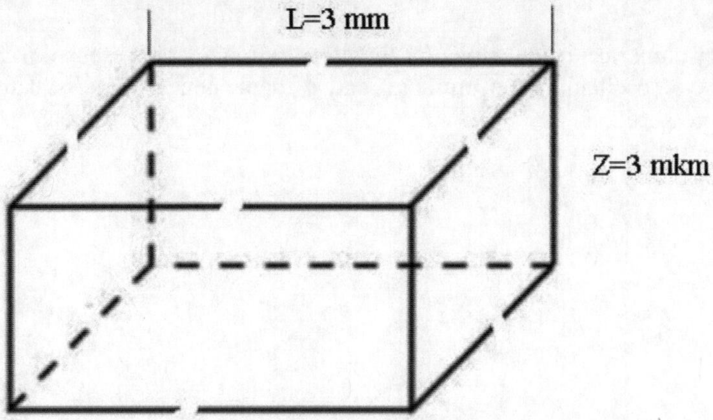

Figure 2. Air gap Z_1.

The total surface area of the parallelepiped describing the air gap:

$$St = 4 * L_1 * h + 2 * L_1 * L_2 = 4 * (3 * 10^{-3} * 3 * 10^{-6}) + 2 * (3 * 10^{-3} * 3 * 10^{-3}) = 1{,}9 * 10^{-5}\, m^2$$

Total area of the side faces of the air gap:

$$Ss = 4 * L_1 * h = 4 * (3 * 10^{-3} * 3 * 10^{-6}) = 36 * 10^{-9}\, m^2$$

The ratio of the area of the side faces to the total area is total 0,2 %.

When moving the moving plate at a relative small cross-section of air gaps (0,2 %), small amplitudes of motion and frequency of 100 Hz or more, the expiration of the gas from or flowing into the gaps Z_1 and Z_2can be neglected. Thus, the number of gas molecules in the gap during the time equal to the period of oscillation can be considered constant.

According to the Boyle-Mariotte law for the case of isothermal process, the dependence of the pressure in the gaps on the movement of the plate is expressed by the formula:

$$\mathrm{P} * \mathrm{V} = \mathrm{const}$$

$$\mathrm{P_1}(dZ) = \mathrm{P} * \frac{\mathrm{Z_1}}{\mathrm{Z_1} + \mathrm{dZ}}$$

$$\mathrm{P_2}(\mathrm{dZ}) = \mathrm{P} * \frac{\mathrm{Z_2}}{\mathrm{Z_2} - \mathrm{dZ}}$$

where $\mathrm{P_1}(dZ)$ и $\mathrm{P_2}(dZ)$ - the current gas pressure in the gaps in depending on the displacement of the movable plate, P - measured gas pressure.

Thus, the differential pressure and pneumatic force acts on the plate from the two gaps:

$$F_p(\mathrm{dZ}) = \left(\mathrm{P_1}(\mathrm{dZ}) - \mathrm{P_2}(dZ)\right) * \mathrm{S},$$

where $\mathrm{S} = \mathrm{L_1} * \mathrm{L_2}$ – area of the movable plate from the air gap.

Pneumatic stiffness of the gas spring $\mathrm{G_p}$ we find from the ratio pneumatic power $\mathrm{F_p}$ of movement dZ:

$$\mathrm{G_p} = \frac{\mathrm{F_p}}{\mathrm{dZ}}$$

From the formula of oscillations we find the cyclic frequency

$$\omega = \sqrt{\frac{G_m + G_p}{\mathrm{m}}},$$

where G_m- the mechanical stiffness of the suspension, m – the mass of the movable plate (m= $\mathrm{L_1} * \mathrm{L_2} * h * \rho$, the density of silicon $\rho = 2200$ kg/m^3).

The square of the resonant frequency of oscillation

$$f_p^2 = \frac{G_m + G_p}{4 * \pi^2 * m}$$

The results of numerical calculations for the pressure range are presented in figures 3 (dependence of the frequency of oscillations on pressure) and 4 (dependence of the square of the frequency of oscillations on pressure).

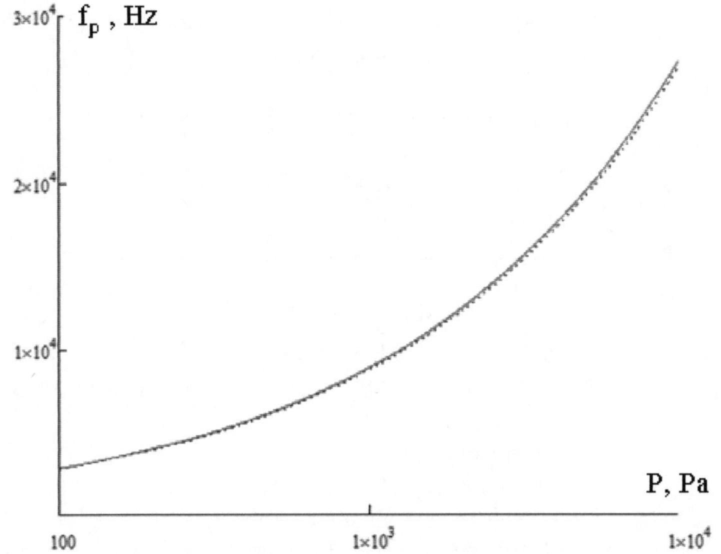

Figure 3. Frequency dependency on pressure.

In figure 3, we can see the coincidence of the results of the numerical calculation of the dependence and the calculation of the analytical formula derived in [1]:

$$P=K*(f_p^2 - f_{p.m.}^2),$$

where $f_{p.m.}$- value of resonance frequency at zero pressure. The discrepancy is about 1 %.

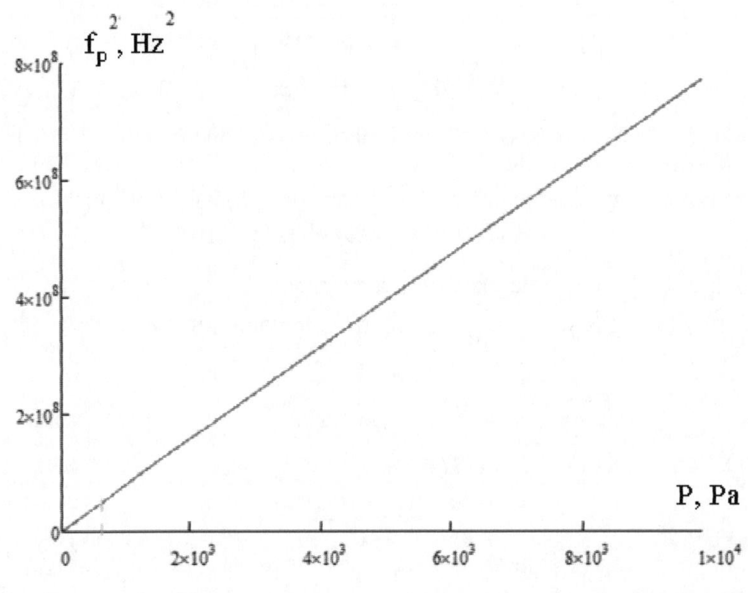

Figure 4. Frequency square dependency on pressure.

IOP Conf. Series: Materials Science and Engineering **387** (2018) 012019 doi:10.1088/1757-899X/387/1/012019

The obtained results confirm the thesis about the high degree of linearity of the dependence of the square of the frequency on the measured pressure (figure 4).

References

[1] Gorobey V, Garshin A and Kuvandykov R 2017 *Vacuum equipment and technology - 2017. Proceedings of the 24th all-Russian scientific and technical conference with international participation*

IOP Conf. Series: Materials Science and Engineering **387** (2018) 012020 doi:10.1088/1757-899X/387/1/012020

Modeling of reactive sputtering of hot titanium target in nitrogen and oxygen

A O Goncharov, E A Minzhulina and V I Shapovalov

Saint-Petersburg State Electrotechnical University "LETI", St. Petersburg, Russia

E-mail: vishapovalov@mail.ru

Abstract. A model for reactive sputtering of a hot titanium target in the environment of oxygen and nitrogen is considered. The target temperature is defined as a function of the ion current density. For the mathematical description of the process, the basic postulate of chemical kinetics including the equation of the Langmuir isotherm and the Arrhenius equation under nonisothermal conditions is used. In addition, the model takes into account thermionic emission and evaporation of the target surface.

Titanium oxynitride films with the composition expressed by the chemical formula TiO_xN_y form a continuous series of compounds with different concentrations of nitrogen and oxygen (from TiO_2 to TiN) [1].

In order to deposit TiO_xN_y films, the methods of reactive magnetron sputtering are most often used [2]. During reactive sputtering, a cold titanium target can operate in reactive or metallic modes [3]. The transition between the modes is controlled by the discharge current and reactive gase flow rate [3]. In a magnetron with a hot target, the physicochemical processes taking place on its surface change [4]. For a correct description of the problem, it is necessary to take into account the effect of the target temperature T.

In the first works devoted to the modeling of the reactive sputtering of a metal target in a mixture of argon, nitrogen, and oxygen, an isothermal chemisorption model for a single reactive gas was developed. A review of this theme is given in [3]. An alternative non-isothermal physicochemical model was originally developed for a single gas environment [5], and later for an environment of two reactive gases [3]. Then this model was developed and used for reactive sputtering of a hot target in an oxygen environment [4].

In this paper, we consider a more complicated problem. Modeling of sputtering of a hot titanium target in argon, nitrogen and oxygen was performed. Due to the limited size of the publication, below we describe the major assumptions, provide a system of algebraic equations in a general form, and present selected results of the solution.

We assume that the reactive sputtering system contains a hot titanium target, an unspecified substrate and a vacuum chamber wall with areas A_t, A_s and A_w, respectively. Elements of the system are located in the gas mixture $Ar + O_2 + N_2$. In this system:

1) the substrate and the wall have constant temperatures T_s and T_w, respectively. The target temperature $T_t = f(j_+)$ depends on the ion current density j_+;

2) we assume that titanium oxynitride is formed on all surfaces in the form of a solid solution of TiO_2 oxide and TiN nitride arising from surface chemical reactions:

$$\mathrm{Ti}+\mathrm{O}_2 \xrightarrow{k_1(T)} \mathrm{TiO}_2, \qquad \mathrm{Ti}+\frac{1}{2}\mathrm{N}_2 \xrightarrow{k_2(T)} \mathrm{TiN}, \tag{1}$$

where $k_1(T)$ and $k_2(T)$ are reaction rate constants (1). Hereinafter indices 1 and 2 refer to oxide (or oxygen) and nitride (or nitrogen), respectively;

1) two processes compete on the target surface:

- formation of a solid solution layer by reactions (1);
- removal of this layer by sputtering and evaporation. The flux density of the evaporated particles J_{ev} from a homogeneous surface gives the Hertz-Knudsen equation:

$$J_{ev} = \frac{p_0}{\sqrt{2\pi m_0 k T_t}}, \tag{2}$$

where p_0 and m_0 are the saturated vapor pressure and the mass of a particle of a material of a homogeneous surface. The concept of "homogeneity" is important. We assume that the target surface in the most general case can have three regions (see fig. 1). One of them, occupying the relative part of the θ_{t1} surface, is coated by an oxide, the second, coated by a nitride, has a relative area θ_{t2}. The remaining part $1-\theta_{t1}-\theta_{t2}$ is an uncoated metal. Each of these areas is called homogeneous. Developing the model, we further introduce refinement indices in (2), where it's required;

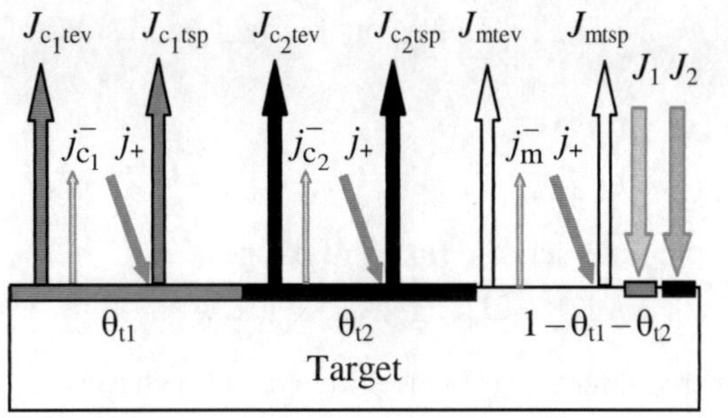

Figure 1. Processes on the target.

2) two stationary modes are possible for a target: metallic or reactive. In the first case, its surface is an uncoated metal and produces only titanium atoms, which create a corresponding film on the surfaces of the substrate and the wall. In the reactive mode, the target generates fluxes of TiO_2 and TiN molecules, which are deposited on the surfaces of the substrate and the chamber walls and create layers of oxynitride;

3) the discharge current j is the sum of the currents that have arisen on parts of the target surface with different physical properties. If the entire surface of the target is homogeneous, then the current density is

$$j = (1+\gamma)j_+ + j^-, \tag{3}$$

where γ and j^- are the coefficient of ion-electron emission and the current density of thermionic emission, respectively, for the target coating material. The value j^- is given by the Richardson-Dushman law:

$$j^-(T_t) = A T_t^2 \exp\left(-\frac{\varphi_t}{k T_t}\right), \tag{4}$$

where $A \approx 120$ A·cm^{-2}·K^{-2}; $k = 1.38\cdot10^{-23}$ J/K is the Boltzmann constant; φ_t is the work function of electrons from the target surface. In order to take into account all possible target coatings (metal, oxide and nitride) in the given task, it is necessary to assign the indices m, c_1 и c_2 in the expressions (3) and (4) to the values γ, j^- and φ_t . Taking into account all components, the expression for the current density of the discharge is now written in the form

$$j=(1-\theta_{t1}-\theta_{t2})[(1+\gamma_m)j_+ + j_m^-]+\theta_{t1}[(1+\gamma_{c_1})j_+ + j_{c_1}^-]+\theta_{t2}[(1+\gamma_{c_2})j_+ + j_{c_2}^-]; \quad (5)$$

4) in order to support the reactions (1), all surfaces consume reactive gases. We denote the oxygen fluxes by Q_{t_1}, Q_{s_1}, Q_{w_1}, and the nitrogen fluxes to these surfaces by Q_{t_2}, Q_{s_2}, Q_{w_2}.

Using assumptions 1–6, the basic postulate of chemical kinetics and the Langmuir isotherm, as it was done in [5], we obtain a system of sixteen algebraic equations. The first six describe the equilibrium on the surface:

- target (see fig. 1)

$$J_{ch_j}[T_t(j_+)]+J_{c_j}(j_+)=0,\ j=1,2, \quad (6)$$

where $J_{ch_j}[T_t(j_+)]$ is the density of the j-chemical flux, which is created by reactions (1) due to the fluxes of oxygen J_1 and nitrogen J_2. The surface of the target, as shown in fig. 1, has three regions. Therefore, it is the source of three fluxes with the density:

$$J_{mt}(j_+)=J_{msp}(j_+)+J_{mev}[T_t(j_+)] \quad (7)$$

for metal and

$$J_{c_jt}(j_+)=J_{c_jsp}(j_+)+J_{c_jev}[T_t(j_+)],\ j=1,2 \quad (8)$$

for oxide (j = 1) and nitride (j = 2). The first terms on the right-hand sides of (7) and (8) reflect sputtering, the second - evaporation;

- substrate and walls

$$J_{ch_j}(T_i)+J_{m_i}(j_+)-J_{c_{ji}}(j_+)=0,\ j=1,2;\ i=s,w. \quad (9)$$

The second group of six equations sets the fluxes of oxygen

$$Q_{i_1}=J_{ch_1}(T_i)A_i,\ i=t,s,w \quad (10)$$

and nitrogen on each of the three internal surfaces of the vacuum chamber:

$$Q_{i_2}=0.5J_{ch_2}(T_i)A_i,\ i=t,s,w. \quad (11)$$

The last group consists of equations, two of which describe the operation of a vacuum pump

$$Q_{p_j}=c^0p_jS_p,\ j=1,2 \quad (12)$$

and two - the balance of gas fluxes:

$$Q_{0_j}=Q_{t_j}+Q_{s_j}+Q_{w_j}+Q_{p_j},\ j=1,2 \quad (13)$$

In expression (12) we use the notation $c^0 = 2.5\cdot10^{25}$, $Pa^{-1}m^{-3}$ – the size coefficient that transfers the flux of Q_p from pV units to the measurement unit, which is more convenient for calculation: the number of particles per time unit; S_p – evacuation speed of the vacuum chamber; p_j - the partial pressure of the j-reactive gas.

The system of algebraic equations (6), (9)–(13) contains three independent variables: the fluxes Q_{0j} (j = 1, 2) and the discharge current density j (5). The dependent variables are the partial gas pressures p_j (j = 1, 2), the gas fluxes Q_{tj}, Q_{sj}, Q_{wj} and Q_{pj} and the fractions of the three surfaces θ_{i1} and θ_{i2} (i= t, s, w), coated by a solid solution of TiO_2 + TiN. The system (6), (9)–(13) was solved numerically. The approximating expression connecting the target temperature in kelvins with the ion current density in amperes per square meter is

$$T_t(j_+)\approx 293+1520(1-\exp[-0.00438\,j_+]), \quad (14)$$

Equations (6), (9)–(13) taking into account (14) were solved in the range of current densities from 200 to 400 A/m^2 and nitrogen flow rates Q_{02} from 0 to 0.3 cm^3/min. Figure 2 shows the S-shaped dependences $p = f(Q_{01})$ outlining that the process is characterized by a hysteresis effect. Figure 2 also shows that the increase in nitrogen consumption leads to the shift of the transition points in the target operating modes, and the derivative dp/dQ_{01} has discontinuities. In addition, an increase in nitrogen consumption leads to a decrease in the width of the hysteresis loop.

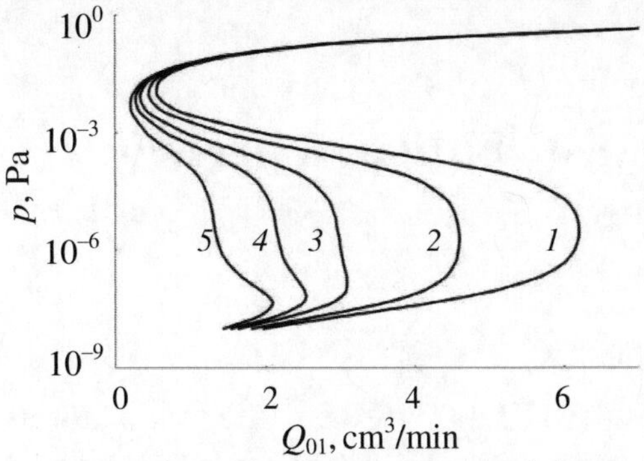

Figure 2. Dependence of the partial pressure of oxygen on the value of Q_{01} at a discharge current density of 200 A/m^2 and the values of Q_{02} (cm^3/min): 1 – 0; 2 – 0.10; 3 – 0.20; 4 – 0.25; 5 – 0.30.

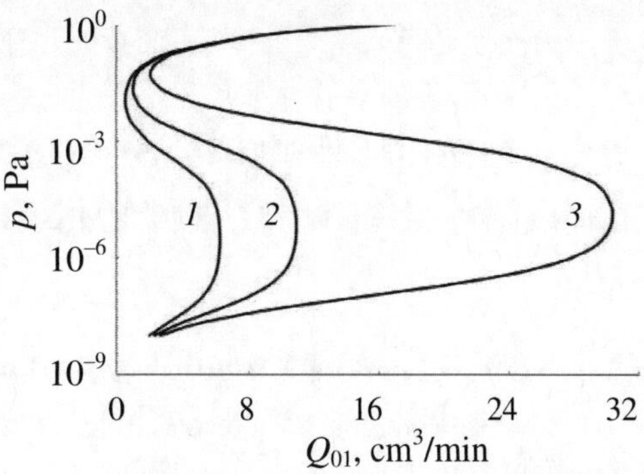

Figure 3. Dependence of oxygen partial pressure on Q_{01} value at Q_{02} = 0 cm^3/min and discharge current density (A/m^2): 1 – 200; 2 – 300; 3 – 400.

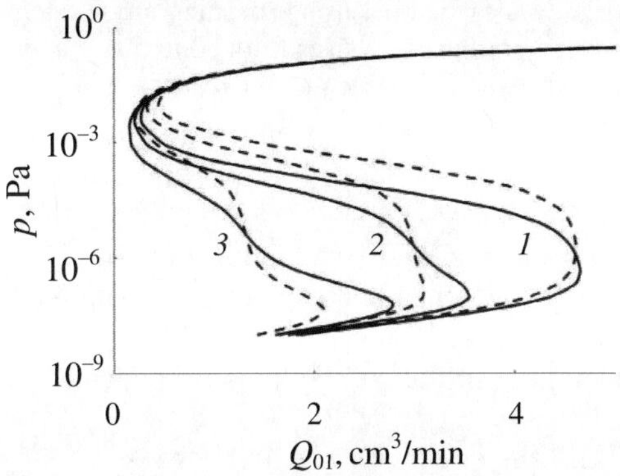

Figure 4. Dependence of oxygen partial pressure on Q_{01} for cold (solid lines) and hot (dashed lines) targets with discharge current density 200 A/m^2 and Q_{02} (cm^3/min): 1 – 0.1; 2 – 0.2; 3 – 0.3

As the current density increases, the transition points are shifted to the region of higher oxygen flow rates (figure 3) and the hysteresis width increases. In addition, the calculations showed that when the hot and cold targets are sputtered, the dependences $p = f(Q_{01})$ are different (figure 4).

Acknowledgment
This work was supported by the Russian Foundation for Basic Research (grant 18-29-11043).

References

[1] Banakha O, Moussa M, Matthey J et al. 2014 *Appl. Sur. Sci.* **317** 986–93
[2] Libardi J, Grigorov K G, Moraes R S et al. 2015 *J. Electron. Mater.* **44** 103–9
[3] Barybin A A, Zavyalov A V and Shapovalov V I 2012 *Glass Phys. Chem.* **38** 396–401
[4] Shapovalov V I, Karzin V V and Bondarenko A S 2017 *Phys. Lett. A.* **381** 472–5
[5] Barybin A A and Shapovalov V I 2007 *J. Appl. Phys.* **101** 054905

The experimental study of spacial distribution of particles produced in the pulse arc discharge

V D Goncharov and R V Yashkardin

St. Petersburg Electrotechnical Univeristy (ETU), 197376, Saint Petersburg, Russia

E-mail: vdgoncharov@rambler.ru

Abstract. The results of experimental researches of special particle psuttering produced due to the energy impact of the pulse discharge plasma moving along the surface of the extended electrodes. The particles distribution study of the substrates placed at the different areas regarding the dis-charge was carried out using scanning probe microscope. It is shown that the major part of the produced particles had sizes from 50 to 500 nm and sputtered orthogonally to the surface of the electrodes and discharge motion vector.

1. Introduction

Using ultradispersed particles (UDP) and materials containing them in technology is very common nowadays. Such materials are widely used in chemical industry, car manufacturing, medicine etc. And there's also a wide variety of methods of producing the UDP [1, 2]. One of such methods involves using pulse arc discharge moving along the surface of the long extended electrodes [3].

2. The device working principles and the order of the experiments

The principles of the device functioning is presented in the figure 1.

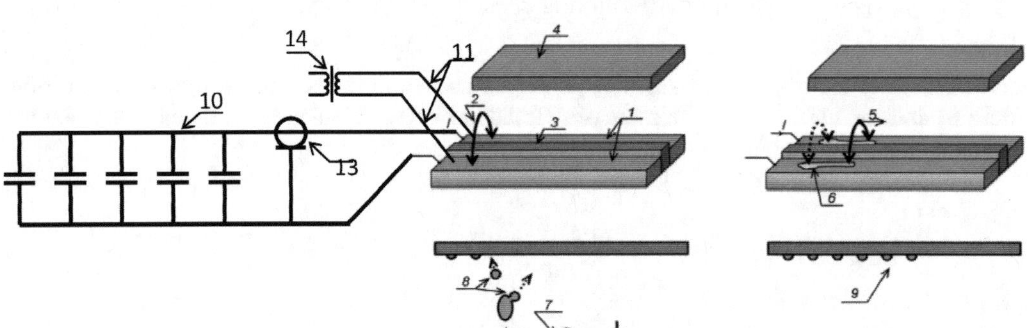

Figure 1. Schematic description of the UDP production device.

The main discharge (2) in the device is in between main electrodes representing two steel beams in parallel to each other (1). In between the main electrodes there's a dielectric inset (3). The capacitive battery (10) and the main electrodes are connected with co-axial power cable (13). The main discharge causing the appearance of the UDP on the surface of the main electrodes in ignited by the additive spark discharge between the main (1) and the additive electrodes (11). The spark is caused by the high voltage impulse on the additive electrodes from the high voltage source (14). The energy accumulated in the capacitive batteries is released in the discharge area. The discharge starts its motion along the main electrodes (5) under the magnetic fields interaction. The interaction of the moving discharge with the main electrodes surface causes the formation (7) sputtering and splitting UDP (8). The detailed description of the processes taking place in the device is presented in the paper [4]. We gave up any hope to study the parameters of plasma [5] experimentally from the very beginning and focused on matching technological parameters, the shape of the voltage pulse and studying special distribution of the UDP.

Experimental researches on sputtering the particles on the glass substrate were carried out using the device in the figure 1. The glass substrates were chosen due to their consistent surface roughness (less than 5nm whereas estimated sizes of the particles were from 10 to 500 nm). So using these substrates allows to determine the number of the UDP quite accurately. The substrates (4) were situated opposite to the electrodes and in front and aside to the anode (1) and cathode (2) (with the semi-conductive inset (3) in between them) (figure 2(a)). There were three studied areas on each substrate: in the beginning (1) in the middle (2) and in the end (3) of the arc motion track (figure 2(b)). On the presented photo of the cathode one can see also the discharge ignition area (4).

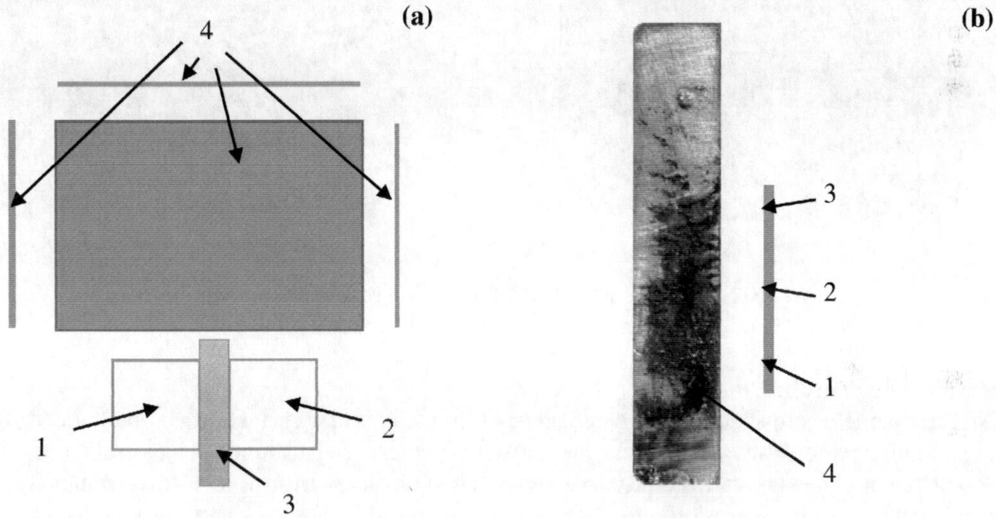

Figure 2. AFM studied substrates placements (**a**); photo of electrode surface with the discharge trace (**b**).

The experiments were made using different types of sputtered targets (both magnetic and non-magnetic (titanium)). The study of the substrates surfaces were made using scanning probe microscope (SPM) Certus Light (NanoScanTech, Russia). We used a probe cantilever NSC 15 with resonance frequency 325 kHz. The maximal scale of the scanning area is 100×100 um. The number of probed points for each dimension can be varied. The more points – the more detailed picture. However, it also dramatically increases the time of scanning and the cantilever wear. That's why we had to make a number of preparatory researches before moving on to the actual study of the particles sputtering distribution. And these experiments showed that:

- There's nearly zero differences between particles distributions between 100×100 and 25×25 um scanning areas. Of course, the picture of smaller scanning scale is 4 times more detailed with the same resolution. That's why further on we used that scale with the resolution 1000×1000 points (the time of one line pass was 1 second). In order to make sure that the scanning area was averagely covered by particles, the area for 25 um square scanning was chosen only after 500×500 points 100 um square preparatory study.

- The results of 25×25 um scanning area can be extrapolated to 5×5 mm square area. After comparing the results of studying five different areas of 5×5 mm square (in each corner and in the center) we found the differences in surface integral parameters (waviness and roughness) never exceed 10 %.
- The information about transverse sizes of the particles can be obtained very easily using Phase Imaging mode. For example, in the figure 3, the results of the substrate scanning in the middle of the discharge trace are presented. One can see the results of two different modes used (amplitude mode (a) and phase imaging (b)). The particles size distribution at the phase imaging mode is very clear and contrast compared to the Amplitude mode, and most of them are around 0.3–0.5 um in diameter.

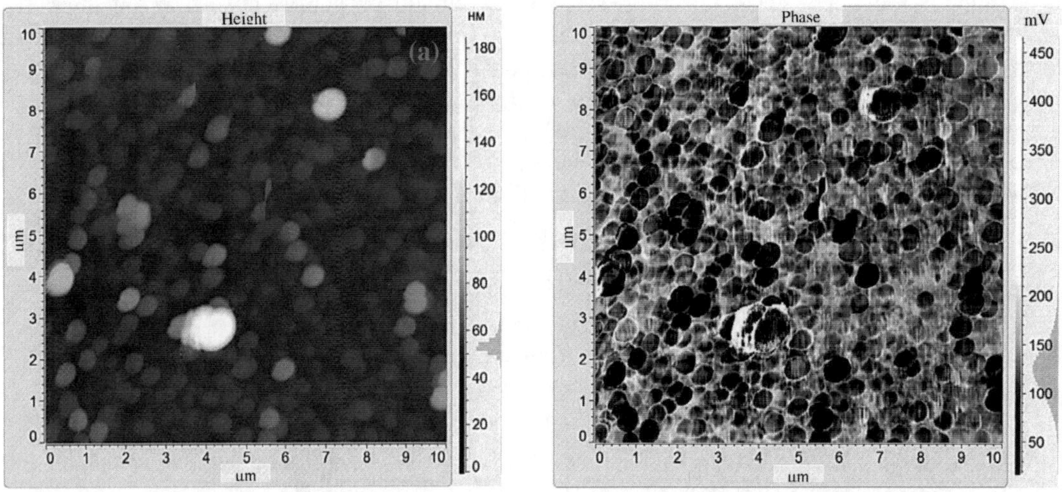

Figure 3. Substrate surface AFM scanning results in the amplitude (**a**) and phase imaging (**b**) modes.

3. Results of researches

Mostly the particles are sputtered perpendicularly to the electrodes surface. It can be noticed from the figures 4 and 5, where we can compare the particles sputtering distribution for the substrates situated in the different places around the discharge area. The substrates from figure 4(a),(b) and figure 5(a) were placed in the area 2 from figure 2 (figure 4(a) – near the cathode, figure 4(b) – distant over the electrodes, figure 5(a) – near the anode). The substrate from figure 5(d) was placed on the way of the discharge motion.

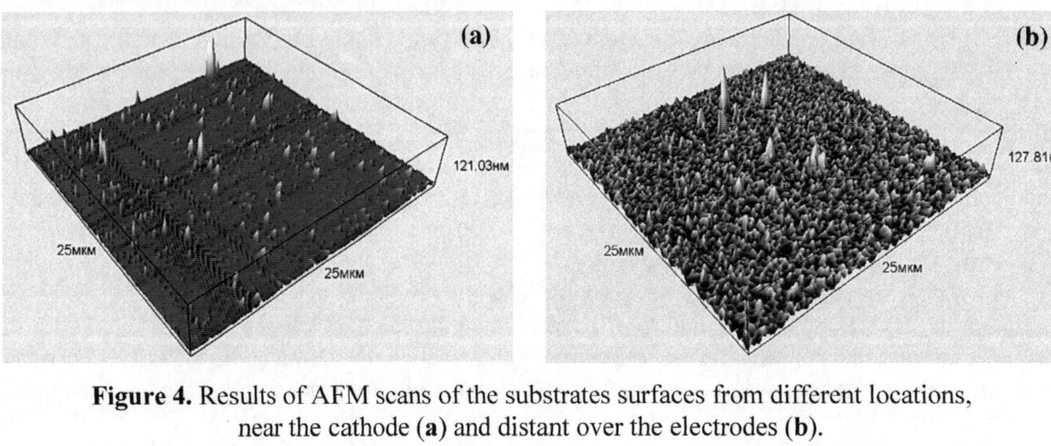

Figure 4. Results of AFM scans of the substrates surfaces from different locations, near the cathode (**a**) and distant over the electrodes (**b**).

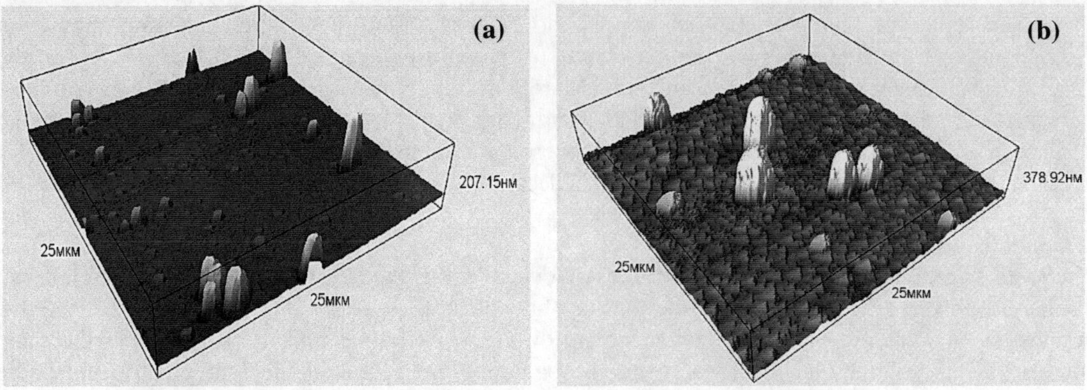

Figure 5. Results of AFM scans of the substrates surfaces from different locations, near the anode (**a**) and place on the way of the discharge motion (**b**).

It's obvious that:

- The particles on the surface of the substrate near the anode (figure 4(a)) are smaller than the ones near the cathode (figure 5(a)).
- Highest amount of particles was found on the surface of substract places right over the electrodes with the major part of particles minimal sizes 0.1–0.5 um. These particles are highly visible In Phase Imaging mode at figure 6 at 10×10 um scale.

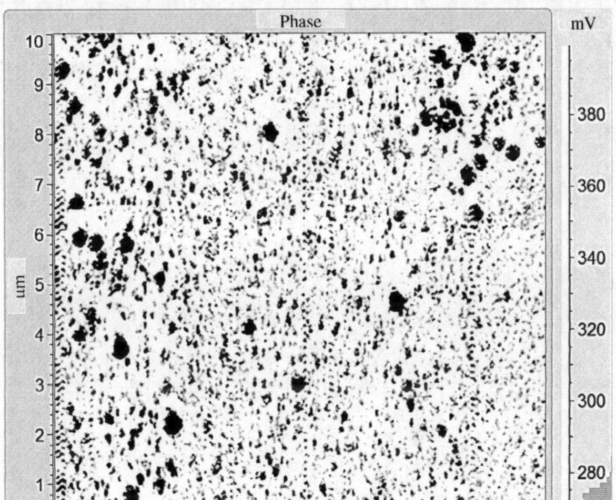

Figure 6. Results of AFM scanning of the substrate surface with particles on it at phase imaging mode.

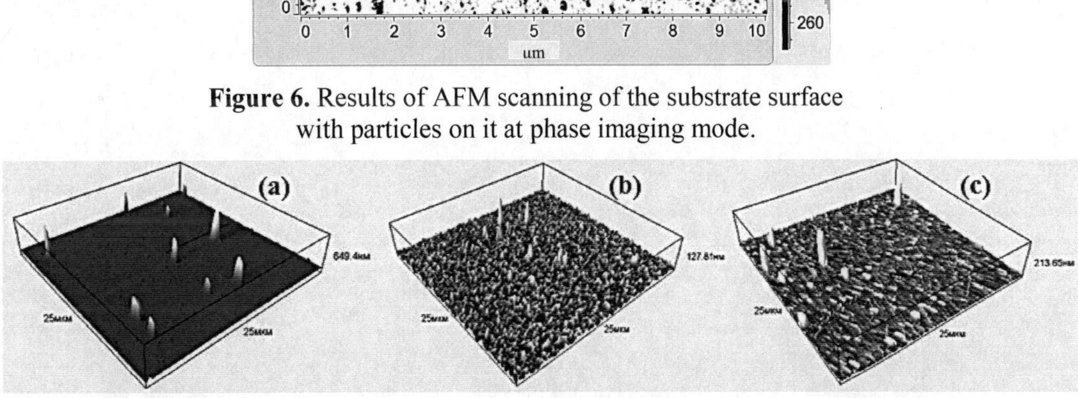

Figure 7. Results of AFM scanning of the substrates surfaces placed over the electrodes.

- The results of scanning also heavily depend on the studied area of the substrate choose. In particular, in the figure 7 the scanning results of the substrate placed over the electrodes. In the area, corresponding the beginning of the discharge trace (a) we find a minor number of particles, which are relatively massive. In the middle (b) a significant number of small particles. Studying their profile leads to the conclusion that they were moving nearly orthogonally to the substrate. By the end of the trace (c) the number of particles decreases and they become bigger once again.

4. Conclusions

The sizes and the number of particles are heavily dependent of the substrate placement. The first two discharges are cleaning and preparing the surface of the electrodes, so the surfaces of the substrates for new and a bit worn out electrodes sputtering are different. After the third discharge under the same conditions (the voltage and the capacity of the batteries, the height of the semi-electric inset) the average number of the particles and their size remains the same in all the experiments. The biggest particles (up to 3 um) were found on the substrate which was placed on the discharge motion direction.

The particles themselves drawn up along the discharge motion which shown that particles were moving angled to the substrate and still were fluid when touched it.

References

[1] Gorokhov M V, Kozhevin V M, Yavsin D A et al. 2012 *Journal of Technical Physics* **6** 135–41
[2] Gubin S P, Koksharov Yu A, Khomutov G B and Yurkov G Yu 2005 *Uspekhi Khimii* **6** 539–74
[3] Goncharov V D and Samsonov D S 2015 *Journal of Technical Physics* **5** 37–42
[4] Yaskardin R V and Sorokin K S 2016 *IEEE NW Russia Young Researchers in the Electrical and Electronic Engineering Conference (2016 ElConRusNW)*
[5] Kostrin D K et al. 2016 *Journal of Physics: Conference Series* **729** 012030

IOP Conf. Series: Materials Science and Engineering **387** (2018) 012022 doi:10.1088/1757-899X/387/1/012022

The possibility of using permanent magnets in planar magnetron installations for sputtering magnetic targets

V D Goncharov, R V Yashkardin and E M Fiskin

St. Petersburg Electrotechnical Univeristy (ETU), 197376, Saint Petersburg, Russia

E-mail: vdgoncharov@rambler.ru

Abstract. The paper includes the results of numerical modeling of the spatial distribution of the magnetic field strength of a magnetron discharge. Special attention is paid to the case where the cathode-target is made of a magnetic material. The paper shows that it is possible to obtain the required value of magnetic field induction above the target surface from a magnetic material either at a target thickness of less than 2 mm or by using a heat removal constraint from the target surface.

1. Introduction

The properties of the discharge depend on the complicated character of the electron motion in the cathode area of this discharge form. Due to the crossed electric and magnetic fields special conditions are created in the cathode area in the way to force, the electrons overcome the distance between cathode and anode couple thousand times. As a result, at large free paths of atoms, electrons and ions, which are characteristics of low gas pressures, it is possible to obtain conditions for efficient sputtering of target materials.

This type of discharge is widely used in magnetron sputtering systems (MSS). These systems have found wide application in manufacturing for the modification of the surface of solids [2]. The paper deals with the most common design of magnetron sputtering systems, containing a magnetic core, permanent magnets and a flat target (figure 1). This model includes “arc” type of the magnetic field distribution and the electric field strength directed normal to the target surface.

The existence of the cathode area of a magnetron discharge is associated with the presence of crossed electric and magnetic fields in it. Knocked from the target surface electrons, accelerating in the electric field, fall into a magnetic field (magnetic trap). The electron moves along the cycloid trajectory. This motion is maintained until the electron experiences a collision with a heavy particle. On the one hand, the cycloid radius should be enough to make the electron reach a certain energy level to make the gas molecules ionization possible and on the other hand, it should not allow the electron to leave the magnetic trap. These conditions define the upper and lower boundaries of the intensity of the magnetic field. Within these boundaries this form of discharge can exist.

If the intensity of the magnetic field created above the target surface is less than the lower boundary, the electron knocked from the cathode surface irrevocably leaves the magnetic trap. The magnetron discharge does not burn at that case. Thus, it is necessary to ensure the value of the longitudinal component of magnetic field induction (MF) above the target surface of at least 0.03 T [3] for the existence of a magnetron discharge. Knowing this value and the conditions under which such a value can be obtained is especially important in the design of MSS with magnetic targets [4]. The task of developing MSS designs can also be solved experimentally. For this, it will be necessary to create a prototype of the system and determine from the spectrum of emission of the discharge its characteristic by changing the thickness of the magnetic target [5].

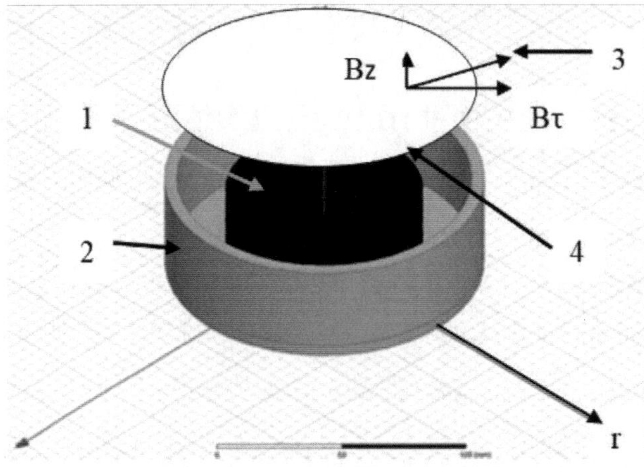

Figure 1. The geometry of studied area.

However, this way requires a considerable amount of time. Therefore, we have previously decided to determine under what conditions the discharge will burn during the design of the MSS with a target of magnetic material. This requires knowing the distribution of the induction of the magnetic field above the target surface. Determination of the induction of MF in a complex con-figuration of the magnetic circuit and permanent magnets is a difficult task. The solution of such a problem is significantly complicated by the fact that the MSSs "work" at high values of the magnetic field strength. In order to obtain adequate information on the field distribution the dependence of the induction of MF on its strength must be taken into account. The task becomes nonlinear and can be solved only by numerical modeling. The paper is dedicated to the development of a mathematical model of the magnetic field in the MSS and to its numerical realization.

2. Modeling the distribution of the magnetic field in the MSS

The real geometry of the MSS magnetic system (schematically shown in figure 1) was used in the calculations. Magnetic field is created by two SmCo magnets (1). The lines of the magnetic field are commutated through the magnetic core made of electrotechnical steel (2). The vector of the magnetic induction might be presented as vector sum of longitudinal (Br) and transversal (Bz) components (3). The target (4) is located in the immediate vicinity from the upper surface of the magnetic circuit (2 mm are necessary for vacuum isolation of the system and its uniform cooling).

The field distribution was defined by fourth Maxwell equation solution:

$$\mathrm{div}\vec{\boldsymbol{B}} = 0, \tag{1}$$

where B – magnetic induction vector.

The induction of the magnetic field was assumed to be zero on all the boundaries at a significant distance from the investigated region. A three-dimensional model was developed. It allows us to investigate not only axisymmetric systems (as most authors), but also the variants for including complex targets, consisting of magnetic and nonmagnetic materials. This task was solved numerically with the help of the created mathematical model in the ANSYS program. The adequacy of the obtained results was determined by comparing the results of physical experiments performed with the help of the magnetometer RSh1-10. The longitudinal component of the magnetic induction experimentally was measured experimentally over the surface of a titanium target 5 mm thick. In addition, it was obtained by numerical simulation in the field of the existence of a magnetron discharge and differed from the experimental one by no more than 5 %. The accuracy of the results of the numerical simulation depended on the domain partition. The magnetic core, the target (especially made of magnetic material), and the region of the formation of the magnetron discharge must be broken down in detail (the maximum number of elements of the partition).

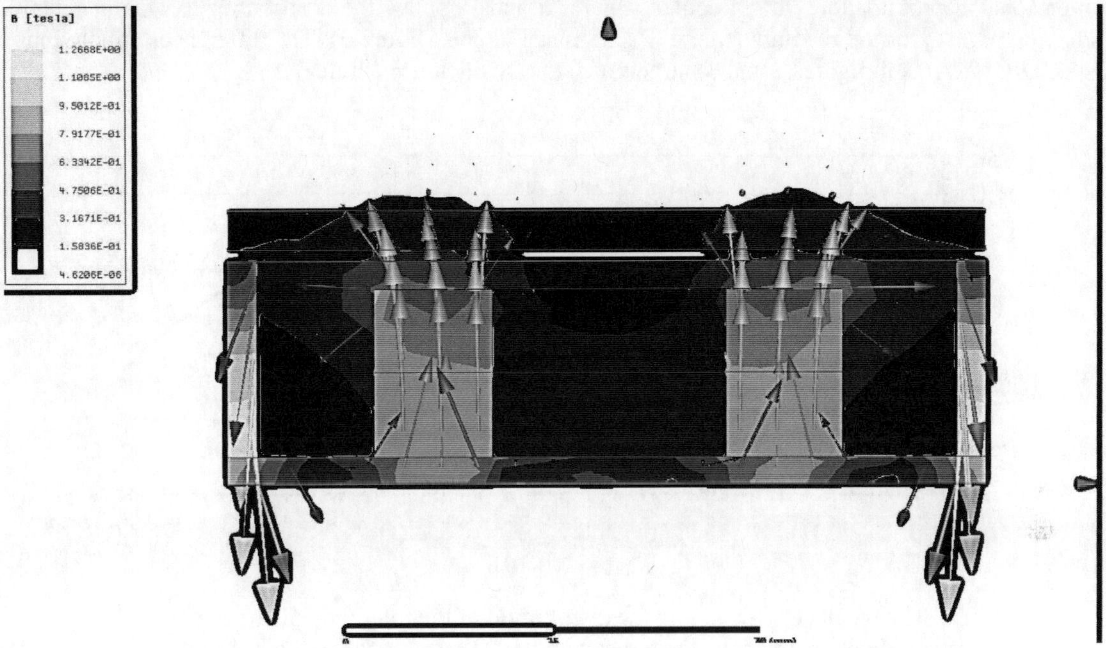

Figure 2. The results of the calculation of the field strength distribution created by the magnetic system of MSS using a non-magnetic target.

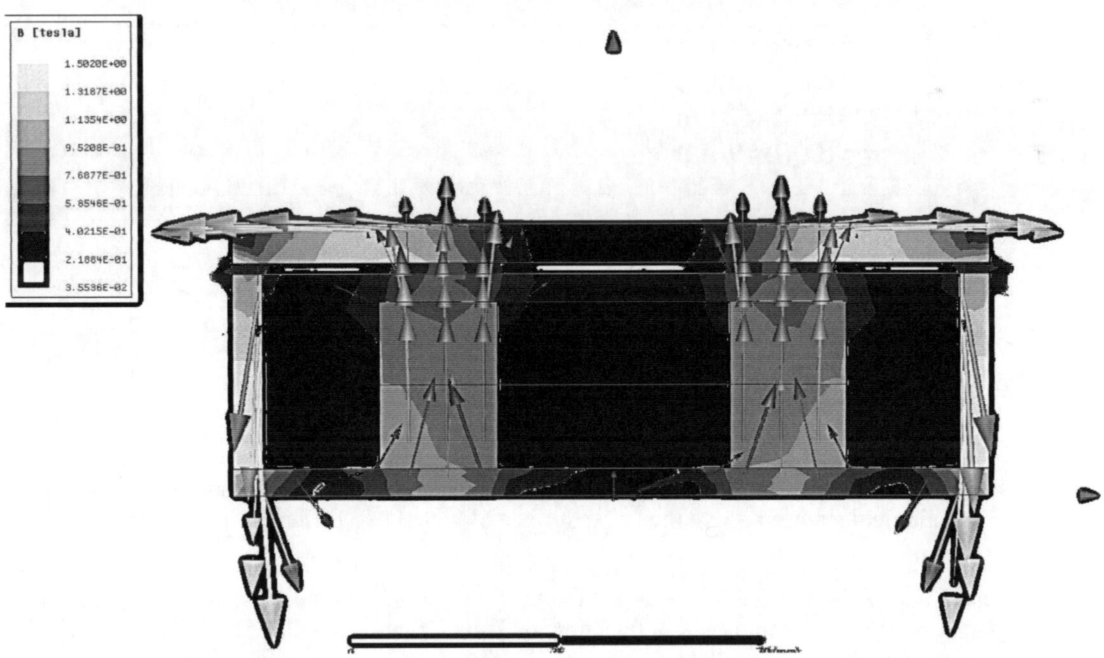

Figure 3. The results of the calculation of the field strength distribution created by the magnetic system of MSS using magnetic target.

The results of the calculations are obtained in the form of a vector distribution of the magnetic induction and can be hardly estimated in practice. These results can be used primarily as illustrations of ongoing processes. It is much easier to estimate the distribution of the MF strength in the two-dimensional approximation. In particular, figure 2 and 3 show the distribution of the magnetic induction module for the titanium (figure 2) and nickel targets (figure 3). The differences are obvious. Especially when (as in figures 2 and 3) targets are equally thick h = 7 mm.

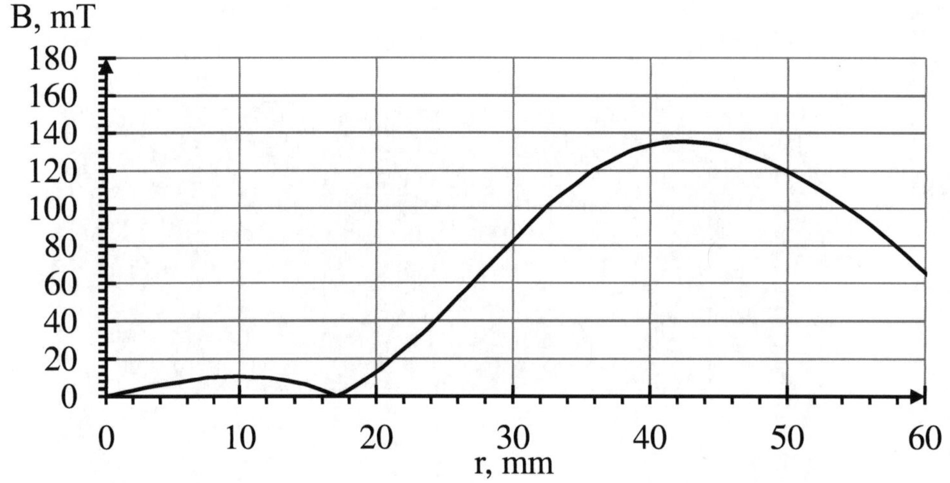

Figure 4. The results of calculating the distribution of the longitudinal component of the magnetic field strength above the target surface of titanium with a thickness of 7 mm.

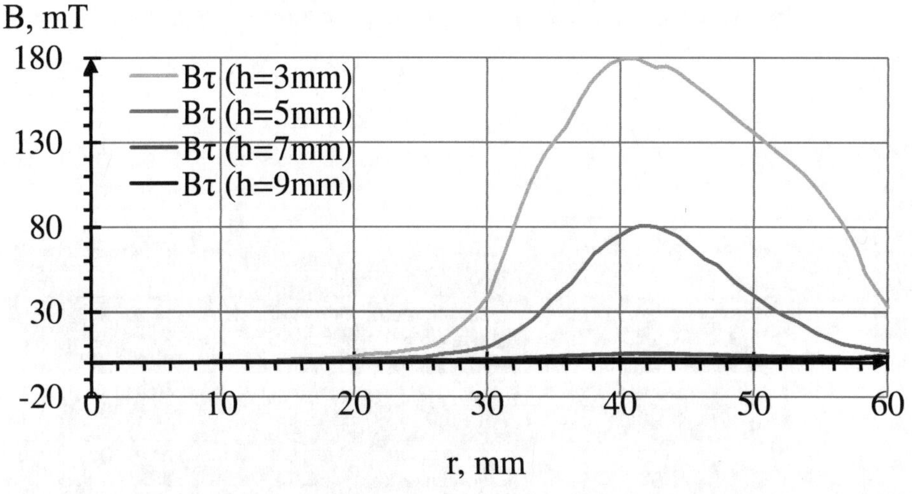

Figure 5. The results of calculating the distribution of the longitudinal component of the magnetic field strength above the target surface of nickel of a different (h) thickness.

However, the highest interest is in tangential component of magnetic induction, directed along the target. Figure 4 and 5 show the distribution of the tangential component of the magnetic induction above the target surface of titanium with a thickness of 7 mm (figure 4) and nickel of different thicknesses (figure 5). These results was obtained when the target was 2 mm from the magnetic system, that is at h = 3 mm the thickness of the target is 1 mm. The comparison of the results shows that it is possible to achieve the desired value of the magnetic induction using permanent magnets only by reducing the thickness of the target from nickel with relatively thin targets (not more than 2 mm thick).

The use of such targets is technologically unfeasible. This is complicated by the need for its frequent replacement. In addition, the use of thin targets is associated with the difficulty of uniform heat removal from their surface. As a result, there is possibility of local overheating of the target above the Curie temperature and the occurrence of temperature instability in the given region. This manifests itself strongly when more than 50 % of the target is generated. The desired value of induction can be obtained either by using combined targets with a ring-shaped liner of magnetic material in a nonmagnetic base, or by heating the nickel target above the Curie temperature. The latter can be achieved by reducing the heat removal from the target, i.e. creating a target design similar to that proposed in the paper [6]. The calculated distribution of the magnetic induction is shown in figure 6 (nickel target of thickness h = 2 mm and the limitation of the heat removal from the discharge burning region). In the calculation of the target temperature distribution, the power deposited in the target surface of 1 kW was assumed to be distributed evenly in the region of the burning discharge (ring with outer and inner diameters of 120 and 110 mm). A heat removal carried out by contacting the edges (5 mm in the outer and 10 mm in the inner diameter) of the target with a copper water-cooled disc. Numerical calculations show that in this case the tangential component of the magnetic induction was sufficient to burn the magnetron discharge.

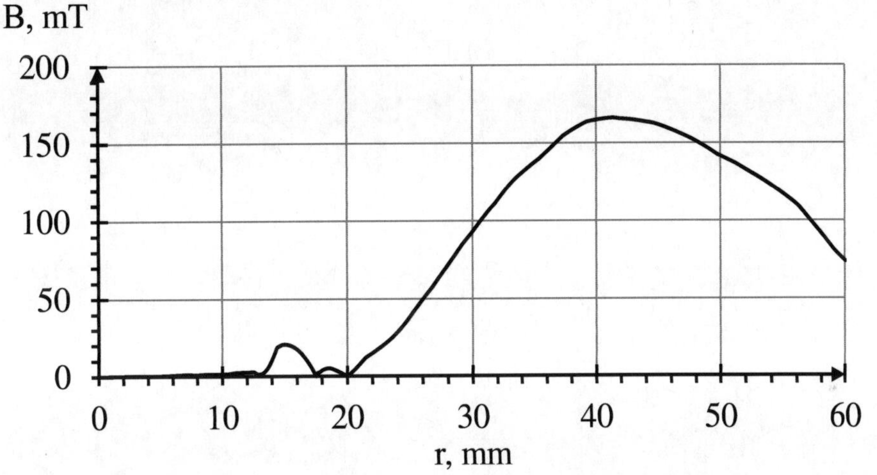

Figure 6. The results of calculating the distribution of the longitudinal component of the magnetic induction with the heat removal above the target surface of nickel with a thickness h = 4 mm.

3. Conclusions

1. The accuracy of the calculation of the MF distribution parameters depends on the domain partition of the individual elements.

2. The desired value of magnetic field induction using permanent magnets can only be achieved by reducing the thickness of the target from nickel (less than 2 mm).

3. The induction value required for the "operation" of the MSS can be achieved by warming up the nickel target above the Curie temperature (the target thickness cannot be more than 5 mm).

The accuracy of the results of the numerical simulation depended on the domain partition.

References

[1] Danilin B S 1989 *Primenenie nizkotemperaturnoj plazmy dlja nanesenija tonkih plenok* (Moscow)

[2] Wiatrowski A, Posadowski W M and Radzimski Z J 2008 *Journal of Physics: Conference Series* **100** 62004–7

[3] Vlasov M A, Zharinov A V and Kovalenko Yu A 2001 *JTF* **12** 34–42
[4] Caillard A, El'Mokh M, Semmar N et al. 2014 *IEEE transactions on plasma science* **42** 1–2
[5] Kostrin D K et al. 2016 *Journal of Physics: Conference Series* **729** 012030
[6] Bleykher G A, Borduleva A O, Krivobokov V P et al. 2016 *Vacuum* **132** 62

IOP Conf. Series: Materials Science and Engineering **387** (2018) 012023 doi:10.1088/1757-899X/387/1/012023

New state primary standard GET 49-2016 of reproduction pressure unit in range 10^{-6}–10^3 Pa

V N Gorobei, E K Izrailov, R E Kuvandykov, D M Fomin and A A Chernyshenko

D. I. Mendeleyev Institute for Metrology (VNIIM), 190005, St. Petersburg, Russia

E-mail: V.N.Gorobey@vniim.ru, ruskuw@mail.ru, fomindm@yandex.ru, vacuum@vniim.ru

Abstract. The article contains data on composition, principle of operation, measuring and functional capabilities of the state special primary standard of pressure unit for absolute pressures in the range 10^{-6}-10^3 Pa.

1. Introduction
The state special standard of pressure unit for absolute pressures in the range 10^{-6}-10^3 Pa (GET 49-2016) was approved in D. I. Mendeleev Institute of Metrology in 2016. It was developed in the course of work to improve the Russian reference base. As a result of work for improving the state standard for absolute pressure the system and some of functional characteristics of the previous standard were changed.

2. Principle of operation and composition of GET 49-2016
The state primary standard GET 49-2016 is a measuring complex designed for reproducing, storing and transferring the pressure unit to the reference and working measurement instruments (MI) in the range of 10^{-6} ... 10^3 Pa. There are several principles of measurements in the standard:

1. Compensation principle, based on balancing between the pressure force and electrostatic force acting on the plates of flat capacitor. This principle is realised in the membrane-capacitive converter of the compensating type. The equation for measurements has the following form:

$$p = kU^2 \quad (1)$$

where p is the reproducible pressure; U - compensating voltage; k is the constant. The constant of the membrane-capacitance manometer is calculated on the base of measurement and capacitance of the flat capacitor.

2. The principle of static expansion, based on isothermal expansion of gas and the Boyle-Mariott Act. The equation of measurement has the following form:

$$p = p_1 (V_1 / V) = k\, p_1 \quad (2)$$

where V_1 is the volume occupied by the gas before expansion (small volume); V is the volume occupied by the gas after expansion (large volume); p_1 is the gas pressure measured by membrane-capacitive converter.

3. The principle of dynamic expansion (reduction), based on the equations of continuum. The equation of measurement has the following form:

$$p_x=p_0+k\Delta p+0.5q\,Q_l, \tag{3}$$

where p_x is the reproducible pressure; p_0 - residual pressure in ultrahigh vacuum chamber of the standart; Δp is the pressure difference between the initial pressure in reference chambers and the measuring chamber. k is the reduction coefficient determined by the ratio of the conductivities of the elements in vacuum system of the standard; $0.5q\,Q_l$ - gas evolution from the surface of the vacuum system of the standard.

These measurement principles are implemented in the standard GET 49-2016 in the form of several reference complexes and installations: figures 1 and 2 shows its structural scheme and the general view of the primary standard.

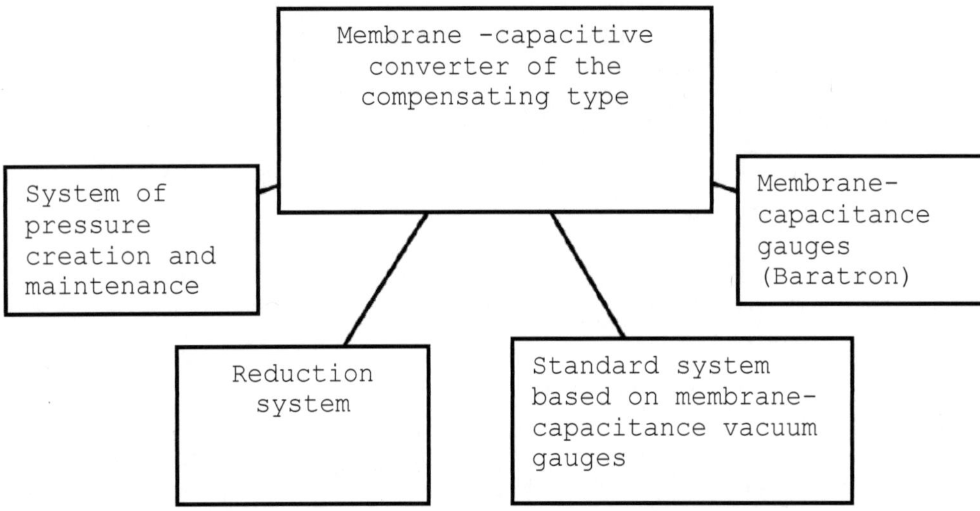

Figure 1. Structural scheme of the state primary standard.

3. Results of testing the state primary standard and metrological characteristics of the standard GET 49-2016

During the development of the GET 49-2016 the practical and theoretical testing was carried out which included the following stages:

1. Theoretical and practical researching of metrological characteristics of reduction installation.
2. Interlaboratory comparisons of the developed standard with the state special standard for pressures difference GET 95-75.
3. The first stage of bilateral comparisons of national standards of Russia and Turkey in the range of low absolute pressures and vacuum according to the instructions COOMET 711/TR/16 and 712/TR/16.

The researches have confirmed theoretical calculations and assumed metrological characteristics of the standard GET 49-2016. The standard has the following metrological characteristics:

Measurement range of absolute pressure, Pa: $1\cdot10^{-6}$ ... $1\cdot10^{3}$.

Number of independent measurements: 10.

The mean square deviation of the measurement result, S_0, not more than:

$0{,}15\cdot10^{-2}$ in the range of $1\cdot10^{-3}$ ... 40 Pa;
$0.30\cdot10^{-2}$ in the range 1 ... $1\cdot10^{3}$ Pa;
$2.0\cdot10^{-2}$ in the range of $1\cdot10^{-6}$ ... $1\cdot10^{-2}$ Pa.

Unexpected systematic error, θ_0, not more than:

$0.17\cdot10^{-2}$ in the range of $1\cdot10^{-3}$ ... 40 Pa;
$0.30\cdot10^{-2}$ in the range 1 ... $1\cdot10^{3}$ Pa;
$2.5\cdot10^{-2}$ in the range of $1\cdot10^{-6}$ ... $1\cdot10^{-2}$ Pa.

Figure 2. General view of the primary standard.

4. The results
The obtained results confirmed the correctness of the technical solutions adopted by the improvement of the reference base in the field of absolute pressures. The development of the reference system was based on the use of modern digital, precision electronics. The standard is build with using the latest vacuum technologies, components and instruments: oil-free pumps for creating and maintaining pressure, electrolytic polishing of internal vacuum surfaces, modern vacuum gauges, fittings and valves.

During the work the measurement range was extended by three orders. The accuracy at the range of high vacuum is increased, and also the functionality of the GET 49-2016 has beem improved.

The state primary standard GET 49-2016 satisfies requirements of industry and allows to solve the problems.

5. Prospects and ways of further improvement
Research work carried out for improving the standard for the range of absolute pressures allowed to formulate and plan further work to improve the Russian reference base for measurement of low absolute pressure and vacuum. In the future the following researching works should be carried out:

- Development and creation of a new membrane-capacitive converter of compensation type with adaptive membrane;
- Development and creation of the standard and reference measurement instruments based on MEMS technologies;
- Development and creation of a natural standard of absolute pressure (the realization of a triple points);
- Development of the system for standard complex in the range og ultrahigh vacuum up to 10^{-10}-10^{-12} Pa;
- Unification of existing pressure standards into one complex of one state standard.

References
[1] Rozanov L N 1990 *Vacuum Technique* (M .: Higher School) 320

[2] Gorobey V N and Izrailov E K 2015 *Vacuum technique and technology* **2** 49–53
[3] Kuzmin V V 1992 *Vacuum measurements* (Moscow: Standards) 228
[4] Gorobey V N and Izrailov E K 2011 *Measuring technique* **4** 70–3
[5] Katys G P and Katys P G 2001 *Nano-and microsystem technology* **11** 3–7

IOP Conf. Series: Materials Science and Engineering **387** (2018) 012024 doi:10.1088/1757-899X/387/1/012024

Production technology of micromechanical vacuum gauge

V N Gorobey[2], S A Konakov[1], R E Kuvandykov[2], I V Popova[1] and R A Teteruk[2]

[1] JSC GYROOPTICS, 190000, St. Petersburg, Russia
[2] D.I. Mendeleyev Institute for Metrology (VNIIM), 190005, St. Petersburg, Russia

E-mail: konakov.st@gmail.com

Abstract. The article describes technological issues of micromechanical vacuum gauge manufacturing, designed for pressure measurements in the range from 10 to 10000 Pa

1. Introduction

One of the important problems in modern industry is development and production of devises and systems for vacuum measurement. It is important to investigate a new methods and approaches for vacuum measurement. JSC GYROOPTICS and VNIIM carry out a project dedicated to development of a micromechanical vibration vacuum gauge [1]. Schema of the vacuum gauge is presented in figure 1.

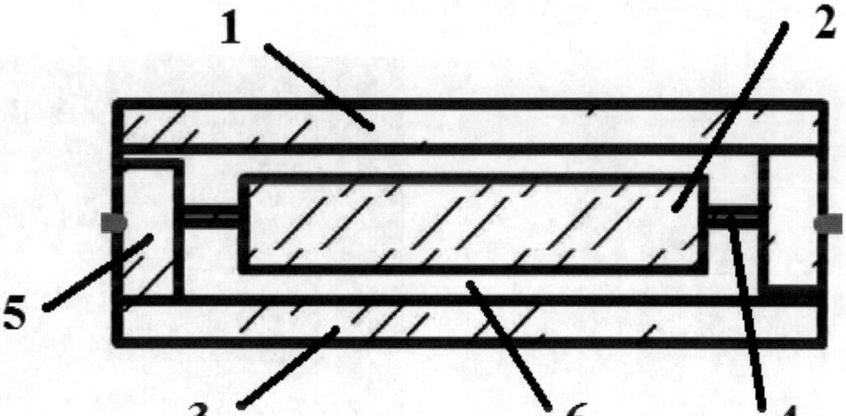

Figure 1. Schema of micromechanical vibration vacuum gauge cross section. 1 is top glass, 2 is silicon plate, 3 is bottom glass, 4 are torsions, 5 is silicon frame, 6 is gap,

Micromechanical vibration vacuum gauge consists of three layers: 75 μm of silicon between 400 μm of glass. Silicon plate with dimensions of 0.9x0.9 mm fixes on torsions which provide vibration motion along oz axis. There is 3 μm gap between silicon and glass. The gap is formed by plasma etching technology. There are metal electrodes on top and bottom glasses. They provide a silicon plate vibration motion by electrostatic force. Also they allow to detect an oscillation amplitude and frequency during vacuum gauge operation.

IOP Conf. Series: Materials Science and Engineering **387** (2018) 012024 doi:10.1088/1757-899X/387/1/012024

Output signal of vacuum gauge is resonance frequency of silicon plate which depends on pressure. This phenomenon based on addition of gas springs in vibration system. Parameters of the spring are function of pressure. In this paper we describe technology issues of micromechanical vacuum gauge manufacturing.

2. Manufacturing technology of micromechanical vacuum gauge

Group technology is used for micromechanical vacuum gauge production. It means that a lot of gauges produced on one wafer at the same time. In manufacturing cycle wafers are treated by standard set of microelectronics and micromechanics processes.

Micromechanical vibration vacuum gauge process flow includes initial preparation of separated wafers: two glass wafers and one silicon wafer. Two thin layers of Al form electrodes on glass wafers by sputtering and lithography processes. 3 μm cavities in silicon wafer are formed by plasma etching process. Top glass wafer is bonded with silicon one by anodic process with alignment. Grinding and polishing processes of silicon layer down to 75 μm thickness are the next steps. Then we carry out two lithography and etching processes on silicon surface. They form the bottom 3 μm cavity and free micromechanical structure of vibration vacuum gauge. Then we carry out the second anodic bonding and form outside metallization and contact pads.

Most of described operations are standard for MEMS production. But vacuum gauge design creates several technological issues, which must be careful development. Also we need investigate a technology tolerance for described process flow.

3. Analyses of technology tolerance

Uniform and flatness of gap between silicon plate and electrode on glass is one of the most important parameter in vacuum gauge design. This 3 μm cavity is formed by plasma etching in silicon. The process has own uniformity of etching rate along wafer. It can lead to parameter distribution for batch of micromechanical vacuum gauges. To evaluate this uniformity we carried out series of tests. 5 wafers with 21 structures on each (total 105 pcs.) have been etched and cavity depth was measured. Obtained data after statistical treatment present in figure 2.

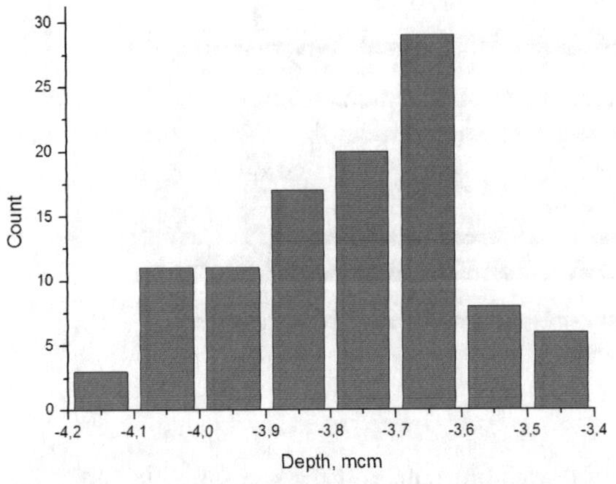

Figure 2. Cavity depth distribution in test structure set.

Figure 3. SEM image of cavity bottom after plasma etching.

It is possible to note that presented distribution has maximum at 3.65 μm. Maximal deviation from this value is plus 0.25 μm and minus 0.55 μm that corresponds 7 and 15 %.

Flatness of cavity bottom was checked on test structures. Etching depth was increased from 3 μm to 15 μm for accuracy investigation. Etched structures were prepared for cross section investigation by SEM. Example of test structure SEM image is presented in figure 3. We can note that cavity bottom has high uniformity and flatness if etching process is tuned accurately. It is possible to conclude that plasma etching process of cavity in silicon plate allow to form a gap between plate and electrodes.

Another property of described production flow of micromechanical vacuum gauge is etching of top gap by plasma etching process in silicon-on-glass wafer. This wafer (after the first anodic bonding process) has different electrical and heat transfer properties then silicon one. It means plasma etching process carry out in different way with increased distribution of etch rate. It is a reason of technological problem of top and bottom equivalent gap production. So we should investigate influence of gap difference on micromechanical vacuum gauge parameters.

To solve this problem we made analytical mathematical model which describe pressure changes in gap during silicon plate oscillation as function of cavity depth. We considered cases with equivalent top and bottom gaps (3 and 4 μm) and case when top gap is 3 μm and bottom one is 4 μm and vise versa. Plots of pressure changes depend on plate displacement for this cases are presented in figure 4.

We can note that gap increasing from 3 μm to 4 μm leads to reducing of pressure difference at set value of silicon plate displacement. The same behavior we obtain when increase only one gap, top or bottom separately.

The first derivative of pressure difference describes sensitivity. This parameter has weak nonlinearity as it presented in figure 5. Ideal vacuum gauge should have high sensitivity independent of oscillation amplitude. Physical principle of vacuum gauge operation could not provide linear characteristic. For low amplitude (0.03) calculated nonlinearity has value around 0.3%. For high amplitude (0.1) nonlinearity increase up to 3 %. This is a reason to restrict maximum displacement of silicon plate. We find interesting effect when top and bottom gap are not equivalent. For separated cases nonlinearity increases to 1.2%. But when we unit this cases, because plate has oscillation motion, nonlinearity are summarized and total value is only 0.5 % (for low amplitude).

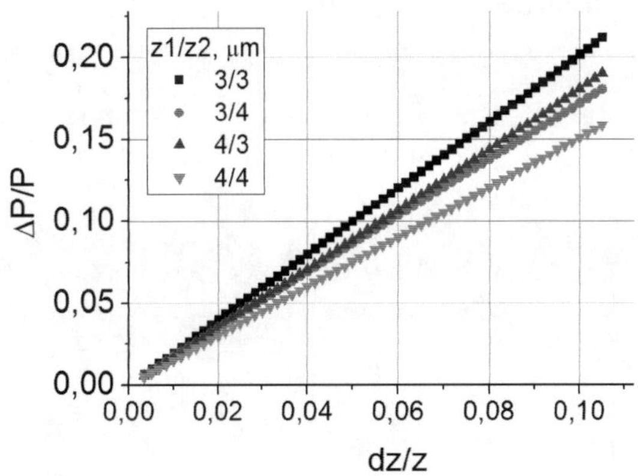

Figure 4. Pressure depends on plate displacement at different gap.

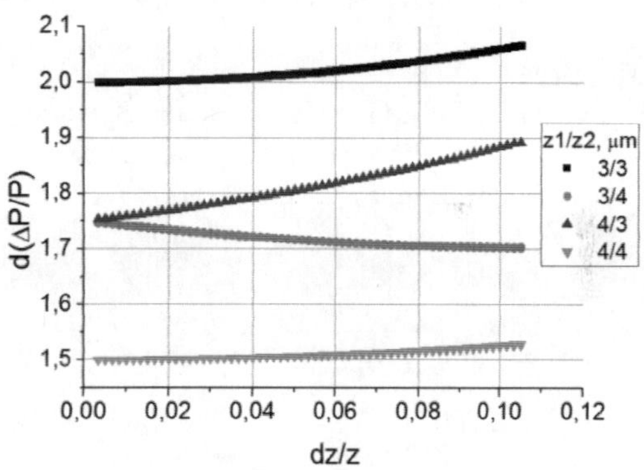

Figure 5. Pressure derivative depends on plate displacement at different gap.

4. Computer simulation

Computer model of micromechanical vacuum gauge was developed for investigation of technological inaccuracy. Influence of natural gravity acceleration (1 g) on silicon plate was investigated. Computer simulation results show that maximum displacement of silicon plate in the case is 0.0684 μm. The value is lower than technology tolerance of gap and lower than plate oscillation amplitude. By the reason we can do not take in account plate displacement under standard gravity acceleration.

In addition we carried out simulation of vacuum gauge operation at absolute vacuum and at pressure in range from 10 to 2000 Pa. In case when pressure does not zero we should include gas springs in vibration system which changes a resonance frequency. These gas springs have different influence on different vibration modes. We can simulate this effect changing one gas spring on many springs, distributed along silicon plate. In figure 1 simulation results of resonance frequency at different pressure are presented.

Table 1. Calculated values of the resonant frequency as a function of pressure.

Pressure, Pa	Stiffness of gas spring, N/m	Frequency, Hz (Simulation)		Frequency, Hz (Analytical model)
		1 gas spring	5 gas springs	
10	5,649	1294,04 (1 mod)	1293,97 (1 mod)	1338
492,127	278	6741,04 (3 mod)	6739,75 (3 mod)	7154
994,876	562	9542,43 (6 mod)	9538,78 (4 mod)	10130
1498	846	11689,8 (6 mod)	11683,1 (4 mod)	12420
2000	1130	13498,9 (6 mod)	13488,6 (4 mod)	14340

These results show that pressure increasing leads to changing of vibration mode of parallel plate motion along oz axis. At zero pressure this motion is the first vibration mode but at 2000 Pa it is the fourth mode. It means that there is some pressure when frequencies of different modes are equivalent. It may lead to problem during vacuum gauge operation and we need take in account this effect.

5. Conclusions

In this work we present results of modeling, simulation and technological tests which dedicate to technological issues of vibration micromechanical vacuum gauge production. Results describe formation problem of equivalent top and bottom gap. Analytical calculations show that technological inaccuracy of cavity depth does not influence on vacuum gauge parameter significantly. Obtained results allow begin a manufacturing process of micromechanical vacuum gauge.

References

[1] Gorobey V N, Garshin A Ya and Kuvandykov R E 2017 *Vacuum Technology and Technologies - 2017. Proceedings of the 24th All-Russian Scientific and Technical Conference with International Participation)*

IOP Conf. Series: Materials Science and Engineering **387** (2018) 012025 doi:10.1088/1757-899X/387/1/012025

Inhomogeneous metal oxide films with a fractal structure as an object for observation of quantum size effects

S E Igoshina[1], I A Averin[1], A A Karmanov[1], I A Pronin[1], N D Yakushova[1] and V A Moshnikov[1,2]

[1]Department of Nano- and Microelectronics, Penza State University, 440026, Penza, Russia
[2]Department of Micro- and Nanoelectronics, Saint-Petersburg Electrotechnical University, 197376, Saint-Petersburg, Russia

E-mail: nano-micro@mail.ru

Abstract. The energy spectrum of charge carriers in inhomogeneous metal oxide films is studied. The principle of specifying the potential on the boundary of nanomaterial/substrate, nanomaterial/atmosphere is demonstrated using the theory of fractals. The energy of the dimension quantization levels in the tin dioxide films with a fractal surface is numerically assessed.

At present, homogeneous semiconductor or semimetal films of nanometer thickness are a classical object for studying various kinds of quantum effects. Theoretical papers devoted to this area of research appeared in the 40s of the XX century [1]. The first convincing experimental proof of the existence of a quasidiscrete spectrum of electrons (on the example of semimetallic Bi films) was presented in the article by Ogrin, Lutsky, and Elinson in 1966 [2]. Since the observation of this effect, a lot of works have been published [3, 4], a stable foundation has been created, suitable for describing and predicting the properties of such films in magnetic and electric fields, under the influence of X-ray, optical and infrared radiation, etc. However, the study of the possibility of existence, observation and practical use of quantum size effects (QSE) in inhomogeneous films of complex composition, for example, based on metal oxides, remains an unresolved problem [5].

Synthesis of such kind of films is possible via various physical and chemical methods, including magnetron sputtering, spray pyrolysis, sol-gel gel technology, etc. Depending on the method of formation, qualitative and quantitative composition, type of substrate used, and a number of other factors the structure of the surface and volume of the film is varied in a wide range [6]. In this connection, there is the question of the conditions for QSE observing, and the influence of the morphology of the film surface on the energy spectrum of the charge carriers. The relevance of this issue is due to the potential possibility for creating elements and devices of nanoelectronics on the basis of such kind of nanomaterials using new physical principles. For example, the possibility of using wide-band gap semiconductor metal oxides with a discrete energy spectrum as highly selective sensing elements of gas sensors is predicted and analyzed in a number of experimental studies [7]. The studies devoted to the use of low-size structures as highly effective photocatalysts are also of considerable interest [8].

The solution of the considered problem for a surface of a random form is difficult, but it is possible to consider some particular cases in the framework of known theories taking into account small corrections. For example, the theory of fractals and multifractals can be considered in the field of quantum mechanics, which allowed forming a new direction of research [9]. The purpose of this paper is to analyze inhomogeneous metal oxide films with a fractal structure as an object for QSE observing. In this case, the quantum-mechanical approach is used to find the energy spectrum and wave functions of electrons, and the fractal approach is used to define the boundary conditions in terms of the potential. The object of the study is tin dioxide films of nanometer thickness synthesized in the framework of the sol-gel technology on the substrates made of oxidized single-crystal silicon (figure 1(a)).

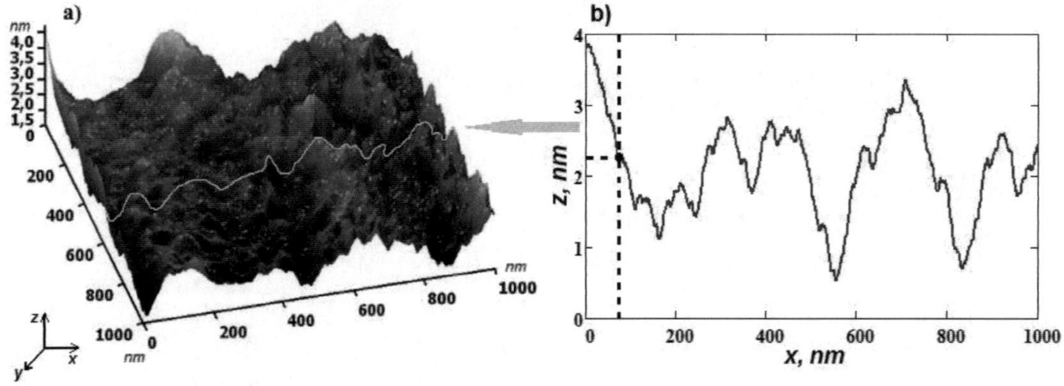

Figure 1. An AFM image of a metal oxide film surface based on tin dioxide (**a**), and a fractal model describing the inhomogeneity of the nanomaterial thickness (**b**).

A self-similar surface structure [10] with a fractal size varying in a wide range from 2.1 to 2.8, depending on the parameters of the synthesis, including the annealing temperature, is reliably proven for such kind of films. The thickness of metal oxide sol-gel nanomaterials is usually tens or hundreds of nanometers, which, taking into account the large effective mass of charge carriers and the high defectiveness of the structure makes a question of QSE observing unreasonable. However, due to limiting the growth of aggregates in the film-forming sol, inhomogeneous in thickness metal oxides can be obtained in the early stages of synthesis, in which quantum-size effects make a significant contribution to electrophysical, photocatalytic, gas sensitive, and other properties. A description of the inhomogeneity of such materials is possible in the framework of the well-known approach outlined in the classical paper [11]. In this case, to describe the given surface the principle of superimposing a small number of fractal transfer surfaces oriented randomly is used. To obtain the height of the surface as a function of the coordinate, the complex Fourier transform of the following form was used:

$$\xi(x) = \sum_j C_{f_j} \exp\left(2\pi i f_j x\right), \quad j = 1,2,... \tag{1}$$

where C_{f_j} are the amplitudes corresponding to the sinusoidal transfer surface, f_j are the spatial frequencies; the smallest ones determine displacement of the entire surface along the height, and the largest one determines scale resolution.

Taking into account the use of the superposition of oscillating functions for constructing the transfer surface, the frequency spectrum is determined by the relation:

$$f_j = f_{j-1}^{\alpha} = f_0^{\alpha^j}, \tag{2}$$

where α is a parameter characterizing the general form of the fractal surface, which value is usually in the range from 0.7 to 1.4.

Within the framework of the present paper, the assumption of the dominant influence of amplitudes at low frequencies on the structure of the fractal surface was used, that is why their scale-invariant form of the Fourier expansion was chosen:

$$C_f \sim f^{-\beta} \tag{3}$$

where β is the proportionality coefficient connected with the scale of the reproduced fractal surface.

The real transfer surfaces were determined taking into account the function (1) as follows:

$$Z(x, y \,|\, \varphi) = Z' \cos\varphi + Z'' \sin\varphi, \tag{4}$$

where $Z' = \mathrm{Re}\,\xi(x)$, $Z'' = \mathrm{Im}\,\xi(x)$ are real and imaginary parts of the function (1), and the angle φ is chosen randomly from the range of 0-2π.

Figure 1(b) demonstrates a particular case of the application of the described approach for inhomogeneity assessment of the nanomaterial thickness. The fractal model corresponds to the curve $\eta(x) = Z(x, y = w_y/2 \,|\, \varphi = \pi/5)$, where w_x, w_y are the geometric dimensions of the metal oxide film along the x and y directions, respectively. It is seen that using the parameter α=1.35 and the coefficient β=1.1 the obtained curve is in good agreement with the relief of the real surface.

The quantum mechanical model describing the analyzed metal oxide films was constructed based on the provisions of motion limitation of charge carriers along one spatial direction (z axis), and free motion along the (x, y) plane. This in turn allows reducing the problem under consideration to the known quantum well model with walls of finite height, one of which (on the silicon dioxide/tin dioxide boundary) is stationary, and the second (on the boundary of tin dioxide/atmosphere-vacuum) is mobile, and its position is determined by a point on the fractal curve (figure 1(b)). Then the particular case corresponding to the inhomogeneity of the thickness of the nanomaterial along the x axis leads to the stationary Schrödinger equation of the form:

$$-\frac{\hbar^2}{2m_k^*}\frac{\partial^2\psi}{\partial z^2} + \left[U(z) - E\right]\psi = 0 \tag{5}$$

where $\hbar$ is the Planck constant, m_k^* (k=1,2,3) is the effective mass of charge carriers in the substrate with z≤0 (silicon dioxide $m_1^* = 0.4m_0$, where m_0 is the electron rest mass), in the nanomaterial with 0<z<η(x) (tin dioxide $m_2^* = 0.275m_0$), and in the atmosphere (vacuum $m_3^* = m_0$) with z≥η(x), and the height of the potential barrier U(z) defined as:

$$U(z) = \begin{cases} U_1 & z \le 0 \\ 0 & 0 < z < \eta(x) \\ U_2 & z \ge \eta(x) \end{cases} \tag{6}$$

where the potential U_1=4.8 eV is determined by the difference of electron affinity in silicon dioxide and in tin dioxide, and the potential U_1=5.7 eV is inherently the work function of the tin dioxide.

In this case, the solution of the Schrödinger equation for the corresponding domains 1, 2, and 3 can be presented in the form:

$$\begin{aligned} \psi_1(z) &= A_1 \exp(\beta_1 z) \\ \psi_2(z) &= A_2 \exp(iK_z z) + B_2 \exp(-iK_z z) \\ \psi_3(z) &= B_3 \exp(-\beta_2 z) \end{aligned} \tag{7}$$

where $K = \sqrt{2m_2 E}/\hbar$; $\beta_1 = \sqrt{2m_1(U_1 - E)}/\hbar$; $\beta_2 = \sqrt{2m_3(U_2 - E)}/\hbar$; A_1, A_2, B_2, B_3 are the amplitudes of the wave function.

Then the energy of size quantization levels and the normalized values of the amplitudes are determined by the condition of continuity of the wave functions and their first derivatives at the quantum well boundary:

$$m_1^{\alpha}\psi_1\big|_{z=0} = m_2^{\alpha}\psi_2\big|_{z=0}$$
$$m_2^{\alpha}\psi_2\big|_{z=\eta(x)} = m_3^{\alpha}\psi_3\big|_{z=\eta(x)}$$
$$m_2^{\alpha+1}\frac{\partial\psi_1}{\partial z}\Big|_{z=0} = m_1^{\alpha+1}\frac{\partial\psi_{21}}{\partial z}\Big|_{z=0} \quad (8)$$
$$m_3^{\alpha+1}\frac{\partial\psi_2}{\partial z}\Big|_{z=\eta(x)} = m_2^{\alpha+1}\frac{\partial\psi_3}{\partial z}\Big|_{z=\eta(x)}$$

In this case, the total energy of the charge carriers in the analyzed film has a discrete-continuous spectrum, representing the sum of the discrete component connected with the motion in the quantization direction, and the continuous one describing the motions in the plane (x, y):

$$E = E_n + \frac{\hbar^2}{2m^*}(K_x^2 + K_y^2) \quad (9)$$

where E_n is the energy of the size quantization levels determined by the fractal surface of the nanomaterial, n = 1,2,3, ... is the principal quantum number, K_x and K_y are x and y components of the wave vector.

Figure 2 shows the results of calculating the size quantization levels in a metal oxide film with fractal structure on the basis of tin dioxide, as well as the spatial distribution of the nanomaterial thickness L taking into account the inhomogeneity. An analysis of the obtained dependence shows that the investigated films are characterized by a complex form of the spatial dependence of energy levels of size quantization, which are fractal curves in the general case. At the same time, minimum of the dependence (4) corresponds to the maxima of the dependences $E_n(x)$, which corresponds to proportionality $E_n \sim \eta(x)^{-D}$, where D is the fractal dimensional characterizing the inhomogeneity of the thickness of the nanomaterial, and is explicitly related to the fractal dimensional of its surface. Mathematically, this relationship is difficult to express, since using even a sinusoidal curve that is not fractal as a transfer surface will lead to the formation of some fractal object.

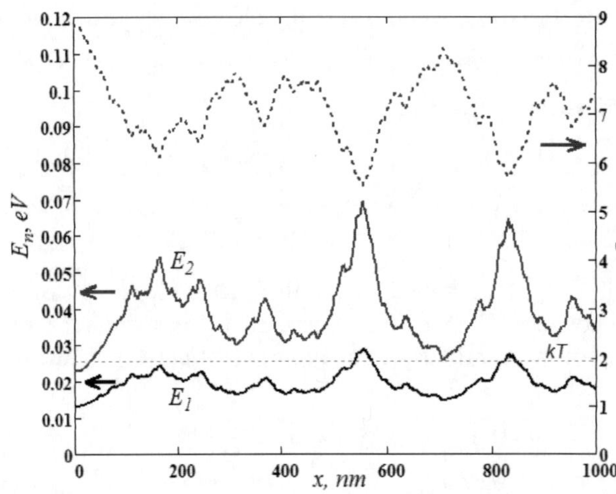

Figure 2. Spatial dependence of the contribution of the size quantization levels in energy spectrum of electrons in a metal oxide film with a fractal structure.

A numerical assessment E_n shows that only small in size sections of the tin dioxide film have a ground-state energy E_1 exceeding thermal energy kT at room temperature. Accordingly, it is possible to predict the QSE manifestation only on local areas of the surface, which size will increase with T

decreasing. However, even at temperature close to room temperature, there will be areas of metal-oxide film having electrophysical, catalytic, gas-sensitive, and other properties, unacceptable to volume material. This phenomenon, in turn, can be used to create elements and devices of nanoelectronics based on quantum-size effects. It should be noted that the proposed model has some drawbacks: it explicitly ignores defects of metal oxide structure, the scattering of charge carriers on the fractal surface and a number of other effects. However, it allows evaluating the contribution of the energy levels of size quantization to the overall energy spectrum of the inhomogeneous film.

Acknowledgment

The work has been financially supported by the Ministry of Education and Science of the Russian Federation within the framework of the project part of the state assignment for Penza State University No. 16.897.2017/4.6, as well as the Scholarship of the President of the Russian Federation No. SP-3800.2018.1.

References

[1] Frohlich H 1937 *Physica* **6** 406–12
[2] Ogrin Yu V, Lutskii V N and Elinson M I 1966 *JETP Letters* **3** 71–3
[3] Samarth N 2017 *Nature Materials* **16** 1068–76
[4] Batista Jr F F, Chavcs A, da Costa D R and Farias G A 2018 *Physica E: Low-dimensional Systems and Nanostructures* **99** 304–9
[5] Mosquera A A, Horwat D, Rashkovskiy A, Kovalev A, Miska P, Wainstein D, Albella J M and Endrino J L 2013 *Scientific Reports* **3** 1714
[6] Pronin I A, Averin I A, Yakushova N D, Karmanov A A, Moshnikov V A, Cho B K and Korotcenkov G 2017 *Arabian Journal for Science and Engineering* **10** 4299–305
[7] Wang L, Li J, Wang J, Yu K, Tang X, Zhang Y, Wang S and Wei C 2016 *Scientific Reports* **6** 35079
[8] Zhu L, Hong M and Ho G W 2015 *Scientific Reports* **5** 11609
[9] Wójcik D, Białynicki-Birula I and Zyczkowsk K 2000 *Physical Review Letters* **85** 5022–5
[10] Moshnikov V A, Maksimov A I, Aleksandrova O A, Pronin I A, Karmanov A A, Terukov E I, Yakushova N D, Averin I A, Bobkov A A, Permyakov N V 2016 *Technical Physics Letters* **9** 967–9
[11] Mandelbrot B B 1975 *Proc. Nat. Acad. Sci. USA* **10** 3825–8

Ab initio modeling of carbon films deposited by laser plasma

V V Ilyasov, O M Holodova, I V Ershov, N V Prutsakova, T P Zhdanova, I P Gritsay and A V Ashkanov

Department of Physics, Don state technical university, 344000, Gagarin Square 1, Rostov-on-Don, Russia

E-mail: viily@mail.ru

Abstract. The results of modeling of atomic and electronic structures of amorphous ultrathin carbon films are presented from the first principles. The calculations of the system total energy were performed within the density functional theory using the Grimm DFT-D2 scheme. The atomic models of amorphous carbon films of a-C and a-C-H types are proposed. The calculated electronic energy spectra of the amorphous carbon films a-C(:H) models are well consistent with UPS and XANES of graphite and the hydrogenated carbon film. On the basis of DFT calculation the energy gaps for the considered models of amorphous carbon a-C(:H) are determined and the mechanisms of its formation are discussed.

1. Introduction

The amorphous carbon unique properties (mechanical durability, transparency in a wide range of spectrum, chemical inertness) open up the prospects of its use in the elements of optical and nanoelectronics devices [1, 2]. The ultrathin films of amorphous carbon a-C (:H) have metastable structure which is formed by mixing of fractions with the sp^3 - and sp^2 - carbon atomic binding. Within J. Robertson's [3] model the lattice sites with an sp^3-binding are put into the clusters with a graphite-like sp^2 - binding. The hydrogen content up to 60% is noted in the films a-C (:H). The ratio sp^3/sp^2 in the hydrogenated carbon films depends on the technological conditions of their synthesis and the choice of a coating process method. In addition, this ratio correlates with the value of the optical gap in the amorphous carbon a-C (:H) energy spectrum.

In this paper, the simulation of atomic and electronic structures of amorphous ultrathin carbon films a-C(: H) synthesized by pulsed laser deposition (PLD) is performed from the first principles. The simulation was performed based on the density functional theory (DFT).

2. Model and method

In DFT calculation the simplest atomic models of ultrathin films of the a-C and a-C-H types: graphene, (graphene, G) and single-side-hydrogenated graphene, SSHG are used. It is known, that in hexagonal graphite structure the interaction between the carbon layers in the direction [0001] is slight (insignificant). The latter allows us to construct a model atomic structure, using, for example, four graphene layers in the Bernal's (ABAB) laying. The quasi-two-dimensional models of the a-C and a-C:H types films are constructed according to the slab scheme. Each slab has been separated from the other with the vacuum area in the direction (0001) of 1.5 nm. The supercell fragments presented in figure 1 are composed from 92 and 105 atoms respectively. DFT calculations of total energy of these systems have shown that these configurations are the most stable. The supercells include graphene unit cells (4x4) in the plane (0001).

The total energy and the system electronic spectrum are calculated by the pseudopotential method within the density functional theory (DFT) similar to works [5, 6]. In particular, nonlocal exchange and correlation functional in parametrization of Perdew-Burke-Ernzerhof (PBE) has been used [7]. The plane wave cutoff energy of 410 eV was used for the self-consistent calculation. The convergence of a cell total energy not less than 10^{-5} Ry/cell is achieved. The functional with the dispersion correction (PBE-D2) [8] is used for the van der Waals interaction accounting.

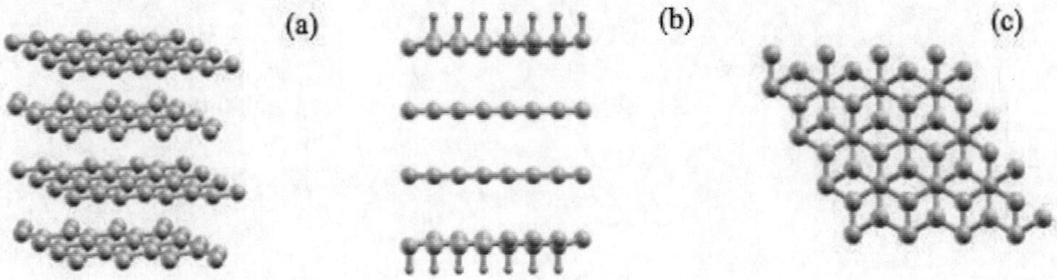

Figure 1. The amorphous carbon film fragments: side view (a) a-C type and (b) the a-C:H type (after relaxation). Top view (c). Carbon atoms (yellow) and hydrogen (turquoise).

3. Results and discussion

Based on the DFT the a-C model, formed by 4 graphene monolayers, atomic structure has been studied. The vertical distance between the monolayers was 0.335 nm. In the graphene layers, the C-C bond length was 0.142 nm. The atomic structure relaxation of a-C model film has been performed. TDOS and partial (PDOS) electronic state densities are calculated for the noted atomic model of amorphous carbon films a-C. In figure 2 TDOS and PDOS of ultrathin a-C films are correlated with a ultraviolet photoelectron spectrum UPS and the near-edge XANES area of pyrolitic graphite. The combination of TDOS and UPS and XANES spectra was performed by the main maxima with energy-6.9 eV and 1.9 eV. It should be noted that the calculated electronic energy spectrum (TDOS) of the offered above model of the amorphous carbon film a-C is well consistent with the graphite spectra UPS [9] and graphite XANES [10] (see figure 2).

In figure 2 projected electronic density on 2s- and 2p-states of carbon are given. PDOS analysis allows to note, that the vicinity of the Fermi level in the energy range from –5 eV to +5 the eV is formed by the filled and free states of carbon 2p-electrons. The valence electron C2s-states are in the energy range (–20÷–6) eV. Thus, the bonding and antibonding states of the carbon C2p-electron states form the energy gap with the value $E_g = 0.4$ eV at the Fermi level. This value is 1.5 times less than that of the data of the theoretical calculation for graphite, which was made by the authors [11]. The obtained energy gap estimation characterizes the 4 graphene sheets model in which there is an interaction between graphene layers. The weak symmetry violation of sublattices in the carbon monolayers may be responsible for this interaction similar to the graphene bilayer (BLG) [12, 13]. The BLG energy spectrum has the gap $E_g = 0.1$ eV the value of which is measured by the infrared spectroscopy method [13].

The atomic structure of the hydrogenated carbon model a-C:H is formed by 4 graphene nanosheets and 2 hydrogen monolayers (see figure 1(b)). The C-H bond length initially was 0.111 nm. As a result of the atomic structure relaxation, the length of C-H-bond was 0.116 nm. The equilibrium lattice parameters, the atomic positions of the atoms of 4 graphene nanosheets and 2 hydrogen monolayers, the vertical distance between graphene atomic layers and the C-C - bond length in the plane (0001) are determined. The vertical distance between the interior monolayers was 0.320 nm, and as a result of relaxation, the external graphene monolayers moved at a distance 0.313 nm from the interior sheets. The C-C length - bonds in the graphene internal layers was 0.142 nm, and it was 0.146 nm in the external (outer) sheets. The later indicates the C-C - bond deformation. For the above considered atomic model of amorphous carbon a-C:H type films, the total (TDOS) and partial (PDOS) density of electronic states were calculated . PDOS are calculated in the spin-polarized variant.

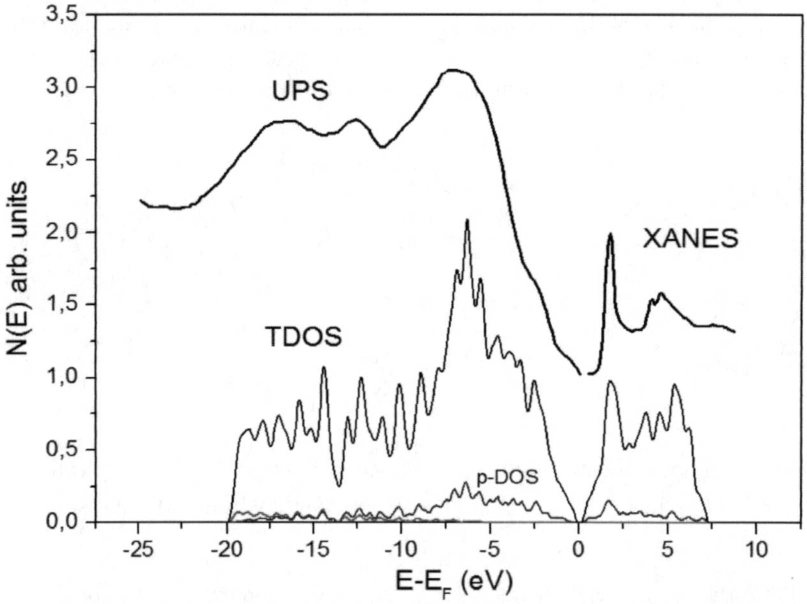

Figure 2. DFT calculation of TDOS and projected s- and p-DOS of the amorphous carbon a-C film in comparison with UPS [9] and XANES [10] spectra of the a-C film.

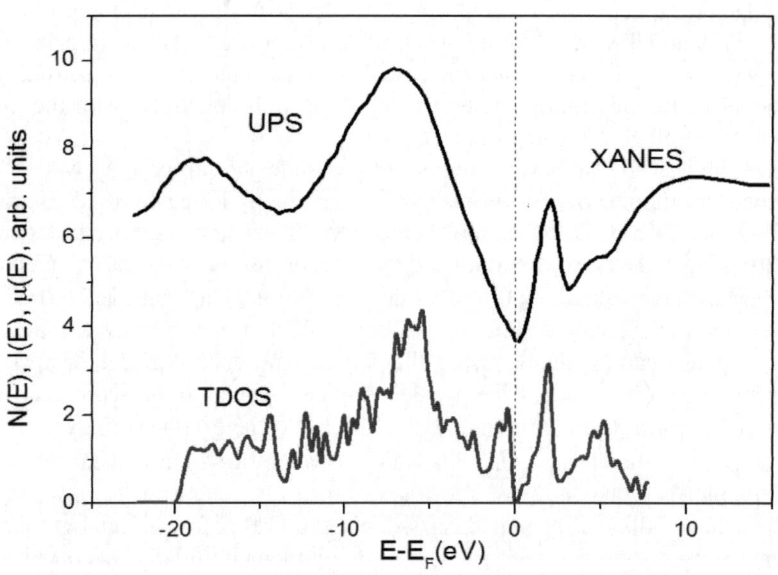

Figure 3. Total density of states TDOS (blue) in comparison with the UPS [9] and XANES [10] spectra (top panel) and projected on atomic orbitals DOS (per atom) for a-C:H - C2p (blue), C2s (red), H1s (purple) - bottom panel.

In figure 3 total DOS of the a-C:H model which is compared with the ultraviolet photoelectron spectrum UPS and the near-edge XANES spectrum of the a-C:H carbon film is shown. The overlaping of TDOS and UPS and XANES spectrum was performed by the main maxima with energy -6.8 eV and 1.9 eV. It should be noted, that the calculated electronic energy spectrum (TDOS) of the considered above amorphous hydrogenated carbon a-C:H model, is qualitatively consistent with the UPS [9] and XANES [10] spectra of the hydrogenated carbon a-C:H (see figure 3). The difference of the theoretical and experimental spectra is the existence of two peaks with the energy -0.9 eV and 1.9 eV, which are located on the Fermi level both sides, in the calculated electron energy spectrum. The existence of these peaks is caused by the rearrangement of the electronic structures at the hydrogenation of the upper and lower graphene monolayers. The peak 0.9 eV is determined by the electronic subsystem with a spin up, and the electron subsystem with a spin down is responsible for a peak with the energy 1.9 eV. In the a-C:H system, the Fermi level energy gap E_g = 0.2 eV is formed by the bonding and antibonding states of carbon C2p-electrons.

4. Conclusions

Based on the density functional theory (DFT), the modeling of the amorphous carbon a-C and the a-C:(H) films deposited by laser plasma is performed. The calculated electron energy spectra of model amorphous carbon films of a-C (: H) are well consistent with UPS and XANES of graphite and the hydrogenated carbon film. The energy gaps for the model amorphous a-C(:H) carbon films are defined.

References

[1] Qindeel R 2010 *Journal of Non-Oxide Glasses* **1** 191–7
[2] Maitre N 2005 *J. NonCrystalline Solids* **351** 877–8
[3] Robertson J and O'Reilly E P 1987 *Phys. Rev. B* **35** 2946–12
[4] Giannozzi P, Baroni S and Bonini N 2009 J. *Phys.: Condens. Matter.* **21** 395502–39
[5] Ilyasov V V, Popova I G, Ershov I V and Nguyen D Ch 2017 *Diamond & Related Materials* **74** 31–9
[6] Ilyasov V V, Popova I G and Ershov I V 2017 *Applied Surface Science* **419** 924–9
[7] Burke K, Perdew J P and Ernzerhof M 1997 *Int. J. Quantum Chem.* **61** 287–7
[8] Grimme S 2004 *J. Comput. Chem.* **25** 1463
[9] Wesner D, Krummacher S, Carr R, Sham T K, Strongin M, Eberhardt W, Weng S L, Williams G, Howells M, Kampas F, Heald S and Smith F W 1983 *Phys. Rev. B* **28** 2152–5
[10] Fink J, Miller-Heinzerling T, Pflüger J, Bubenzer A and Koidl P 1983 *Solid State Communications* **47** 687–5
[11] Alisultanov Z 2013 *Semiconductors* **47** 954–7
[12] Kuzmenko A B, van Heumen E, van der Marek D, Lerch P, Blake P, Novoselov K S and Geim A.K 2009 *Phys. Rev. B* **79** 115441–5

IOP Conf. Series: Materials Science and Engineering **387** (2018) 012027 doi:10.1088/1757-899X/387/1/012027

Vacuum pumping system for the multi-aperture low-energy continuously-operated electron accelerator with a high-density beam current

A G Ivanov, D A Karpov, S L Kosogorov and N A Uspensky

JSC "NIIEFA", Doroga na Metallostroy, 3, 196641 Saint Petersburg, Russia

Abstract. The vacuum pumping system for an electron accelerator has been developed on the basis of arc sources of getter film and ion getter pumps. The contradictory problem has been solved, that is, it was necessary to provide, on the one hand, the maximum efficiency of the pumping speed and, on the other, a minimal return flow of the metallic plasma shunting the accelerating gap. The program and procedure have been elaborated for the numerical simulation of getter pumping taking into account the mass-transfer non-uniformity. The experimentally obtained vacuum characteristics of the developed pumping system for the accelerator are presented.

1. Introduction

The multi-aperture low-energy electron accelerator (EA) under development at the JSC "NIIEFA' uses filament-type tungsten-thorium thermal cathodes as an electron emitter, with the beam extracted through the output foil window (OFW). The accelerator operates continuously with the following parameters of the electron beam: beam cross-section – 2000 cm^2, electron energy – 200 keV, beam current density behind the foil window – 100 μA/cm^2. To provide normal operation of the thermal cathodes and high-voltage gaps in the accelerator chamber the operating pressure should be maintained at a level of $1.3 \cdot 10^{-4}...6.7 \cdot 10^{-4}$ Pa (at the maximum allowable operating pressure of the thermal cathodes amounting to $2.6 \cdot 10^{-3}$ Pa).

The purpose of the study was to develop the vacuum pumping system (VPS) for an electron accelerator, with the VPS operating at widely varying temperatures and at extreme vibrations. Series-produced vacuum pumps (turbomolecular, cryosorption, diffusion oil-vapor pumps) are inoperable under such conditions. In these cases, high vacuum can be efficiently produced by arc sources (AS) of a getter film [1]. This pumping system is based on continuous or periodic deposition of a getter film (titanium) on the cooled surfaces and sorption of active gases by this film. Inert gases and hydrocarbons are pumped by an additional pump, as a rule, by an ion getter pump. These pumps have no moving parts and can operate under the above-mentioned environmental conditions. But, for normal operation the high-voltage gaps are to be protected against metal plasma flows generated by the arc sources (with an ionization degree of about 80 %), while providing the required pumping of gas release flows, including electron-stimulated gas release during extraction of a powerful electron beam through OFW. This paper presents the main results of calculations, design and tests of this vacuum pumping system.

2. Description of the vacuum pumping system

The vacuum chamber (CV) of the electron accelerator is made of stainless steel and has a volume of 0.35 m^3. The electron beam leaves through OFW located on the vacuum chamber wall. The thermal desorption flux in the electron accelerator chamber, including the electron-stimulated gas release, amounted, as calculated, to $1.2 \cdot 10^{-3}$ m^3Pa/s, but with the system of thermal cathode being switched-on and without beam extraction it amounted to $8.6 \cdot 10^{-4}$ m^3Pa/s. In the operation pressure range of the electron accelerator $(1.3–6.7)\cdot10^{-4}$ Pa the vacuum pumping system should provide the effective pumping speed of the chamber from 1.8 to 8.8 m^3/s. Such high pumping speed is ensured by the arc sources (AS1, AS2) of a getter film developed at the NIIEFA (figure 1), which is switched on immediately prior to generation of the electron beam. The vacuum arc sources do not pump out inert gases. In this connection, ion getter pumps (NM1, NM2) are used to maintain the background pressure and pumping of inert gases. Two ion getter NMD-0,4 pumps (with a pumping speed of 0.4 m^3/s each) ensure effective pumping speed of the electron accelerator chamber amounting to (0.4…0.5) m^3/s, which, as calculated, ensures a pressure of $(1.7…2.1)\cdot10^{-3}$Pa with thermal cathode glow in the electron accelerator chamber. The forevacuum system consists of the pump (NL1) NVR-16D, valve (VP1) and nitrogen trap (BS1) and is connected to the vacuum pumping system through the valve (VM1). Two wide-range vacuum gauges (PMT1, PMT2) serve for measurement of pressure.

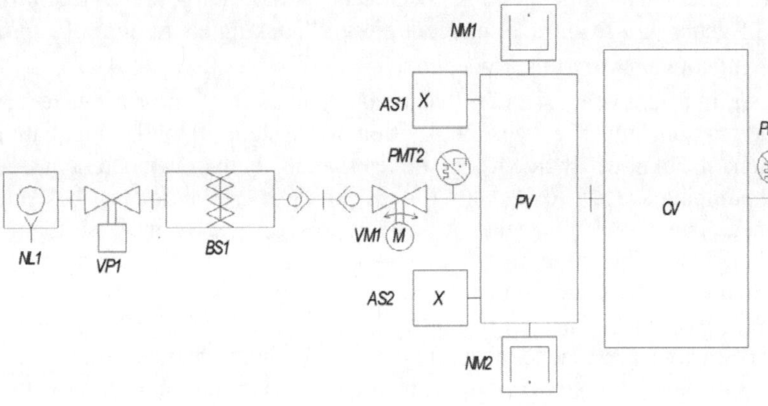

Figure 1. Basic diagram of the vacuum pumping system.

3. Analysis of design versions for the vacuum pumping system

To attain the optimal pumping parameters, to ensure reliability and serviceability several design versions of the vacuum pumping system of the electron accelerator were analyzed. The analysis was carried out both for continuous getter evaporation and for periodic mode. But, only the continuous operation mode of the arc sources turned out to be acceptable in all analyzed versions at the specified level of vacuum required for accelerator operation and at the calculated gas release fluxes.

The pumping speed S_{0ni} of the getter film as to i_{th} (nitrogen, oxygen, hydrogen) is defined by the sorbing surface area F_{cn}, m^2, and its specific pumping speed s_{0ni}, $m^3/(c\cdot m^2)$ [2, 3]:

$$S_{0ni} = s_{0ni} \cdot F_{cn} = 36.38 \cdot \beta_{0i} \cdot (T_i/M_i)^{0,5} \cdot F_{cn}, \quad (1)$$

where: $\beta_{0i} = 0.35$ – the initial coefficient of gas sticking to titanium (value for nitrogen) [2]; $T_i = 293K$ – the gas temperature; $M_i = 28$ g/mole – the molar mass of gas (value for nitrogen).

At the film saturation degree less than the critical one ($Z_{cr} = 5 \cdot 10^{20}$ molecule/g, value for nitrogen on a titanium film) the sticking coefficient β_i is practically constant and amounts to the initial β_{0i}. With the saturation degree exceeding the critical value, the sticking coefficient is exponentially decreased. The basic values of Z_{cri} and β_{0i} for calculation of getter pumping are reported in [2–5], but they vary in a wide range. In this study, use was made of the refined numerical values of Z_{cri} *and* β_{0i} obtained experimentally [6].

The mass transfer non-uniformity of the getter and the evacuated gas can cause local changes in the sticking coefficient. To account for this, a program based on angular coefficient method was created in the Octave GNU UNIX software. In this case, the titanium plasma was conventionally replaced by neutral particles with a sticking coefficient of titanium to the walls of the vacuum chamber α=1. All surfaces with a getter film are divided on the surface, within which the distribution of the getter will be considered homogeneous. The program calculates the incident and pumped out fluxes of the getter and gas, the sticking coefficients of the selected surfaces according to a given inlet gas flow. First, according to the cosine law of reflection, the mass transfer coefficients (probability of a particle from one surface site to another) between the selected surfaces are calculated. The fluxes of the getter are calculated on the selected surfaces at a known rate of evaporation of the getter (2…5 mg/s).

The gas incident flux on the i-th surface and its sticking coefficient are determined from:

$$G_i = \sum_{j=1}^{n} W_{ji} = \sum_{j=1}^{n} (w_{ji} + G_j \cdot \varphi_{ji} \cdot (1 - b_j)) \tag{2}$$

$$b_i = f_{app}(Z_i) = f_{app}\left(Q_i / \overline{m_{geti}}\right) = f_{app}(b_i \cdot G_i) \tag{3}$$

where W_{ji} –outgoing flux from j-th surface to i-th surface, m^{3}Pa/s; $w_{ji} = w_j \cdot \varphi_{ji}$ – outgoing flux from *j*-th surface to *i*-th surface due to the intrinsic gas evolution of the surface ($w_j \neq 0$ only for gas inlet) *j*, m^{3}Pa/s; Q_i –the pumped out flux by *i*-th surface, m^{3}Pa/s; G_j – the total incident flux to *j*-th surface, m^{3}Pa/s; φ_{ji} – mass transfer coefficient from *j*-th surface to *i*-th surface; $f_{app}(Z_i)$ – function approximating the dependence of the sticking coefficient on the saturation degree of the getter; Z_i – saturation degree of getter on *i*-th surface, molecul/g; $b_{i,j}$ – sticking coefficient of *i-th* or *j*-th surface; $\overline{m_{geti}}$ – deposition rate of getter to *i*-th surface, g/s.

A system of ***2 · n***, in the general case non-linear, equations is created for ***n*** selected surfaces by the equation (2, 3).The program solver realizes the solution of a system of nonlinear equations of the form F (x) = 0 in which: ***n*** unknown incident gas fluxes G_i and ***n*** unknown sticking coefficients $b_i$of surfaces.The total pumping speed is calculated for the averaged gas parameters based on the sticking coefficients obtained. This procedure makes it possible to significantly increase the accuracy of the calculation, especially with increased fluxes of evacuated gas or an unsuccessful design of inner surfaces resulting in a local oversaturation of the getter film.

The qualitative assessment of the return plasma flow into the electron accelerator chamber was performed with the sticking coefficient of titanium to the vacuum chamber walls equal to α = 0.99. This assessment has made it possible to compare the return plasma flows into the electron chamber for different design versions of the vacuum pumping system.

The analyzed design versions of the vacuum pumping system differ by arrangement of the ion getter pumps, arc sources, as well as by the design of the water-cooled inner shields I, II (figure 2).

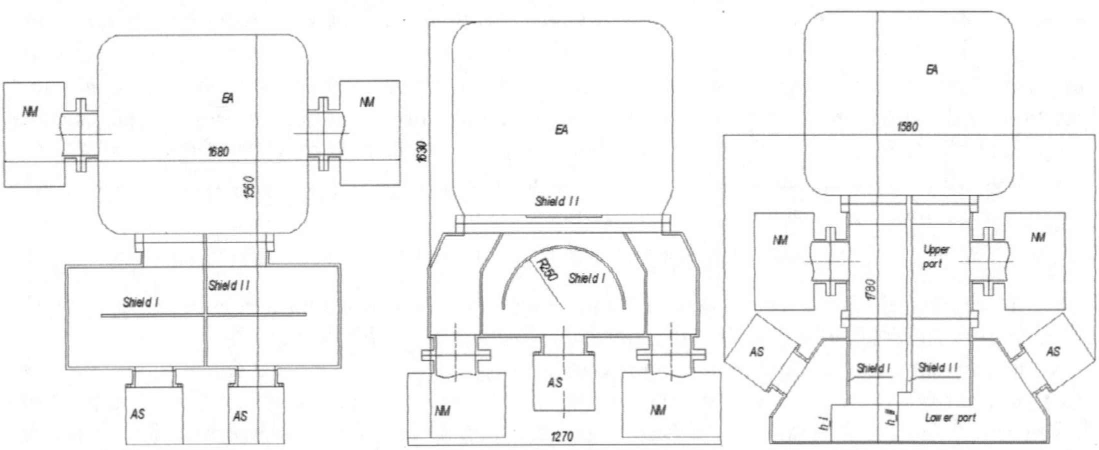

(**a**) (**b**) one VAS not shown (**c**)

Figure 2. Design versions of the vacuum arc source.

The overall dimensions and serviceability of the accelerator determine the arrangement of the ion getter pumps and arc sources, as well as influence the conductivity of the vacuum pumping system pipeline. The water-cooled inner shields are intended to provide a sufficient area of the getter film and construction of a labyrinth, which protects the electron accelerator vacuum chamber with high-voltage electrodes against ingress of the metal plasma of getter material into the chamber, which might cause shunting of the high-voltage gap and a disruption in the normal operation regime of the electron accelerator.

With the NMD-0,4 pumps (NM) mounted on the electron accelerator (figure 2(a)), their stray magnetic fields might affect the accelerator operation. With the NMD-0,4 pumps or arc sources (AS) located vertically at the bottom (figure 2(a), (b)), generated dust and getter particles might cause shorting of their electrodes and make inoperative the vacuum pumping system.

The scheme in figure 2(c) provides ease of operation and the highest reliability of the vacuum pumping system. This design version contains two demountable water-cooled shields (shield I and shield II). With two shields being mounted, conductivity from the electron accelerator chamber to the volume with the getter film is minimal, but protection against the return plasma flow is maximal. With shield I removed and only shield II remained, the conductivity is markedly increased, but protection against the return plasma flow is reduced. As the numerical experiment revealed, while evaporation rate of the getter equal 4 mg/s, the particle flux into the accelerator chamber through the inlet flange amounted to $1.1 \cdot 10^{15}$ 1/s without the shield I (figure 2(c)) and to $1.0 \cdot 10^{14}$ 1/s with the shield I (h_I = 14 cm) (figure 2(c)). In both cases, conductivity of the pipeline to the chamber with sorbing surfaces ensures the required effective pumping speed. The final optimal design can be chosen only experimentally. The calculated characteristics of 4 versions of the vacuum pumping system are summarized in table 1.

Table 1. Main design characteristics of the vacuum pumping system

Parameter	Version *a*	Version *b*	Version c (h_I=14 cm)	Version *c* without shield I
Conductivity of 2 ASs, m^3/s	3.70	4.92	3.36	5.77
Getter area of 2 ASs, m^2	1.64	0.66	1.25	1.50
Pumping speed of 2 ASs for nitrogen, m^3/s	67.6	27.2	51.4	62
Effective pumping speed of 2 ASs, m^3/s	3.5	4.16	3.16	5.28
Conductivity of 2 NM pumps, m^3/s	1.88	1.48	1.62	1.62
Effective pumping speed of 2 NMs, m^3/s	0.56	0.52	0.54	0.54
Effective pumping speed of EA, m^3/s	4.06	4.68	3.70	5.82
Operation pressure of EA, 10^{-4} Pa	3.5	3.3	4.0	2.4

In the final design version chosen by the calculation results (figure 1(c)) the vacuum pumping system is mounted on the lower wall of the electron accelerator chamber and comprises two ports. Two NMD-0,4 ion getter pumps are symmetrically mounted on the upper port, and two arc sources are mounted on the lower "port-chamber". The inner water-cooled surfaces of the lower port with the installed shield are made as distributed pumping surfaces, which are dusted with getter. The demountable shields I are of different length in order to vary the size h_I from 14 to 40 cm, the demountable shield II is h_II = 15 cm in size. Figure 3 shows the calculated diagram of pressure in the electron accelerator chamber of the chosen design version.

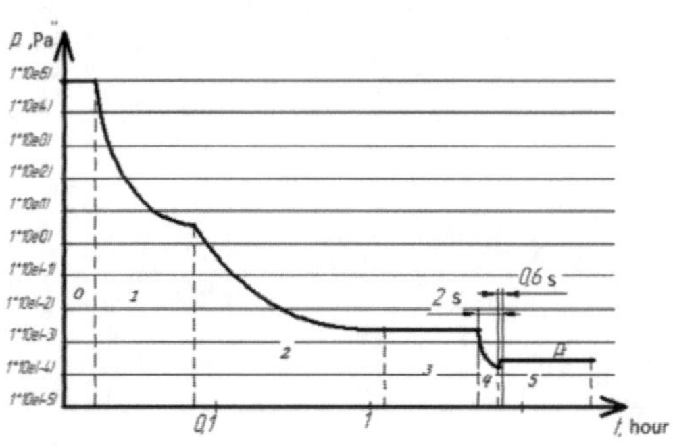

Figure 3. Time diagram of pressure in the EA chamber (without training regime): 0 – atmospheric pressure; 1 – forevacuum pumping; 2 – pumping of NM with a short-term start of AS; 3 – NM ensures background pressure; 4 – joint operation of AS and NM; 5 – extraction of electron beam through OFW.

4. Testing of the vacuum pumping system

The vacuum pumping system was tested in the process of the adjustment and start-up of the electron accelerator. The background pressure (without generation of the electron beam) ensured by the ion getter pumps was maintained at the level of about $5 \cdot 10^{-4}$ Pa. Pressure during generation of the electron beam provided by the joint operation of the vacuum arc sources and ion getter pumps amounted to about $2 \cdot 10^{-3}$ Pa, which is higher than the calculated pressure. The reason might be that the training time of the electron accelerator with electron beam generation was not sufficient. Still, this pressure level provided the normal operation regimes of the high-voltage gap and thermal cathodes. With the inner shield I removed, the shunting current in the high-voltage gap of the electron accelerator prevented the electron beam with the required parameters from being obtained, while the shield, when being mounted (h_I = 14 cm), provided the normal operation of the electron accelerator. This design version of the electron accelerator was accepted as optimal.

5. Conclusion

The vacuum pumping system for the multi-aperture low-energy continuously-operated electron accelerator with a high-density beam current operating in the wide range of temperatures and extreme vibrations has been developed. This vacuum pumping system allows one to pump out high gas flows ensuring vacuum required for operation of the electron accelerator. The chosen scheme with the demountable shields makes it possible to adjust the system conductivity and distribution of getter fluxes to provide the normal operation of the accelerator.

References

[1] Karpov D A 1995 *Vacuum* **8–10** 825–6
[2] Gurevich L S 1978 *Engineering method for calculation of the main vacuum characteristics of evaporation getter pumps* (L. NIIEFA Preprint A-0389)
[3] Saksagansky G L 1988 *Electrophysical vacuum pumps* (M. Energoatomizdat)
[4] *VACOM Product Catalog* 2013 (presented at the Vacuum Tekh Expo Conference)
[5] Kontor E I 1977 *Getter and ion-getter pumps* (M. Mashinostroenie)
[6] Ivanov A G and Karpov D A 2017 *Proceedings of the 13th International Conference "Films and Coatings - 2017" (SPb)* pp 511–5

IOP Conf. Series: Materials Science and Engineering **387** (2018) 012028 doi:10.1088/1757-899X/387/1/012028

Study of the waveguide elements for mm-wave traveling wave tube

A S Ivanov, O N Mikhailova, V N Titov and A D Tupitsyn

School of Electronics, Saint Petersburg Electrotechnical University "LETI", 197376, Saint Petersburg, Russia

E-mail: asivanov@etu.ru

Abstract. The use of a millimeter waves for future mobile networks demands a creation of broadband transceivers with output power more than 10 W. Only vacuum electron devices, in particularly, traveling wave tubes can provide those parameters. Because of the short wavelength, the reference dimension of the interaction system becomes close to 100 microns and skin layer depth reaches 0,4 micron. Its fabrication comes across with low surface roughness challenge. In this paper, we evaluate interaction system elements fabricated by electrical discharge machining. The surface roughness of the microdimensional slot in a brass plate is determined experimentally. Based on the experimental roughness value, a computer simulation of a folded waveguide, a perspective type of an interactive system for a millimeter band traveling wave tube has been carried out. It is found that the loss value on the 40-period waveguide due to the roughness increases by 80 % in comparison with the ideal surface of the waveguide walls.

1. Introduction

Transition to the next generation of mobile communications (5G) includes a development and implementation of several key technical directions. One of them is a use of a millimeter wave frequency band providing peak data rate up to 20 Gbps [1]. Taking into account an availability of a low atmospheric attenuation window for 92–95 GHz frequencies, this W-range opens a new prospective for mobile networks in the future.

The use of a millimeter wave band for mobile networks demands a creation of broadband transceivers with output power more than 10 W. In this frequency band it can be provided only by vacuum electron devices. More and more attention is being given to a development of traveling wave tubes (TWT) [2–8], which fully meet the requirements of wide frequency bandwidth and output power.

Numerous publications on simulation results of separate TWTs blocks have determined their optimum design features. First, it is a use of folded waveguide [2, 4] as interaction structure between travelling wave and sheet electron beam. The wavelength about 3 mm results in decrease of TWT reference dimensions, particularly the minimum size of waveguide cross section, which can make only 100 microns, and a skin layer depth – 0,4 microns (for brass). The losses in interaction structures produced by the roughness of conducting surfaces significantly influence TWT gain. These factors determine way of fabrication for folded waveguides.

The minimum roughness of conducting surface is reached at milling [9] of the waveguides, however this way of processing demands the use of unique microdimensional milling cutters and is applicable for waveguides with cross sectional dimensions of more than 500 microns. Electrical discharge machining (EDM) [10–11], widely applied in electronic devices parts manufacturing, succumbs in surface finish fineness, but allow to produce more complex shapes of parts.

The work purpose is an estimation of a surface roughness and finding ways of its decrease in microdimensional slots formed by EDM in a brass plate.

2. Experimental data

For surface roughness estimation experimental model was made as a brass plate with a 350 microns width through slot (figure 1) using EDM. Thickness of dark layer on the inner side of a slot gives one an opportunity to visually evaluate maximum depth of surface layer as a half difference of measured width of a slot (PL2-PL1) /2=28 a micron.

a) b)

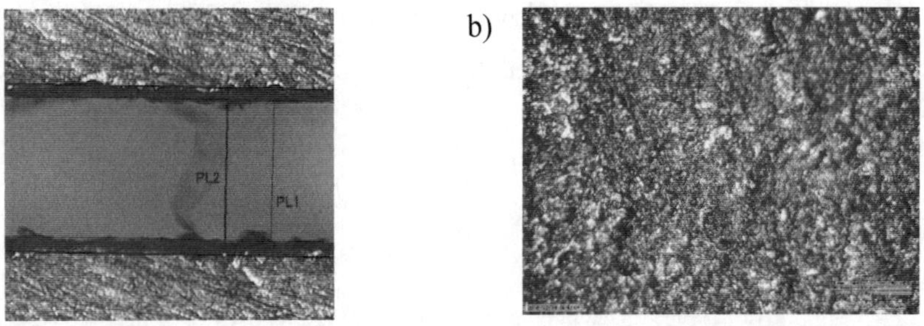

Figure 1. Microphotographs (**a**) of a through slot (dark horizontal band in centre) in a brass plate before polishing PL1=354 microns, PL2=410 microns; side slot surface, magnification x420 (**b**).

Later the experimental model was electrochemically polished to decrease roughness (figure 2). As a result, width of a slot increased at 43 microns.

a) b)

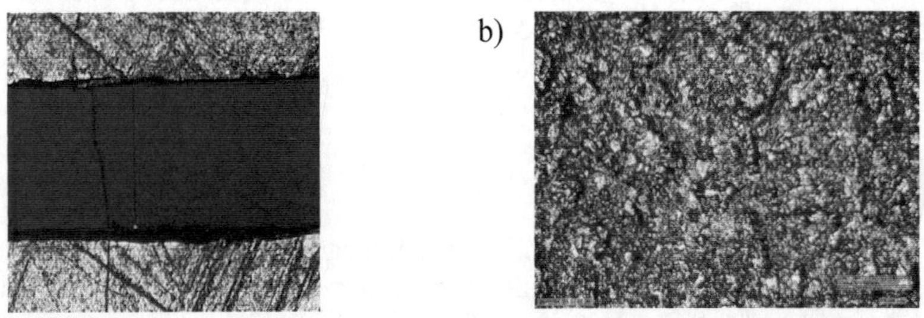

Figure 2. Microphotographs (**a**) of a through slot (dark horizontal band in centre) in a brass plate after polishing PL1=397 microns; side slot surface, magnification x420 (**b**).

For visualization of the objects, geometrical measurement and relief depth definition high-resolution microscope Hirox KH-7700 with a set of lenses for wide range of magnifications was used as well as for contactless geometrical measurements of experimental model. The software and hardware complex HIROX KH-7700 synthesize three-dimensional model of object using a series of captured frames at different levels of a focusing (up to 100 slices).

Resolution of focusing (altitude of irregularities) reaches 0,25 microns. In cases, if the depth of a relief does not exceed microns, the model is synthesized under in a semiautomatic mode with a preliminary manual measurement of depth of a relief. The three-dimensional models of objects allow to thoroughly studying a relief of a surface.

The map of three-dimensional model of a surface before polishing is shown in figure 3. It is found that an average profile of a surface has maximum depth of 20 microns. The map of three-dimensional model of a surface after polishing is shown in figure 4. It is found that an average profile of a surface has maximum depth of 7 microns.

a)

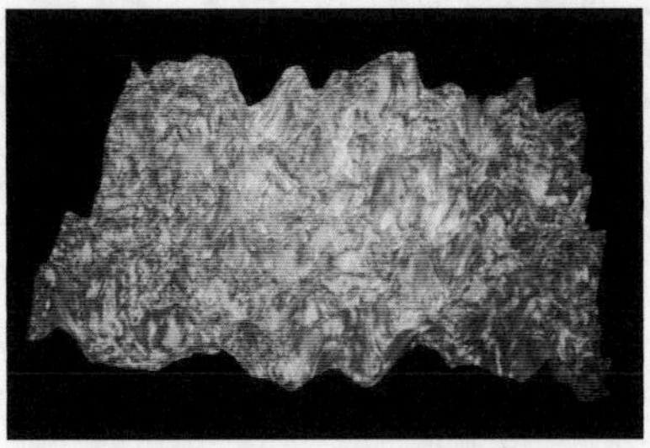

b)

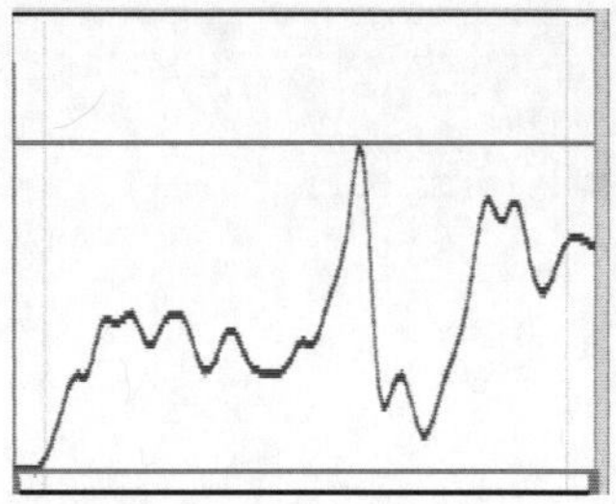

Figure 3. Map of three-dimensional model of a surface before polishing (**a**); an average profile of a surface with maximum depth of 20 microns (between horizontal lines) (**b**).

a)

b)

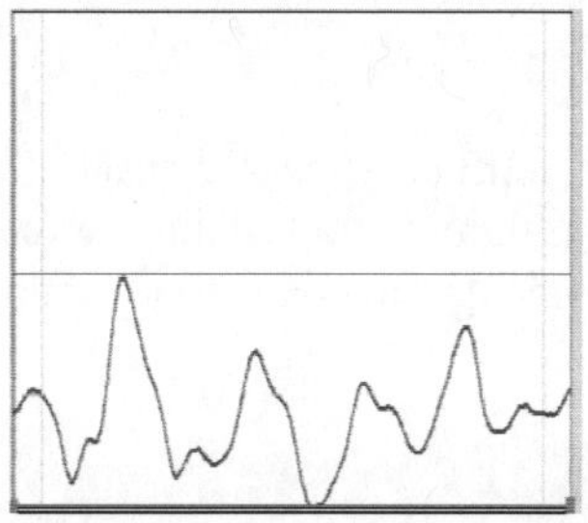

Figure 4. Map of three-dimensional model of a surface after polishing (**a**); an average profile of a surface with maximum depth of 7 microns (between horizontal lines) (**b**).

3. Simulation

Simulation of an ohmic loss in the 40-periods folded waveguide, a four period segment of which is shown in figure 5, has been carried out in CST Studio Suite with surface roughness of metal taken into consideration. The gradient model of surface losses [12] was used, making possible a correct description of an ohmic loss in case of a considerable surface roughness, when the depth of a skin layer appears to be comparable and even less than typical metal roughness. Single-parametric description of a roughness by means of route mean square deviation of a profile parameter Rq is used. Its value can be determined experimentally.

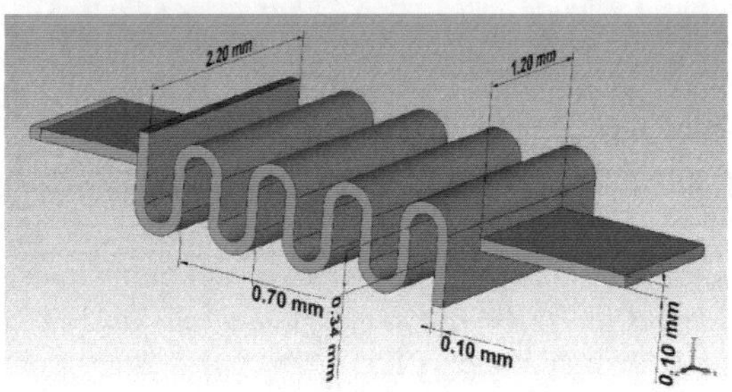

Figure 5. Geometry of four periods of folded waveguide with channel for sheet electron beam [5].

4. Simulation results

The simulation results as relations of s-parameters across a surface roughness of the 40-periodic folded brass waveguide are represented in figure 6. In case of folded waveguide, fabricating by milling the increase of losses on the 40-periodic waveguide at the expense of a roughness does not exceed on an order of magnitude maximum value of a factor of increase of losses, which one can be obtained using conventional approaches [13–14]. At increase of a roughness, more than 1 micron losses in the waveguide essentially increase.

a)

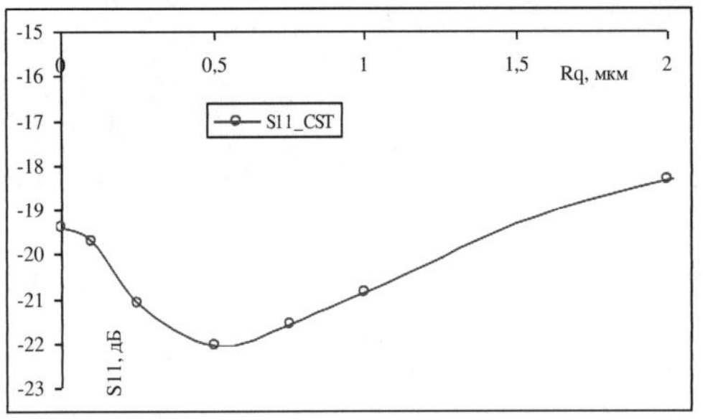

b)

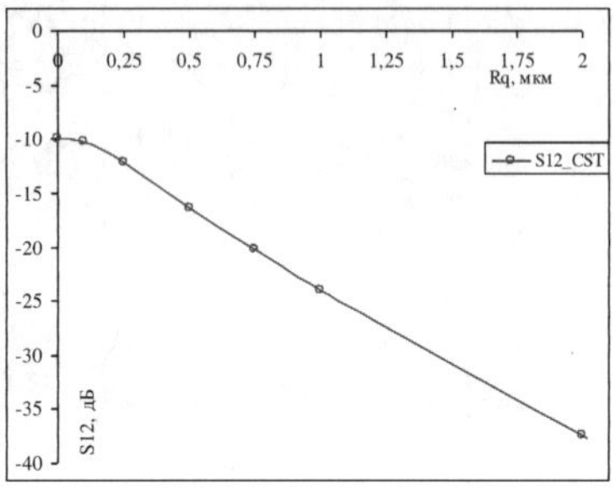

Figure 6. Relation of S-parameters to a surface roughness of the brass folded waveguide (40 periods): (**a**) – S11; (**b**) – S12.

5. Conclusions

The standard electrical discharge machining for the multiperiodic brass folded waveguides create a surface with a roughness producing increase of losses up to 80 % in comparison with ideal surface. The electrochemical polishing allow reducing a roughness almost three times, however loss value remain very high.

References

[1] Rappaport T S, Sun S, Mayzus R, Zhao H, Azar Y, Wang K, Wong G N, Schulz J K, Samimi M and Gutierrez F 2013 *IEEE Access* **1** 335
[2] Cai J, Feng J, Li B and Liao F 2005 *IEEE International Symposium on Microwave, Antenna, Propagation and EMC Technologies for Wireless Communications* DOI: 10.1109/ MAPE.2005.1618016
[3] Ryskin N M, Rozhnev A G, Karetnikova T A, Torgashov G V, Sinitsyn N I, Shalaev P D and Bourtsev A A 2013 *IEEE International Vacuum Electronics Conference* DOI: 10.1109/IVEC.2013.6571022
[4] Sharma R K, Grede A and Chaudhary S 2014 *IEEE Transactions on Plasma Science* **10** 343
[5] Grigoriev A D, Ivanov A S, Ilyin V A, Luchinin V V and Titov V N 2015 *Electron Engineering, ser. 1, HF Engineering* **4** 26–32
[6] Ryskin N M, Karetnikova T A, Rozhnev A G, Torgashov G V, *Bushuev* N A and Shalaev P D 2015 *IEEE International Vacuum Electronics Conference* DOI: 10.1109/IVEC.2015.7223784
[7] Paoloni C, Letizia R, André F, Kohler S, Magne F, Rocchi M, Marilier M, Zimmerman R, Krozer V, Ulisse G, Ramirez A and Vilar R 2016 *IEEE International Vacuum Electronics Conference* DOI: 10.1109/IVEC.2016.7561865
[8] Gamzina D, Li H, Himes L, Barchfeld R, Popovic B, Pan Pan, Letizia R, Mineo M, Feng J, Paoloni C and Luhmann Jr. N C 2016 *IEEE Transactions on Nanotechnology* **1** 85
[9] Stil I, Fontana A L, Lefranc B, Navarrini A, Serres P and Schuster K F 2012 *22nd International Symposium on Space Terahertz Technology Proceedings* pp 151–3
[10] Chung D K, Lee K H, Jeong J and Chu C N 2014 *International Journal of Precision Engineering and Manufacturing* **9** 1785
[11] Liu Q, Zhang Q, Zhu G, Wang K, Zhang J and Dong J 2016 *MMP* **43** 391
[12] Gold G and Helmreich K A 2012 *European Integrated Circuits Microwave Conference Proc.* 42
[13] Hammerstad E O and Bekkadal F 1975 *Microstrip Handbook* (Norwegian Institute of Technology)
[14] Groiss S, Bardi I, Biro O, Preis K and Richter K R 1996 *IEEE Transactions on Magnetics* **3** 894–7

IOP Conf. Series: Materials Science and Engineering **387** (2018) 012029 doi:10.1088/1757-899X/387/1/012029

Investigation of gas discharge features during HiPIMS using optical emission spectroscopy

V V Karzin and K I Karapets

Saint Petersburg Electrotechnical University "LETI", 197376, Saint Petersburg, Russia

Email: karzin.ru@gmail.com

Abstract. The paper presents the results of the investigation of the emission spectra of a gas discharge during high-power impulse magnetron sputtering (HiPIMS) of a titanium target at different power supply parameters. In such mode, the fraction of ionized atoms of the target material in the discharge increases significantly, what affects the properties of the deposited coatings. The characteristic features of the operation of the system in HiPIMS mode were singled out, and also the discharge spectra were compared at high-power impulse and conventional (sputtering on a direct current) operation modes of the magnetron. As a result of the study, it was shown that the use of optical emission spectroscopy during HiPIMS makes it possible to effectively determine the parameters of the stable operation of the system and determine the boundary at which a transition into high-power regime occurs.

1. Introduction

One of the most promising methods for applying conductive and dielectric coatings is HiPIMS (High-Power Impulse Magnetron Sputtering) [1–5], the first mention of which was made in the late 90's. During two decades, this technology has overcome the long path of development and adaptation and now occupies a significant place in industrial production (microelectronics and machine building). Using this method, the percentage of ionized atoms of the sputtered material can reach 90 % [1–3]. This and a number of other parameters significantly improve such properties of the formed coatings as their adhesion to the substrate, density and uniformity.

In this paper, a study of the gas discharge of a magnetron during HiPIMS by optical emission spectroscopy is presented. This tool of diagnosing the discharge parameters allows determining its component composition and other properties quite accurately [4–5]. The measurements were carried out at different settings of the high-power unit: frequency, peak voltage, and pulse duration. When a high-power pulsed magnetron power supply unit operates at low voltages (pulse amplitude up to ~500 V) in argon medium, necessary for the synthesis of high-density coatings discharge conditions might not occur. The reason for this is the low gas and the sputtered substance ionization (these parameter corresponds to the discharge of a classic magnetron which is operating on a direct current). Optical emission spectroscopy of the discharge is an effective way to determine the transition to the HiPIMS regime.

2. Experimental

Experiments were carried out at the upgraded vacuum deposition unit UVN-71. The residual pressure in the chamber was 0.02 mTorr, the working pressure of argon (purity 99.95 %) was 0.6 mTorr. A planar 6 mm thick magnetron with a diameter of 115 mm with a cooled titanium target (99.9 % purity) was used as the sputtering system. To operate the magnetron in high-power pulsed mode, a specialized power unit with a peak voltage value of 1000 V and current of 1000 A was used (the average power was not more than 5 kW).

To perform the spectral analysis, the ISM3600 spectrophotometric attachment with the special software complex Aspect 2010 was used, with the help of which the data obtained during the experiments were processed.

For the study, the most intense spectral lines corresponding to the emission of titanium and argon atoms and ions were chosen. In figure 1(a), (b) discharge emission spectra corresponding to magnetron sputtering at direct current (DC mode), direct impulse current (DC-pulse (PM)) and high-power impulse mode (HiPIMS) are shown.

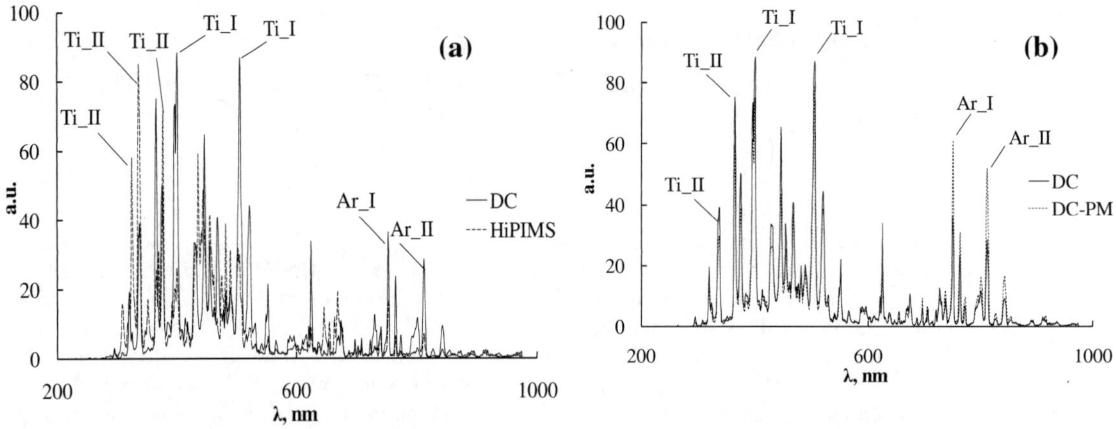

Figure 1. Spectra of the gas discharge of a magnetron during sputtering in HiPIMS and DC (**a**) and DC and DC-PM (**b**) modes.

In figure 2(a)–(c) the dependences of the intensities of the spectral lines Ti I (atom) on the peak voltage of the pulse at its various durations are shown. At pulse duration of 50 μs, the radiation intensity is small regardless of frequency, what may indicate a low deposition rate in this mode. At pulse duration of 100 and 150 microseconds, a significant increase in the intensities of the spectral lines is observed. At a voltage of 540–560 V, the increase of the growth rate of the intensities of the lines under consideration is well seen.

In figure 2(d)–(f) analogous dependences for Ti II (single-charged ion) are presented. It can be seen from the dependences that the degree of ionization at a pulse duration of 50 μs is small. With a further increase in the pulse duration, it significantly increases. In this case, the intensification of the growth of spectral lines occurs even at a voltage of 520–530 V.

Figure 3 shows the graphs for Ar I ((a)–(c)) and Ar II ((d)–(f)). The dependences for argon shown in figure 3 correlate with analogous ones for titanium - with the increase of maximum pulse voltage, the intensity of all lines increases. There is also a critical value of this voltage (inflection point) at which the growth rate of spectral lines increases.

In figure 4 the ratio of the main spectral lines as a function of the frequency and duration of the pulse for voltages of 500, 540, and 580 V is shown in the form of column diagrams.

Figure 4 clearly represents the distribution of the emission intensities of various particles in the discharge. At a frequency of 100 Hz and a pulse duration of 50 μs (figure 4(a)), the dependence is similar to sputtering in the DC mode – the lines corresponding to the emission of titanium atoms predominate. With further growth of the pulse frequency and duration, redistribution of the intensities of the atomic lines (Ti I) and ions (Ti II) of the target material occurs - the fraction of the ionized component noticeably increases. The spectral lines corresponding to the emission of titanium ions show the highest intensity, what indicates a transition to the real HiPIMS regime. At the same time, with the increase of the frequency and duration of pulses, the intensities of emission lines for all groups of charged particles increase, without a significant redistribution of their ratio.

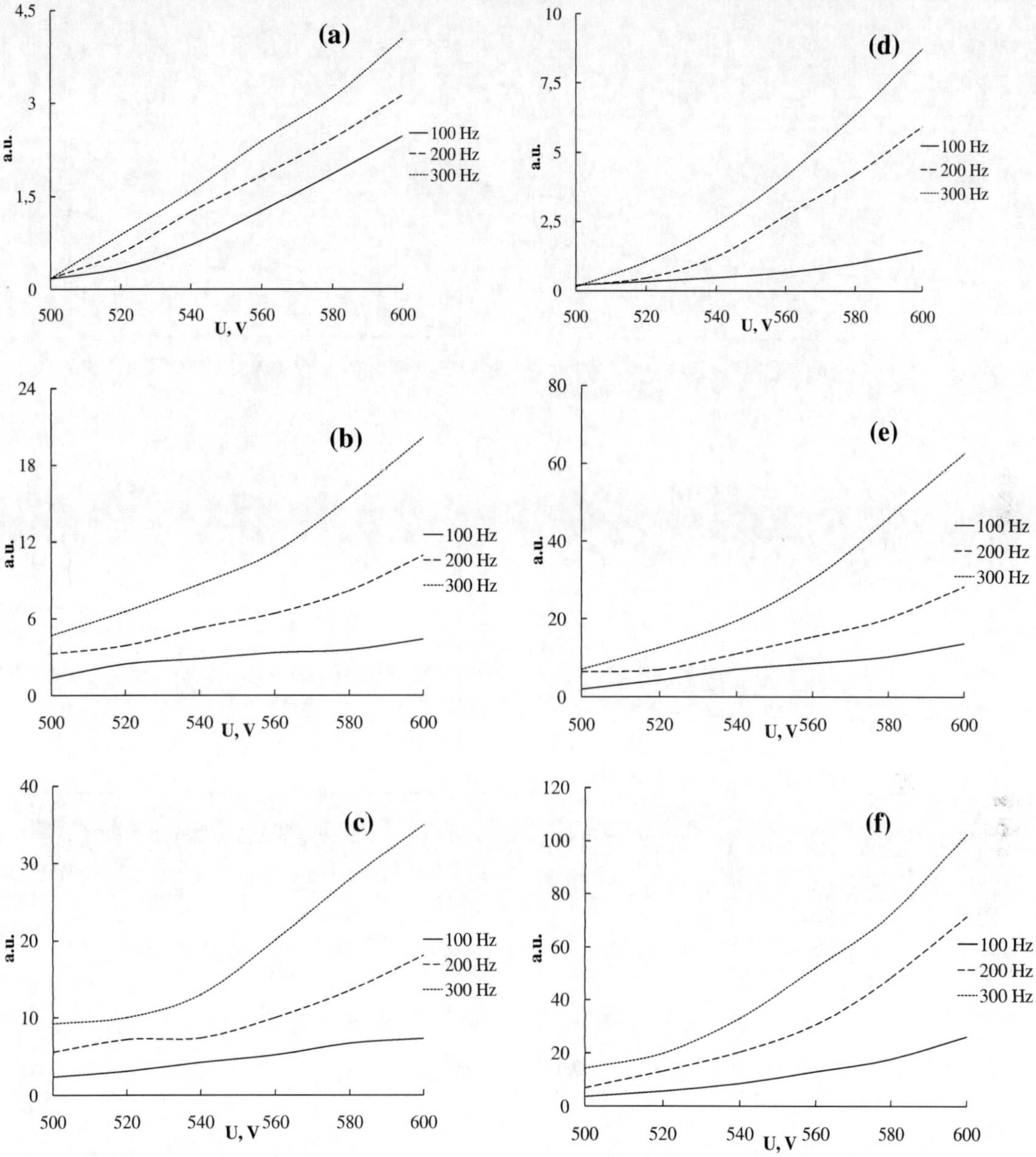

Figure 2. Intensities of the spectral lines Ti I and Ti II at a pulse duration of 50 (**a**), (**d**), 100 (**b**), (**e**) and 150 (**c**), (**f**) microseconds.

Figure 4 clearly represents the distribution of the emission intensities of various particles in the discharge. At a frequency of 100 Hz and a pulse duration of 50 μs (figure 4(a)), the dependence is similar to sputtering in the DC mode – the lines corresponding to the emission of titanium atoms predominate. With further growth of the pulse frequency and duration, redistribution of the intensities of the atomic lines (Ti I) and ions (Ti II) of the target material occurs - the fraction of the ionized component noticeably increases. The spectral lines corresponding to the emission of titanium ions show the highest intensity, what indicates a transition to the real HiPIMS regime. At the same time, with the increase of the frequency and duration of pulses, the intensities of emission lines for all groups of charged particles increase, without a significant redistribution of their ratio.

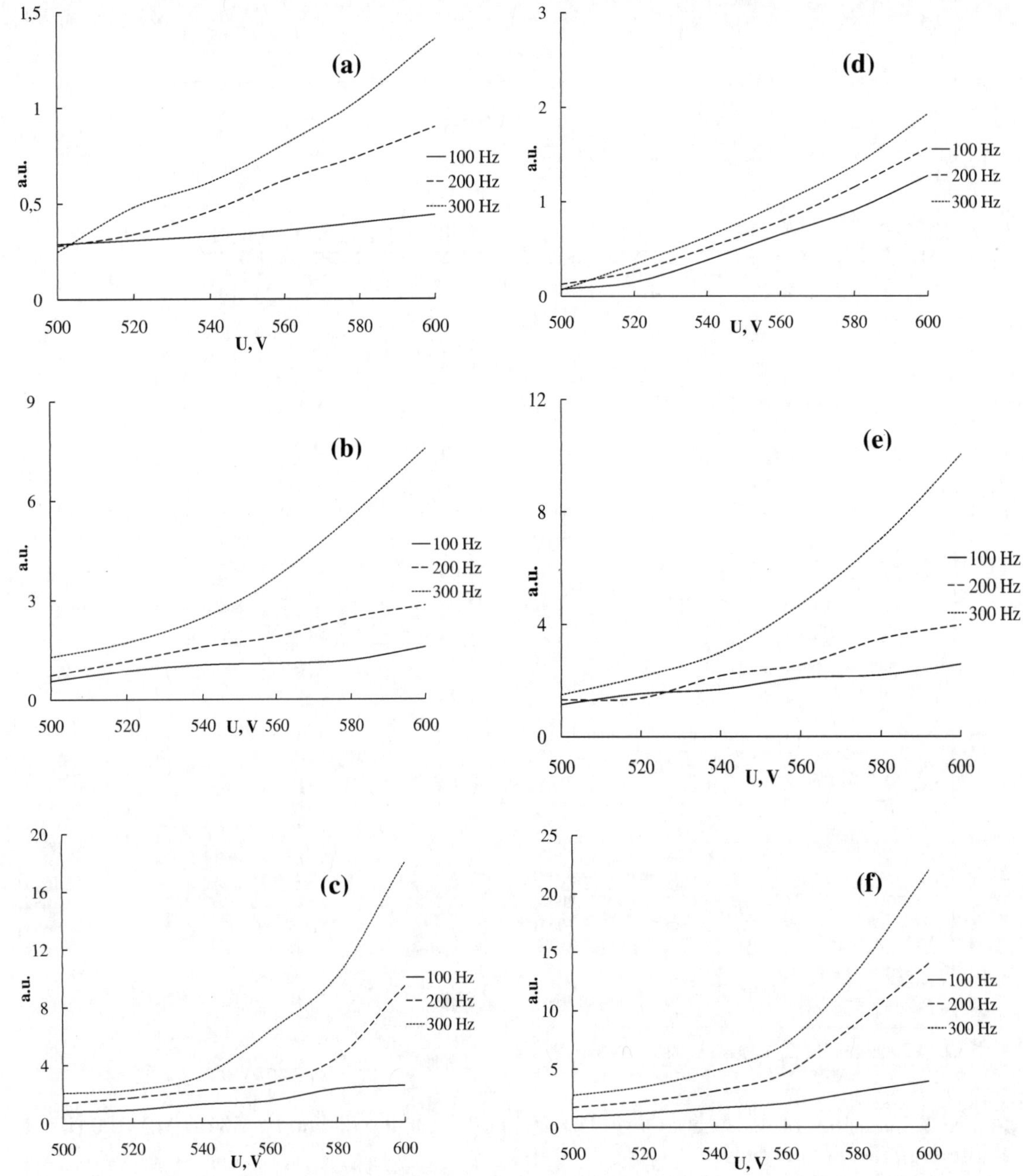

Figure 3. Intensity of the spectral lines of Ar I and Ar II at a pulse duration of 50 (**a**), (**d**), 100 (**b**), (**e**) and 150 (**c**), (**f**) μs.

Figure 5 shows the emission spectra of a gas discharge measured at a voltage of 600 V and a pulse duration of 50 μs, but at different frequencies which were 100 Hz (HiPIMS (–)) and 200 Hz (HiPIMS (+)). "–" indicates the absence of the HiPIMS mode, the amplitude of the Ti I spectral lines predominates over Ti II; "+" – the inverse mode, in which the spectrum takes the opposite form and corresponds to the HiPIMS mode. At a frequency of 100 Hz, the lines corresponding to the emission of neutral titanium atoms have the highest intensity, which is typical for operation in the classical DC mode of the magnetron, while when increasing the frequency up to 200 Hz, the line of ionized titanium acquires the highest intensity. This indicates a transition from low-power to high-power mode. For the technology of manufacturing film structures, a precise determination of the magnetron operation modes is an important task, the solution of which allows achieving efficient deposition of high-quality coatings with specified characteristics.

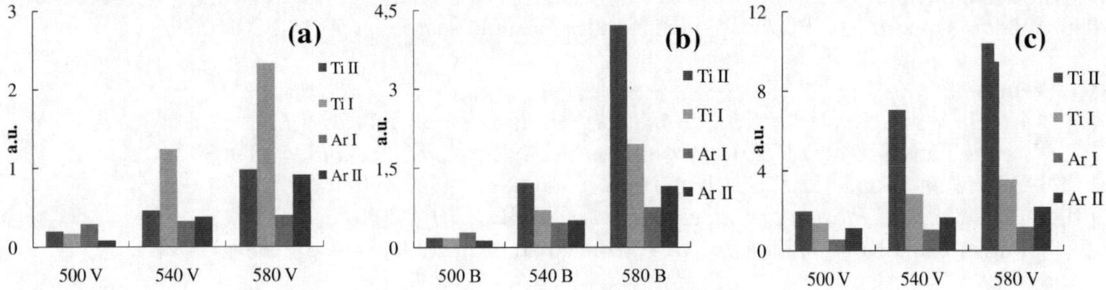

Figure 4. The ratio of the intensities of the spectral lines in the discharge for frequencies and pulse widths of 50 μs and 100 Hz (**a**), 50 μs and 200 Hz (**b**), 100 μs 100 Hz (**c**), respectively.

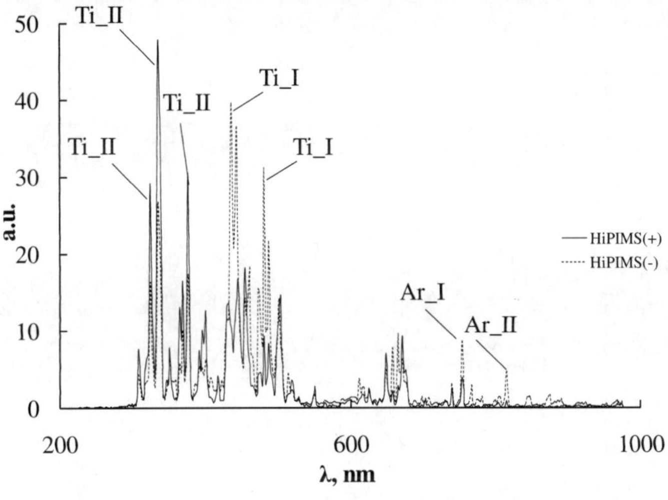

Figure 5. Spectra of the gas discharge of a magnetron under different operating conditions.

From the above data follows the fact that optical emission spectroscopy is an effective way of determining the modes of operation of a magnetron during the usage of HiPIMS technology. With the help of this diagnostic method it is possible not only to determine the state of the gas discharge, but also to estimate the ratio of charged particles in it.

3. Conclusion

A high-power gas discharge of a magnetron was studied using optical emission spectroscopy. To achieve the most efficient operation in the HiPIMS mode, optimum frequency and pulse width settings of a high-amplitude power supply are required. During magnetron sputtering of a titanium target at frequencies below 100 Hz and pulse duration less than 50 microseconds, at voltages no higher than 580 V, the qualitative picture of the distribution of the intensities of the spectral lines is similar to the magnetron sputtering at a direct current (DC mode). At a relatively low peak voltage, short pulse duration and their frequency, it is not possible to achieve theefficient ionization of sputtered target and gas atoms in a vacuum chamber, what doesn't allow using the advantages of HiPIMS technology. Optical emission spectroscopy is an effective tool for determining the transition point to a high-power pulse mode.

The possibility of qualitative analysis of the discharge was determined on the basis of measuring the spectral characteristics of the radiation. This makes it possible to quickly determine the mode in which the magnetron sputtering system operates at the given moment and, if necessary, make the appropriate adjustment, by changing the parameters of the power unit.

Acknowledgment
This work is supported by Russian Foundation for Basic Research (grant 18-32-01063).

References

[1] Karzin V and Komlev A *Surface and coatings technology* **344** 269–73
[2] Arcos T and Schroder R 2014 *Plasma sources science and technology* **23** 963–73
[3] Breilmann W and Maszl C 2013 *Journal of physics: applied physics* **46** 22–34
[4] Britun N 2015 *Plasma processes and polymers* **12** 1010–27
[5] Barker P and Konstantinidis D 2014 *Surface and coatings technology* **258** 631–8

IOP Conf. Series: Materials Science and Engineering **387** (2018) 012030 doi:10.1088/1757-899X/387/1/012030

Study of nanoscale metal oxide coatings produced by magnetron sputtering of copper followed by heat treatment

I N Kashkul, L D Bakhchova, K G Elanskaia and D N Redka

Saint-Petersburg State Electrotechnical University "LETI", 197376, St. Petersburg, Russia

E-mail: rd89@bk.ru, k.elans@yandex.ru

Abstract. The composition and microstructure of nanoscale metal oxide coatings deposited by magnetron RF sputtering copper on the quartz glass with followed heat treatment were investigated. The dependence of the film thickness and the content of oxides CuO and Cu_2O in it depending on the temperature and time of heat treatment conditions were obtained. Studies have shown that, depending on the mode of receipt of copper oxide films may be multiphasic and include various oxide phases, predominantly, CuO and Cu_2O with different bandgap and p-type conductivity.

1. Introduction

Copper oxides are widely used in solid state electronics and optoelectronics [1]. One important advantage of using the copper oxide lies in non-toxicity and widely availability [2]. Various methods for the obtaining of copper oxide, such as thermal oxidation, a reactive high-frequency magnetron sputtering, hydrothermal synthesis, electrodeposition, thermal evaporation, and others are known [3]. It is also known that copper oxide has two phases with different stoichiometry, Cu_2O and CuO [3]. Interest in the study of copper oxide films of different stoichiometries connected with the possibility of their use in solar cells [4, 5]. Application of this type of photovoltaic constrained insufficient knowledge of the physical processes at the interface of oxide structures with developed surface morphology.

Theoretical efficiency conversion of solar energy into electrical energy for copper oxide is 9–12 %, but still the experimental oxide thin film heterostructures CuO_x/ZnO, obtained by different research groups [4–8], have low efficiency. In addition to the n-ZnO/p-CuO heterojunctions, the heterostructures of copper oxide with PZT are also of interest for photovoltaic applications [9] due to the ferroelectric photovoltaic effect and significant electric fields at the interface of the PZT and different semiconductors [9–13]. Thus, copper oxide can be considered as a promising photosensitive layer in combination with ZnO or PZT. The main method of increasing the efficiency of solar cells based on copper oxide is to improve the quality of the oxide layers, which requires a detailed examination and developing of processes for obtaining oxide layers.

This is a continuation study for the works [3] about development and investigation of thin-film layers of copper oxides for photovoltaic applications. As a convenient method of depositing layers, we use magnetron sputtering.

2. Methods of obtaining and studying samples

On a purified quartz and glass substrates have deposited at room temperature copper thin film, 100 nm thick by magnetron sputtering on installation Plasmalab System 100 (at a pressure of 5 mbar and a charge power of 100 mW). Substrates were cut mechanically into rectangular samples of size 10x20 mm. Then, the samples were placed in electric furnace PT-1,2-70 and have suffered thermal oxidation in air at temperatures ranging from 220 to 500 °C and the time (t) from 5 to 60 minutes.

Evaluation of the film thickness were performed at the facility AlphaStep D-120. The values of film thickness by temperature and annealing time of copper film are shown in table 1. As seen from the profilometry data, metal film 100 nm thick is fully oxidized at 500 °C for 30 minutes.

Table 1. The values of film thickness after oxidation in air.

Temperature, °C	Thickness, nm						Phase composition
	t= 0 min	t= 5 min	t= 10 min	t= 20 min	t= 30 min	t= 60 min	($t \geq 30$ min)
250	100	110	115	128	140	165	Cu_2O
300	100	135	145	160	175	185	$Cu_2O + CuO$
350	100	140	155	170	180	190	$Cu_2O + CuO$
500	100	190	200	210	215	215	CuO

The prepared samples were investigated by Raman spectroscopy, X-ray analysis and atomic force microscopy. Optical and electrical properties have been studied.

Survey of the phase composition of the films by Raman spectroscopy performed with the setup LamRam HR800 using green laser with wavelength 532 nm, 2 mW beam power, accumulation time 30 seconds and the number of scans 10. The surface morphology was studying by the atomic force microscope "Integra Thermo" with a resolution of 10 nm in the horizontal plane, and the vertical better than 1 nm. The optical properties of the films were studied by Varian Cary® 50 UV-Vis Spectrophotometer. Measurements of the changing electrophysical parameters were carried out on the installation ECOPIA HMS-5000.

3. Results and discussion

Studying the phase composition of the films formed by oxidation at various temperatures of nanoscale copper layers on the air, carried out independently in two ways – by Raman spectroscopy and X-ray analysis (figure 1). Analyzing of Raman spectra identified lines Cu_2O (98 cm^{-1}, 145 cm^{-1}, 217 cm^{-1}) and CuO (296 cm^{-1}, 342 cm^{-1}, 627 cm^{-1}). Results showed that structure of the copper oxide film varies depending on the temperature and time of annealing in air. Therefore, Cu_2O phase formed by isochronous annealing for 30 minutes at temperatures of 220–250 °C. With further increasing the temperature to 350 °C is obtained the film as a mixture phases Cu_2O and CuO, but at a temperature above 500 °C formed a single phase CuO.

X-ray analysis allowed to draw similar conclusions about the change in the phase composition of the copper oxide films with increasing annealing temperature. In a manner, by varying the annealing temperature in the range of 220 – 500°C we are able to manage phase composition and to obtain copper oxide as single-phase CuO or Cu_2O and multiphasic Cu_2O + CuO films.

Morphology of obtained surface layers were examined by atomic force microscopy (AFM) in tapping mode on a Nano-lab NTEGRA-Therma. Analysis of the data showed particles ranging in size from 100 to 300 nm are evenly distributed on the sample surface. The surface of obtained layers rough, while the scope of heights at the characteristic surface topography length of 10 microns. For samples subjected to thermo processing at higher temperatures, this scope is larger. Investigation results of copper oxide films surface morphology can be concluded that the oxidation process leads to increase surface roughness.

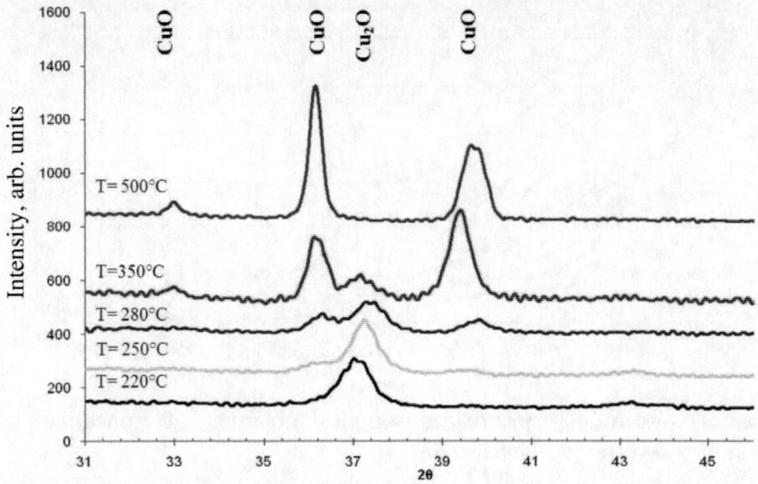

Figure 1. X-ray diffraction patterns copper oxide film obtained by oxidizing in air at various temperatures for 30 min.

According to the theory, in the process of annealing molecular oxygen at atmospheric pressure depending on the processing temperature may oxidize copper fully, partially or in two stages:

$$Cu + 0{,}5O_2 \leftrightarrow CuO;\ 2Cu + 0{,}5O_2 \leftrightarrow Cu_2O;\ Cu_2O + 0{,}5O_2 \leftrightarrow 2CuO.$$

Due to the polycrystalline structure of copper, according to [9], the rate of oxidation of copper with the bulk diffusion of deposits in the lattice and grain boundary diffusion can be expressed as

$$\frac{d\left(x^2\right)}{dt} = 2D_\upsilon V\Delta c + \frac{4D_b V\Delta c\delta}{d_t} = K_\upsilon + K_b f,$$

wherein x – thickness of the oxide layer at time t; D_υ – bulk diffusion coefficient; V – molar volume of the oxide; Δc – gradient of the concentration of point defects across the oxide layer; D_b – the grain boundary diffusion coefficient; δ – the width of the grain boundary; d_t – the average grain size at time t; K_υ – component of parabolic constant oxidation associated with the bulk diffusion in the crystal lattice; K_b – component associated with the grain boundary diffusion; $f=2\delta/d_t$ - the proportion of all diffusion regions disposed along the grain boundaries.

Analysis of experimental data have shown that the data in table 1 are well described by semiempirical equation:

$$h(t) = h_m \int_0^{Dt/h_0^2} e^{-k^2} dk,$$

where h_m and h_0 – maximal and the initial film thickness of the copper oxide, t - time, D - the effective diffusion coefficient.

Study optical properties of the films obtained at temperatures of 220–250 °C showed that samples have high transmittance (50%). Whereas the films obtained at temperatures of 350–500 °C have a high absorption coefficient in the visible spectrum and substantially transparent in the infrared spectrum (figure 2). By transmission and absorption spectra were calculated band gap of 1.2 and 2.28 eV for CuO and Cu_2O respectively. Based on the differences in transmission for samples more promising for use in photovoltaics is Cu_2O. Investigation of electrophysical parameters showed that the samples have p-type conductivity and high surface resistance.

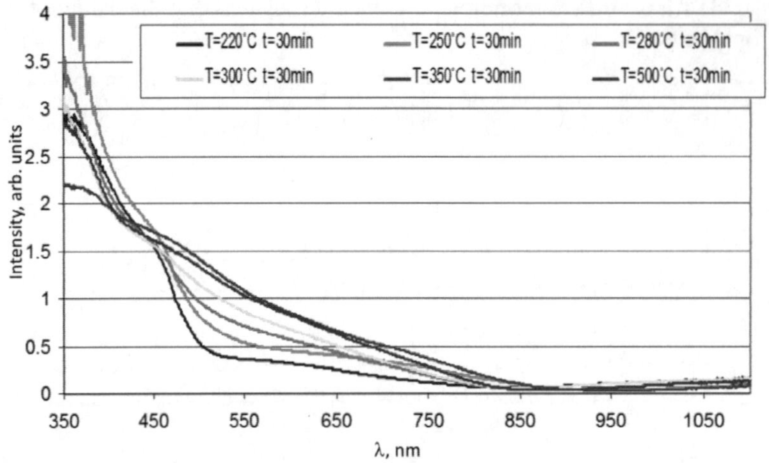

Figure 2. Absorption spectra of samples obtained at annealing for t=30 minutes at different temperatures.

4. Conclusions

As a result of experimental studies were obtained dependences of the oxide film thickness and the percentage content of CuO and Cu_2O oxides upon the annealing temperature-time conditions. Thus, at isochronous annealing for 30 minutes at temperatures of 220–250 °C formed Cu_2O, by further increasing the temperature to 350 °C obtained mixture phase Cu_2O and CuO and at a temperature above 500 °C formed a single phase CuO.

Acknowledgment

This work was supported by the grant of The Ministry of Education and Science (Minobrnauka) of the Russian Federation (project Goszadanie № 3.3990.2017/4.6).

References

[1] Jia W, Reitz E, Shimpi P, Rodriguez E G, Gao P X and Lei Y 2009 *Materials Research Bulletin* **8** 1681–6
[2] Oral A Y, Mensur E, Aslan M H and Basaran E 2004 *Materials Chemistry and Physics* **1** 140–4
[3] Adilov Sh R, Afanasiev V P, Kashkul I N, Kumekov S E, Mukhin N V and Terukov E I 2017 *Glass Phys. Chem.* **3** 272–5
[4] Adilov Sh R, Kumekov M E, Kumekov S E and Terukov E I 2013 *Semiconductors* **5** 655–6
[5] Omayio E O, Karimi P M, Njoroge W K and Mugwanga F K 2013 *Int. J. Thin Film Sci. Tec.* **1** 25–8
[6] Redka D N, Mukhin N V and Zakharov I G 2016 *Technical Physics* **61** 1744–6
[7] Afanasjev V P, Konoplev G A, Chigirev D A, Mukhin N V, Terukov E I and Terukova E E 2016 *Proc. of SPIE* **9896** 98980V
[8] Afanasjev V P, Mukhin N V, Redka D N, Rudenko M V, Terukov E I, Oseev A and Hirsch S 2017 *Ferroelectrics* **508** 124–9
[9] Cao D, Wang C, Zheng F, Dong W, Fang L and Shen M 2012 *Nano Letters* **6** 2803–9
[10] Afanasjev V P, Chigirev D A, Mukhin N V and Petrov A A 2016 *Ferroelectrics* **1** 170–6
[11] Agafonova D S, Kozyrev A B, Mukhin N V, Red'ka D N, Elanskaya K G, Rudenko M V, Oseev A, Hirsch S. 2018 *Glass Phys. Chem.* **1** 15–20
[12] Mukhin N V 2016 *Glass Phys. Chem.* **1** 64–6
[13] Afanas'ev V P, Vorotilov K A and Mukhin N V 2016 *Glass Phys. Chem.* **3** 295–301

IOP Conf. Series: Materials Science and Engineering **387** (2018) 012031 doi:10.1088/1757-899X/387/1/012031

Specifics of oil vacuum pumps operation in handling radioactive materials

N T Kazakovsky and V A Korolev

Federal State Unitary Enterprise "Russian Federal Nuclear Center – All-Russian Research Institute of Experimental Physics" (FSUE "RFNC-VNIIEF"), 607200, Sarov, Russia

E-mail: kazakovsky@cam.vniief.ru

Abstract. Some problems of operation of oil pumps in handling radioactive materials are addressed. Simple and safe methods of spent oil liquid radioactive waste treatment in-situ are proposed.

1. Introduction

Different types of oil pumps are widely used in operation of vacuum systems and apparatus containing radioactive materials including tritium. Urgency of using oil vacuum pumps remains despite a large number of recently developed new types of dry pumps with rather high dynamic behavior. Wide range of application of this type of equipment is stimulated by several positive factors: relatively low initial cost, long service life, maintainability in-situ, etc. Selection of oil pumps is often determined by the fact that the Russian enterprises have successfully mastered manufacture of the required spectrum of such pumps. The scope of this article considers the problems which arise in the process of operation of oil vacuum pumps in handling radioactive materials including those with tritium.

2. Research done and results obtained

Periodic routine efforts on changing oil are required during operation of vacuum pumps in order to maintain an adequate level of specs. In some cases radioactive materials are accumulated in vacuum oil of the pump. For example, in handling tritium-containing materials, tritium is accumulated in oil as a result of reactions of isotopic exchange, oxidation and tritium water sorption. When specific activity of tritium in replaceable oil exceeds 100 Bq/g, it turns to a category of liquid radioactive waste (LRW). LRW of vacuum oil can be formed when handling other radioactive materials especially if they are fine-dispersed and volatile. The Federal Law "On Radioactive Waste Management" [1] binds all entities whose activity entails generation of radioactive waste (RW) to carry out recovery operations for their own expense. Maintaining a required safety level constitutes a key aspect of LRW management as a source of ionizing radiation. One of the methods of solving this problem lies in decreasing a class of LRW hazard through extraction of insoluble high-level solid components with subsequent solidification of a liquid phase. Conditioning of LRW in-situ appears to be the most reasonable measure from the viewpoint of ensuring environmental safety and economic effectiveness.

If LRW of vacuum oil include insoluble fine-dispersed suspended solid high-radiated radioactive components than separation of phases is necessary since such components may initiate significant radiolysis of the liquid phase accompanied with release of hydrogen during storage. Filtering seems to be a principal approach to separate liquid from insoluble impurities distributed in it. Though use of filters for separation of the fine-dispersed phase has some disadvantages – short life and large number of secondary RW being in the first place. Methods of physical and chemical binding of the dispersed phase with follow-on detachment of formed deposit became also predominant. However, deposition leads to concentration of solid components in one part of the fluid which does not solve the problem completely.

The authors propose spinning as a new technique for extraction of the fine-dispersed phase from LRW of vacuum oils [2]. The proposed technique implies feeding of the second fluid, which can be conditionally called "spacer fluid", to LRW of oil prior to spinning. Density of spacer fluid should be higher than that of LRW base and should not mix with them. Spinning of the LRW – spacer fluid mixture involves concentration of suspended fine-dispersed particles into the bulk of the latter. The authors suggest using water as the spacer fluid for LRW of vacuum oil with particles distributed in it. The two fluids do not practically mix. Upon localization of fine-dispersed particles in water, they can be safely separated and bound into a solid matrix by applying the known technique.

Water and oil are separated after suspended fine-dispersed particles have been transferred into water. To do this, water with fine-dispersed particles and refined oil (contains neither water nor fine-dispersed particles) are exposed to negative temperature up to water freeze-up. Remaining liquid oil is removed and, as it might have residual activity, is transferred into a solid aggregate state. Water can be used to spin next portions of LRW of oil.

A hardening agent consisting of a mixture of saturated hydrocarbons – paraffin, stearic acid and ceresin – was proposed in research for oil solidification [11]. Upon transfer into liquid state at T <70 °C, this hardener mixes well with vacuum oil and, after solidification, forms solid homogeneous hydrophobic compound which ensures safe storage. Filling of the final compound with vacuum oil comes to up to 60 % of the final mass. Radiation safety of the proposed compound was verified experimentally by performing comparative studies of extraction of tritium from tritium-containing oil specimens prior and upon solidification. A laboratory installation in the form of a looped circuit was assembled to do the research. The circuit comprised a cell to locate the specimen, a membrane pump and an ionization chamber. The membrane pump pushed gas medium through the cell with the specimen and the ionization chamber which recorded tritium volumetric activity. Volumetric activity of gas medium in the installation grew continuously during blowdown of tritium-containing medium-activity vacuum oil specimen. It increased 10-fold within an hour as against background level. Periodic control exercised within 2 months revealed no tritium extraction when the same oil specimen was blown down after solidification.

In order to assess applicability of the proposed hardening agent for conditioning of LRW on the basis of several commercial organic solvents (solvent, xylene, kerosene, white spirit, rubber solvent) we defined their solubilities. According to the results obtained, concentration of solvents in the final compound makes up, at least, 10 wt. %. Solubility limit for these materials was not defined as chemical composition of materials can vary over a wide range within one and the same standard. Therefore, preliminary prototype tests are needed for practical solidification of LRW of a particular organic solvent.

3. Conclusions

It should be mentioned that the class of waste-posed hazard goes down upon solidification. This factor significantly affects radiation safety in further newly formed RW handling. The developed technologies rely on methods never proposed before. Their uniqueness lies, primarily, in simple implementation. This enables to harden LRW in-situ under any in-lab or production conditions without applying high-tech dedicated equipment. Introduction of the proposed technologies will permit (1) to rule out necessity of LRW storing in-situ, thus, improving production radiation safety; and (2) to significantly cut down costs of the enterprise for transportation to a recovery site and LRW conditioning by the third party. No secondary RW is generated which is very important for ensuring environmental safety of the proposed methods.

IOP Conf. Series: Materials Science and Engineering **387** (2018) 012031 doi:10.1088/1757-899X/387/1/012031

The proposed methods are of general purpose, can be applied to a large spectrum of materials, do not require any special labware and are capable of reaching high practical efficiency.

Possibility of carrying out the entire process of extraction and subsequent fixation of a dispersed phase in a stable matrix material in the absence of any contact with environment (in a hermetically sealed box) is a major factor in handling such radionuclides as tritium.

Radiation monitoring performed at all stages of operation demonstrated their safety for personnel and environment. Values of volumetric activity in the air of a workroom and at the external surface of a container with produced compound as well as internal radiation doses of personnel involved are tens times lower than existing reference levels.

References

[1] Federal Law *"On Radioactive Waste Management and Making Amendments into Separate Legislative Acts"* 190-FZ of 11.07.2011
[2] Kazakovsky N T and Korolev V A *Patent RU 2637811 G21F9/04* 07.12.2017
[3] Application for Invention 2017124109 of 07.06.2017 (decision of patent issue of 13.03.2018)

Research of acoustically stimulated thermal desorption in electrovacuum microwave devices

I F Khanbekov[1,2], I P Li[1], V S Petrov[1], V P Mikhaylov[2] and E A Deulin[2]

[1]JSC Pluton, 105120, Moscow, Russian Federation
[2]Bauman Moscow State University, 105005, Moscow, Russian Federation

E-mail: ivan.khan@ya.ru, mikh-vp@yandex.ru

Abstract. The article describes the results of the research of gas desorption from internal surfaces and from the thickness of metal and non-metal parts of electrovacuum microwave devices into free gas due to the simultaneous thermal and acoustic effects on the device body. There are adduced the comparison of the rate of gas release at different technologies of electric devices pumping. Also, it's given the physical explanation of the mechanism of thermoacoustic desorption of gas components in a vacuum. Also, it's given the physical explanation of the mechanism of thermoacoustic desorption of gas components in a vacuum.

A degassing process of electrovacuum microwave devices is the most prolonged technological process, which is connected with high energy consumption and high responsibility. During a degassing process the working capacity of the whole product is being formed. Nowadays not enough attention is paid to degassing process of electrovacuum microwave devices. Often the technology hasn't changed since the moment of the development, despite that the last 30–40 years the perceptible progress in the field of vacuum technology has occurred. The electrovacuum microwave device degassing process occurs according to classical scheme with using a step heating of product housing and with isothermal holding at a temperature about 550 °C. The duration of this process could reach several dozens of hours. Obviously, it is impossible to accelerate the degassing process of electrovacuum devices using the classical technology inasmuch that thermal desorption capabilities are exhausted in this cause and the further increasing of a temperature will case a damage or destruction of electrovacuum microwave elements. An application of pumps which are more powerful will increase the sizes of operating units. It is also connected with larger financial investments. Herewith, using more powerful pumps will allow increasing degassing velocity insignificantly, which is limited with exhaust tube capacity in this case. An application of non-thermal desorption types could be more effective way to accelerate degassing process of electrovacuum microwave devices. It allows to intensify gas separation. Considering all existing types of desorption we can distinguish an acoustic desorption, more precisely – an acoustically stimulated thermal desorption. This method is the most appropriate. It is easy to implement and doesn't require large financial costs. That why it allows to increase the efficiency of degassing of devices significantly.

The mechanism of an acoustically stimulated thermal desorption is rather complicated and consists of several simultaneous processes, with the advantages desorption acceleration and acceleration of a vacuum pumping process of gas components from a vacuum volume of electrovacuum microwave devices. An increasing thermal desorption intensity with using acoustic fluctuations could occur due to stimulation of a separation colloidal particles from surfaces inside the vacuum volume at the expense of its mechanical acceleration. It is known that the process of degassing device walls is parallel to the process of shaking off microscopic coarse-colloid particles from the inside walls of the device to a certain degree. Actually a significant quantity of foreign particles is located on all components of inside fittings of an item. They don't only slow down vacuum degassing process by closing surface areas by themselves, but they also appear as sources of gas flows. Another factor of accelerating degassing of electro vacuum microwave devices is directional speed up of desorbed molecules motion of a non-associated gas by an acoustic-surface-wave (ASW) which is spreading from degassed volume to lower pressure side. This effect is described in the sources [2, 3].

Presumably, the main mechanism of intensifying the gas components desorption is the compression and tension of the metal crystal lattice and deformation of intercrystalline gaps, caused by acoustic vibrations of the polycrystalline metal structure.

The one mechanism of the previously described acoustically stimulated thermal desorption is the closing and opening of intercrystalline gaps, which are "open" to atoms (primarily hydrogen) in the case when the walls of intercrystalline gaps are covered with at least one monolayer of sorbate with high volumetric concentration CS1 (figure 1(a), right). Acoustic oscillations are closing the intercrystalline gaps (figure 1(b), right) and causing instantaneous connection of sorbate layers on the gaps walls, simultaneously causing heating in the area of the newly formed contact (figure 1(b), left) up to temperature TS1=TS2, because sorption is an exothermic process. In our case we are dealing with "collective sorption" and the contact of "groups" of molecules located on opposite surfaces [5].

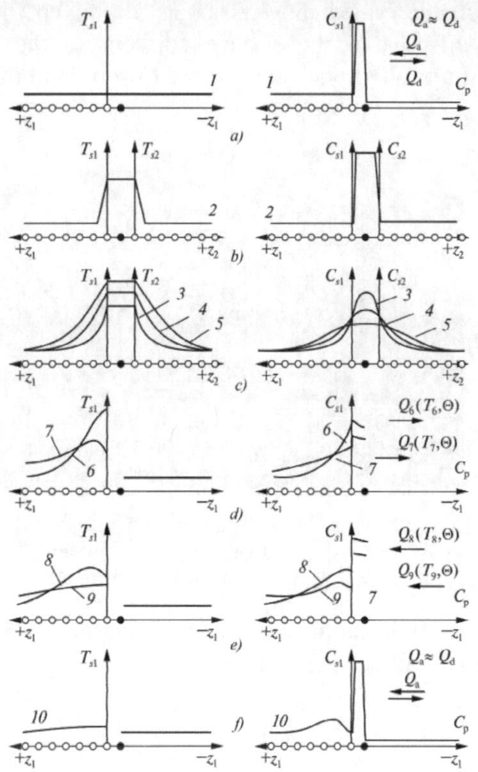

Figure 1. View plots of the distribution of temperatures T_S and volume concentrations C_S of the gases atoms in the acoustic vibration intercrystalline gaps.

This "collective sorption" process occurs at ultrasonic frequency and at the nanoscale distances. The result of this process is instant local heating of the polycrystalline structure of the acoustically stimulated material (figure 1(c), (d), left). Herewith, the diffusion processes take place both deep in the metal crystal lattice ntothe and on the surfaces of the intergranular boundaries, by stretching and compressing the crystal lattice. Finally, the described repeated a high frequency process with leads to the appearance of pulses of the gas flow (Qd) from the gap (figure 1(f), right). At this point, the material is not heated (temperature TS curve 10 in figure 1(f), left), because the desorption process comes with heat absorption.

The result is the appearance of a gas release flow (Qd) from the gaps between the crystallites, as shown in figure 2. Hence, schematically the acoustically stimulated thermodesorption can be represented as the process movement of thermally activated atoms in shrinking and stretching crystal lattice of metal, located in the acoustic field, and process "collective" sorption, causing the pulsed local heating of the crystallites surfaces.

Consider the scheme of hydrogen gas evolution from the surface layer of the polycrystalline structure of corrosion-resistant steel (figure 2) when exposed to ultrasonic vibrations [6, 7]. As shown in figure 2, crystallite sizes are in a fairly wide range from units to tens of micrometers, the intergranular boundaries width is from tens to tens of angstrems. Precisely in these intergranular boundaries, the gas sorbed on the surfaces of crystals and dissolved in the metal volume is mainly located [8–10]. Under the influence of longitudinal and transverse ultrasonic vibrations on crystallites there are forced elastic vibrations. In this case, the width of the intercrystalline boundaries also changes with the frequency of forced oscillations. Under certain conditions, there is a sharp increase in the amplitude of the elastic oscillations of the width of the boundaries between the crystals. In this case, the pulsed local heating of the surface of crystallites occurs and the mobility of "dissolved" hydrogen atoms is increased, as described above. The gas "dissolved" in the intergranular boundaries and the volume of grains diffuses to the surface adjacent to the vacuum medium, where its concentration in conditions of ultra-high vacuum is close to zero, and then it is desorbed and pumped out by the vacuum system.

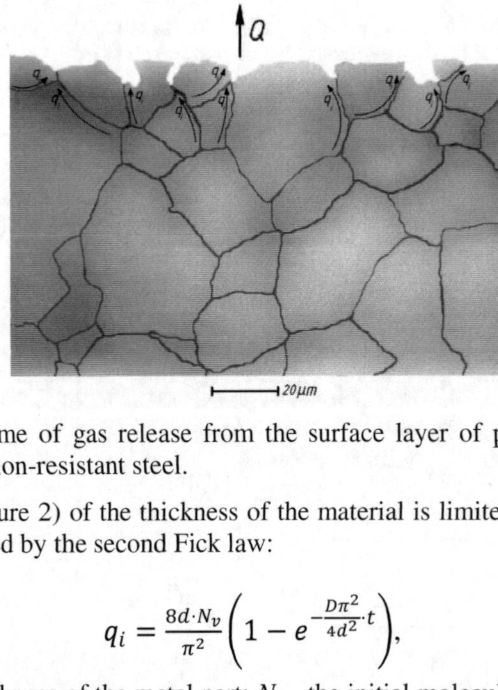

Figure 2. Scheme of gas release from the surface layer of polycrystalline structure corrosion-resistant steel.

The specific gas flow q_i (figure 2) of the thickness of the material is limited by the diffusion process, which in this case is described by the second Fick law:

$$q_i = \frac{8d \cdot N_v}{\pi^2}\left(1 - e^{-\frac{D\pi^2}{4d^2} \cdot t}\right), \tag{1}$$

where d – is a half the thickness of the metal part; N_v – the initial molecular concentration of the gas in the metal; D – the diffusion coefficient of the gas in the metal.

The gas diffusion coefficient in the metal is determined as follows:

$$D = \frac{1}{6} a \cdot \overline{u} \cdot e^{-\frac{E}{\mathrm{RT}}}, \tag{2}$$

where a – constant of a crystal lattice of metal; $\overline{u}$ – average thermal velocity of atoms; E – heat or activation energy of gas diffusion.

The efficiency of an acoustically stimulated thermal desorption has been proved experimentally. The essence of the experiments was in registration a gas components pressure increasing in a previously degassed volume of electrovacuum microwave device till $1*10^{-7}$Torr. In the first case trials were conducted without a heating of product housing. On the graphic chart (figure 3) there is a time dependence of residual gases pressure increasing in a volume of magnetron after a vacuum pump disconnection. The curve 1 shows a pressure increasing without an acoustic influence, the curve 2 – with an acoustic influence, accordingly. Point a and b denotes conditionally an identic saturation of vacuum volume with desorbed gases for the different time.

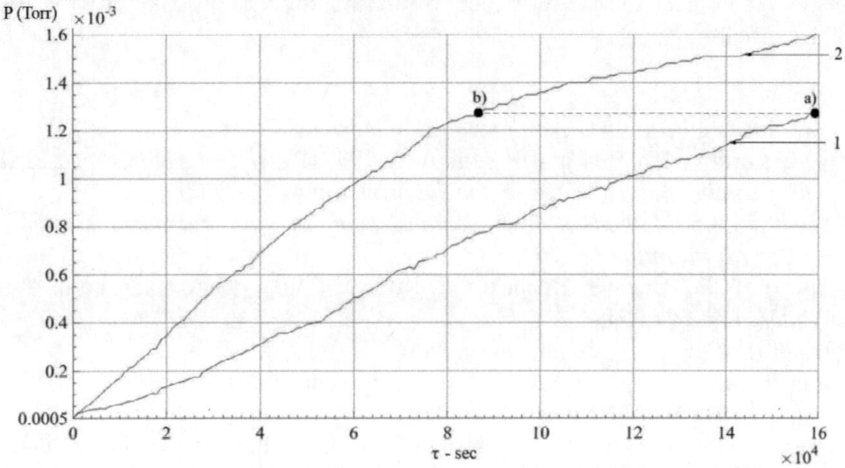

Figure 3. Graphic chart of a time dependence of residual gases pressure increasing in a volume of a magnetron: 1 – graph without ultrasonic; 2 – graph with ultrsonic.

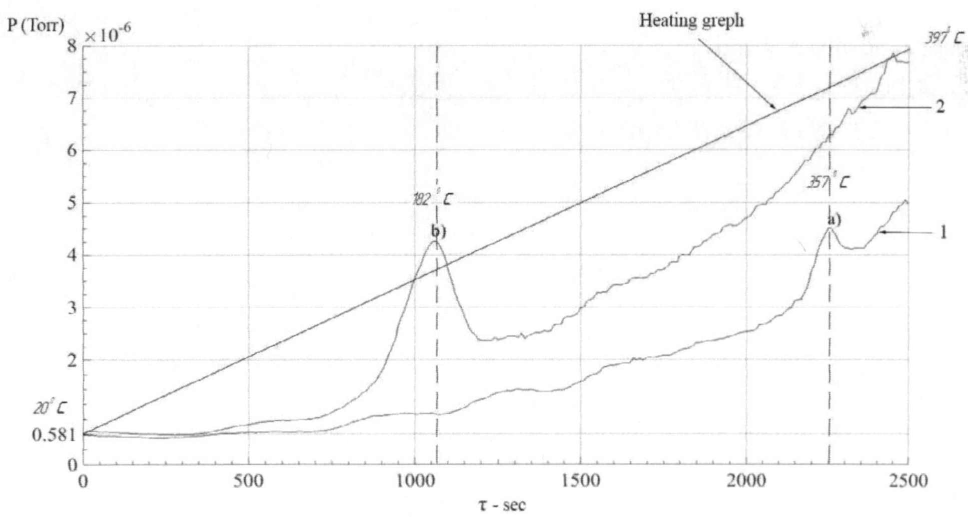

Figure 4. Graphic chart of a time dependence of residual gases pressure increasing in a volume of magnetron with a heat up: 1 – graph without ultrasonic; 2 – graph with ultrasonic.

In the second case a heat up of magnetron housing was added to a similar trial. On the graphic chart (figure 4) a residual gases pressure increasing in a volume of a magnetron is shown with increasing the temperature according to the linear law with a speed of 9 °C/min in the range of 20÷393 °C. The peak emission of desorbed molecules during a thermal desorption, peak a) will appear after 37,5 min. at the temperature 357 °C (1 – graph without ultrasonic). And during the acoustically stimulated thermal desorption peak b). will appear after 17,7 min. at the temperature 182 °C (2 – graph with ultrasonic). Therefore the duration of vacuum pumping out is reduced 2,12 times and the temperature of the effective water molecules desorption is reduced by 175 °C or 2 times [4].

Thereby we can draw a conclusion about the efficiency of an acoustically stimulated thermal desorption. Obviously suchlike decision allows to increase an intensity of gas components desorption and diffusion from the inside parts of machined product, reduce the electrovacuum devices degassing time duration and therefore to increase volumes of finished products output.

Acknowledgment
The authors express their grace to Maxim A Naumov and Alexey V Glushenkov for the assistance in the preparation of this article.

References

[1] Ostroumov G A and Fedotov G A 1973 *Acoustic journal* 1 76–9
[2] Makarov A, Luskinovich P N and Ryzhikov A 1997 *Method of pumping gases and vacuum pump for pumping gases. RF patent for the invention № 2079000*
[3] Aitkhozhin A 1995 *Method of high vacuum pumping and pump for its implementation. Application for invention № 95105038*
[4] Lee P, Lifanov N D, Petrov S, Hanbekov F, Kotlicki I Yu, Alenkov I, Zabelin A N, Sugars A 2018 *Patent RU 2 644 553*
[5] Deulin E A 2018 *Physics of vacuum mechanics*
[6] Kudryashov B A, Grib V V, Gaevskii V V, Dem'yanushko I V, Kartsov S K 2017 *Russian Engineering Research* **12** 1081–3
[7] Dem'yanushko I V, Kazantsev V F, Karagodin V I, Luzhnov Y M, Prikhod'ko V M 2017 *Russian Engineering Research* **12** 1084–6
[8] Komshina A V, Pomel'nikova A S and Fetisov G P 2015 *Russian Metallurgy* **13** 1158–60
[9] Alexander P, Andrey P, Alexander N and Elena N 2015 *Defect and Diffusion Forum* **363** 142–7
[10] Kurkin A S, Makarov E L, Kurkin A B, Rubtsov D E and Rubtsov M E 2017 *Metal Science and Heat Treatment* **3–4** 250–4

IOP Conf. Series: Materials Science and Engineering **387** (2018) 012033 doi:10.1088/1757-899X/387/1/012033

Investigation of a small-sized source parameters for magnetron sputtering with liquid-phase target

A A Khramova, A R Matanin, D A Polin, D D Vasil′ev and K M Moiseev

Bauman Moscow State Technical University, 105005, Moscow, Russia

E-mail: annakhramova3@gmail.com, k.moiseev@bmstu.ru

Abstract. The article describes an investigation of a small-sized source called NMSA-52M for magnetron sputtering with liquid-phase target. Causes and features affecting the discharge power required for the target melting are considered. Methods of increasing the discharge power of a given source are proposed. The design of the crucible reducing the intensity of the heat sink is given.

1. Introduction

Thick conductive films (from 20 to 50 μm) are widely used in electronics and power engineering: they are used in power microelectronics, in thermoelectric modules production, in hybrid and traction engines manufacture, as well as in wind power and rocket engineering [1–2, 14]. There are various methods of forming such coatings: copperplating, DBC technology, vacuum arc deposition and magnetron sputtering (MS). For a number of factors, such as the rate of deposition, the cleanliness of the coating, the economy and environmental friendliness, MS exceeds the other methods by 30–100 % [1–5]. However, during the MS process the biggest amount of energy transferred to the target goes to its heating. As a development of this method MS with liquid-phase target was invented. To implement this method, the target is placed in a refractory crucible and is insulated from the cooling system, due to what the target material melts, and thermal evaporation is added to the sputtering. As advantages of this method comparing to standard MS it is possible to list 5- to 10-fold increase in deposition rate, an improved cleanliness of the coating by a factor of 20 to 50, an energy consumption reduction to 50–100 V/Atom [1, 5–10].

To melt the target and initiate the "self-sputtering" process mode, it is necessary to achieve a certain discharge power. The main parameters affecting the discharge power are: the discharge voltage, the working pressure, and the value of the of the magnetic field induction on the target surface. The first two parameters are limited by the plant technological capabilities, while the magnetic induction tangential component value on the target surface is determined by the magnetron and magnetic system constructional features. Nowadays, the MS with liquid-phase target process is investigated using magnetrons with overall dimensions from 80 mm to 150 mm [8, 16]. The magnetron presented in this paper is small-sized (its target diameter is 50 mm), which is one of the reasons for requiring a larger magnetic field induction tangential component value to achieve the required discharge power for the target melting. Also, the design and the crucible material influence the discharge power supplied to the target [6, 10–14]. The aim of this paper is to investigate a small-sized source for magnetron sputtering with a liquid-phase target.

2. Magnetron NMSA-52M construction

The target diameter of small-sized magnetron NMSA-52M is 50 mm (2"). In the magnetron frame there are holes for water cooling with a diameter of 6 mm (figure 1(a)). The magnetic system consists of cylindrical and circular magnets made of SmCo material (figure 1(b)).

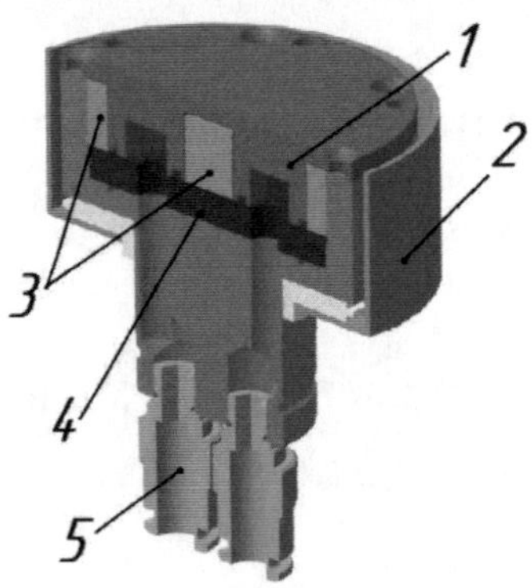

(a)

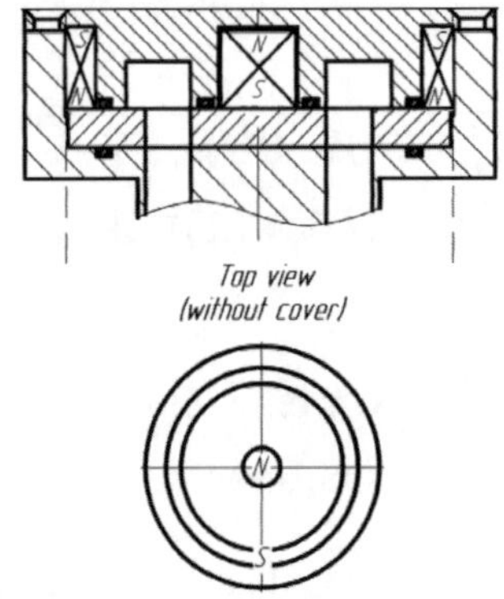

(b)

Figure 1. Magnetron NMSA-52M: section of the body (**a**): 1 – cover, 2 – frame, 3 – magnets, 4 – magnetic core, 5 – holes for cooling; general view (**b**).

During the magnetron sputtering with liquid-phase target process, the target material is often placed in a refractory crucible made of molybdenum which is located at the magnetron surface. The gap between the crucible and the cover is maintained by crucible legs (figure 2), which ensures thermal insulation of the crucible from the cooled magnetron cover [15].

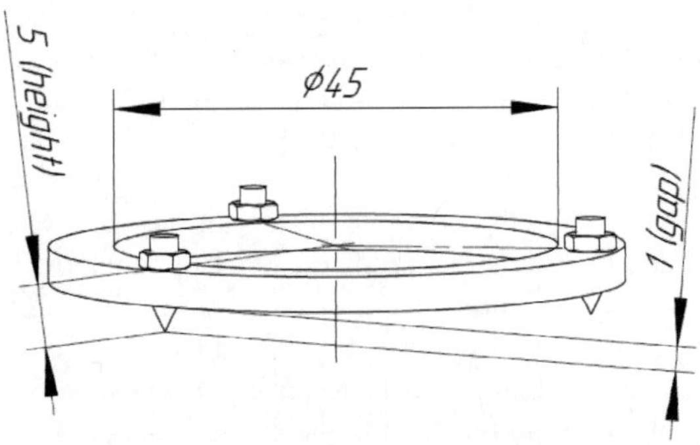

Figure 2. Molybdenum crucible with legs.

3. Magnetron NMSA-52M construction features analysis

3.1. Small overall dimensions

The magnetron for MS with liquid-phase target process presented in this article is small-sized, which allows its use it as a technological source in laboratory vacuum plants. However, in case of a small-sized source a larger value of the magnetic field induction on the target surface is needed to achieve a discharge power sufficient to melt the target. This may be due to the fact that the value of the Lorentz force sufficient to hold electrons in the plasma region is inversely proportional to the electron trajectory radius. Earlier studies were carried out for the presented magnetron [17], where a magnetic system with two plasma regions was modeled and implemented to increase the value of the magnetic field induction tangential component (figure 3). When the magnetron was powered, discharge combustion was observed only in the magnets outer the ring region, in spite of the fact that the value of the magnetic field induction tangential component on the target surface in both regions was equal (figure 4).

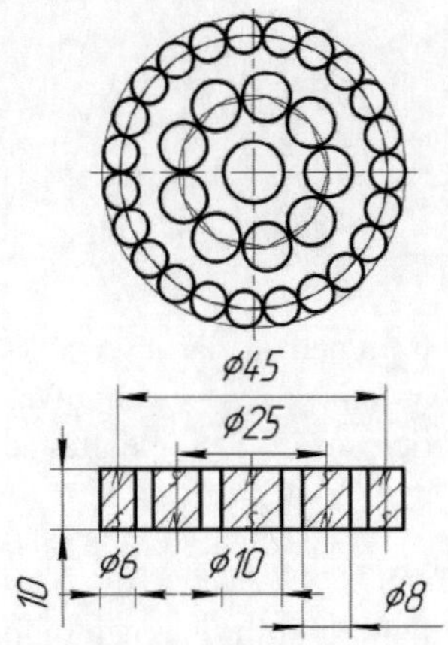

Figure 3. The magnetic system construction of the magnetron NMSA-52M with two plasma regions.

Despite the increase of the magnetic induction tangential component value, the use of such magnetic system did not lead to a discharge power increase. It can be concluded, that the discharge power is affected both by the magnetic field induction tangential component value on the target surface and the magnetron discharge plasma region geometric parameters. Experiments carried out to study the ring magnet diameter influence on the discharge power showed that a decrease in the ring magnet diameter leads to a decrease in the discharge power, despite the increased tangential component of the magnetic field induction on the target surface. Thus, the small overall magnetron dimensions prevent achieving larger discharge power values.

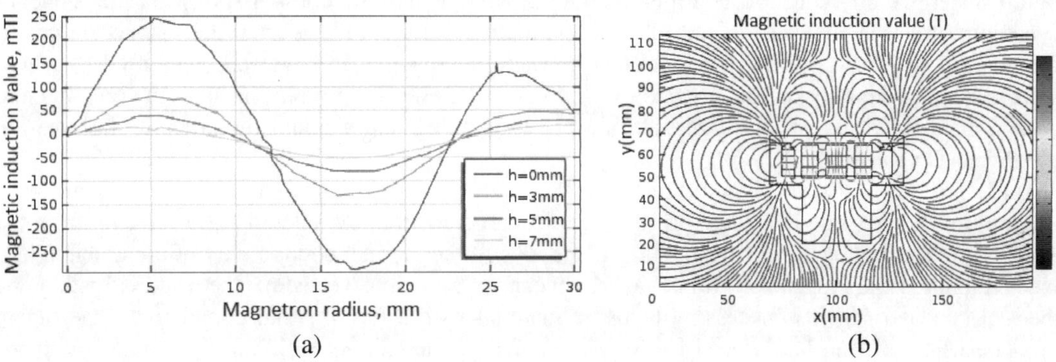

Figure 4. Magnetic system with two plasma regions magnetic field characteristics: the dependence of the magnetic field induction tangential component on the distance to the magnetron center (**a**); the magnetic field distribution (**b**).

3.2. Magnetic system

To obtain the required discharge power value, it is not sufficient to increase only the magnetic field induction tangential component on the target surface. It is also necessary for the magnetic system to create a magnetic field uniform distribution in the plasma region [17]. For this purpose, in the described case of a planar magnetron with axial symmetry, the magnetic system must also possess axial symmetry. For magnetron NMSA-52 such a system is an annular magnetic system (figure 5), consisting of a central cylindrical and peripheral annular magnets mounted on a magnetic core made of ferromagnetic material.

In addition, the strongest magnet material should be used. In the original construction magnets made of SmCo material are used, its residual magnetic induction value is 1.07 T. To increase the magnetic field induction tangential component, one can use, for example, magnets made of NdFeB N45 material, which has a higher induction value of 1.35 T.

IOP Conf. Series: Materials Science and Engineering **387** (2018) 012033 doi:10.1088/1757-899X/387/1/012033

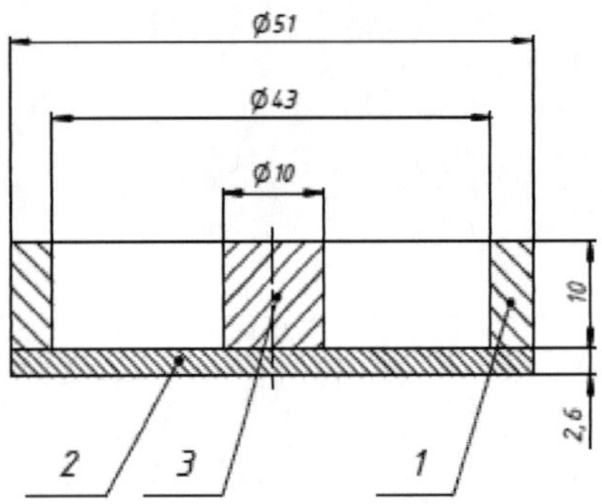

Figure 5. Magnetron NMSA-52M annular magnetic system: 1 – annular magnet; 2 – magnetic core; 3 – central cylindrical magnet.

3.3. Crucible material and construction

The crucible must be made of refractory material, for which the melting point is much higher than the target material melting point. Therefore the most suitable materials for the crucible are molybdenum and tantalum. The latter is a more technological material for making crucibles by means of cold stamping.

The crucible thickness, the gap size when installed on the magnetron cover, and the target thickness (or material granules) contribute to the magnetic field induction value on the target surface. To ensure the discharge power required for the target melting and initiate the process in "self-sputtering" mode, the crucible thickness and the gap size should be minimal. A new construction of crucible shown on the figure 6 was developed to increase discharge power. The crucible is made from a 0.5 mm thick tantalum sheet which reduces the crucible thickness by 2 mm in comparison with the standard molybdenum crucible showed in figure 2. The tantalum crucible is screwed to the magnetron cover, which ensures a more stable electrical contact between the magnetron cover and the crucible. It was established experimentally that a gap size of 1 mm is sufficient to thermally insulate the crucible and provide a guaranteed gap in the target melting process, taking into account the crucible deformation due to thermal expansion.

In order to solve the problem of target melting using a small-sized magnetron, a decrease is considered in the necessary discharge power value for target melting by reducing heat removal from the crucible. Two heat removal mechanisms take place: due to the thermal conductivity of the crucible legs connected to the cooled magnetron frame, and by means of radiation. To reduce the heat dissipation due to thermal conductivity, it is necessary to use a material with a low thermal conductivity coefficient value. The heat sink can be reduced by increasing the legs length, as well as their contact with the crucible at points, where the temperature differs the most from the molten target temperature, which explains the unusual crucible legs shape (figure 6). To reduce the heat removal intensity due to radiation, it is necessary to use a material with a low radiation coefficient value [8, 16].

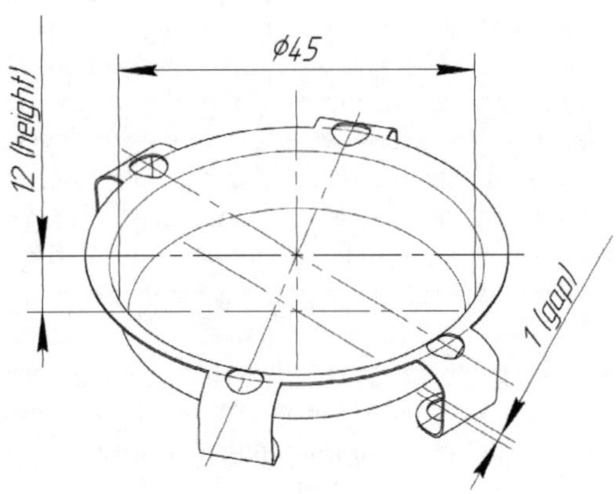

Figure 6. Tantalum crucible with legs.

IOP Conf. Series: Materials Science and Engineering **387** (2018) 012033 doi:10.1088/1757-899X/387/1/012033

4. Results and conclusions

Varying the voltage and working gas pressure in the range determined by the technological capabilities of the installation does not allow to achieve the required discharge power value when using a small-sized source for MS with liquid-phase target. A discharge power increase in this case is possible due to a magnetic field induction increase on the target surface, at the same time a decrease in the source size for MS with liquid-phase target leads to an increase in the magnetic field induction necessary value to achieve the required discharge power value. The magnetic field above the magnetron surface must be uniform to guarantee axial symmetry of the magnetic system, provided the magnetron is co-planar and axially symmetric. The discharge power is influenced both by the magnetic field induction tangential component value on the target surface, and by the discharge plasma region geometric parameters, the small dimensions of the magnetron hinder the achievement of significant discharge power values. Another way to achieve a sufficient power to melt the target is to reduce the heat removal intensity from the crucible, which leads to a decrease in the discharge power required for the target melting. The proposed crucible design makes it possible to reduce the heat sink intensity.

References

[1] Sidelev D V, Yuryeva A V, Krivobokov V P, Shabunin A S, Syrtanov M S and Koishybayeva Z 2016 *Journal of Physics: Conference Series* **741** 012193
[2] Asainov O H, Bainov D D, Krivobokov V P, Romanenko S E and Chernyatina A A 2011 *Russian Physics Journal* **54** 158–61
[3] Erofeev E V, Kagadei V A, Kazimirov A I and Fedin I V 2015 *Proc. Int. Sib. Conf. on Control and Communications SIBCON (Omsk)* 7147054
[4] Kashkarov E B, Nikitenkov N N, Syrtanov M S, Sutygina A N, Shulepov I A and Lider A M 2016 *Applied Surface Science* **370** 142–8
[5] Bleykher G A, Krivobokov V P and Yuryeva A V 2015 *Russian Physics Journal* **4** 431–7
[6] Danilin B S, Kakurin M V and Minaichev V E 1978 *Elektr. Tekh. Ser. Mikroelektr.* **24** 84–7
[7] Bleykher G A, Krivobokov V P and Tret'yakov R S 2011 *Izv. Vyssh. Uchebn. Zaved. Fiz.* **11/2** 148–53
[8] Musil J, Lestina J, Vclek J and Tolg T 2001 *Journal of Vacuum Science and Technology A* **19** 420–4
[9] Yuryeva A V, Shabunin A S, Korzhenko D V, Korneva O S and Nikolaev M V 2017 *Vacuum* **141** 135–8
[10] Bleykher G A, Krivobokov V P, Yurjeva A V and Sadykova I 2016 *Vacuum* **124** 1–7
[11] Breauer G, Szyszka B, Vergohl M and Bandorf R 2010 *Vacuum* **84** 1354–9
[12] Bleykher G A and Krivobokov V P 2012 *Nucl. Instr. Meth. Phys. Res.* **292** 34–9
[13] Sakurai Y, Nakajima R and Nakamura H 2009 *Solid State Phenomena* **154** 175–9
[14] Khramova A A, Matanin A R, Polin D A, Vasiliev D D and Moiseev K M 2018 *13th Int. Scientific and Technical Conf. "Vacuum Technique, Materials and Technology"* ed S B Nesterov (Moscow: NOVELLA) pp 124–8
[15] Akishin M Y, Rogozhin A A, Vasiliev D D and Moiseev K M 2016 *The Future of Mech. Eng. in Russia: a collection of works of the 9th All-Russian Conf. of Young Scientists and Specialists* (Moscow: BMSTU Publishing) pp 290–2
[16] Musil J, Rajský A, Bell A J, Matouš J, Čepera M, and Zeman J 1996 *J. Vac. Sci. Technol. A* **4** 2187–91
[17] Khramova A A, Matanin A R, Vasiliev D D and Moiseev K M 2017 *Materials of the 24th Scientific and Technical Conf. with the participation of foreign specialists "Vacuum science and technology"* ed S B Nesterov (Moscow: NOVELLA) pp 187–92

Influence of hydrogen content in working gas on growth kinetics of hardened layer at ion nitriding of 16MnCr5 and A290C1M steels

Y G Khusainov, R S Esipov, K N Ramazanov, E L Vardanyan, P V Tarasov and S R Shekhtman

Ufa State Aviation Technical University, 450008, Ufa, Russia

E-mail: ramazanovkn@gmail.com

Abstract. The impact of hydrogen content from 10 to 30% in the mixture of operating gas environment on the growth kinetics and structure of a hardened layer during ion nitriding of martensite and pearlite steels has been studied. It is stated that hydrogen in the vacuum chamber in the range from 10 to 30% contributed to a 1.2-time increase of nitrogen diffusion rate in 16MnCr5 steel and 1.3-time increase in A290C1M steel. The dependence of the hardened layer thickness and microhardness distribution over the depth on the gas environment composition is stated. The structure of the nitrided layer of steels at different hydrogen contents during ion nitriding is studied.

1. Introduction

At present most critical parts of machines are manufactured from high-strength alloyed structural steels. An example of such a part is a gear wheel, the working surface of which is a tooth bearing. This tooth bearing of the part is subjected to alternating and contact stresses, which determines the application of surface hardening techniques in production. One of the widely used hardening techniques applied to the surface to structural steels is glow discharge nitriding (ion nitriding) [1]. However, the significant disadvantage of ion nitriding is the long duration of processing (25–30 h) [2].

It is stated in [3–5] that during glow discharge nitriding the composition of the gas environment significantly effects the structure, properties and growth rate of the hardened layer. Use of operating environments, which contain nitrogen and hydrogen (up to 5 %), allows increasing the rate of diffusion saturation in the surface due to chemical etching of oxides on the processed surface [3]. Use of argon (up to 80 %) in the operating environment also enables considerably increasing the growth rate of the diffusion layer, as under the effect of electrical field argon ions diffuse the nitride layer on the surface of a cathode-part, which results in the reduction of the barrier-layer effects [4, 5]. However, at present there are no published data on the impact of multi-component saturating environments on the rate of diffusion growth and mechanical properties of the strengthened layer.

Thus, the study on the impact of the operating gas environment composition on the growth kinetics of the nitride layer and regulation of the ratio between the gas mixture components with the aim to enhance the efficiency of ion nitriding of structural steels is an urgent problem.

2. Experimental procedures

Heat-treated structural steels 16MnCr5 and A290C1M were subjected to ion nitriding. Low-carbon steel

16MnCr5 of the martensite type is widely used to produce critical parts of machines, for example, high-loaded gear wheels. Nitride forming elements allow using it as a nitride one to increase the wear resistance of machine parts. Steel A290C1M is the pearlite one and is used to produce gear wheels, cylinder barrels, worms, spindles, barrels and other parts operating at temperatures up to 450 °C.

Ion nitriding was performed on the modernized set ELU-5M designed to conduct thermal and chemical and thermal treatments in the vacuum. Before nitriding the samples were subjected to ionic clearing for 10 minutes at P=10 Pa in the argon environment, the surface temperature did not exceed T=250 °C. The diffusion saturation was performed in the operating mixture of nitrogen N_2, argon Ar and hydrogen H_2. The operating gas environment composition and nitriding regimes are listed in table 1.

Table 1. Operating gas environment composition and ion nitriding parameters.

№ treatment regime	Composition of gas, %			Temperature, °C	Pressure, Pa	Duration, h
	Ar	N_2	H_2			
1	60	30	10			
2	55	30	15			
3	50	30	20	550	150	6
4	45	30	25			
5	40	30	30			

The microhardness measurements of the surface were carried out on the microhardness tester Struers Duramin-1. The statical load value applied to the diamond indenter for 10 seconds was 980.7 mN (100 g). The microstructure studies of the samples were performed with the help of an optical microscope Olympus GX-51. Prior to metallographic studies the cross section samples of steels were etched in 3 % solution of HNO_3.

3. Results and discussion

Figure 1 displays the optical images of the surface layer microstructure of steel 16MnCr5 samples after ion nitriding with the hydrogen content in the operating gas environment in the range from 10 to 30 %.

The microstructure analysis of the 16MnCr5 steel samples after nitriding shows that the microstructure of all the samples consists of three zones: I – zone of nitrides, II – diffusion zone, III – matrix. The transition between the diffusion zone (II) and matrix (III) is smooth. The zone of nitrides (I) presumably consists of iron nitrides and alloying elements, and the diffusion layer consists of solid solution of nitrogen in α-Fe [7, 8]. It is stated that during nitriding of 16MnCr5 steel samples (when the hydrogen content changes in the operating gas environment in the range from 10 to 25 %) (figure 1(a)–(c)) a modified layer forms with a thickness (*h*) of 185 μm. This is explained by the fact that the nitrogen content in the gas mixture is constant of 30 % (table 1). However, after nitriding of the 16MnCr5 steel sample with the hydrogen content in the operating gas environment of 30 % (figure 1(e)), there is no nitride layer (I), the thickness of the diffusion zone is h ~180 μm. The absence of the nitride layer is possibly connected with its spalling during specimen preparation. Spalling of a nitride later is conditioned by its hydrogen embrittlement, as the hydrogen content in the saturating atmosphere is about 30 %. The reduction of the diffusion zone thickness is caused by a low argon content in the operating gas environment (40 %), which results in the efficiency loss of the barrier nitride layer diffusion.

Figure 2 displays the optical images of the microstructure of the surface layer in steel A290C1M samples after ion nitriding at different contents of the operating gas environment. The microstructure analysis shows that in all the samples the transition between the diffusion zone (II) and the matrix (III) is smooth. The nitride zone (I) consists of iron nitrides, the diffusion zone (II) is the solid solution [7,

8]. There were no microcracks and cleavages on the surface. It is stated that during nitriding of A290C1M steel with the hydrogen content in the operating gas environment of 10, 15, 20, 25, 30 % modified layer with a thickness of h ~195, 230, 240, 250 and 230 μm, forms, respectively. The decrease of the hardened layer thickness in the sample after nitriding with a hydrogen content of 30 % in the operating gas environment is explained by a low volume of argon in the saturating atmosphere.

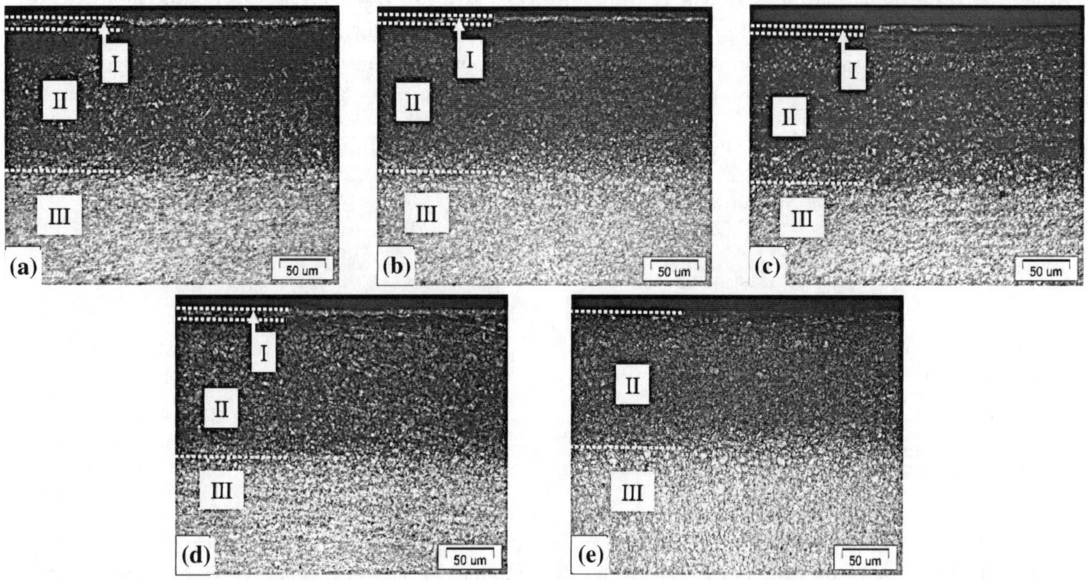

Figure 1. Microstructure of 16MnCr5 steel samples after ion nitriding with different content of hydrogen in the operating gas environment: **(a)** – 10 % H_2; **(b)** – 15 % H_2; **(c)** – 20 % H_2; **(d)** – 25 % H_2; **(e)** – 30 % H_2.

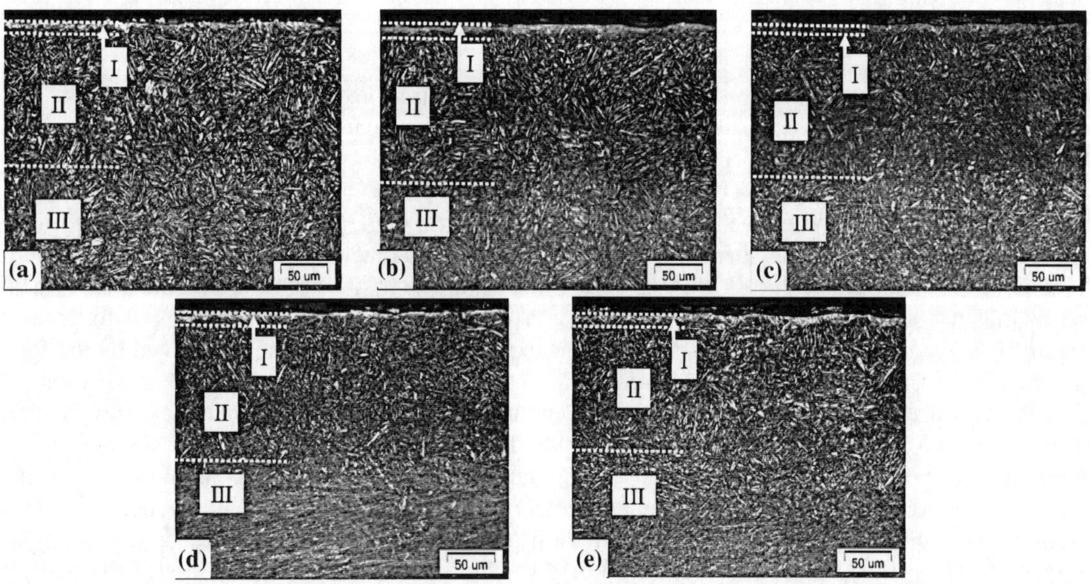

Figure 2. Microstructure of A290C1M steel samples after ion nitriding at different contents of hydrogen in the operating gas environment: **(a)** – 10 % H_2; **(b)** – 15 % H_2; **(c)** – 20 % H_2; **(d)** – 25 % H_2; **(e)** – 30 % H_2.

Figure 3 displays the dependence of the microhardness distribution over the thickness of the surface layer of 16MnCr5 and A290C1M steel samples after ion nitriding with different contents of hydrogen in the gas environment.

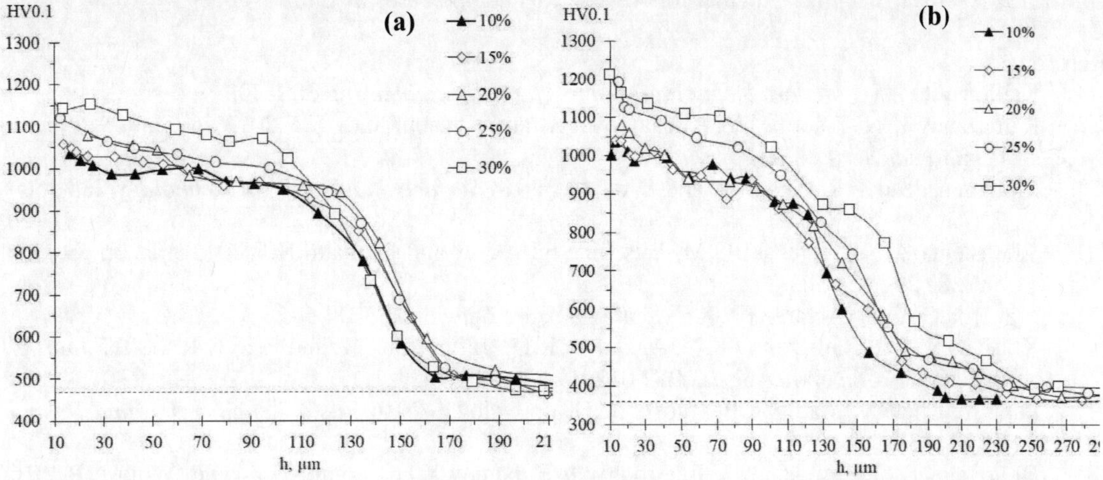

Figure 3. Dependences of microhardness distribution over the thickness of the surface layer after ion nitriding with different contents of hydrogen in the gas environment: 16MnCr5 (**a**), A290C1M (**b**).

The microhardness distribution graphs over the depth of steel 16MnCr5 after ion nitriding (figure 3(a)) show a smooth change of the microhardness over the thickness of all the studied compositions of the operating gas environment. When the hydrogen content in the vacuum chamber changes from 10 to 30% some the hardened layer thickness increases from 170 to 190 µm and the surface microhardness enhances from 1025 to 1150 $HV_{0.1}$. The microhardness increment is conditioned by the increase of microstresses in the modified layer under the impact of enhanced concentration of dissolved nitrogen in α-Fe [9, 10]. Thus, the hydrogen content increase in the operating gas environment from 10 to 30 % contributes to hardness enhancement of the hardened layer of 16MnCr5 steel by 2–2.5 times.

The dependence graphs of microhardness distribution over the depth in steel A290C1M after ion nitriding (figure 3(b)) demonstrate a smooth change in the microhardness over the thickness in all the studied compositions of the operating gas environment. When the hydrogen content increases in the vacuum chamber from 10 to 25 %, the hardened layer thickness enhances from 195 to 240 µm, and the surface microhardness grows from 1000 to 1175 $HV_{0.1}$. When the hydrogen content in the gas environment is about 30 %, the modified layer thickness decreases to 240 µm, and the hardened layer microhardness is 1200 $HV_{0.1}$. Thus, the hydrogen content increase in the gas environment from 10 to 30 % contributes to enhancement of the hardened layer hardness in steel A290C1M by 2.5–3.5 times.

4. Conclusions

As a result of the studies on the impact of the hydrogen content in the operating gas environment on the growth kinetics of the hardened layer during ion nitriding of steels 16MnCr5 and A290C1M, it was states that:

– the growth kinetics of the hardened layer during the hydrogen content increase in the gas environment from 10 to 25 % for 16MnCr5 steel practically does not change, and that for A290C1M steel increases by 1.3 times. The use of the operating gas atmosphere with the hydrogen content of about 30 % and argon of 40 % for the studied steels results in the reduction of the growth kinetics of the hardened layer, as in the mixture of gases the argon content reduces, and the barrier effect takes place due to formation of a nitride layer. The nitride layer prevents nitrogen atoms to penetrate the matrix.

– the hydrogen content increase in the gas environment from 10 to 30 % results in the enhancement of the efficiency of oxides etching on the steel surface and increase of the dissolved nitrogen concentration in α-Fe. As a result the microhardness increment occurs in the hardened layer for 16MnCr5 steel by 2–2.5 times and that for A290C1M steel increases by 2.5–3.5 times.

References

[1] Kaplun V G 2007 *Modern technologies in mechanical engineering* 388–403
[2] Ramazanov K N, Zolotov I V, Khusainov Y G and Khusnutdinov R F 2015 *Journal of Physics: Conference Series* **652**
[3] Mohammadzadeh R, Akbari A and Drouet M 2014 *Surface and Coatings Technology* **258** 566–73
[4] Allensteina A N, Lepienski C M, Buschinellic A J A and Brunatto S F 2013 *Applied Surface Science* **277** 15–24
[5] Figueroa C A and Alvarez F 2006 *Applied Surface Science* **253** 1806–9
[6] Khusainov U G, Ramazanov K N, Agzamov R D, Vardanyan E L and Esipov R S 2017 *Journal of Physics: Conference Series* **872** 012017
[7] Fang Y, Chen X, Madigan B, Cao H and Konovalov S 2016 *Fusion Engineering and Design* **103** 21–30
[8] Konovalov S V, Kormyshev V E, Gromov V E, Ivanov Y F, Kapralov E V and Semin A P 2016 *Journal of Surface Investigation* **10** 1119–24
[9] Zuev L B, Danilov V I, Konovalov S V, Filip'ev R A and Gromov V E 2009 *Physics of the Solid State* **51** 1137–41
[10] Gromov V E, Gorbunov S V, Ivanov Y F, Vorobiev S V and Konovalov S V 2011 *Journal of Surface Investigation* **5** 974–8

Investigation of current-voltage characteristics of a glow discharge in molecular gases

A S Kiselev, E A Smirnov and A A Zagoskin

Department of electronic instruments and devices, Saint Petersburg Electrotechnical University "LETI", 197376, Saint Petersburg, Russia

E-mail: askiselev@etu.ru

Abstract. In this work current-voltage characteristics of a glow discharge flowing in narrow long tubes are investigated. The studies were carried out on a specially developed layout, which includes discharge tubes of various diameters. The model was connected to the pumping and filling system, which allows varying the gas content and working pressure. The discharge was excited in pure carbon dioxide gas, as well as mixtures based on it: carbon dioxide with nitrogen in the ratio 1:1, carbon dioxide with helium in the ratio 1:8, and also carbon dioxide, nitrogen and helium in the ratio 1:1:8. An approximating expression is obtained for the CVC in the form of a power function including the discharge current, the length of the discharge gap, and the longitudinal potential gradient calculated in the automated mode.

The glow discharge is widely used in various fields of science and technology. One of the possible variants of using a glow discharge is the excitation of the active gas medium of lasers. The parameters of radiation of glow discharge lasers are significantly influenced by the electrical characteristics of gas-discharge tubes. First of all, this refers to the type of current-voltage characteristics (CVC) of extended discharge gaps of lasers. The current-voltage characteristic of glowing lasers makes it possible to judge the state of their active medium, including the degree of leakage of atmospheric gases [1]. The impedance of the plasma, which affects the dynamic stability of the glow discharge lasers current circuit, depends to a considerable extent on the values of the static and differential resistances determined by the CVC. In addition, the form of the current-voltage characteristic has a significant effect on the dynamic properties of the discharge, which determine the stability of the current-carrying chain of glow discharge lasers [2].

Knowledge of the behavior of the CVC of the discharge tubes of the glow discharge lasers makes it possible to control the deviation of the parameters of the active elements of lasers from the nominal values and to carry out the necessary correction. However, the individuality of technical solutions implemented in the glow discharge lasers makes it difficult to assess and predict the effect of changes in the discharge conditions on the CVC of lasers. To exclude the need for additional experiments for the specific discharge conditions of a laser, it is proposed to use approximating expressions for the current-voltage characteristic of a glow discharge lasers, taking into account the variety of discharge conditions for different types of lasers.

Layout for the study of discharge characteristics is four discharge tubes with immobile molybdenum anodes (figure 1). The length of the positive column is fixed and is 350 mm. The internal diameters of the tubes are 3; 5; 10; 15 mm. Such dimensions are typical for CO_2-lasers. Before filling the tubes with a working gas mixture they were carefully pumped out. The pressure control during evacuation was made by a thermocouple sensor connected to a vacuum gauge [3].

Experiments to study the discharge were carried out for different gas contents: pure carbon dioxide, carbon dioxide with nitrogen in the ratio 1:1, carbon dioxide with helium in the ratio 1:8, and also carbon dioxide, nitrogen and helium in the ratio 1:1:8. The pressure of the gas mixture ranged from 0.5 to 10 Torr and was controlled with an oil U-shaped manometer.

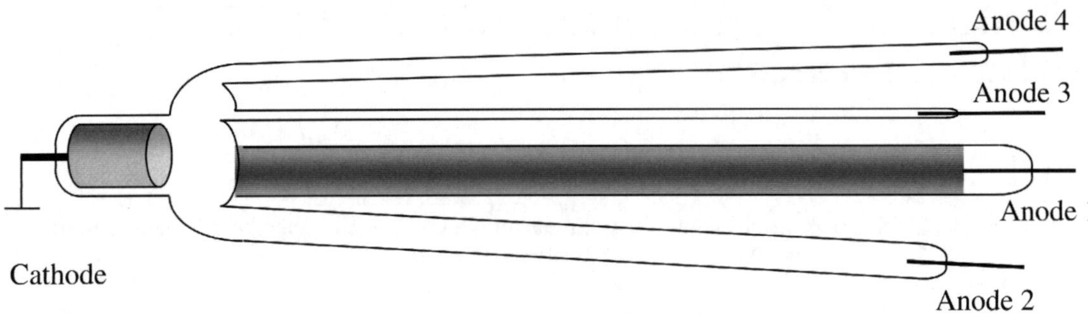

Figure 1. Layout for the study of discharge characteristics.

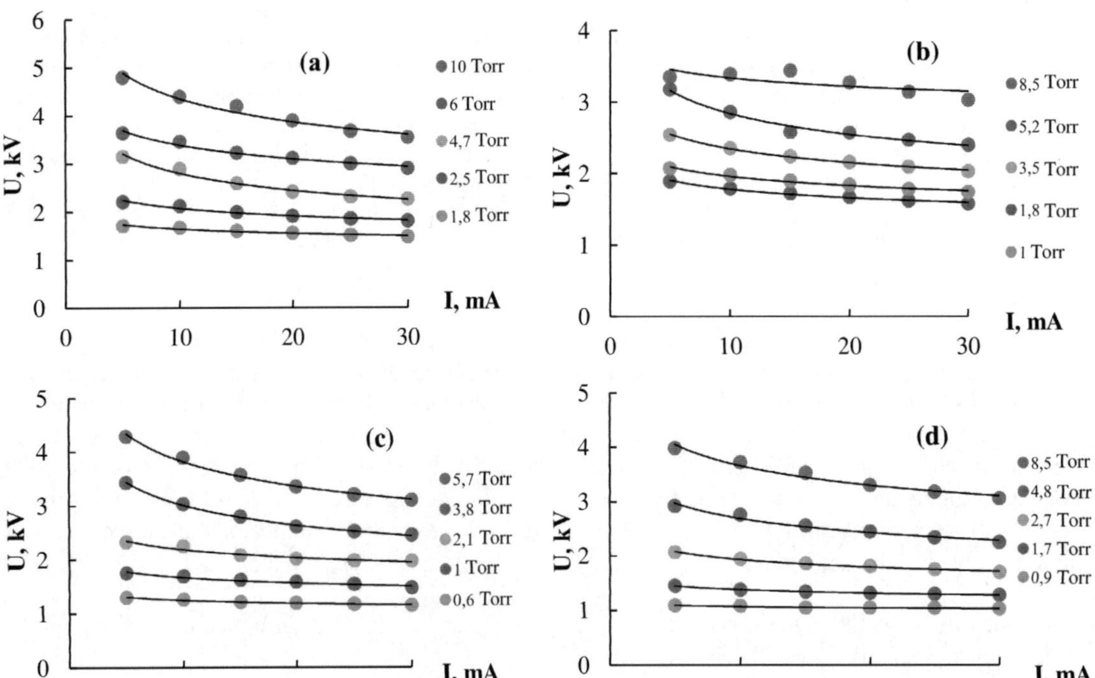

Figure. 2. Current-voltage characteristics: **(a)** – mixture CO_2:N_2:He=1:1:8, tube diameter 3 mm; **(b)** – CO_2:N_2 =1:1, tube diameter 15 mm; **(c)** – pure CO_2, tube diameter 5 mm; **(d)** – CO_2:He =1:8, tube diameter 10 mm.

In the ranges of working currents, the current-voltage characteristic of a glow discharge in long tubes is of the same type of falling character. Figure 2 presents examples of the current-voltage characteristic for various fillings and diameters of discharge tubes. The general tendency is an increase in the discharge voltage with an increase in the total pressure p and a decrease in the diameter of the discharge tube *d*. This is due to the increase in the gradient of the potential of the positive column E_z [4].

For the approximation of the working sections of the falling CVC, the functional relationship between the voltage drop across the tube *U* and the discharge current *I* are expediently displayed in the form of a power law:

$$U - U_k = aE_z LI^b, \quad (1)$$

where U_k is the cathode potential drop; *a*, *b* is approximation coefficients calculated on the basis of a set of experimental current-voltage characteristics; E_z is the longitudinal gradient of the potential in the positive column; L is the length of the positive column. When choosing the form of expression (1), it was assumed that the near-anode potential drop is close to zero [5]. The cathodic drop in the potential depends on the material and design of the cold hollow cathode and the kind of gas. In the normal glow discharge mode, typical for glow discharge lasers conditions, U_k is 100–250 V and is not a current function. In the narrow extended discharge capillaries used in the glow discharge lasers, E_z turns out to be high and amounts to a few–tens of kilovolts per meter [6]. Therefore, the main part of the total voltage drop falls on the positive column, the longitudinal gradient of the potential in which depends on the discharge current.

The value of E_z in the approximating expression was determined as:

$$E_z = \frac{3kT_e\sqrt{\chi}}{2e\lambda_e},$$

where χ is coefficient of accommodation of electrons; λ_e is the mean free path of electrons in the gas mixture; T_e is the electron temperature; *k* is the Boltzmann constant [7].

During the experiments under certain discharge conditions, standing striations were observed in the tubes. They are alternating regions in which either gas excitation or ionization predominates. Strata were predominantly observed at low pressures. The spatial period of the stratum (the distance between the luminous regions of the discharge) depended on the diameter of the tube, and also on the discharge current. Under certain conditions (the growth of the discharge current, the heating of the tube), there were situations when the strata ceased to fill the entire discharge gap, or disappeared altogether. The conditions for the existence and appearance of strata are certainly of interest and require the carrying out of separate full-scale studies.

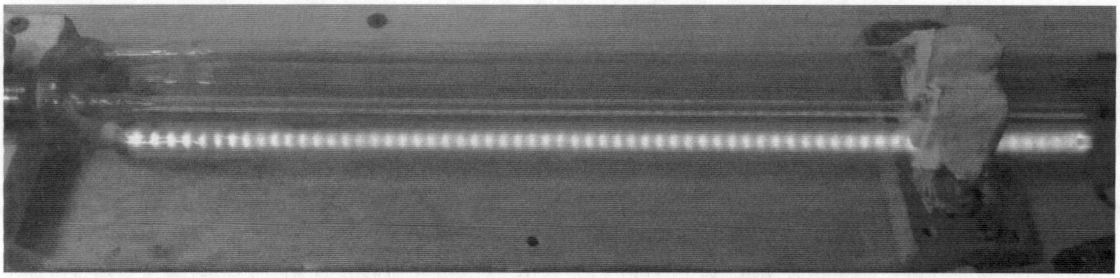

Figure 3. The emergence of striations in the discharge of pure carbon dioxide.

References

[1] Kiselev A S, Kostrin D K, Lisenkov A A and Smirnov E A 2017 *Journal of Physics: Conference Series* **789** 012027
[2] Chernigovskiy V V, Kostrin D K, Martsinukov S A and Lisenkov A A 2016 *Vakuum in Forschung und Praxis* **6** 34–7
[3] Smirnov E A 2013 *Proceeding of Saint Petersburg Electrotechnical University "LETI"* **10** 6–10
[4] Kiselev A S and Smirnov E A 2014 *Vacuum technique and technology* **1** 56–9
[5] Granovsky V L 1971 *Electrical current in gas* (Moscow: Science)
[6] Kiselev A S and Smirnov E A 2017 *Journal of Physics: Conference Series* **929** 012053
[7] Kiselev A S and Smirnov E A 2017 *Journal of Physics: Conference Series* **872** 012092

Phase difference meter between the alternating current and voltage components for determining the impedance of plasma of glow discharge lasers

A S Kiselev, E A Smirnov and A S Fadeev

Department of electronic instruments and devices, Saint Petersburg Electrotechnical University "LETI", 197376, Saint Petersburg, Russia

E-mail: askiselev@etu.ru

Abstract. In this work meter for the phase difference of the variable components of the discharge current and the voltage drop was developed and tested. The operation of the meter is based on the transformation of harmonic signals into rectangular pulses, and further determination of the time shift between them. The possibility of using this meter for monitoring the dynamic resistance of glow discharge lasers is shown.

Gas-discharge devices are widely used in various fields of science and technology. Among the widest range of such devices, a special place is occupied by instruments with an extended positive column of a glow discharge, to which, in particular, glow discharge lasers belong. Glow discharge lasers is actively used for technological and metrological purposes [1, 2]. However, due to physical and structural features, the laser radiation has a temporary instability caused by the influence of various factors. This, in turn, involves the use of glow discharge lasers in conjunction with systems to stabilize the power of their radiation [3].

Methods of stabilizing the power of laser radiation are divided into passive and active. In practice, active methods are often used, since passive methods do not take into account all the factors that affect the instability of the laser radiation power. Active stabilization methods suggest negative feedback. One of the options for stabilizing the power of laser radiation is to adjust the pump level. In glow discharge lasers, this process reduces to the modulation of the discharge current. Such a system is called a current power stabilization system. Let us assume that at a time t the laser had a power level P_0 at a current I_0, which corresponds to point A (Figure 1). However, due to external influences, the energy characteristic can shift and the power $P_0 + \delta P$ (point B) will correspond to this current. The principle of current stabilization of power consists in forcibly changing the discharge current by such a value δI so that the radiation power level again is equal to P_0 (point C). In this case, the output power level of the laser radiation will be at a stabilized level [4].

The discharge conditions realized in glow discharge lasers correspond to the region of sharp drop in the current-voltage characteristics characteristic of the gas gap, when there is a transition stage from free electron diffusion to ambipolar. Since glow discharge lasers operate on direct current, one can use the method of small perturbations to obtain expressions for the total dynamic resistance. In accordance with this, the current and voltage of the discharge will, respectively, have the form: $U = U_0 + u$, $I = I_0 + i$. In these expressions I_0, U_0 are stationary values of the current and voltage drop at the discharge, i and u are the variable components of the discharge current and voltage, where $u << U_0$, $i << I_0$. Since the glow discharge plasma has reactive properties, there is a phase shift between the values of *u* and *i*. In this case, the dynamic resistance of the discharge will be defined as the total resistance to alternating current [5]:

$$Z_d = \frac{u}{i} e^{j\Delta\varphi} = R + jX,$$

where *R* and *X* are the active and reactive components of the discharge impedance; $\Delta\varphi$ is the phase shift between the alternating current and voltage components [2].

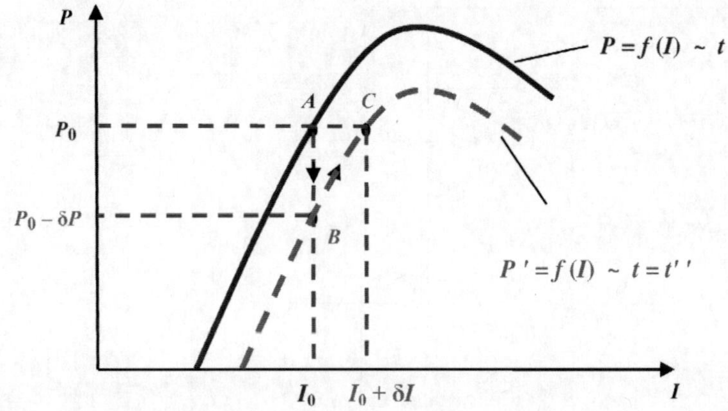

Figure 1. Explanation of the principle of the current stabilization of radiation power.

The basis of the experimental study of the dynamic discharge resistance is the measurement of the values of the alternating current and voltage components at a fixed frequency, as well as the phase shift between them. In the circuit, the discharge burns at a constant current, and is modulated by an external sinusoidal signal fed from a high-voltage oscillator with tunable frequency. Variable components of current and voltage are taken from the measuring resistors and amplified. Next, the phase shift between the signals is determined using a phase difference meter [6].

To register the difference between the phases of the current and the voltage drop of the discharge gaps of lasers, it is necessary to simultaneously measure and compare them with each other. The paper suggests the following version of this task. First, it is necessary to convert the sinusoidal signal into rectangular pulses of the same frequency (figure 2). For this purpose, the chip MAX942EPA + serves, in one case it contains two comparators.

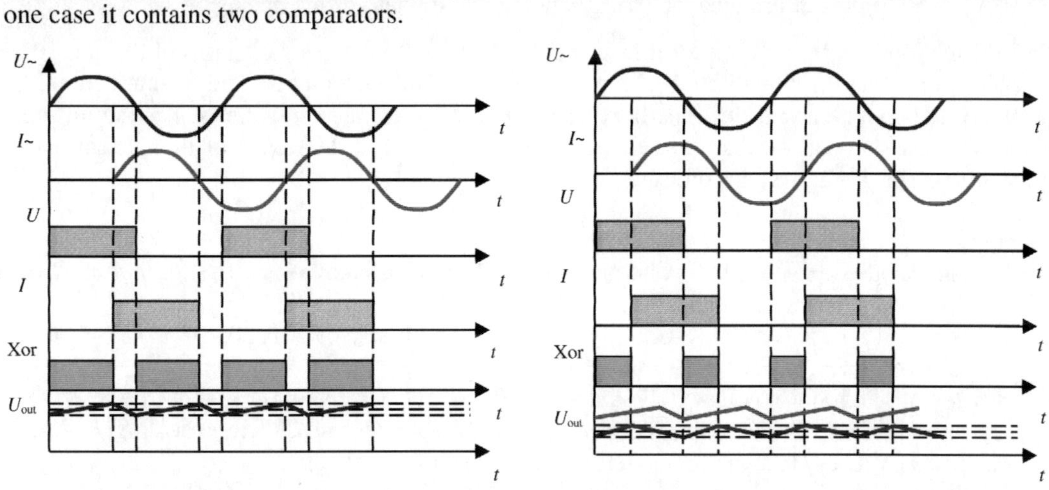

Figure 2. Explanation of the principle of operation of the phase difference meter.

The input of the comparators is fed with signals proportional to the alternating components of the discharge current and the voltage drop. Then the signals from the comparators go to the logical element "excluding or", at the output of which rectangular pulses are learned. This scheme operates on the principle of pulse width modulation: the greater the phase shift between current and voltage, the lower the duty cycle of the pulses obtained. The output signal from the logic element can be fed to the oscilloscope, or smoothed using an RC filter. The level of the smoothed signal will be proportional to the phase difference between the signals [7].

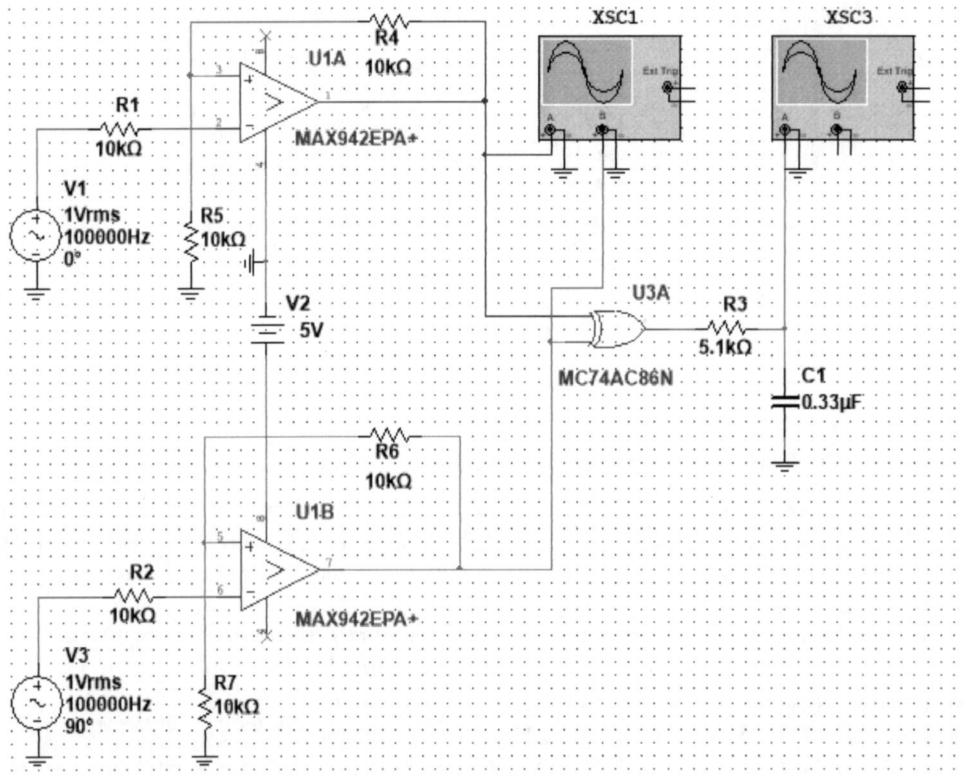

Figure 3. Principal electrical scheme of the phase difference meter.

In the course of this work, a circuit for measuring the difference between the phases of the current and the voltage drop across the discharge gap of the laser was developed and manufactured, and preliminary tests were carried out. The next stage of work is planned to connect a phase difference meter to mock-ups of gas-discharge lasers and to study the phase difference of the current and the voltage drop of the discharge gaps of lasers.

References

[1] Martsinukov S A, Kostrin D K, Chernigovskiy V V and Lisenkov A A 2016 *Journal of Physics: Conference Series* **735** 012049
[2] Martsinukov S A, Kostrin D K, Chernigovskiy V V and Lisenkov A A 2016 *Journal of Physics: Conference Series* **729** 012023
[3] Kiselev A S and Smirnov E A 2017 *Journal of Physics: Conference Series* **929** 012053
[4] Smirnov E A 2005 *Proceeding of Saint Petersburg Electrotechnical University "LETI"* **2** 43–8
[5] Granovsky V L 1971 *Electrical current in gas* (Moscow: Science)
[6] Kiselev A S and Smirnov E A 2017 *Journal of Physics: Conference Series* **872** 012092
[7] Kiselev A S and Smirnov E A 2014 *Vacuum technique and technology* **1** 56–9

IOP Conf. Series: Materials Science and Engineering **387** (2018) 012037 doi:10.1088/1757-899X/387/1/012037

Influence of the technological development on the adhesion increment of metallic thin film coatings to ceramic substrates of aluminum oxide

L L Kolesnik, Myo Kyaw Hlaing and Zaw Phyo Aung

Bauman Moscow State Technical University

E-mail: kolesnik@bmstu.ru

Abstract. In this journal are considered that the comparing of the adhesion strength with and without Ti layer between substrate and current layers. Comparing of the DC magnetron and pulse magnetron with various pressure in vacuum chamber. With Ti sub layer adhesion strength 10 times more than without adhesion layers. The best value of adhesion strength was obtained by using with a direct current magnetron and pressure in the chamber 8.8×10^{-2} mbar. A series of experiments was conducted with differences substrate position and maximum and minimum distance between substrate and target to prevent the destruction of the substrate when coating with a hot cathode target, it is necessary to limit the thermal effect on it.

1. Introduction

Power electronic modules producing by the domestic industry are contained in the structure of metal-ceramic compound elements. These compounds are required for adhesion strength, sustainability to temperature changes. Typical applications: aircraft navigation systems, elements of the commutations board and etc. A typical module of power devices shown in figure 1 and consists of a commutations board on which conductive paths are formed, electronic components and cooling radiators are installed.

Literature analysis requires the necessary parameters to the power modules, the main important are the working temperature and the adhesion of the films to the substrate. During the exploitation of power modules, they are exposed to thermal and mechanical stresses, including vibration. This leads to the destruction of components from overheating and mechanical damage. Often damages occur in the substrate-conductive coating structure. So we need to choose the substrate material for removing the heat and preventing the damage which was describes above.

At the moment, these connections are performed on the basic of thick film technology, especially uses direct bonded copper DBC-technology and other various atmospheric methods. These modules are used in the field of technology associated with the need to transmit high-current signals. Thin films technology ensures generally the absence of the several types of thick film technologies and on the other indicator surpasses them. Transition to thin film technology for creating the basis soldering elements of vacuum devices and power modules will be reduced the percentages of rejects and will be improved the quality of these connections.

Figure 1. Typical commutation board and view of the power electronics module.

Since 2016 at Bauman Moscow State Technological University has been taken modernization of laboratory apparatus for processing thin film coating technology into vacuum. As a result, the installation was received two independent sources: thermal evaporator and magnetron spraying system. The modernization process consisted of the mounting of gas and magnetron spraying system, which provided the technology base for processing the development of thin film coating. This makes it possible to form multilayers film coatings in a single vacuum.

Special flange was designed and manufactured for apparatus of magnetron system. Cooling water continuously supplied to magnetron system along a close circuit and we have been used that cooling system for prevention of overheating and its demagnetization. However, the cooling system is realized so that was cooled magnets, but the target of magnetron by itself was not cooled and the process was heated temperature about 900 °C.

At present, various types of ceramics are used as substrate materials for the fabrication of the circuit board of power modules. In the following table show the properties of the basic materials for substrate selection. Important characteristics are mechanical strength, thermal conductivity and coefficient of thermal expansion. The coefficient of thermal expansion should be close to the film coating in order to avoid the destruction of the metal-ceramic structure when the temperature changes. Substrate of aluminum nitride is the most promising candidate for improving heat dissipation and has high thermal conductivity 10 times higher than aluminum oxide (Al_2O_3), thermal expansion coefficient close to the coefficient of thermal expansion of silicon, good mechanical properties and non-toxicity, environmentally friendly and relatively inexpensive specific durability is 5–7 times lower than the specific stability of ceramics from beryllium oxide). In terms of these important characteristics, the preferred material is aluminum nitride. However, it shows much worse adhesion performance compared to aluminum oxide.

The purpose of this work is to determine the temperature dependence of the substrate position in the chamber and the angle of spraying with magnetron sputtering. Increasing the adhesion strength of ceramic substrates for the manufacture of power electronics devices and electro vacuum devices.

2. Experimental details

A series of experiments was conducted to investigate the application of coatings in a laboratory installation. In this paper, alumina substrates are used. We chose Ti material for the adhesive layer from many popular materials, for example, Cr, Ta and Al and. etc. The adhesive layer was applied using the method of magnetron sputtering of direct current and in the pulsed mode. The coating of current films was preliminarily coated by the impulse magnetron sputtering with different frequencies and filling coefficients. The copper layer was formed by thermal evaporation, a power mode of 0.20 kV was determined for each experiment. In these regimes, a series of experiments was performed to apply multilayer $Cu/Ti/Al_2O_3$ films by two methods in a single vacuum cycle. Before coating, the substrate surface was cleaned in the ultrasonic bath for about 10 minutes, and the temperature was 25–27 °C. With ultrasonic cleaning, large grease dirt is removed due to intense local mixing under the action of shock waves created in the solvent. The disadvantage of chemical surface preparation is also the fact that this treatment does not in principle ensure the removal of physically adsorbed moisture, the more chemisorbed. Multilayer film on ceramic substrate was coated with the following conditions: Argon gas – 30 sccm; Voltages – 650 V; Coating time – 5 min; Distance – 75 mm; Power – 380 W.

3. Results and discussion

First sires of experiment is single layer current film coating on aluminum oxide substrate and then multilayer was coated on the same substrates for comparing these two films for adhesion strength. The obtained samples were investigated for adhesion strength by peel test method.

Adhesion strength with Ti layers was 10 times greater than the single-layer copper coating. It was shown that Ti inserting as adhesion layer which are used for soldering components (copper, tin-gold etc.).

To determine the influence of the application regimes of the adhesive sub layer, studies were carried out under various conditions of the power source and pressure during application. For the Ti sub layer with DC magnetron 650 V and pulse mode magnetron was chosen.

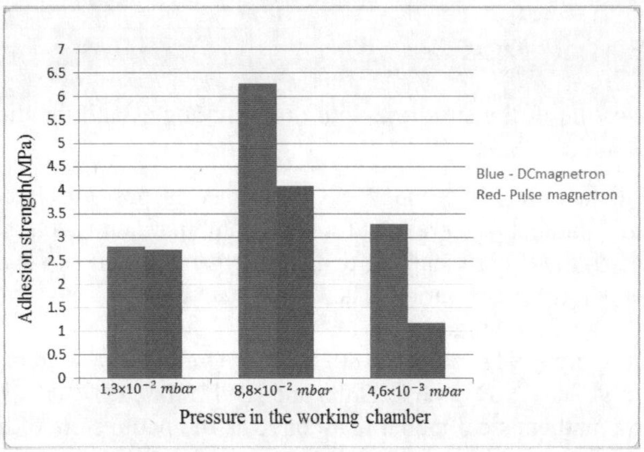

Figure 2. Comparing the adhesion strength with DC magnetron and pulse magnetron various pressures in working chamber.

The best value of adhesion strength was obtained by using with a direct current magnetron and pressure in the chamber 8.8×10^{-2} mbar.

A series of experiments was conducted to study the deposition of coatings with the temperature regime. Different frequencies and coefficient filling with pulse magnetron sputtering f -30 kHz D- 50%, f -30 kHz D -70%, f -90kHz D-70 and f-70kHz D-55 was chosen. The magnetron operated in the following mode: power - 380 W, Working gas-Argon 30 sccm, Target - titanium. The distance from the target to the substrate is 75 mm. The substrate temperature was measured with a thermocouple for every 20 seconds. At a frequency of 20 kHz and a fill factor of 80 %, the highest substrate temperature is obtained. It is determined that the substrate is heated to a temperature of 220 °C or higher, and at the initial stage the heating rate reaches up to 40 °C/min.

During the processing of regimes, cases of specimen destruction were observed. Since the deposited material comes from a hot Ti target heated to 1000 ℃, it has been suggested that the destruction is due to thermal effects. It was found that due to the diffuse scattering of the material, due to the high pressure in the deposition area, the back side of the substrate is dusted. High heating rate at the initial stage, leads to the destruction of films coatings. The substrate temperature was measured for different positions of the substrate holder. The maximum distance from the target to the substrate was 110 mm, and the minimum distance between them was 75 mm.

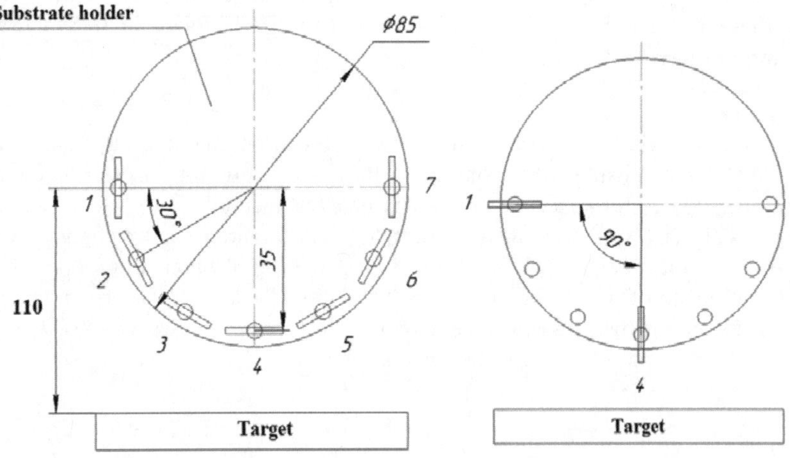

Figure 3. The location of the substrate with different angles and the distance from the target to the substrate.

In these experiments, a full factorial experiment was performed at positions 1 and 4. Two experiments were carried out, each with 3 parallel observations for each combination of factors in order to improve the accuracy of the experiment. In the experiment, the distance between the substrate and the target varies, for each distance value the spraying angle is varied. The output parameter is the temperature of the substrate. Deposition regimes describe the following.

The substrate temperature are (d=110mm, α=180° is 75 Degree Celsius), (d=75mm, α=180° is 93 Degree Celsius), (d=75mm, α=90° is 109 Degree Celsius) and (d=110mm, α=90° is 153 Degree Celsius).

It is necessary to build a mathematical model that connects the parameters of the heat flux, which will allow to determine the optimum position of the substrate to achieve the required film growth rate with minimal thermal impact.

4. Conclusion

The insertion of an adhesive titanium sub layer leads to an increase in the adhesive strength of a copper film on an alumina substrate by nine or more times.

The best adhesion strength is achieved when the adhesive under layer of titanium is applied by DC magnetron at pressure of 8.8×10^{-2}mbar.

To prevent the destruction of the substrate when coating with a hot cathode target, it is necessary to limit the thermal effect on it.

The magnitude of the thermal impact is determined by the geometric characteristics of the equipment (distance) and the technological parameters of the process (temperature, pressure, evaporation rate

The developed mathematical model will allow to determine the optimum position of the substrate during the deposition from the hot cathode to ensure the required minimum thermal effect on the substrate.

References

[1] Kolesnik L L, Zhuleva T S, Predtechenskiy P O, Myo Kyaw Hlaing and Zaw Phyo Aung *Journal of Physics: Conference Series* **872**

[2] *Handbook of deposition technologies for films and coatings: science, applications and technology* ed. P. M. Martin (Burlington; Oxford: William Andrew/Elsevier) 2010 912
[3] Luchkin A G and Luchkin G S *Cleaning substrate surfaces for coating deposition by plasma methods in vacuum*
[4] Kolesnik L L, Dejev B A and Seleznev A B 2013 *Vacuum science and technology: Materials of the 20th scientific and technical conference with the participation of foreign specialists* 165–7
[5] *US Patent 5571332 Electron Jet Vapor Deposition System* 1996
[6] Danilin B S and Syrchin V K 1982 *Magnetronnyye raspylitel'nyye sistemy* (Radio i svyaz') 72
[7] Marakhtanov M K 1990 *Magnetronnyye sistemy ionnogo raspyleniya (Osnovy teorii i rascheta)* (M.: BMSTU) 76

Features of magnetron sputtering of a doubled Ta/Ti target

A E Komlev, E S Shutova and A A Komlev

St. Petersburg Electrotechnical University “LETI”, 197376, St. Petersburg, Russia

E-mail: a.komlev@mail.ru

Abstract. In this work, investigation of magnetron sputtering process of the doubled Ta/Ti target is presented. Such a construction allows to simultaneously sputter two materials, forming complex film structures on a substrate. The results of computer simulation and experimental investigation of the thermal behavior of the target, current-voltage and spectral characteristics of discharges, and characterization of the fabricated coating samples are given.

1. Introduction

The modern development of micro- and nanotechnologies opens new perspectives and possibilities for creating multicomponent film structures that significantly expand the functionality of the raw materials. Adding to a TiN film materials such as Si, Al [1] makes it possible to increase microhardness by more than 3 times, to the value of 42 GPa [2]. Heusler’s alloys such as Fe_2NbSn, $Co_2Cr_{0.6}Fe_{0.4}Al$, PtMnSb, Co_2FeSi, Co_2FeAl [3] are actively created and studied, as they define a new stage of spin electronics development.

One of the traditional ways of forming such structures is the method of magnetron sputtering by several confocal sources. Schematically, this process is shown in figure 1. The method of co-sputtering allows to accurately control the elemental composition of the deposited films.

As an alternative method, a number of authors consider the possibility of magnetron sputtering of composite ("mosaic") targets [4–6]. The mosaic target is a matrix of a water-cooled cathode made from a material A (figure 2), in which the body has varying geometry inserts made of matter B (figure 2). The dimensions of the target (radius R_t), insert parts (radius R_s) and their relative positions, e.g., circle radius R_v, are determined based on the requirements to the structure and composition of the coating. Using reactive gases makes possible to form films compounds, for example, metal oxides (Ti-VO [7]), which are actively used in optoelectronics.

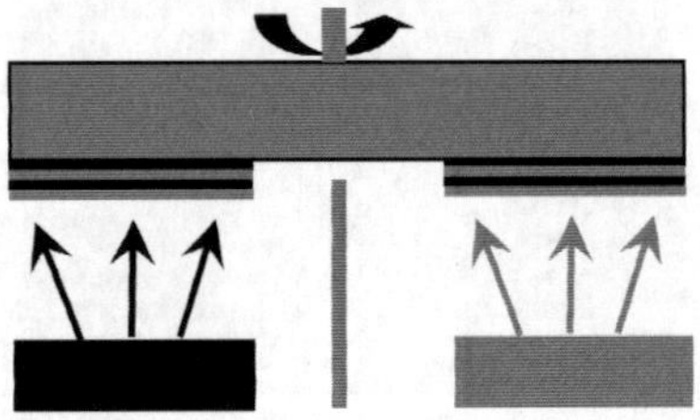

Figure 1. Sputtering scheme with two sources.

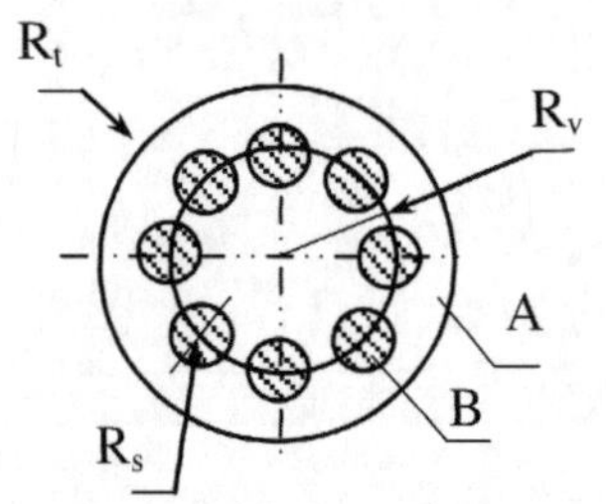

Figure 2. Geometric model of the "mosaic" target.

The application of such structures is challenging because of the need for accurate preliminary selection of the geometry of the insert parts to achieve the required composition of the deposited film. In the article [4], authors state that properly selected materials and their parameters allow to obtain uniform coatings over a large area. The advantage of using a "mosaic" target is especially evident in cases where it is necessary to obtain films containing elements with low mutual solubility or a large difference in the melting points.

As an alternative to the "mosaic" target the authors [8] suggest using a more efficient and simple doubled-target design: bottom water-cooled target of material A and the top one made of material B with apertures of a predetermined size and geometric shape in the erosion zone. This design allows the simultaneous sputtering of two materials. By varying the gap between the targets, it is possible to vary the temperature of the top plate, thereby realizing the "hot" target mode, which leads to an increase in the effects of thermionic emission and evaporation.

2. Investigation of the gas discharge emission and volt-ampere characteristics during sputtering of the hot double target in an inert medium

To study the physico-chemical processes that occur during sputtering a double target with a hot top cathode, a magnetron sputtering system was developed (figure 2). To implement the hot regime, the bottom target, made from tantalum sheet (99.99 %) with a diameter of 130 mm and 1 mm thickness, was fixed with a gap of 0.2 mm on a metal water-cooled base. The top target, which was a titanium sheet with 130 mm diameter and 1 mm thick, was fixed above the tantalum sheet with a gap of 1 mm. To realize the possibility of simultaneous sputtering of 2 targets, 8 holes with a diameter of 17 mm were cut out in a titanium disk and located along a radius of 29.2 mm corresponding to the middle of the erosion zone with an internal diameter of 40 mm and an external one of 76.6 mm. The experimentally determined ratio of the sputtered areas of the bottom tantalum (2054 mm^2) and the top titanium target (1786 mm^2) is 1.15 to 1. The sputtering coefficients of the selected metals are close to each other: 0.69 for tantalum and 0.71 for titanium.

The residual pressure in the chamber was less than 10^{-2} mtorr, the argon pressure during sputtering was 1 mtorr, the current density varied from 13 to 123 mA/cm^2.

As a result of studying the composition of the discharge, emission spectra were obtained by sputtering the target in an inert argon atmosphere. During the processing of these spectra, the most informative lines of the emission spectrum of the discharge were chosen and their intensity was measured. The dependence of the emission line intensity on the current density is shown in figure 3a. The corresponding I-V characteristics are given in figure 3(b).

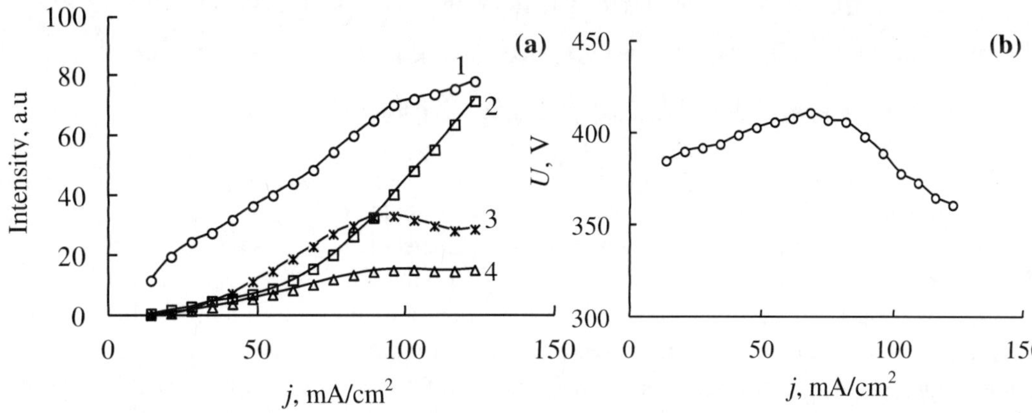

Figure 3. Dependence of the intensity lines of 1 – Ar, 2 – 656.2 nm line, 3 – Ti, 4 – Ta on the discharge current density (**a**); I-V characteristic upon sputtering of the doubled Ta + Ti target (**b**).

The obtained I-V characteristic corresponds to the standard form of the curve when a hot metal target is sputtered in an argon atmosphere. When the current density exceeds 80 mA/cm^2, the voltage decreases, which is possibly due to the increase in thermionic emission. The gas composition of the discharge, which can be characterized by the results of optical emission spectroscopy (figure 3(a)), also has peculiarities at 80 mA / cm^2 current density. If the intensity of the neutral argon, tantalum and titanium emission lines increases gradually with increasing current density from 13 to 80 mA/cm^2, after reaching 80 mA/cm^2 saturation of the Ta line with a simultaneous decrease in the intensity of the Ti line is observed . When analyzing the emission spectra of the discharge, a line with a sharp increase of intensity in the current density range of 70–123 mA/cm^2 was observed at 656.2 nm. According to the reference data, this wavelength corresponds to neutral hydrogen, which can be present in the chamber in minimum volumes in the form of residual gases. Therefore, the observed effect requires further detailed study.

Thus, the analysis of the gaseous atmosphere composition through the results of the optical emission spectroscopy and the I-V characteristics of the discharge indicates the characteristic features observed in a gas discharge at a current density of more than 80 mA/cm^2 and reaching an top target temperature of 1400–1500 K.

3. Modeling of the thermal regime of the "hot" composite target

At the initial stage, the modeling of thermal processes during sputtering of a doubled target in the COMSOL Multiphysics software package was performed. This software allows to consider the dependence of the material parameters from temperature in the form of analytical, piecewise continuous functions, as well as interpolation of a number of data during the calculation.

Figure 4 shows a general view of the developed 3D model of the magnetron. The main elements are:

- the top titanium target of a given thickness with holes in the erosion zone;
- bottom tantalum target of a given thickness, with an annular zone of erosion;
- cooled stainless steel base;
- stainless steel fasteners (screws and washers that determine the size of the gap).

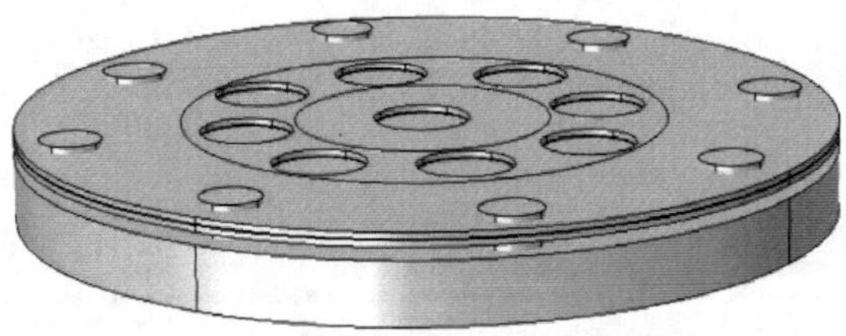

Figure 4. 3D model of a magnetron with a hot composite target.

A one-dimensional problem is described by a system of homogeneous differential equations:

$$\rho_i c_i \frac{\partial T(x,t)}{\partial t} = \lambda_i \frac{\partial^2 T(x,t)}{\partial x^2}, \quad i = 1, 2, 3, 4 \tag{1}$$

where: T – temperature; c_i, ρ_i, λ_i – heat capacities, densities and coefficients of thermal conductivity of the target materials (i = 1, 2), cooled base (i = 3), fastener and chamber elements (i = 4), and x is the spatial coordinate. For calculations were used material parameters from the embedded library.

The problem was solved for various values of the discharge current density taken from the experimental I-V characteristic. One of the boundary conditions implies a cooled wall of the base (T = 283 K). A typical example of the calculation result for the target is shown in figure 5. The maximum target temperature is reached at a current density of 123 mA/cm^2 and is 1631 K (figure 6).

4. Experimental determination of target temperature

The target temperature was determined experimentally from the results of measuring the total emission spectrum of a gas discharge and the radiation from a heated target.

IOP Conf. Series: Materials Science and Engineering **387** (2018) 012038 doi:10.1088/1757-899X/387/1/012038

The ISM3600 spectrometer was used to measure the emission spectra of a gas discharge. As can be seen from figure 7, when the power is increased, the spectrum of the gas discharge undergoes a significant change. The detected optical spectrum recorded by sputtering a "hot" target is a combination of the spectra of glow discharge and thermal radiation of a target, the intensity of which increases with increasing current density.

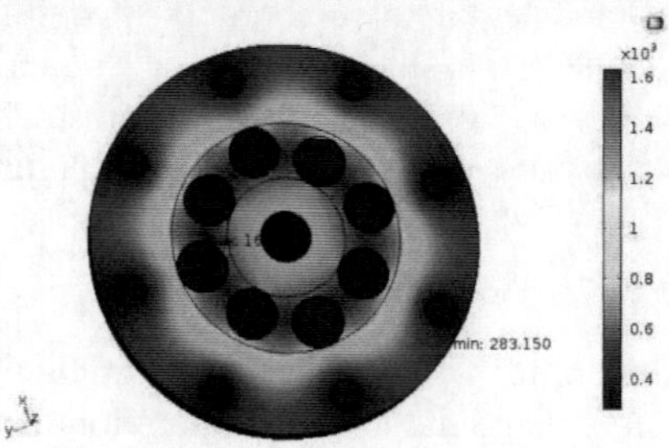

Figure 5. Temperature distribution on the surface of the target.

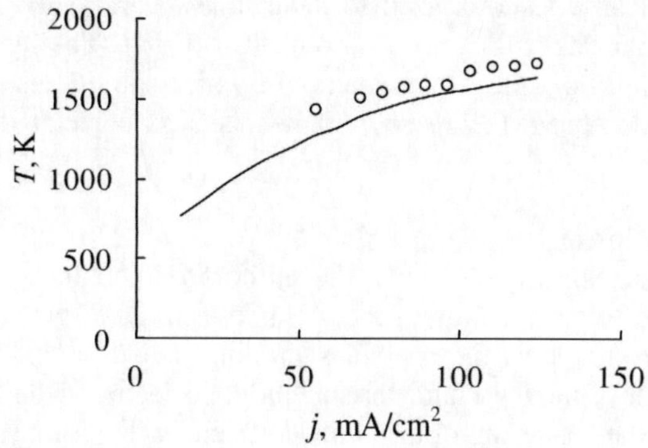

Figure 6. Dependence of the temperature of the composite target on the current density, the solid line corresponds to the model and points are from experiment.

Temperature calculation method is based on comparison of the theoretical emission spectrum of an absolutely black body at a given temperature with the position of minima on the total spectrum. Calculation of the temperature is carried out according to the well-known Planck's formula:

$$I(\lambda) = \frac{2\pi \mathrm{h}\mathrm{c}^2}{\lambda^5 \left[\exp\left(\frac{\mathrm{hc}}{\lambda \mathrm{k} T} \right) - 1 \right]}, \tag{2}$$

where $I(\lambda)$ – spectral intensity at the given wavelength λ; h – Plank's constant; k – Bolzman's constant; T – object's temperature (K); c – speed of light in vacuum.

The results of determining the temperature from the discharge emission spectra are given in figure 6. The discrepancy between the calculated and experimental data does not exceed 6%, which indicates the adequacy and correctness of the developed model.

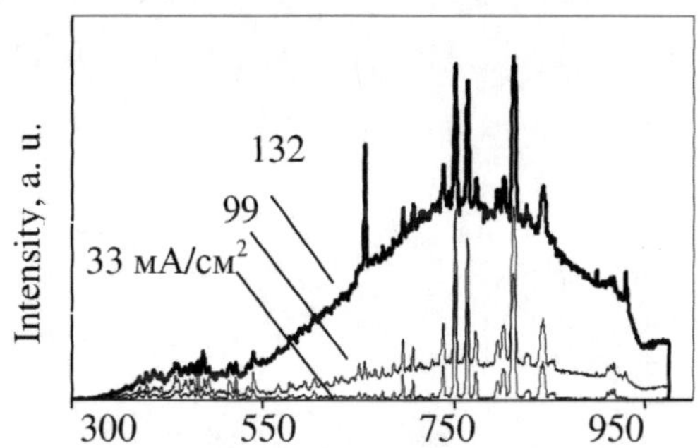

Figure 7. Spectra of plasma emission in an Ar atmosphere at a pressure of 1 mtorr for different current densities on a hot target.

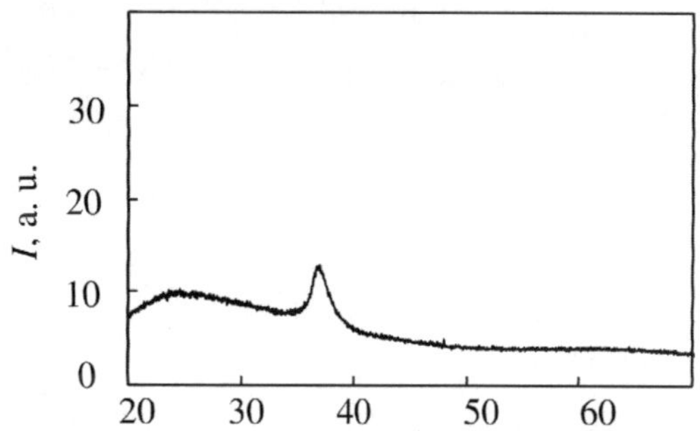

Figure 8. XRD spectra of the manufactured coating sample.

IOP Conf. Series: Materials Science and Engineering **387** (2018) 012038 doi:10.1088/1757-899X/387/1/012038

Figure 8 shows the XRD spectra of the manufactured coating sample. It has one broad maximum near $2\theta = 38$ deg. This angle corresponds to the cubic body-centered modification Ta with the preferential orientation (110) at 37.6 degrees and to the hexagonal Ti with orientation (111) at 38.0 degrees, coherent-scattering region compile 4.5 nm. The width of the peak does not allow one to give an uniquely identify one or two phases are present in the sample. The data of the electron probe X-ray spectral analysis was used to determine the ratio of titanium and tantalum in the deposited films: 45 atomic % of Ti and 55 atomic % of Ta). Taking into account the similarity of the sputtering coefficients of the selected materials, which should ensure the deposition of the film on the substrate at the same speed, the result obtained is well correlated with the ratio of the areas of the sputtered surfaces.

5. Conclusions

The studies carried out by the authors demonstrated the possibility of practical application of a doubled target. Change of the gap between the targets allows to change the mode of operation of the magnetron sputtering system, switching it to the "hot" mode. Nevertheless, the peculiarities of the behavior of the I-V characteristic and the spectra of the gas-discharge emission detected by the authors during the sputtering of the "hot" doubled Ta/Ti target require additional investigation.

References

[1] Xu Y, Li L, Cai X, Chu P K Chadrasekhar Loka, Ho Tak Yu and Kee-Sun Lee 2007 *Surface & Coatings Technology* **201** 6824–7

[2] Shukla K, Rane R, Alphonsa J, Maity P and Mukherjee S 2017 *Surface & Coatings Technology* **324** 167–74

[3] Viglin N A, Marchenkov V V and Milyaev M A 2013 *The Physics of Metals and Metallography* **114** 1003–8

[4] Vidyarthia V S, Suchanecka G, Gerlacha G, Levinb A A, Meyerb D C and Grötzschel R 2010 *Thin Solid Films* **518** 4106–12

[5] Dolique V, Thomann A L, Brault P, Tessier Y and Gillon P 2009 *Materials Chemistry and Physics* **517** 142–7

[6] Nakano J, Miyazaki H, Kimura T, Goto T and Zhang S 2004 *Journal of Ceramic Society of Japan* **112** 908–11

[7] Karolina Sieradzka, Michal Mazur and Damian Wojcieszak 2012 *Thin Solid Films* **520** 3472–6

[8] Shapovalov V I, Kozin A A, Minzhulina E A and Smirnov V V 2018 *Int. Conf. on Vacuum Technique and Technology* (St. Petersburg: St. Petersburg Electrotechnical University) pp 92–5

IOP Conf. Series: Materials Science and Engineering **387** (2018) 012039 doi:10.1088/1757-899X/387/1/012039

Possibility of measuring range extension of growing thin films thickness control, based on surface plasmon resonance effect

A E Komlev, R V Babinova and R V Dyukin

St. Petersburg Electrotechnical University “LETI”, 197376, St. Petersburg, Russia

E-mail: a.komlev@mail.ru

Abstract. Dependence of excitation angle of surface electromagnetic waves on HfO_2 film thickness was investigated. Modeling for different wavelengths showed the method’s applicability for real-time and high-resolution film thickness control. Proposed method can be used for development of fundamentally new nanoscale films thickness control systems.

1. Introduction

Precise control of thin films thickness is especially important task in optical devices manufacturing. For example, over a hundred of alternating layers with different refractive indexes can be used in narrow-band dielectric optical filters. Depending on operating wavelength, thickness of individual layer can varies from tens to hundreds nanometers. Moreover, thickness of layers has to be controlled with angstrom accuracy. For this purpose, two control systems are mainly used: quartz resonator and photometry. However, both these methods have some limitations. Authors have developed a method based on surface plasmon resonance, which allows calculating in real-time optical thickness of growing nanoscale film [1].

2. Theory

High sensitivity of surface electromagnetic waves (SEW) methods is based on physical principles of surface plasmons excitation [2]. Surface plasmon resonance (SPR) appears during excitation of SEW by light, which falls on layers boundaries with opposite dielectric permittivity signs. Even minor change of electro physical properties of analyzed medium causes noticeable change of excitation conditions and thus, shift of the resonance angle.

Such effect has acted as a cause to investigate the possibility of the SPR-based method applicability for thin film thickness control.

SEW was excited in Kretschmann geometry [3]. As a plasmon-generating layer, 50 nm Au film was used. Reflected from Au surface laser light intensity has a pronounced minimum, which is SPR excitation angle. During deposition of a matter on the gold surface, excitation angle increases.

As a first step, to determine the maximum thickness of HfO_2 film on sensor surface, at which resonance angle doesn’t exceed 80 degrees, modeling in WinSpall program was performed. Three wavelengths of laser were studied: 632.8 nm, 1064 nm and 2200 nm. $HfO_2$was chosen as analyzed matter because of its wide utilization in dielectric optical filters from UV up to near IR regions.

For verification of the modeling results, test samples were produced and SEW excitation experiments in “glass – Au - HfO_2 – air” structures were performed.

3. Experiment

For the verification tests, multi-layers structures on BK7 glass were fabricated. Plasmon generating layer (Au 99.99 %, 50 nm thickness) was deposited by magnetron sputtering. HfO_2 films with 1, 5 and 20 nm thickness were deposited on Au layer by electron beam evaporation. Thickness of HfO_2 films was controlled by quartz resonator system.

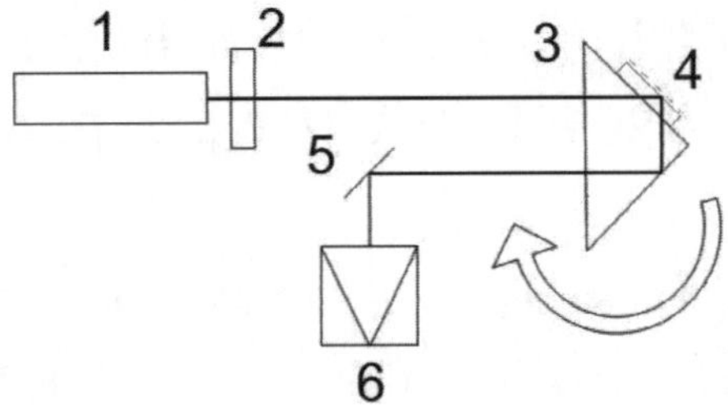

Figure 1. Scheme of SEW excitation angle measurement experimental setup: 1 – He-Ne laser, 2 – attenuation optical filter, 3 – prism on angle rotary stage, 4 – test sample, 5 – mirror, 6 – photodetector.

Sample was fixed on prism by oil immersion, which was mounted on rotary stage with 1′ angle step. Dependencies of reflected laser light intensity on incident angle to Au surface were measured for pure Au, Au + 1 nmHfO_2, Au + 5 nmHfO_2 and Au + 20 nmHfO_2.

4. Modeling

Experimental results are in a good agreement with data from theoretical modeling in WinSpall software.

As a light source was chosen p-polarized He-Ne laser with wavelength632.8 nm. Sample matter was BK7, $\varepsilon_3 = 2.2952$.Dielectric permittivity of Au$\varepsilon_1 = -10.5 + i1.2$.

From figure 2 it is visible, that resonance angle shift, caused by deposition of 1 nm HfO_2 is about 0.3 degree in both theory and experiment.

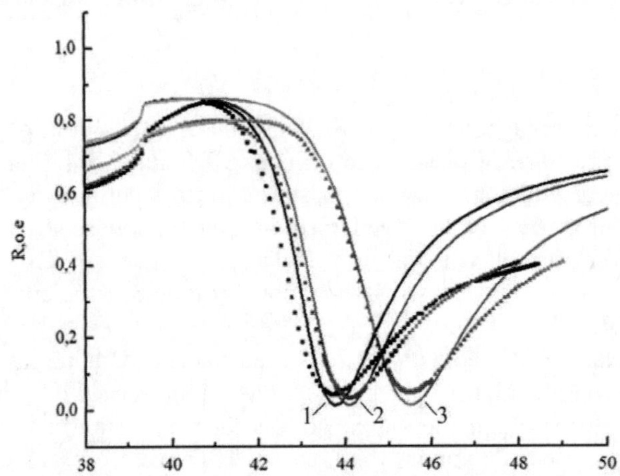

Figure 2. Calculated (solid curve) and experimental data comparison: 1 – sample Au/Air, 2 – sample Au/1 nm HfO_2/Air, 3 – sample Au/5 nm HfO_2/Air.

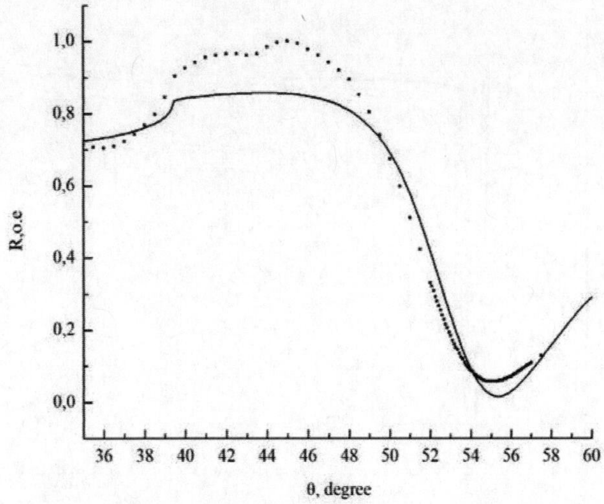

Figure 3. Calculated (solid curve) and experimental data comparison for Au/20 nm HfO_2/Air sample.

Existing modern ways of signal detection allows to register even smaller angle shifts [4]. Thus, experimental results confirm the possibility of detection an angstrom optical thickness changes.

Nevertheless, modeling results showed applicability limit of the method: when HfO_2 film thickness reaches 32 nm, resonance angle becomes over 80 degree (figure 4). Further thickness increase leads to SPR disappearance and makes detection impossible.

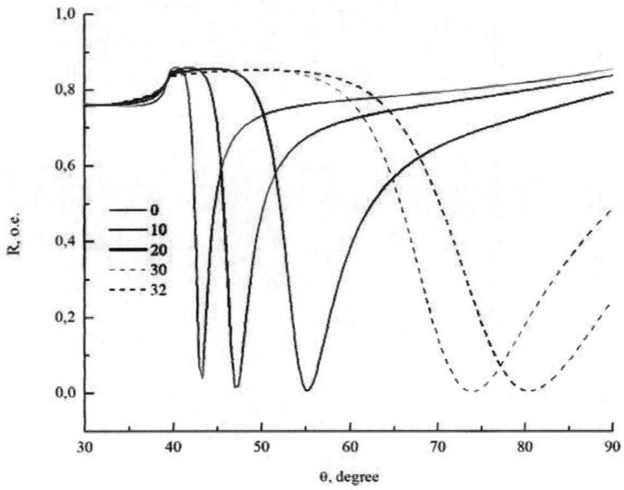

Figure 4. Dependence of reflectance coefficient on incidence angle to Au surface for "BK7/Au/HfO_2/Air" multilayer system. HfO_2 thicknesses are 0, 10, 20, 30, 32 nm. Laser wavelength is 632.8 nm.

The easiest available way of thickness measuring range extension is utilization of longer laser wavelength [5].

Modeling of emergence conditions of surface plasmon resonance in system glass–Au–HfO_2–air was also done for 1064 and 2200 nm lasers. These wavelengths exists in commonly used fiber and solid state lasers, including DPSS.

As seen in figure 5, with 1064 nm laser maximum thickness of HfO_2 layer rises up to 87 nm. At the same time, remains high sensitivity and possibility of precise layer thickness control.

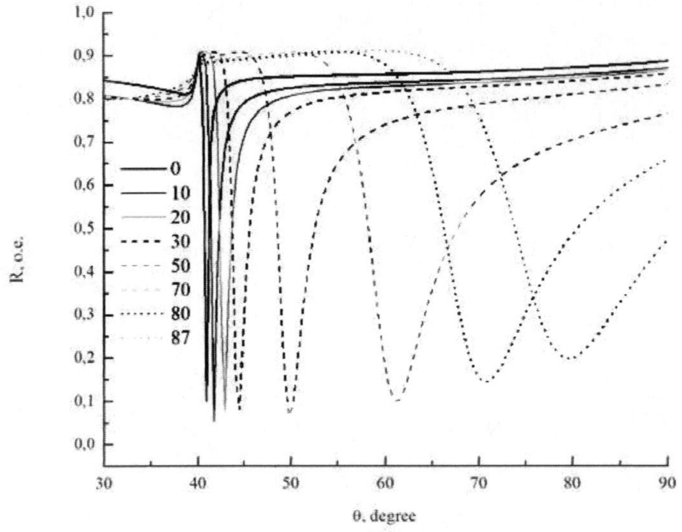

Figure 5. Dependence of reflectance coefficient on incidence angle to Au surface for "BK7/Au/HfO_2/Air" multilayer system. HfO_2 thicknesses are 0, 10, 20, 30, 50, 70, 80,87 nm. Laser wavelength is 1064 nm.

At laser operating wavelength of 2200 nm, thickness control limit rises up to about 220 nm (figure 6).

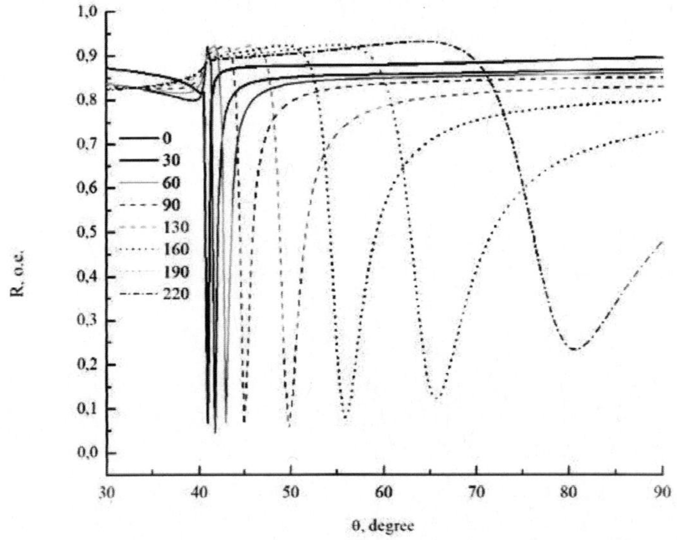

Figure 6. Dependence of reflectance coefficient on incidence angle to Au surface for "BK7/Au/HfO_2/Air" multilayer system. HfO_2 thicknesses are 0, 30, 60, 90, 130, 160, 190, 220 nm. Laser wavelength is 2200 nm.

Modeling results for different wavelengths and 1 nm thickHfO_2film are listed below (table 1).

Table 1. Summarized SPR modeling results for "BK7/Au/HfO_2/Air" for different laser wavelength.

Wavelength (nm)	632.8	1064	2200
θ_{res}, thickness HfO_2=0nm	43.16	40.95	41.07
θ_{res}, thickness HfO_2=1nm	43.45	41.00	41.08

IOP Conf. Series: Materials Science and Engineering **387** (2018) 012039 doi:10.1088/1757-899X/387/1/012039

5. Conclusion

It was found out that increase of laser wavelength extends SPR angle shift measuring range. Thus, the maximum film thickness detection range changes from tens to hundreds of nanometers. However, with increase of wavelength, the sensitivity of the method goes down: corresponding to SPR (minimum of reflection) angle shift between 0 and 1 nm thick HfO_2 is 0.29°, 0.05°, 0.01° for wavelengths of 632.8 nm, 1064 nm, 2200 nm respectively.

References

[1] Komlev A E, Dyukin R V and Shutova E S 2017 *Journal of Physics: Conf. Series* **872** 012042
[2] Sharlandjiev P and Gushterova P 2003 *Vacuum* **69** 21–5
[3] Liang H Q, Liu B and Hu J F 2017 *Optik* **149** 149–54
[4] Sotnikov D V, Zherdev A V and Dzantiev 2015 *Advances in Biochemistry* **55** 391–420
[5] Nelson B P, Frutos A G, Brockman J M and Corn R M 1999 *Analytical Chemistry* **71** 3928–34

IOP Conf. Series: Materials Science and Engineering **387** (2018) 012040 doi:10.1088/1757-899X/387/1/012040

Metallic coatings for MEMS structures

A V Korlyakov, O N Mikhailova and A V Serkov

Saint Petersburg State Electrotechnical University «LETI»

E-mail: novaja.tasamaya@yandex.ru

Abstract. This paper presents the results of mechanical stresses measurement in thin metal films in the composition of MEMS structures. The main requirements for metal films for MEMS are presented, recommendations for choosing methods of film deposition are given.

1. Introduction

Thin metal films are a part of microelectromechanical and micro-optomechanical devices, switches, sensors, actuators. They are used to form conductors, contacts and reflective coatings, so as in microelectronics technology. However, their area of use in MEMS is wider. Metals are often used as structural layers, as well as sacrificial or protective layers, which are subsequently removed.

It should be noted that mechanical properties play an important role for MEMS applications, and therefore, much attention in this work has been paid to them.

2. The application of metal films in MEMS

Table 1 shows the main areas of application of metal films for micro-optomechanical structures.

Table 1. The main application of metal films in MEMS.

Main applications in MEMS	Metal films
Structural (mechanically movable, functional) layers	Au, Cr, Ni, W
Adhesive and contact sublayers	Cr, Ti, Ta, W, V
Resistive and thermoresistant films	Ni, Pt,W
Form memory	Alloy Ti-Ni
Magnetic films	Fe, Alloy Fe, Co, и Ni
Reflective coatings	Au, W, Pt, Cr, Al
Conductive coatings	Au, Ni, Al
Sacrificial layers	Al, V, Cu, W
Masks for wet and dry etching	Al, Ni, Cu, Cr

The application of metal films in MEMS can be different. The following requirements are putted on films, depending on the area of their use:

- Resistance to certain etchants;
- Absence of pores and defects;
- High reflection coefficient;

IOP Conf. Series: Materials Science and Engineering **387** (2018) 012040 doi:10.1088/1757-899X/387/1/012040

- Minimum mechanical stresses;
- Good adhesion;
- High conductivity;
- Ohmic contact to the underlying conductive layer.

Compositions of metal films are often used to solve all these problems. The properties of metal films are strongly dependent on the methods and conditions of their deposition. A wide range of deposition methods makes thin metal films one of the most versatile classes of materials used in MEMS devices.

3. Methods of deposition metal films for MEMS

There are a lot of methods of producing metal films. They differ in the temperature of deposition, the degree of vacuum and other parameters. These parameters determine the properties of the produced films. The process temperature is one of the most important parameters of the film producing for several reasons. Firstly, this is the occurrence of thermomechanical stresses. They can lead to bending of the substrate or film peeling. Furthermore, they can strongly influence on the mechanical characteristics of the MEMS device. In addition, MEMS structures are often products in which the film is freed from the substrate. And if the value of the mechanical stresses is significant or they are compressive, this leads to the bending of such structures when they are freed from the substrate and, as a consequence, structures become unsuitable for use. Secondly, MEMS structures include multilayer compositions, and at high temperatures there can be interdiffusion of the layers, which leads to a change in the properties of the films. For example, during annealing of membrane elements with a composition of gold-chromium layers, chromium diffused to the surface, which led to a change in the mechanical properties of the composition of the layers and a change in the characteristics of the device. The application of metallization, as a rule, is one of the final stages of structure producing. Methods of depositing metal films at lower temperatures such as magnetron deposition should be chosen to avoid layer interdiffusion. Also, it is possible to use barrier layers.

A low vacuum in the chamber during deposition can result in the film saturation with impurities from the residual atmosphere. This is not desirable for micromechanical structures because the impurities from the residual atmosphere can strongly change the value of the mechanical stresses of the metal film, and as a result, the mechanical characteristics of the entire structure.

In addition, since the mechanical characteristics of MEMS are controlled not only in the process of producing structures, but also in the process of modification and operation of the device, it is important that it is possible to form metal films on already finished structures (for example, to produce electrodes). The following requirements are often made for MEMS due to the features of technology and design:

- Low film deposition temperatures;
- High vacuum;
- Possibility of deposition metal on the already produced device;
- The required value of mechanical stresses.

However, not all technological processes can meet these requirements. Therefore, the choice of the method of forming metal films depends on the specific task. The required characteristics of the film are taken into account.

4. Methods for measuring the mechanical characteristics of thin films

Measurement of mechanical stress in microelectronics technology is mainly carried out by measuring the bending of the substrate under the action stresses of the film or by X-ray diffraction [1]. MEMS technology uses other methods related to the design features and the application area of the device. For example, a method of measuring the deformation of a fixed–fixed film is used. Registration of the deflection is often carried out by optical methods. In the structures obtained by the technology of bulk micromachining, methods based on forced deformation of the film under the action of external forces are used: pressure or electric action. These methods are considered in detail in [2]. In this work we used a technique developed for micromechanical membranes based on measuring the deflection of a membrane under the action of air pressure. The measurements were held by means of a Fabry–Perot interferometer formed by the reflective end of the optical fiber and the reflecting surface of a metallized membrane [3].

5. Results and discussion

Mechanical stresses of many metal films obtained by magnetron deposition can be both tensile and compressive depending on the process conditions [4]. The sign and magnitude of the stresses play an important role for MEMS devices. Full compensation of mechanical stresses can be achieved by using film compositions with opposite mechanical stresses. And, as a result, the sensitivity of the structure increases.

Thin films of tungsten, obtained by magnetron deposition, have compressive mechanical stresses from the measurements. Therefore, they compensate the tensile stresses. The fact that the stresses are compressive can be explained by the oxidation of the film in the first stages of deposition. Thin films of titanium, obtained by the magnetron method, also have compressive mechanical stresses. High compressive stresses often lead to their detachment from the substrate in local areas. In addition, a thin layer of titanium oxide is observed on the surface due to the ability of titanium to strongly oxidize in air. Therefore, titanium films are recommended as a sublayer, and the deposition of the base material should be carried out without depressurizing of the vacuum chamber. The compressive stresses of titanium can compensate the tensile stresses of the structural layers.

Mechanical stresses in nickel films obtained by the method of magnetron deposition, are near zero, according to the measurements.

The results of measuring mechanical stresses in thin chromium films obtained by the magnetron method in various technological regimes were obtained. The feature is the high content of chromium oxide in the used target. Mechanical stresses in chromium films of thickness 100 nm, obtained in the investigated range of temperature and pressure of the working gas, are tensile and tend to decrease during deposition with increasing pressure of the working gas in the chamber (Figure 1, a). The mechanical stresses of the resulting films also change when the substrate temperature changes during deposition. The obtained dependence is complex (figure 1(b)).

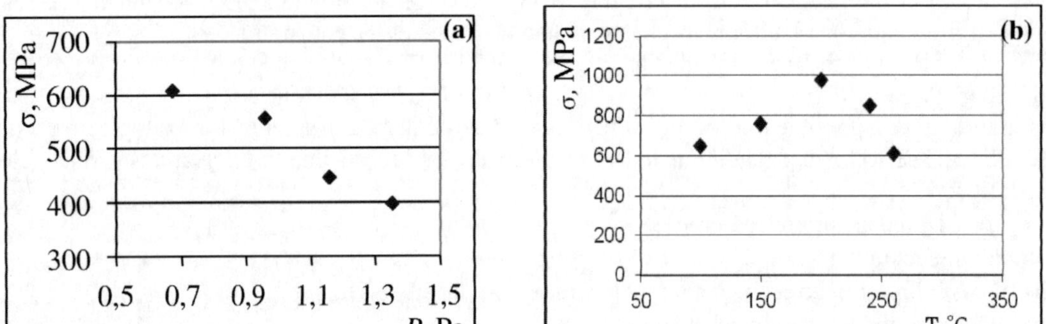

Figure 1. Dependence of mechanical stresses in chromium films on the working gas (Ar) pressure in the chamber during deposition processes by the magnetron method (**a**) Dependence of mechanical stresses in chromium films on substrate temperature during deposition processes by the magnetron method (**b**).

Table 2 presents the results of measuring mechanical stresses in various metal films deposited by magnetron method and shows how they were used in MEMS structures.

The obtained results show that not only the choice of the method, but also the choice of parameters of the deposition process have a significant effect on the mechanical properties of metal films. Layers with opposite mechanical stresses should be chosen to minimize the total stresses for MEMS compositions, which include sublayers, structural materials, and reflective or conductive coatings.

Table 2. Results of measuring mechanical stresses in various metal films (Me), deposited by the magnetron method.

Me	Film thickness, nm	Stress sign	Mechanical stresses, MPa	Application in MEMS
Cr	100	tensile	450 – 650 (depend on the conditions of the deposition process)	Reflective coatings and structural layers for sensors
Ti	100	compressive	-200	Sublayer for reflective and conductive coatings
W	80	compressive	-150	
Ni	100	–	Near zero	Reflective coatings and conductive layers for sensors
Au	90	change sign	From -190 to 560	Reflective coatings and structural layers for sensors, electrode material of activated membrane elements, wiring and commutation layers for MEMS switches

6. Conclusions

The values of mechanical stresses in thin films of various metals used in MEMS structures for various purposes deposited by a magnetron method are obtained.

The mechanical stresses in metal films depend on the conditions and methods of their deposition. They depend not only on the difference in the temperature coefficients of linear expansion of the film and substrate. Mechanical stresses are formed as a result of oxidation processes during growth, as well as at higher temperatures due to interdiffusion of materials used as structural layers. The layers of metals used for adhesion are easily oxidized and often form compressive stresses. This is especially true for thin layers. Such metal films like Ni or Au oxidize to a lesser degree and can have mechanical stresses near zero or change the sign of mechanical stresses depending on the deposition conditions.

References

[1] Gottfried K, Kriz J, Werninghaus T, Thumer M, Kaufmann Ch, Zahn D R T and Gressner T 1997 *Science and engineering* **B46** 171–5

[2] Astashenkova O N and Korlyakov A V 2013 *Nano- and microsystems technology* **2** 24–9

[3] Astashenkova O N 2015 *Ph.D. thesis* (Saint Petersburg Electrotechnical University, Saint Petersburg)

[4] Tamulevicius S 1998 *Vacuum* **2** 127–39

IOP Conf. Series: Materials Science and Engineering **387** (2018) 012041 doi:10.1088/1757-899X/387/1/012041

Application of a vacuum-arc discharge for the production of biocompatible coatings

D K Kostrin and A A Lisenkov

Department of electronic instruments and devices, Saint Petersburg Electrotechnical University "LETI", 197376, Saint Petersburg, Russia

E-mail: dkkostrin@mail.ru

Abstract. Vacuum-arc discharge is an effective method of forming pure metal and metal-like coatings of complex composition. In this paper, the possibility of its use for the production of biocompatible coatings on the example of synthesis of titanium nitride and titanium carbide on metal substrates is evaluated.

1. Introduction

Biomedical coatings are used to change the properties of contact between implants and the human body, to achieve the desired properties, such as wear resistance in moving parts of the prosthesis and corrosion protection under the influence of biological fluids of the body, as well as to reduce the diffusion of the implant material into the living tissue. Titanium nitride and carbide coatings received special attention [1], the advantage of their biomedical application is due to the properties of biological compatibility, which depend on their composition and structural properties, such as stoichiometry, surface morphology, microhardness, corrosion and wear resistance. Such coatings are used in the manufacture of prostheses hip and knee joints, as well as screws and nails for bone synthesis (figure 1) [2, 3].

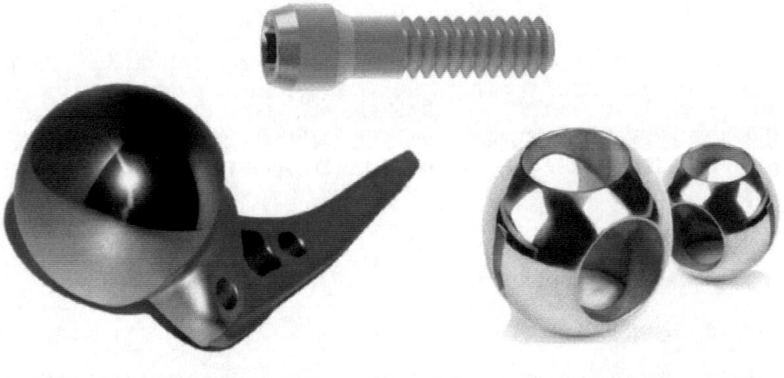

Figure 1. Examples of medical devices with applied protective biocompatible coating.

Biomedical materials must meet a number of often mutually exclusive requirements, and different structural elements may require different properties of the materials used. The use of metal coatings is possible if the surface material is extremely hard and durable. The manufacture of implants entirely from materials that meet these requirements is not possible, as such hard materials will be too fragile to use. Thus, the implant is made of available materials with the required characteristics, and the surface properties are achieved by applying biocompatible metal or metal-like coatings. The main requirements for the applied coatings are that they should not be brittle or prone to cracking, should be sufficiently rigid and durable, and their material should be compatible with the substrate material [4].

In turn, the requirements for the coating technology are the possibility of obtaining not only simple metal, but also dense composite coatings, ensuring a stable connection between the coating and the substrate material, minimal heat dissipation, so as not to cause excessive microstructure damage to the substrate material, as well as adjustable and sufficient deposition rate of the material, capable of achieving the required coating thickness in a reasonable time.

2. Technology of obtaining coatings

The flow of the process in vacuum at a pressure of not more than 10^{-3} Pa, provides both purity and high adhesion of the coatings to materials with different physicochemical characteristics. The vacuum-plasma technology is based on the formation of coatings due to the deposition of ions with high kinetic energy. Changing the accelerating potential U_{bias} allows to regulate the energy of precipitating ions, and, therefore, to control the flow of the technological process. In this regard, the vacuum coating process consists of two stages.

At the first stage, the surface is prepared for processing: external pollution, adsorbed gases and oxide films are removed from it [5]. At the same time, the ions of the applied material are introduced into the substance of the treated surface, which leads to the formation of a pseudo-diffusion layer, which is of great importance in improving the adhesive properties of the resulting coating. At the second stage, the coating is formed, the properties of which on different substrates are determined by the heating temperature of the treated surface. Vacuum-arc plasma sources make it possible to perform plasmachemical synthesis of nitride and carbide compounds [6, 7], for which the reaction-capable working gas is introduced into the generated plasma flow (figure 2).

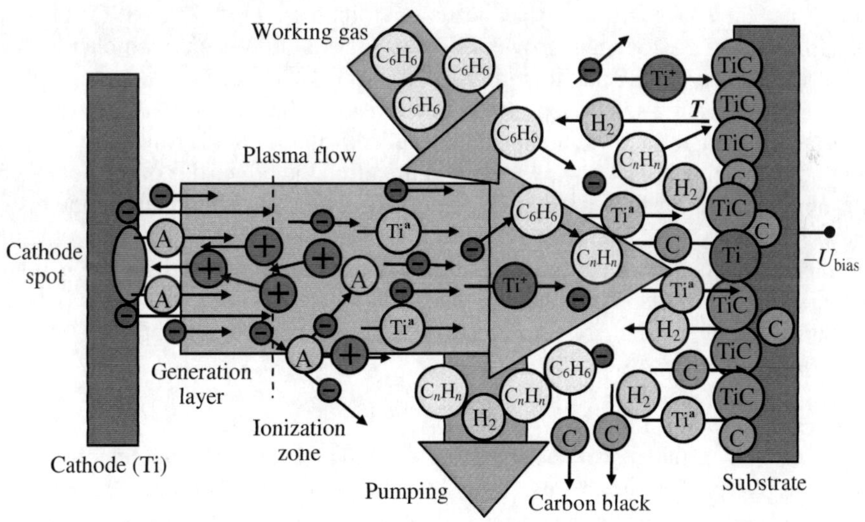

Figure 2. Schematic representation of the processes in the plasma flow of vacuum-arc discharge during deposition of titanium carbide.

The rate of the reaction gas supply is based on the coordination of the flows of the interacting metal ion particles with the gas molecules. Gas pressure, discharge current and the distance between the cathode and the substrate lead to a change in the ratio of chemical elements in the coating and its phase-structural composition, allowing obtaining coatings with desired properties [8].

The peculiarity of the generated metal plasma flow is the presence in it of a noticeable number of cathode material particles: droplets having rectilinear trajectories and leading to a decrease in the quality of the coating due to the deterioration of microrelief and the appearance of porosity in the formed structure. This disadvantage is the root cause of premature failure, corrosion and decrease in the performance properties of coatings. By lowering the discharge current the total number of droplets in the plasma flow cannot be significantly reduced.

For cleaning the plasma flow from the droplet fraction is used a separation system of the "jalousie" type creating a solid, impenetrable barrier to droplet formation, having the straight path of movement and deposited on the separator surface facing the working surface of the cathode or a special plasmaoptical system for transporting the plasma flow providing a reversal of the flow by 90°. By changing the magnetic field in time, it is possible to control the thickness of the formed coating with a predictable degree of unevenness not exceeding one percent, and in addition to obtain the required thickness at any point on the surface of the processed parts.

3. Synthesis of nitride compounds

For the synthesis of nitride compounds – metal-like formations with high hardness chemically active molecular nitrogen N_2 with a strong triple bond *d* (N–N) is used as a reaction gas. The destruction of the triple bond is hindered by the unusually strong first of the disrupted bonds in the N_2 molecule. Dissociation of nitrogen leads to the reaction of interaction of nitrogen atoms with the sprayed metal ions (Me) with the subsequent formation of a compound Me_xN_y:

$$Me_{\text{substrate}} + (2Me_{\text{plasma}} + 2N_2\uparrow)_{\text{vacuum chamber}} \rightarrow [Me_{\text{substrate}} + (2Me_{\text{plasma}}N)_{\text{coating}}] + (N_2\uparrow)_{\text{vacuum chamber}}.$$

Nitrogen atom in nitrides having an isolated state s^2p^6-configuration of valence electrons can be both a donor and acceptor. In the formation of compounds, depending on the electronic structure of the elements, the system can be transformed both into a stable s^2p^6- and decomposed sp^3-configuration.

Efficiency of the plasmachemical reactions of compounds synthesis depends both on the degree and nature of molecules excitation and on the kinetic energy of ions. The increase in the energy of charged particles bombarding the condensation surface leads to an increase in the surface diffusion mobility of adatoms, thereby causing a predominant growth of crystallites in the condensation plane. Chemical bond in metal-like nitrides is very strong due to participation in bonds between metal atoms and nitrogen not only of external electrons, but also of deep-seated electrons of unfinished shells. As a result, the nitrides of transition metals are very refractory and have high hardness and chemical resistance.

The area of homogeneity of TiN is very wide and therefore the properties of the compound strongly depend on the amount of nitrogen in the nitride, which affects the microhardness of the resulting films, which varies widely from 20 to 40 GPa. High hardness of the condensed TiN is a consequence of the high level of internal stresses. The dependence of the microhardness on the pressure is nonmonotonic. For stoichiometric TiN films, the main phase is the phase with a face-centered cubic lattice. The value of the lattice parameter is determined by a number of factors: nitrogen content, film thickness, as well as the presence of internal stresses.

4. Synthesis of carbide compounds

To carry out the plasma chemical synthesis of carbide compounds, a carbon-containing working gas is introduced into the plasma flow. In this case, the decomposition of hydrocarbon compounds is carried out both by their collisions with particles of the plasma flow and by thermal decomposition, which is a complex process consisting of a set of elementary reactions occurring both simultaneously and sequentially, with the formation of a large number of intermediate products [9]. In practice, benzene C_6H_6, an aromatic hydrocarbon, whose molecules contain stable cyclic groups of atoms (benzene nuclei) with a closed system of conjugated bonds, is often used as a working substance.

Thermal transformations of benzene can proceed both at 800...1100 K, and at higher temperatures. At low temperatures, there is a rupture of the C–H bond and the formation of diphenyl $C_{12}H_{10}$ and, to a lesser extent, polyphenols. C_6H_7 radical reactions play a significant role, which occurs both in the initiation processes and as a result of the addition of atomic hydrogen to benzene. At higher temperatures there are complex and non-selective processes of splitting the phenyl nucleus. The decay rate of C_6H_5 and C_4H_3 is so high that the kinetics of benzene decomposition under these conditions can be described under the assumption of direct decay reactions, such as $C_6H_6 \rightarrow C_2H_2 + C_4H_2 + 2H$. The formation of a compound that occurs on the condensation surface can occur according to the following scheme of splitting hydrocarbons with long bonds into molecules of smaller length:

$$C_6H_6^{\uparrow} \xrightarrow{W} C_nH_{xn}^{\uparrow} + C_xH_{yn}^{\uparrow} + H_2^{\uparrow} \xrightarrow{W} C + C_xH_y^{\uparrow} + H_2^{\uparrow};$$

$$Me_{\text{substrate}} + (Me_{\text{plasma}} + C_6H_6)_{\text{vacuum chamber}} \xrightarrow{T} (Me_{\text{substrate}} + Me_{\text{substrate}}C + Me_{\text{plasma}}C + C) + C_{\text{carbon black}} \downarrow + C_nH_n^{\uparrow}.$$

Titanium carbide TiC_{1-x} belongs to the group of non-stoichiometric compounds and has a wide area of homogeneity within which the carbon content ($TiC_{0.48}...TiC_{0.95}$) changes without rebuilding the crystal lattice. It is experimentally established that trigonal and cubic ordered phases of Ti_2C type are observed in TiC_x carbide in the region of $0.5 < x < 0.7$.

5. Results and discussion

The phase composition of the formed coatings was studied by X-ray phase analysis on the X-ray diffractometer DRON-3 in the filtered $Cu_{K\alpha}$ radiation. The thickness of the coatings was controlled by means of metallographic sections. For the analysis of coatings was used scanning electron microscope JSM-35CF. Figure 3 shows the surface of titanium carbide coatings with and without the use of plasma cleaning system, and also the metallographic section of the obtained coating.

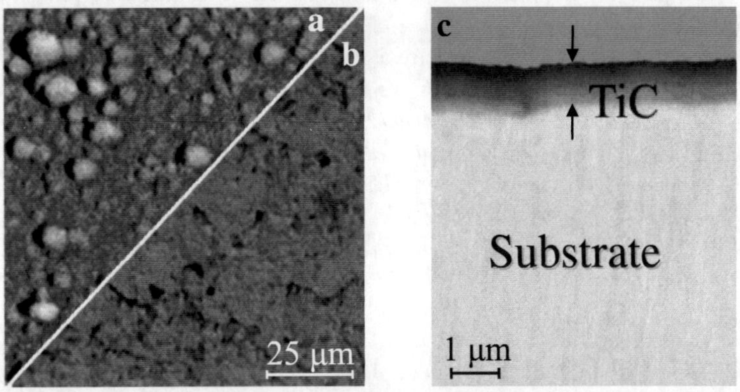

Figure 3. Surface of titanium carbide coatings without the use of a plasma cleaning system (**a**) and with the use of a plasma cleaning system (**b**), metallographic section of the obtained coating (**c**).

Improving the quality of formed coatings can be achieved not only due to the technological conditions of deposition, but also by reducing the thickness of the applied layers. It is known that the hardness of nanometer films is several times higher than the hardness of conventional films with micrometer thicknesses. The effects of hardening of thin surface films at the present time have not received an unambiguous physical interpretation and this explains the interest in research in the field of strength physics of composite materials.

When forming multilayer coatings, the determining factor is the correct choice of materials that make up the composite layer – in case of unsuccessful choice of materials, the disintegration of individual layers in the coating and the reduction of its mechanical characteristics occurs. Double and triple systems of carbides, nitrides, and carbonitrides of transition materials can be used as multilayer coatings [10]. High thermodynamic stability, hardness and strength of these compounds are due to the similarity of structures and the close size of atoms.

6. Conclusion

The developed technology of plasmachemical synthesis of titanium-based compounds in the flow of metallic plasma generated by a vacuum-arc discharge allows regulating the characteristics of the resulting compounds. Studies have shown that the main parameters affecting the structure and properties of coatings are: gas pressure responsible for the elemental and phase composition of the coating; discharge current, which determines the composition of the plasma flow and the growth rate of the coating; bias potential, which determines the growth rate of the coating, microstructure and hardness; substrate temperature, which determines adhesion, microstructure and residual stresses.

It is shown that in the presence of droplet formations, coatings with a developed surface are obtained, and when cleaning the plasma flow, it is possible to effectively control the composition and structure of the formed coating.

References

[1] Vladescu A, Cotrut C, Braic V, Balaceanu M and Braic M 2004 *Romanian Reports in Physics* **3** 460–5

[2] Markatos K, Tsoucalas G and Sgantzos M 2016 *Acta Medico-Historica Adriatica* **1** 161–76

[3] Understanding Orthopedic Hardware - not just screws and wires! URL: https://posterng.netkey.at/esr/viewing/index.php?module=viewing_poster&task=&pi=119719.

[4] Ratner B D, Hoffman A S, Schoen F J and Lemons J E 1996 *Biomaterials science: an introduction to materials in medicine* (San Diego: Academic Press)

[5] Bystrov Yu A, Kostrin D K, Lisenkov A A and Vetrov N Z 2015 *Vakuum in Forschung und Praxis* **2** 22–5

[6] Bystrov Yu A, Vetrov N Z, Lisenkov A A and Kostrin D K 2014 *Vakuum in Forschung und Praxis* **5** 19–23

[7] Vinogradov M L, Kostrin D K, Smelova V V, Trifonov S A and Lisenkov A A 2016 *Proceedings of the 2016 IEEE North West Russia Section Young Researchers in Electrical and Electronic Engineering Conference* 729–30

[8] Kostrin D K and Lisenkov A A 2016 *Materials Science Forum* **843** 278–83

[9] Kostrin D K, Pikus M I, Trifonov S A and Lisenkov A A 2017 *Journal of Physics: Conference Series* **789** 012029

[10] Kostrin D K and Lisenkov A A 2016 *Materials Science Forum* **870** 371–6

IOP Conf. Series: Materials Science and Engineering **387** (2018) 012042 doi:10.1088/1757-899X/387/1/012042

Numerical study of rarefied gas flows in microchannels with oscillating elements

V V Kosyanchuk[1,2,3] **and A N Yakunchikov**[1,2,3]

[1] Faculty of mathematics and mechanics, Lomonosov Moscow State University, 119991, Moscow, Russia
[2] Institute of Mechanics of Lomonosov Moscow State University, 119192, Moscow, Russia
[3] Mechanical Engineering Research Institute of the Russian Academy of Sciences, 101990, Moscow, Russia

E-mail: vasiliy_ksnk@mail.ru

Abstract. This paper present a numerical study of a free-molecular (Kn >10) gas flow in two devices with high-frequency oscillating boundary: (1) flow in plane channel with a series of oscillating barriers. (2) flow in a channel with wall performing forced harmonic motion. All calculations were performed using simplified collision-free version of Event-Driven Molecular Dynamics method (EDMD). It was shown that high-frequency oscillations have a significant influence on the gas flow, and this influence depends on molecules characteristic thermal speed. Thus it was shown, that this influence will be different for gases with different molecular masses, which leads to gas separation effect.

1. Introduction

Constant development of microelectronics and microtechnology has lead to the emergence of many new applications of microelectromechanical systems (MEMS). Microdevices with moving parts can be distinguished as separate subclass among all MEMS devices with applications in various areas of microfluidics [1–3]. Recently a series of works appeared devoted to microdevices utilizing surface acoustic waves (SAW) to separate solid microparticles with different sizes [4–6]. Meanwhile, applications of MEMS with moving parts to rarefied gas dynamics problems are very rare.

Authors of this work were the first to discover a series of novel effects appearing for rarefied gas flow in microstructures with moving elements [7–11]. The main idea here is that for gas flowing in free-molecular regime, the only thing affecting gas molecules motion is the number and pattern of gas-surface collisions. For instance, it was noted that for gas flowing in plane straight channel with diffuse reflection law used on the walls, the probability of molecule to pass through the channel decreases exponentially with the number of collisions. Thus for gas flows in microstructures with moving parts molecules passing probability (and, hence, mass flux) can change significantly when the ration of characteristic oscillation speed is compatible with characteristic thermal speed of molecules

In this work we study the free-molecular gas flow in two devices: in microchannel with a series of oscillating barriers and in the channel with walls moving according to the harmonic law. It is shown that one can influence the conductivity of the channel for a given gas due to forced oscillations of parts the boundary. It is also explained how this effect can be used to separate gas mixtures.

2. Problem statements

First problem studies flow in a plane channel with diameter D and a series of oscillating barriers (figure 1). Barriers have distance L between them and oscillate with frequency ω. Initial phases of barriers motion are chosen in such a way that common motion of barriers lowest point represent a travelling "pseudo-wave" with length λ. Characteristic wave speed k of such motion is the product λω.

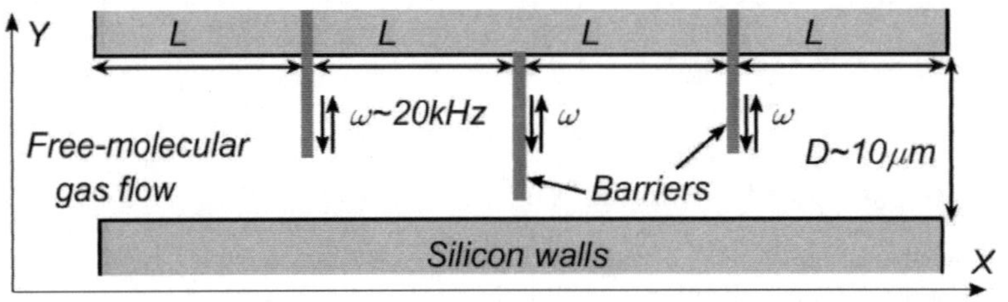

Figure 1. Geometry of the first problem. Channel with series of oscillating barriers.

Second problem studies rarefied gas flow in a curving channel (figure 2), having length L and width D in motionless state (i.e. when there are no oscillations and channel walls are straight). Channel wall are curving according to harmonic law with amplitude A, wave length λ and wave speed k:

$$y_{\pm} = y_{0,\pm} + A\sin\left(2\pi\frac{x-kt}{\lambda} + \psi\right), \quad y_{0,+} - y_{0,-} = D, \tag{1}$$

index ± corresponds to upper and lower surfaces.

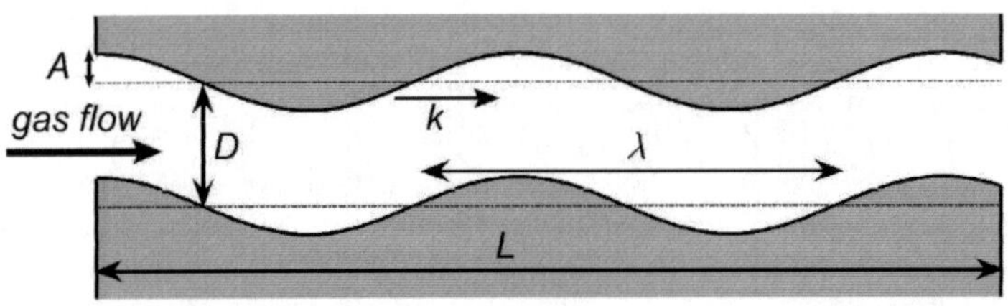

Figure 2. Geometry of the second problem. Channel with wall curving according to harmonic law.

Interaction between gas molecule and solid surface was described in terms of Maxwell diffuse scattering model (model with full energy and momentum accomodation). This means that molecule comes in thermal equilibrium with the wall after collision and molecules velocity after collision has equlibrium Maxwell distribution with temperature of the wall. The problem is studied in isothermal case – temperature of the wall and gas at inlet are constant in space and time. The flow is considered free-molecular which means that collision between molecules are neglected. Molecules were simulated as material point with no internal degrees of freedom. Mass forces were not taken into account.

3. Solution procedure

Numerical algorithm was thoroughly discussed in authors works [7–11]. Assumption about free-molecular flow regime allows to ignore collisions between molecules and calculate molecules trajectories independently. This significantly simplifies numerical algorithm which can be treated as either collision-free version of event-driven molecular dynamics method (EDMD [12, 13]) or of direct simulation Monte-Carlo (DSMC) method.

Calculations for each molecule consisted of three stages. First, a molecule was created at inlet cross section with coordinates, velocity and starting time, sampled from corresponding physical distributions. Point of next collision with surface was calculated after that. New velocity was calculated at collision point according to scattering kernel and then a search for new collision point

IOP Conf. Series: Materials Science and Engineering **387** (2018) 012042 doi:10.1088/1757-899X/387/1/012042

started. Collision-reflection process was repeated until molecule leaved the channel through either inlet or outlet cross-section.

As a result of calculations for great number ($N = 10^7$) of molecules one obtains passing probability P as ratio of number of molecules that passed through the channel to the number N of all probe molecules. Evaluating P for different values of k – characteristic oscillations speed and $c = \sqrt{2k_BT/m}$ – molecules characteristic thermal speed (here T is temperature, k_B – Boltzmann constant, m – mass of the molecule) one obtains profile $P(k/c)$. Based on profile $P(k/c)$ one can then evaluate $\gamma(k)$ – coefficient of separation enhancement for a given pair of gases (for He-Ar mixture in this case).

4. Channel with barriers

Results of calculations for first problem are presented in Figure 3. Characteristic speed of oscillations in this case is the product of pseudo-wave length λ and barriers oscillation frequency ω. The main parameters of the study are also the distance between the barriers L and wavelength λ (determined under fixed L by the number of barriers and phase shift $\Delta\psi$ between their motion laws). Results in figure 3(a) indicate that for small number of barriers separation enhancement can occur only for low values of L/D. However, such close distances between neighboring barriers cannot be achieved in practice. Meanwhile increasing number of barriers (figure 3(b)) one can achieve enhanced separation even for higher values of L/D. Moreover, increasing number of barriers and, hence, increasing pseudo-wave length λ one can achieve values of characteristic speed λω comparable with characteristic thermal speed of molecules even for comparatively low values of frequencies ω (about kHz). This fact means that it is possible to realize this effect in practice.

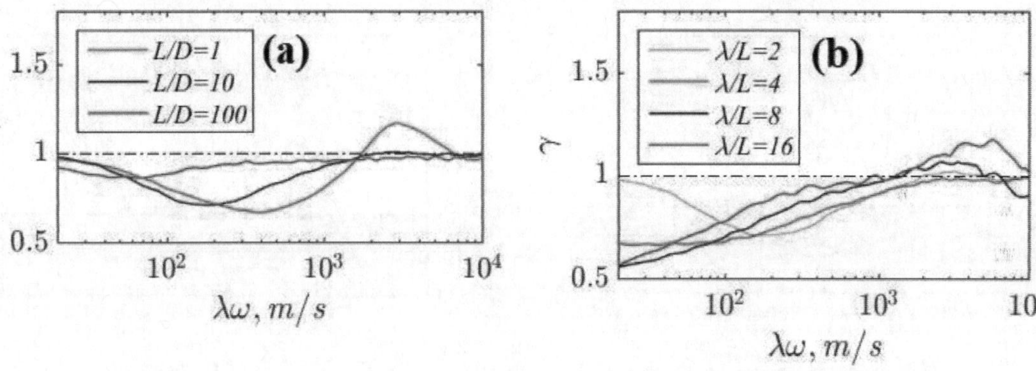

Figure 3. Simulation results in first problem. Dependence of coefficient γ on characteristic speed λω for different values of ratios L/D (**a**) and λ/L (**b**).

5. Channel with curving walls

Second problem has more parameters and only main results are presented in Figure 4. It is shown that separation effect (nonmonotonic $P(k/c)$ profile) is observed only at ratios A/D (oscillation amplitude to channel diameter) around 0.2 (figure 4). Values of separation enhancement coefficient γ in case $A/D = 0.2$ are presented in figure 5(a). For higher and lower values of A/Dseparation effect vanishes (profiles become monotonic) but pumping effect is still observed, which means that for low and high values of A/D this device can be used as a vacuum gas pump. It is noted that with increasing wavelength λ separation effect is getting stronger but requires higher values of wave speed k to reach maximal effect (figure 5(a)). It is also noted that using longer channel also leads to more effective separation (figure 5(b)).

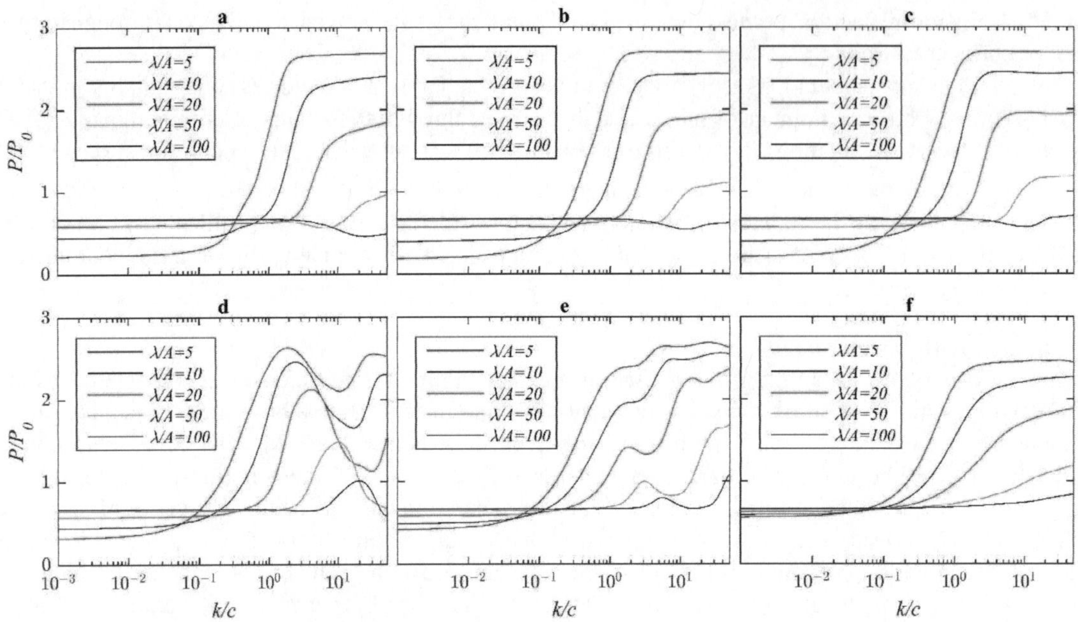

Figure 4. Simulation results in second problem. Profiles $P(k/c)$ for L/D =10 and different values of λ/A (given in legend) and A/D. **(a)** – $A/D = 1.0$. **(b)** – $A/D = 0.5$. **(c)** – $A/D = 0.4$. **(d)** – $A/D = 0.2$. **(e)** – $A/D = 0.1$; **(f)** – $A/D = 0.01$.

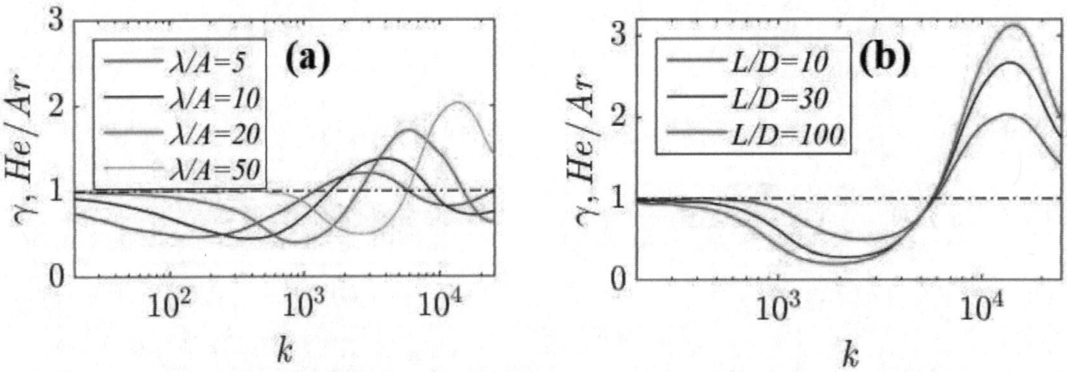

Figure 5. Dependence of coefficient γ on characteristic wave speed k for different values of ratios λ/A **(a)** and L/D **(b)**.

Acknowledgements
The research is carried out using the equipment of the shared research facilities of HPC computing resources at Lomonosov Moscow State University [14] and Joint Supercomputer Center of the Russian Academy of Sciences. This work is supported by the Russian Science Foundation (Grant number 17-71-10227).

References

[1] Friend J and Yeo L Y 2011 *Reviews of Modern Physics* **2** 647
[2] Nisar A, Afzulpulkar N and Mahaisavariya B 2008 *Sensors and Actuators B* **2** 917–42
[3] Iverson B D and Garimella S V 2008 *Microfluidics and Nanofluidics* **2** 145–74
[4] Shi J, Ahmed D, Mao X, Colletti A and Huang T J 2008 *Lab on a Chip* **2** 221–3
[5] Shi J, Ahmed D, Mao X, Lin S, Lawit A and Huang T J 2009 *Lab on a Chip* **20** 2890–5
[6] Devendran C, Gunasekara N R, Collins D J and Neild A 2016 *RSC Advances* **7** 5856–64

[7] Kovalev V L, Kos'yanchuk V V and Yakunchikov A N 2014 *Fluid Dynamics* **49** 524–9
[8] Yakunchikov A, Kovalev V and Kosiantchouk V 2015 *Microfluidics and Nanofluidics* **18** 1039–43
[9] Kosyanchuk V V, Yakunchikov A N, Bryukhanov Y A and Konakov S A 2017 *Microfluidics and Nanofluidics* **21** 116
[10] Kovalev V, Yakunchikov A and Kosiantchouk V 2015 *Acta Astronautica* **116** 282–5
[11] Kosyanchuk V V and Yakunchikov A N 2018 *Microfluidics and Nanofluidics* **22** 60
[12] Bannerman M N, Sargant R and Lue 2011 *Journal of Computational Chemistry* **15** 3329–38
[13] Yakunchikov A N and Kosyanchuk V V 2018 *Computers and Fluids* **170** 121–7
[14] Sadovnichy V, Tikhonravov A, Voevodin V and Opanasenko V 2013 *Contemporary High Performance Computing: From Petascale toward Exascale* 283–307

IOP Conf. Series: Materials Science and Engineering **387** (2018) 012043 doi:10.1088/1757-899X/387/1/012043

Target heating in sputtering unit of magnetron

A A Kozin

Department of Physical Electronics and Technology, Saint Petersburg Electrotechnical University "LETI", 197376, Saint Petersburg, Russia

E-mail: kozin.alexander.andreevich@gmail.com

Abstract. The thermal process in a sputtering magnetron unit with a sandwich-target was studied. The unit contains two targets located on the same axis. The inner target is effectively cooled by running water; the outer target is cooled through the fastening elements and by radiation. The study was performed by modeling in the COMSOL Multiphysics using the Heat Transfer Module. A 3D-model of the sputtering unit has been developed in order to thoroughly analyze the thermal processes in a unit with a cold chromium target and a hot titanium one. The influence of the target thickness and magnetron discharge power on the outer target temperature is found. The model can be used to study a magnetron with any metallic target.

Magnetron sputtering is a commonly used technique to deposit films of metal alloys or solid solutions of simple compounds (nitrides, oxides, etc.). In order to create two fluxes, various sputtering systems are used; the simplest one contains two magnetrons with targets of different metals [1, 2] or one magnetron with a composite target [3, 4]. Traditionally, the target is cooled effectively in these devices. Besides, sputtering systems with hot targets are being studied [5–7], where a single target can be heated up to the melting point and higher [8, 9]. The hot target mode is maintained by heat transfer through a gap (up to 1 mm thick) and by fastening elements [10].

Heating the target sputtered in inert and reactive gas environments changes the physicochemical processes on the target surface, in the gas and on the substrate. In order to understand the influence of the target surface temperature on the composition, crystal structure, and properties of the films, the value of the temperature should be determined. Besides that, it is used as a parameter in modeling of the reactive sputtering process [7, 11, 12].

Direct measurement of the target surface temperature is a complex technical task. To that end, non-contact pyrometric methods [10, 12–14] are used most often. But they give a significant error, since the pyrometer integrates the discharge optical spectrum in the IR region, which is a combination of the glow discharge emission and target thermal radiation spectra. Another approach is based on modeling in various software packages, where a thermal problem is solved numerically for a given model of a device with given boundary conditions.

In this paper, we describe a new type of magnetron with a sputtering unit, which contains two targets located on the same axis. The inner metal target is sputtered through the slots in the outer target and operates in a cold mode. The outer target made of another metal operates in a hot mode. The sputtering unit will be referred to hereinafter as the "sandwich-target".

In order to solve this problem, a 3D-model of a magnetron with a sandwich-target was developed in the COMSOL Multiphysics (figure 1). In the model, the sputtering unit consists of: an inner 4 mm thick target *3* made of chromium; an outer target *1* made of titanium and fastening elements *2* and *4*. In the area of the outer target erosion, an even number of slots *5* of a predetermined diameter are arranged symmetrically about the center of the target. In the calculations, it was assumed that the discharge power is uniformly distributed in the annular region where the greatest erosion is observed. The power incident on the target, the outer target thickness, and the diameter of the slots in the outer target may vary in the model.

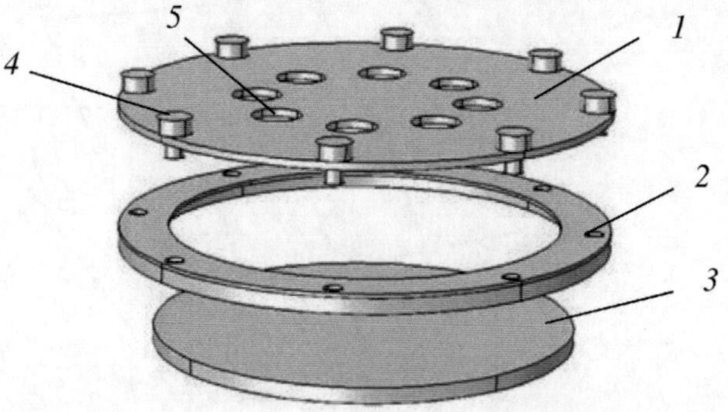

Figure 1. 3D-model of a magnetron with a hot sandwich-target: *1* – outer target; *2* – fastening ring providing a gap between the targets; *3* – inner target cooled by running water; *4* – the bolts; *5* – slots.

Figure 2 shows the temperature distributions over the surface of the outer target in the longitudinal and transverse directions, obtained as a result of calculations. Since Figure 2 reflects the results including those for sections containing slots, discontinuities are observed in the temperature distribution at the edges of the slots.

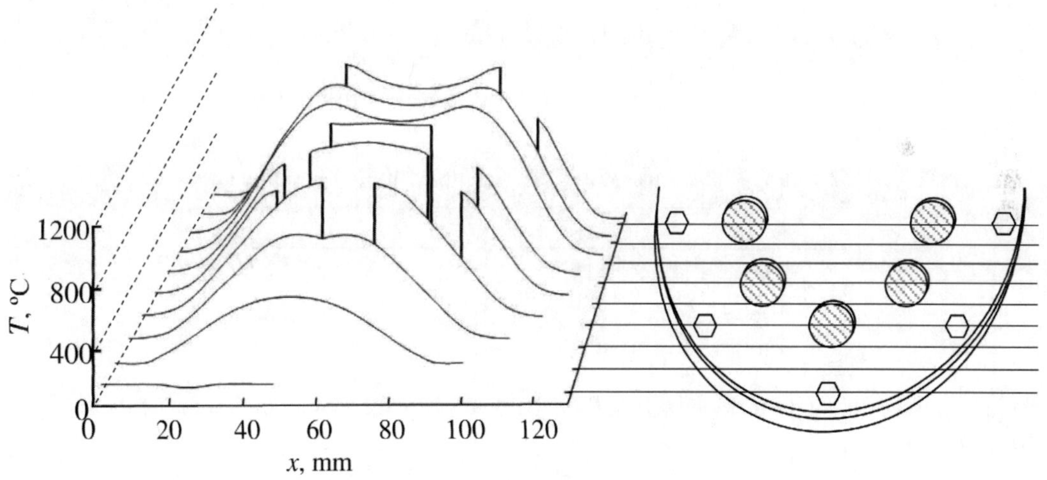

Figure 2. Temperature distribution over the surface of the outer target. Only half of the target is considered.

Since the temperature is distributed unevenly over the target surface, the concept of the effective temperature T_{eff}, which determines its average heat flux, was introduced. It can be calculated by averaging the distribution over the entire surface of the outer target:

$$T_{eff} = \frac{1}{S}\int_{-\infty}^{\infty}\int T(x, y)dxdy, \tag{1}$$

where $T(x, y)$ – is the surface temperature distribution, S – is the surface area of the target, and x and y are the transverse and longitudinal coordinates, respectively.

Figure 3 shows the dependence of the effective temperature on the thickness of the outer target at different powers. Figure 3 demonstrates that an increase in the thickness of the outer target leads to drop of the effective temperature(at fixed power).

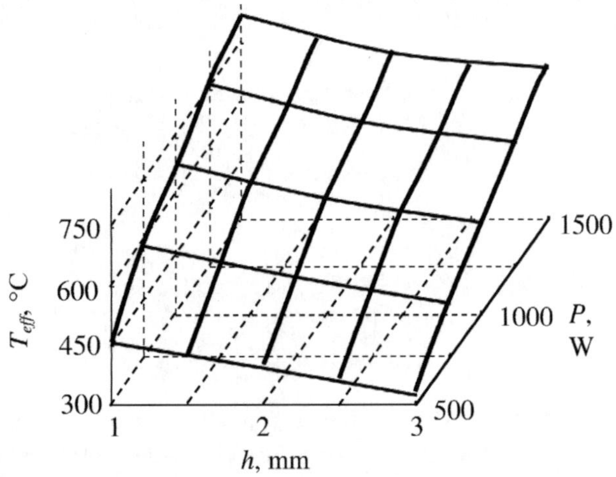

Figure 3. The dependence of the effective temperature of the outer target on its thickness and the incident power.

Based on the received data, it took two steps to develop the model $T = f(P, h)$:

1. determination of the dependence $T = f(h)$ for P = const in the form of a polynomial of the second order:

$$T = b_0 + b_1 h + b_2 h^2. \tag{2}$$

The values of the coefficients b_0, b_1 and b_2 in (2), calculated by the Ordinary Least Squares (OLS), are summarized in table 1.

Table 1. The values of the coefficients of the polynomial (2).

P, W	b_0	b_1	b_2
500	519	–59.0	–2.0
1000	804	–133.5	14.5
1500	980	–188.0	30.0

2. the definition of the dependencies $b_i = f(P)$, $i = 0, 1, 2$ in the form of a polynomial of the second order:

$$b_0(P) = (125 + 0.897P - 2.18 \cdot 10^{-4} P^2); \tag{3}$$
$$b_1(P) = (35.5 - 0.209P - 4 \cdot 10^{-5} P^2); \tag{4}$$
$$b_2(P) = (-19.5 + 0.036P - 2 \cdot 10^{-6} P^2), \tag{5}$$

the parameters of which are also determined by the OLS.

Using (2)–(5), we write the general expression describing the change in the effective temperature caused by the power and thickness of the outer target:

$$T_{eff}(P, h) = b_0(P) + b_1(P)h + b_2 h^2. \tag{5}$$

Table 2 summarizes the results of calculations in the COMSOL package and based on the model (5). Table 2 confirms the relevance of the model.

Table 2. Comparison of the calculation results.

h, mm	*P*, W	*T*, °C	
		COMSOL	Model
1	500	458	458
2	500	393	393
3	500	324	324
1	1000	685	685
2	1000	595	595
3	1000	534	534
1	1500	822	822
2	1500	724	724
3	1500	686	686

As a result of the performed research:

- a 3D-model of a sputtering unit containing a cold chromium and hot titanium target was developed;
- it is revealed that the temperature over the surface of the target is distributed unevenly and strongly depends both on the design parameters and on the power incident on the target;
- a relevant mathematical model has been created in the form of a second-order polynomial, which determines the dependence of the effective surface temperature of the outer target on its thickness and incident power.

Acknowledgment
This work was funded by the Russian Foundation for Basic Research (grant 18-29-11043).

References

[1] Stone D S, Gao H, Chantharangsi C et al. 2014 *Sur. Coat. Technol* **244** 37
[2] Greczynski G, Lu J, Jensen J et al. 2014 *Sur. Coat. Technol.* **257** 15
[3] Zhang W, Li Y, Zhu S and Wang F 2003 *Vac. Sci. Technol.* **21** 1877
[4] Zaleska A 2008 *Recent Patents on Engineering* **2** 157
[5] Tesař J, Martan J and Rezek J 2011 *Sur. Coat. Technol.* **206** 1155
[6] Mercs D, Perry F and Billard A 2006 *Sur. Coat. Technol.* **201** 2276
[7] Shapovalov V I, Karzin V V and Bondarenko A S 2017 *Phys. Lett. A.* **381** 472
[8] Bleykher G A, Krivobokov V P and Yuryeva AV 2015 *Techn. Phys.* **60** 1790
[9] Bleykher G A, Borduleva A O, Krivobokov V P and Sidelev D V 2016 *Vacuum* **132** 62
[10] Lapshin A E, Levitskii V S, Shapovalov V I et al. 2016 *Glass Phys. Chem.* **42** 359
[11] Bondarenko A S, Kolomiytsev A A and Shapovalov V I 2016 *J. Phys.: Confer. Series.* **729** 012006
[12] Shapovalov V I and Smirnov V V 2017 *J. Phys.: Confer. Series.* **857** 012039

Mechanisms of epitaxial growth of SiC films by the method of atom substitution on the surfaces (100) and (111) of Si single crystals and on surfaces of Si films grown on single crystals Al_2O_3

S A Kukushkin and A V Osipov

Institute of Problems of Mechanical Engineering, RAS, St.-Petersburg, 199178 Russia

E-mail: sergey.a.kukushkin@gmail.com

Abstract. The structure and morphology of silicon carbide (SiC) films formed by a new method of atom substitution on the surfaces (100) and (111) of silicon substrates (Si) were investigated. It is shown that the singular faces (100) of Si in the process of substitution of silicon atoms for carbon atoms are transformed into a face of SiC, consisting of an ensemble of facets resembling saw-tooth structures, the side surfaces of which are covered with planes (111) and (110) and (210). Theoretically, a fundamentally different mechanism of atom substitution on the faces (111) and faces (100) is described and experimentally confirmed. It was found that on the vicinal surface of Si, deviated 4° or more from the singular face (100), during the synthesis of SiC, an ordered phase of SiC is formed, the surface morphology of which appears as of facets (scales) consisting of faces (111). It is shown that the method of topochemical substitution of atoms can grow epitaxial SiC layers not only on a silicon substrate, but also on substrates of other materials, in particular, on single-crystal sapphire substrates.

1. Introduction

In papers [1–4] a new method of obtaining of nano-SiC epitaxial films on Si, the so-called method of atom substitution, was theoretically predicted and experimentally discovered. This method of growth of nano-SiC is fundamentally different from all existing methods and technologies of growing single crystals, films and nanostructures. This method is based on the " assembly " of a new silicon carbide matrix on the basis of the old silicon matrix by partial substitution of silicon atoms in the crystalline matrix of the substrate with carbon atoms. The "assembly" of the SiC matrix is carried out by chemical reaction (1)

$$2Si(cr) + CO(gas) = SiC\ (cr) + SiO\ (gas)\uparrow\ . \qquad (1)$$

In [2–4] it was proved that the reaction (1) proceeds in two stages. In the first stage, an intermediate compound is formed – activated complex "silicon vacancy-carbon atom-matrix-silicon" ($C{-}V_{Si}$). Carbon atoms at this stage are located in the interstitial positions of the silicon matrix. At the second stage of the reaction, the activated complex passes into silicon carbide, and the released vacancies merge into pores formed under a layer of silicon carbide. In a cubic symmetry crystal, these two centers of dilatation (a carbon atom and a silicon vacancy) interact elastically with each other. If dilatated dipoles are perpendicular to the plane (111) of silicon, they are attracted to each other [2–5]. At the same time, almost all dilatated elastic energy, which arose due to the introduction of a carbon atom and the formation of a vacancy, relaxes [2–5].

It should be noted that the strict quantum mechanical calculation of the rate of chemical reaction (1) and the interaction of a silicon vacancy with a carbon atom was described in [6, 7]. After the chemical transformation is complete, the mechanical dipoles that have fulfilled their role disintegrate into a film of silicon carbide and pores under its surface. The orientation of the film is determined by the "old" crystal structure of the original matrix of Si, and not only the surface of the substrate, as it usually occur in traditional methods of films growing.

2. Structural and morphological features of formation of SiC films on singular surfaces Si (111) and Si (100)

A distinctive feature of the method of substitution of atoms is that, regardless of the initial crystallographic plane of Si on which SiC was synthesized, a plane [111] is necessarily formed as one of the planes. This effect is due to the fact that the elastic dipoles carbon atom – silicon vacancy, which was mentioned above, are always perpendicular to the direction [111] and lie in the plane [110] perpendicular to the plane (111).

Thus, according to the work [3], the nucleation rate of SiC embryos on a smooth densely packed face $I(n_{cr}, \beta_{cr})$ can be found from the expression (2)

$$I(n_{cr}, \beta_{cr}) = \frac{2D_{cr}}{\sqrt{\pi}}(N_d + N^{Si})\exp(-R_{cr}/k_B T), \tag{2}$$

and the rate of SiC nucleation on the Si vicinal face $I(n_{cr}, \beta_{cr})$ from the expression (3)

$$I(n_{cr,V}) = \frac{2D_{cr,V}}{\sqrt{\pi}} N_d \exp\left(-\frac{R_{cr,V}}{k_B T}\right), \tag{3}$$

where $R_{cr}(\cos\varphi(x,y,z)) = \frac{\Gamma_1\Gamma_2(4\pi\Gamma_2-\Gamma_3\varepsilon)}{[\zeta-\Lambda(\cos(\varphi(x,y,z))]^2} + \frac{\pi\Gamma_2\Gamma_3^2}{\mu^{Si}\varepsilon[\zeta-\Lambda(\cos(\varphi(x,y,z))]}$ is the work of formation of the embryo SiC containing n_{cr} molecules of SiC and etching pits on the closely packed faces, depending on the location of dilatation dipoles, relative to the crystallographic coordinate of the cubic crystal, i.e. to the cos((x,y,z)); $R_{cr,V} = \frac{4\pi\Gamma_1''\Gamma_2^2}{[\zeta-W(\cos((\varphi(x,y,z))\,N_d]^2} + \frac{4\pi a\Gamma_2^2(\frac{\Gamma_2'}{\Gamma_2}+\cos\theta/\sqrt{3})}{[\zeta-W(\cos((\varphi(x,y,z))\,N_d]}(R_{cr,V}) = 4\pi\Gamma_1$ is the work of formation of the embryo SiC containing n_{cr} molecules on SiC vicinal faces, which depends both on displacement of dilatation dipoles, which depends on its location relative to the crystallographic coordinate of the cubic crystal cos(x,y,z), and the angle of deviation of vicinal faces from faces (111) θ ; $\beta_{cr}=R_{cr}/H_{cr}$, the ratio of the critical dimension of the formed shrinkage pore to the critical layer thickness of SiC; $\zeta = k_B T\frac{(V^{SiC}+V^{Si})}{(V^{SiC}V^{Si})}\ln\left[\frac{P_{CO}K^{eq}}{P_{SiO}}\right]$; $\Lambda(\cos(\varphi(x,y,z)) = W(\cos((\varphi(x,y,z))\,N_d + Y_{SiC}$; V^{Sic} , volume per atom in a cubic lattice of SiC; V^{Si} , volume per atom in the cubic lattice of Si; k_B is the Boltzmann constant; T, temperature of synthesis of SiC, depends on the process and lies in the range 1000–1400 ^{0}C (that not to exceed the melting point of silicon); K^{eq} is the equilibrium constant for the formation of SiC from Si under its processing by gas CO; P_{CO}, pressure of carbon monoxide in the vapor phase over the surface of the Si substrate , and P_{SiO} , pressure of silicon monoxide in the reaction zone; W (cos((x,y,z)), elastic energy of the crystal in the formation of dilatation dipoles in its volume, depending on their location relative to the crystallographic coordinates of the cubic crystal cos ((x,y,z)).

The energy W (cos((x,y, z)) can take both positive and negative values depending on the direction of the face; approximately it equals zero for the direction W ([111]) $\approx$ 0 eV at the arrangement of elastic dipoles along the direction [111] and is maximal at the arrangement of dipoles along the direction [100] W ([100])~1 eV; N_d is the density of elastic dipoles in the volume of the intermediate complex in silicon, the rate of chemical reaction of SiC formation is determined, on average, $N_d \approx 1{,}6 \cdot 10^{28} m^{-3}$; $Y_{SiC} = \mu^{Si} K^{SiC} \varepsilon^2 / 2(K^{SiC} + \mu^{Si})$; $\mu^{Si} = 0{,}52 \cdot 10^{11}$ N/m^2 is a shift module of Si; K^{SiC}=2,5·10^{11} N/m^2 is the compression module of SiC; ε is the strain in the surface layer of the substrate that occurs due to the difference in distance between the atoms in Si and SiC along the planes (111), ε ≈ 0,2; since $K^{SiC} \gg \mu^{Si}$, then $Y_{SiC} \approx \mu^{Si}\varepsilon^2/2$;, *a* is the interatomic distance in silicon matrix; Γ_1 describes the change of the total surface energy of the upper and lower faces of the disc-shaped SiC embryo and the pore in the conversion of Si in SiC [8]; Γ_2 is the total surface energy of the side faces of the embryo and of etching pit, bordering the shrinkage pore; Γ_3 is the total surface energy of the lateral surfaces of the embryo of SiC and of the pore in the Si; Γ_1" is the total surface energy of the embryo of SiC and of the pore on the vicinal crystal face; Γ_2' is the total surface energy of the embryo and of the etching pit, formed on vicinal faces; specific estimates of values Γ_1, Γ_2, Γ_1" and Γ_2' can be found in [3]; D_{cr} and $D_{cr,V}$ are pre-exponential kinetic coefficients calculated in [3]. From the formulas (1) and (2) it follows that the nucleation of SiC can occur only if ζ>Λ(cos(φ (x, y, z)) or, what is the same, the pressure of CO must satisfy the condition (4)

$$P_{CO} > \frac{P_{SiO}}{K^{eff}}, \tag{4}$$

where $K^{eff} = K^{eq} \exp\{-[\Lambda(\cos(\varphi(x,y,z))V^{SiC}V^{Si}/(V^{SiC}+V^{Si})k_B T]\}$ is the generalized equilibrium constant K^{eq}, taking into account the elastic energy Λ(cos(φ(x,y,z)).

An important feature of the formulas (2) - (4) describing the nucleation of the SiC layer on Si is that the work of the formation of the SiC embryo significantly depends on the direction along which the chemical bonds in the intermediate complex are broken. According to the formulas (2) and (3), along the direction [111], the term $\Lambda(\cos((\varphi(x,y,z)) = [W(\cos((\varphi(x,y,z)) N_d + Y_{SiC}]$ in the denominator of the work of formation R_{cr} (cosφ(x,y,z)) in the formula (2) or the term $[W(\cos(\varphi(x,y,z)) N_d]$ in the denominator of the work of formation $R_{cr,V}$ in formula (3), take the mimimal values . That means that the nucleation rate will be maximal only along the direction [111]. Namely, these faces (111) are formed primarily in the process of SiC synthesis by the substitution method.

In the case of a vicinal surface deviated from the plane (111) by a small angle, as follows from equation (3), the growth begins already at pressure P_{CO} at which $\ln\left[\frac{P_{CO}K^{eq}}{P_{SiO}}\right] > 0$; for example, for T= 1250°C this pressure must be $P_{CO} \geq 0.03 \cdot P_{SiO}$. In the case of other low-index faces, the situation is changing essentially. According to the calculations carried out in [3] W([100]) ~1. It follows that the value Λ(cos(φ(x, y,z)) for the face (100) will be Λ(cos(φ(x, y,z)) ≈ 3.4·10^9 J / m^3. This value of Λ (cos (φ (x,y,z)) leads to the fact that the formation of the SIC film by the method of atom substitution on the face (100) becomes possible only if CO pressure in the reaction zone is $p_{CO} > 0.5 \cdot p_{SiO}$, i.e. ten times higher than the stable film growth pressure [3] $P_{CO} \geq 25$ Torr. That is why, SiC can't nucleate on the smooth surface (100) of silicon. However, the nature bypasses such a high value of Λ(cos(φ (x,y,z)) transforming the face (100) into a SiC face consisting of a set of facets resembling saw-tooth structures, the side faces of which are covered with planes (111) and (110), and (210). The angle between the face direction (100) and the faces (111) is 54^0 44'. Any small deviation of the face (100) from this direction makes it vicinal, which leads to the instantaneous formation of a very thin layer (several atomic layers) of the film. The growth of thick SiC layers at this pressure is not possible, since a thin SiC layer will close all main pores, through which CO enters the inner layers of Si substrate during the synthesis.

3. Structural and morphological peculiarities of formation of a film of SiC on a vicinal surface of Si (100)

A completely different picture will be observed if SiC is grown by the method of substitution of atoms on the surface (100), disoriented to 2–7° in the direction <011>. If you heat the surface to temperatures in excess of 600 °C, the plane (100) of silicon, according to thermodynamics, will be covered with steps. The terraces of this structure will be planes (100), and the steps will be limited by planes (011). Along the directions <011 > there are "channels" in the silicon lattice, which are associated with the peculiarities of the crystallographic structure of the lattice. Therefore, the CO molecules are rushing in direction perpendicular to the steps inside Si. The Si surface is saturated with CO and, as described above, the interaction reaction between Si and CO occurs, as a result of which the intermediate state is formed that is converted to SiC. In this transformation the part of Si from the step (011) is evaporated and forms the step (111) of SiC. This process removes the "degeneracy" inherent to the non-deviated face (100) and leads to the formation of facets consisting of faces (111) of Si, which are located on the former steps (011), but not randomly displaced as on the face (100). Naturally, these faces make up the same angle of ~55° with the face (100) and an angle of ~35°with the face (011).

Since the attraction between the silicon vacancy and the carbon atom in the silicon matrix is maximal along the direction <111>, part of the step (011) of Si can turn into a fracture (112) of SiC. As is known, the angle between the planes (112) and (100) in a cubic crystal with a diamond lattice is ~35 degrees. As a result, a lateral wedge-like protrusion of silicon carbide is formed, which has an apex protruding above the site of the step and an tilted face reaching the site of the lower step, with the formation of a slope. The surface of Si (100) is covered by an array of wedge-liked parallel steps representing themselves triangular prisms (lateral faces of pyramids) [3, 12].

4. Structural and morphological features of the SiC film formation on the surface (100) of the epitaxial Si film grown on Al2O3

Using the method developed in [1–4], one can obtain the epitaxial layers of SiC not only on a silicon substrate, but also on substrates of other materials. To do this, a single-crystal layer of Si must be deposit on the working surface of the substrate, and then synthesized the SiC layer using the reaction (1). If the initial silicon carbide is epitaxial, the SiC layer would be epitaxial as well. Figure 1 shows the image of the surface and of the cleavage of a single-crystal substrate Al_2O_3 with a layer of epitaxial SiC obtained with a scanning raster microscope (SEM). This structure was obtained as follows. At the beginning, an epitaxial Si layer with a thickness of the order of 610–630 nm and of orientation (100) was deposited on the surface of the R-cut sapphire by CVD. Further, the sapphire substrate with a layer of Si was placed in a special device for the growth of SiC and was annealed at temperature 1250°C in the atmosphere of CO. As a result, an epitaxial layer of SiC was grown. The structure of the SIC layer is clearly visible in figure 1. It consists of a solid layer of SiC of a thickness about 58 nm and of the porous Si layer covered with a layer of SiC. In addition, it can be seen that the SiC layer has a greater degree of roughness.

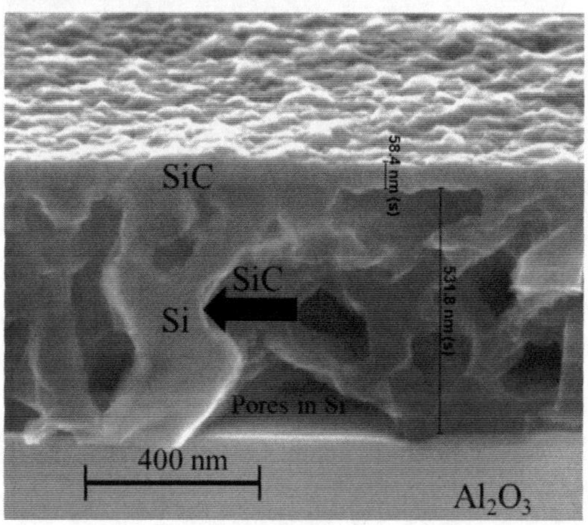

Figure 1. SEM-image of a cross-section cut of sapphire with a formed SiC layer and porous silicon covered with SiC with a thickness of about 532 nm.

The increased degree of SiC surface roughness is due to the fact that, as shown in the above, during the transformation of Si (100) into SiC, the SiC layer of orientation (100) is covered with saw-tooth faces (111). This is that is observed as SiC is growing on sapphire by the method of atom substitution.

5. Conclusions

Thus, it is theoretically and experimentally shown that SiC layers are growing as epitaxial layers on the orientation (111) of Si as well as on (100) orientation. However, the growth mechanism of SiC is fundamentally different. This leads to a significant difference in the surface morphology of SiC layers grown on Si substrates of these orientations. The surface (111) of Si remains smooth when transforming into a surface of SiC, while the surface (100) is covered with facets consisting of faces (111). It is shown that a similar mechanism is observed as SiC is growing from the Si layer grown on sapphire.

Acknowledgements

This work was supported by the grant of the RPF (grant No 14-12-01102). The authors are grateful to A. S. Grashchenko, N. Ah. Feoktistov and S. D. Fedotov for conducting experimental studies on SIC synthesis on sapphire. The work was performed using the equipment of the unique stand (UNO) "Physics, chemistry and mechanics of crystals and thin films" of IPM RAS.

References

[1] Kukushkin S A and Osipov A V 2008 *Phys. Solid State* **50** 1238–45
[2] Kukushkin S A and Osipov A V 2012 *Dokl. Phys.* **57** 217–20
[3] Kukushkin S A and Osipov A V 2014 *J. Phys. D: Appl. Phys.* **47** 313001–42.
[4] Kukushkin S A, Osipov A V and Feoktistov N A 2014 *Phys. Solid State* **56** 1507–35
[5] Kukushkin S A, Osipov A V and Telyatnik R S 2016 *Phys. Solid State* **58** 971–80
[6] Kukushkin S A and Osipov A V 2017 *Phys. Solid State* **59** 1238–41
[7] Kukushkin S A and Osipov A V 2017 *J. Phys. D: Appl. Phys.* **50** 464006–13

IOP Conf. Series: Materials Science and Engineering **387** (2018) 012045 doi:10.1088/1757-899X/387/1/012045

Improvement of performance properties of products by surface treatment of materials with cathode spots of vacuum arc discharge

V G Kuznetsov and T A Kurbanov

Institute of Problems of Mechanical Engineering, Russian Academy of Sciences, 199178, Saint Petersburg, Russia

E-mail: kvgipme@gmail.com

Abstract. It is shown that the impact of cathode spot of vacuum-arc discharge on the plastically deformed surface may be partially or fully recovered her form. This effect was detected during vacuum-arc cleaning of the inner surface of the cylinder. Mathematical and experimental modeling of the process of restoring the inside diameter of the cylinder confirmed the effect of shape recovery upon exposure of the cathode spots of a vacuum arc on the surface.

Fundamental and applied research in the field of mechanics and physics of interaction of plasma jets from the cathode spots of a vacuum arc with a solid surface in recent years greatly intensified in connection with the development of new applications of vacuum-arc discharge. In addition to the widely used direction of vacuum-arc coatings deposition [1], this type of discharge is mastered for hardening of the surface of metals [2], for vacuum arc cleaning [3] of the surface from scale, rust, oil film and other contaminants in mechanical engineering, metallurgy and other fields, for defectoscopy of surface layers of metal products [4], for protect the metal from corrosion by passivation of the surface [5] and in other directions.

When developing cleaning technologies of plastically deformation of inner surface of the cylinder by the cathode spots of vacuum arc discharge was discovered the effect of reducing of the internal diameter. By removing of dirt and minor sputtering surface of the metal layer the diameter of the cylinder must be increased. As the cylinder was used the housing case of axle box of railway wagon. Housing case of the axle box perceives and transmits to wheelset gravity of the loaded wagon, as well as dynamic loads that occur when driving. In the process of operation the axle box is subjected to heavy loads, causing the appearance of plastic deformations in the areas of contact of the housing with the bearing bushing [6]. As these loads are distributed unevenly on the inner surface of the cylinder inner radius of surface due to plastic deformation changes non-uniformly, this creates ovality of the inner surface.

In figure 1 are presented experimental data of changes of the internal diameter of the 190 cylinders when vacuum-arc cleaning.

To develop methods for the measurement of thermal fields and deformations during vacuum-arc treatment of the inner surface of the cylinder was created the experimental apparatus. Figure 2 shows a photograph (figure 2(a)) and scheme of the experimental equipment and the measurement procedure (figure 2(b)).

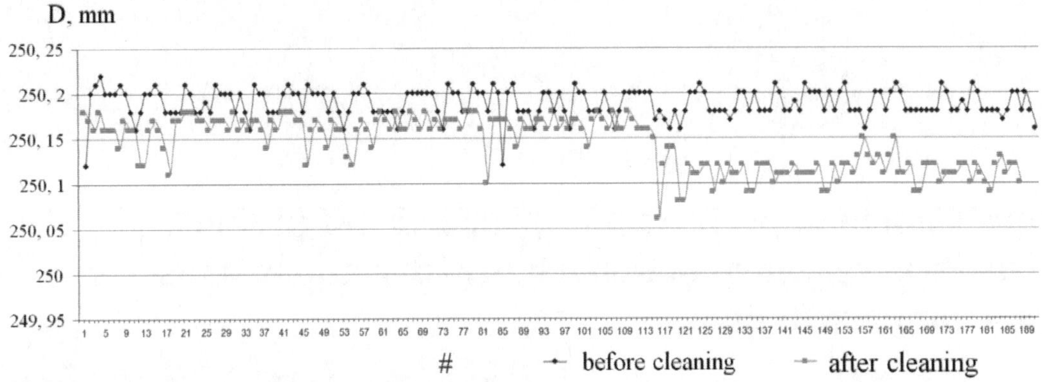

Figure 1. The change in the internal diameter of the 190 cylinders after vacuum-arc cleaning the surface (D – diameter, # – identification number).

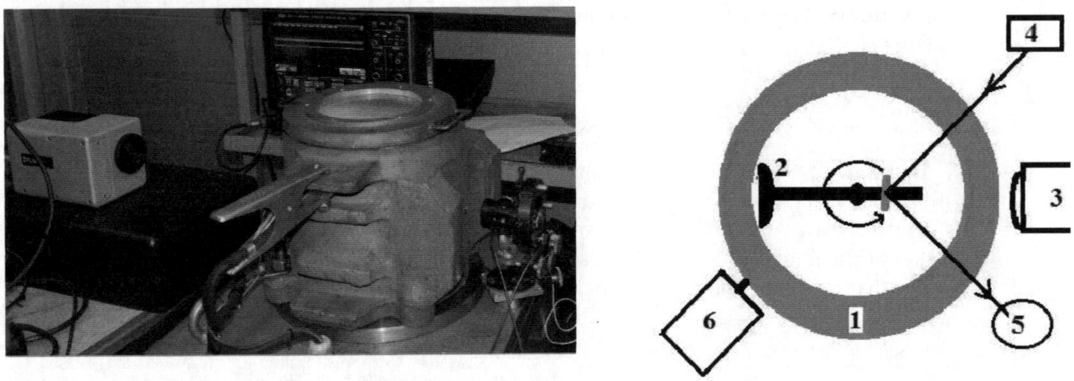

a) b)

Figure 2. Photograph (**a**) and diagram (**b**) measurement of temperature and displacement of the external surface of the cylinder by vacuum-arc heating (1 – cylinder-cathode, 2 – electrode-anode, 3 – thermal imaging camera, 4 – semiconductor laser, 5 – photo detector, 6 – interferometer).

Cylinder is installed on the vacuum post and is closed by a lid with a viewing window. Temperature measurement was carried out by thermal imaging camera (3) ThermaCAM™ SC 3000 (figure 2(b)). For studies of the movements of the outer wall of the cylinder in the process of vacuum-arc heating was used a laser interferometer with a sensitivity of better than 5 nm. In the interferometer implements the classical scheme of the Michelson. One of the mirrors of the interferometer was placed on an elastic membrane with a central steel probe ∅1×5 mm, which was in contact with the wall of the cylinder. In the measurement counted the number of photoelectron bands. Synchronization of the measurements with the movement of the rotating electrode was carried out by the signal from the photodetector (5), arising in the penetration of the laser radiation (4), reflected from a mirror, rigidly connected to the axis of rotation of the anode (figure 2(b)).

Vacuum arc discharge was ignited between the axle box housing (cathode) 1 and rotating along the surface the electrode 2 (anode). The cathode spots of a vacuum arc, followed by the anode, carried out the surface cleaning and at the same time heated surface layer of the cylinder.

Obtained the combined in time dependencies (figure 3) of temperature change (T) and displacement (U) of the outer wall of the cylinder in one pass of the electrode by heating on 2 °C. Shows a qualitative coincidence time dependencies corresponding to the classical model of thermoplasticity.

Developed experimental methods of measuring the temperature and movement of the outer surface of the cylinder possible to obtain data for verification of boundary conditions for solving nonstationary problems of thermoplasticity by finite element method. Conducted finite-element analysis of transient heating of a cylindrical pipe with an initial ovality generated by the uneven pressure on the inner surface of the cylinder.

Consider the finite element model. Owing to the symmetrical geometry of the problem consider one quarter of the cylinder. Thereby neglecting the asymmetry due to local heating by a heat source moving on a circle with angular velocity of 2 revolutions per minute. After processing of the inner surface by means of uneven pressure occurs ovality.

Figures 4 and 5 shows the picture of radial displacements after the removal of the internal pressure and a graph of radial displacement.

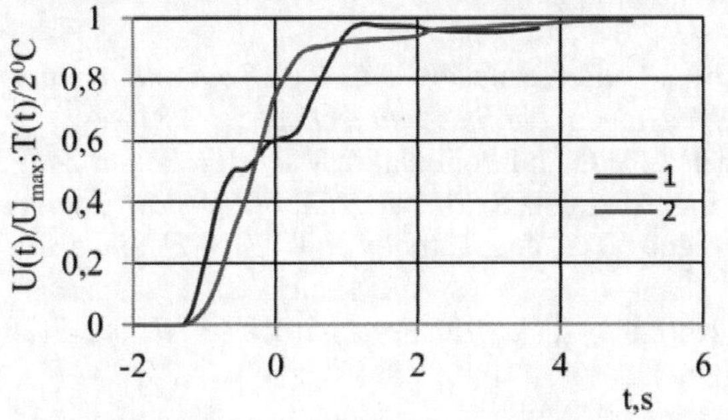

Figure 3. Combined time dependencies of temperature change (1) and moving of the outer wall of the cylinder (2) for one pass of the electrode.

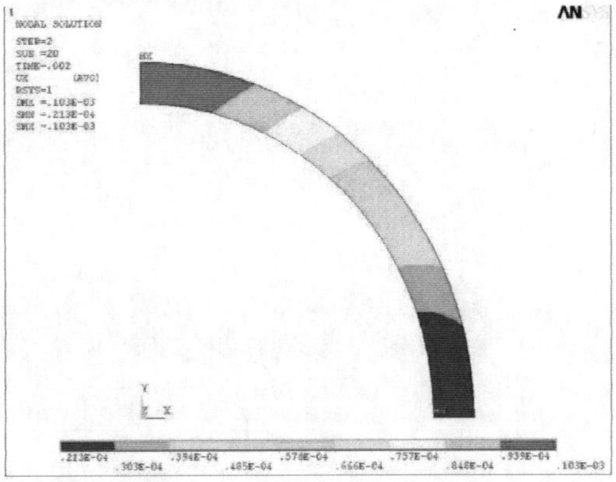

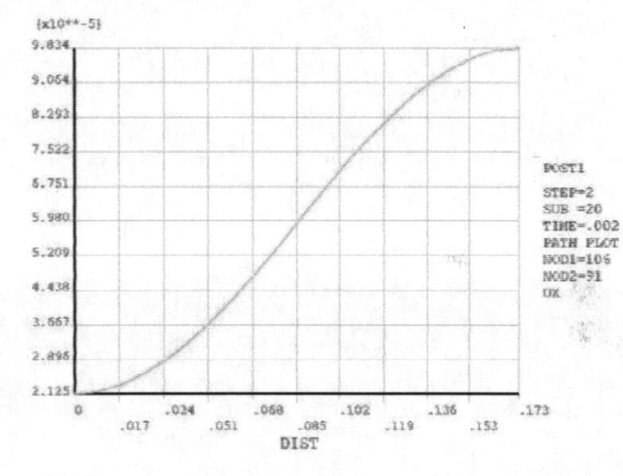

Figure 4. Picture of radial displacements after the removal of the internal pressure.

Figure 5. Radial displacements after the removal of the internal pressure at the circumference.

In the presence of initial ovality of the inner surface, that is, the difference of the values of the radii is equal to 77 μm after two passes of the electrode is reduced ovality, and the difference of the values of the radiuses decreased to 40 microns.

In the result of experimental and theoretical studies on the effect of vacuum-arc exposure on the inner surface of the housing case axle-box of railway wagon has been shown the reducing of out-of-roundness acquired during a maintenance period.

When the load is removed on the inner surface arise the compressive circumferential stress, the intensity of which depends on the magnitude of applied normal pressure. Intense local heating of the surface compression zone creates a thermal expansion and, as a consequence, plastic deformation, depending on the pre-compression: the more pre-compression, the greater the deformation, which explain the effect of reducing the internal diameter and ovality.

Heating of the surface and restoring its shape was carried out by the discharge from the surface of the cylinder to the rotating electrode. Pattern of radial displacements after the removal of the internal pressure and graph of radial displacement was obtained.

Comparison of radial displacements of the inner surface after removing the pressure and after final cooling to room temperature shows the real effect of reducing the induced ovality during the operating period. It is shown that in addition to reducing of ovality due to non-stationary localized heating of the inner surface of the cylinder decreases and her inner diameter. With the help of experimental studies and mathematical calculations on the developed models confirmed the possibility to recover the plastically deformed surface by means of the cathode spots of a vacuum arc at currents of 100-200 A up to 70 μm. When increasing the current of the vacuum arc discharge and adjusting the speed of movement of cathode spots can significantly reduce the ovality

References

[1] Valuev V P, Rybnikov S I and Kuznetsov V G 2004 *Tool and technology* **17–18** 33–40
[2] Kuznetsov V G 2009 *Vacuum technique and technology* **2** 81–4
[3] Indeitsev D A, Kuznetsov V G and Ponimatkin V P 2011 *Patent 2457282 RU* 27.07.2012
[4] Kuznetsov V G and Babushkina E S 2016 *Journal of Physics: Conference Series* **729** 012003
[5] Kuznetsov V G, Rurbanov T A and Tatonov V K 2018 *Engineering technology and materials science* **2** 80–3
[6] Kuznetsov V G and Kurbanov T A 2017 *Patent 2635203 RU* 09.11.2017

IOP Conf. Series: Materials Science and Engineering **387** (2018) 012046 doi:10.1088/1757-899X/387/1/012046

Improving the adhesive strength of the babbit layer of sliding bearings

V G Kuznetsov and T A Kurbanov

Institute of Problems of Mechanical Engineering, Russian Academy of Sciences, 199178, Saint Petersburg, Russia

E-mail: kvgipme@gmail.com

Abstract. The possibility of increasing the adhesion of the babbit coating applied by means of casting on the bearings inserts by preparing of the surface of the bearings inserts by processing with help of cathode spots of vacuum arc discharge is shown. At the same time, the vacuum purity of the surface is provided and a roughness with a partially-regular micro relief is created. In addition, it is possible to increase the adhesive properties of the coating by applying a vacuum arc method of deposition of a thin sublayer of the coating material.

The development of modern turbine construction requires new technical solutions that determine the possibility of increasing the efficiency, reliability and technological effectiveness of turbine plants. This is especially true for power engineering, where sliding bearings are widely used as turbine bearings – the most responsible products, the failure of which entails the failure of the entire unit. To reduce wear, and therefore increase the service life of the bearing, an anti-friction material is applied to its surface. Babbits are used as an anti-friction material, for example, B83 [1]. Babbit B83 stamps consists of 6 % copper, 10 to 12 % antimony, the rest is the tin. The basic technology of coating deposition of babbit on the bearings is casting. In the manufacture of bearings babbits is poured to the heated inserts (250 ^{0}C) at a temperature of 450–480 ^{0}C. The level of casting technology is low, as evidenced by statistics on unscheduled and emergency failures of sliding bearings [2].

When using casting technologies (siphon casting, centrifugal casting) to obtain an anti-friction layer, the following problems arise: the liquation of alloying elements, progressing with increasing of layer thickness, a high tendency to defects such as shells and pores. Also, due to the low cooling rates, an unfavorable structure is formed during the crystallization of babbit [3]. Some improvements are observed when using turbulent casting. However, this technology does not solve the problem with low strength of babbit layer adhesion to the base. To increase the adhesion strength, it is known of using such technological methods as milling of end or spiral grooves, recesses, cuts, application of copper coating to the surface of the bearings inserts before tinning with tin, use of alternative methods of applying babbit coatings, for example, plasma spraying and surfacing [4]. A significant disadvantage of babbits is also the fact that with increasing temperature they reduce all the indicators of mechanical strength, especially fatigue resistance.

The aim of this work is to develop vacuum ion-plasma methods for preparing of the surface of the bearings inserts to provide high adhesion of the babbit coating applied by casting.

With proper preparation of the surface of the bearings inserts and its filling with babbit, a strong connection of babbit and metal should occur along the entire surface of the bearings inserts, which significantly increases the adhesive strength of the coating with the substrate. Therefore, before applying babbit must be removed from the surface of the bearings inserts all existing contamination. Another factor that increases the adhesive strength of the coating with the substrate is to provide a given surface roughness with a regular or partially-regular microrelief. In addition, high adhesive properties of the coating can be provided by creating a thin diffusion layer at the interface substrate-coating and a subsequent thin coating from the coating material (babbit). The implementation of the above-mentioned directions of increasing the adhesive strength of babbit coatings can be provided by the treatment of the surface of the bearings inserts by the cathode spots of vacuum arc discharge and subsequent treatment of the surface by the accelerated flows of metal plasma.

The method of vacuum arc cleaning of the metal surface or the method of cleaning the surface by the cathode spots of the vacuum arc provides a complete cleaning of impurities such as rust, scale, residues of technological grease and others and allows for precision dimensional surface treatment, and contributes to effective visual and technical surface defectoscopy. Cleaning is carried out by means of a vacuum arc discharge burning between the surface of the processed product, which is a cathode, and an anode in a vacuum chamber, in which a lot of fast-moving cathode spots are formed, localized in the places of accumulation of contaminants, oxide films and removing them until the main material of the processed product is completely exposed.

Studies have shown that under the processing conditions typical for surface cleaning, there are no structural-phase changes in the properties of the product material. At the same time, a change in the nature of the surface roughness relief was found (figure 1). In figure 1(a) is shown the initial profile of the surface of detail from steel 20GL, and figure 1(b) – profile of the surface after vacuum-arc cleaning. The arithmetic mean deviation of the Ra profile for the initial surface was 2,944 μm, for the surface after cleaning – 2,561 μm.

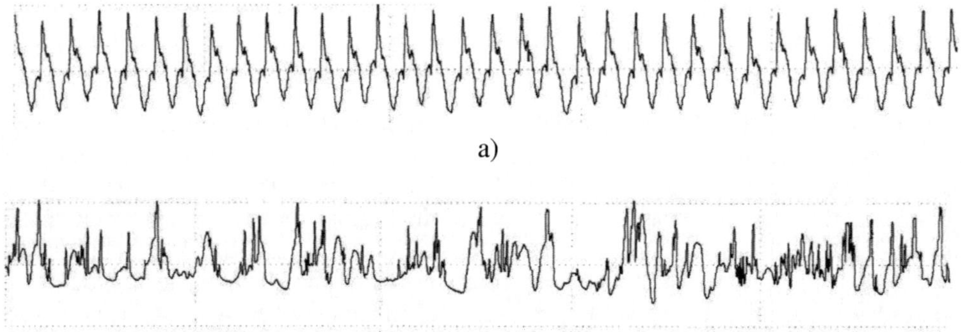

a)

b)

Figure 1. Surface roughness profile: (**a**) – initial; (**b**) – after cleaning.

The initial sample (without cleaning) has a regular profile with approximately equal dispersion of protrusions and depressions. The surface subjected to vacuum-arc treatment is uniformly covered with depressions (figure 2(a)).

At higher magnification (figure 2(b)) it is show that depressions have the shape micro-pocket. This surface is consistent with the definition of micro-relief elements of concave shape. The studies show that the vacuum arc treatment of the surface of materials due to the impact on it of the cathode spot of the vacuum arc can be used to create a regular micro-relief in the form of micro-pockets to improve the adhesive properties of babbit coatings applied to such a surface. As the babbit has been used the babbit B83 stamps.

As a bearings inserts for the application of babbit a sleeve made of steel 20 with an external diameter of 100 mm and an internal diameter of 70 mm was used. Cleaning of the inner surface of the sleeve with the simultaneous creation of a microrelief was carried out in the vacuum chamber of the installation. The bushing was a cathode and was connected to the negative pole of the welding rectifier. Inside the bushing was inserted graphite rod, which served as the anode. Surface cleaning was carried out in the pressure range from 1 to 10 Pa at a vacuum arc discharge current of 100 A.

a) b)

Figure 2. The topography of the surface after vacuum-arc cleaning: the increase in (**a**) – ×100, (**b**) – ×25000.

The creation of a diffusion layer and a transition layer with a thickness of about 5 μm between the substrate and the babbite coating, which was subsequently applied by casting, was carried out with the using of the traditional technology of vacuum arc deposition. As the cathode of the vacuum arc evaporator, a babbit cylindrical ingot was used, which was placed inside the bushing. Thus, the so-called cylinder vacuum-arc evaporator was created.

The adhesion of babbit layer to the substrate was determined by the ring shear method [5]. Pressure on the babbit ring was carried out by a punch. Pressure on the babbit ring was carried out by a punch. Measurement of forces was carried out using a mechanical dynamometer. The results of the researches are presented in the table 1.

Table 1. Results of measurement of adhesion properties.

Surface preparation and material of deposition layer	Average load to failure, N
Chemical cleaning - tinning by tin - babbit	5400
Cleaning by cathode spots with the creation of microrelief - tinning by tin - babbit	6800
Cleaning by cathode spots with the creation of microrelief – babbit	6600
Cleaning by cathode spots with the creation of microrelief - diffusion layer of babbit - tinning by tin – babbit	7500
Cleaning by cathode spots with the creation of microrelief - diffusion layer of babbit - transition layer of babbit – babbit	7200
Cleaning by cathode spots with the creation of microrelief - diffusion layer of babbit - transition layer of babbit - tinning by tin - babbit	7800

The results presented in table 1 allow us to draw the following conclusions:

1. Cleaning of the surface by the cathode spots of the vacuum arc with simultaneous creation of a microrelief on the surface in comparison with the chemical preparation of the surface provides an increasing of the adhesive properties of the babbite coating by about 26 %.

2. Babbit, deposited on the surface of the bearing insert, treated by the cathode spots of a vacuum arc, without tinning by tin slightly inferior in the adhesion of the babbit deposited on the surface with pre-tinning.

3. The diffusion layer produced during ion bombardment of the surface is promotes the increase of babbit adhesion.

4. Tinning by the tin slightly improves the adhesion properties of a babbit coating in comparison with the case where the babbit is deposited on the transition vacuum-arc layer of babbit.

5. The maximum adhesion is obtained for the case of "Cleaning by cathode spots with the creation of microrelief - diffusion layer of babbit - transition layer of babbit - tinning by tin - babbit ", but it is necessary to take into account the economic efficiency of the process. Perhaps, the process of "Cleaning by cathode spots with the creation of microrelief - diffusion layer of babbit - tinning by tin – babbit" may be more economically feasible.

References

[1] Potehin B A, Glushenko A N and Ilushin V V 2006 *Metal technology* **3** 17–22
[2] Mihailova M A 2005 *Bulletin AGTU* **2** 135–40
[3] Potehin B A, Ilushin V V and Hristolubov A S 2009 *Metallurgy and heat treatment of metals* **8** 16–21
[4] Hristolubov A S, Potehin B A, Mihailov B S and Skvortsov A A 2008 *Bulletin IGTU* **4** 33–5
[5] Terelnik V B, Martsinkovskiy V S, Belous A V 2010 *Bulletin HNTUA* **94** 102–8

IOP Conf. Series: Materials Science and Engineering **387** (2018) 012047 doi:10.1088/1757-899X/387/1/012047

System for deposition of diamond like *a*-C:H films in Ar-C_2H_2 plasma with ion beam assistance

I V Lopatin, Yu H Akhmadeev, D Yu Ignatov, S S Kovalskiy and E A Petrikova

Institute of High Current Electronics, Siberian Branch, Russian Academy of Sciences (IHCE SB RAS), 634055, Tomsk, Russia

E-mail: lopatin@opee.hcei.tsc.ru

Abstract. The paper presents a system which comprises a gas plasma source and an ion source and allows surface treatment of dielectrics due to ion charge compensation by plasma electrons. The ion energy corresponds to the discharge voltage of the ion source. The hardness of a-C:H coatings deposited by the method is up to 40–45 GPa.

1. Introduction

The high wear resistance, high hardness, and low friction coefficient of diamond-like carbon (DLC) coatings make them promising for deposition on cutting tools. DLC coatings can be formed by physical vapor deposition, e.g., vacuum arc [1] or magnetron sputtering [2], and by plasma-enhanced chemical vapor deposition in different discharges, e.g., rf discharge [3] or glow discharge [4]. The formation of carbon and hydrocarbon films with desired parameters depends strongly on the ion energy delivered to the surface of a growing film. In pulsed vacuum arc deposition, this energy is specified by the parameters of the discharge and plasma filtering system, and in rf plasmas, by the self-bias of a substrate. Such methods of controlling the ion energy severely limit the operating conditions of discharges, which produce carbon-containing plasmas, and narrow the range of operating parameters. Often, a bias is applied to a substrate with a growing DLC film [4], making it possible to vary the ion energy independently of the plasma generation conditions. However, DLC coatings are non-conducting, and the bias pulse should thus be short (up to tens of microseconds), which limits the coatings thickness. Another shortcoming of such systems is the impossibility to treat dielectric materials. In this context, ion beam technologies appear promising for the formation of DLC coatings [5, 6]. In these technologies, the energy to the surface of a growing film is delivered by ions which are accelerated by a separate device with independently controlled parameters. The use of only ion beams for the formation of dielectric coatings is problematic because of the necessity to remove the positive charge from the surface of a growing film. Therefore, it is reasonable and promising to use an ion beam jointly with a plasma whose electrons can compensate the ion charge at the surface of a coating. The main discharge plasma present in the region of substrates can serve not only as a carbon ion source but also as an electron source to provide the operation of an auxiliary non-self-sustained discharge. The auxiliary discharge can serve as a source of accelerated ions which deliver a desired energy to a coating.

IOP Conf. Series: Materials Science and Engineering **387** (2018) 012047 doi:10.1088/1757-899X/387/1/012047

The non-self-sustained mode of auxiliary discharges allows one to control the discharge parameters and hence the energy and current density of ions over a wide range.

Here we propose a system based on a hollow-cathode non-self-sustained glow discharge for the generation of an ion beam and present research data on its capabilities for depositing DLC coatings.

2. Experimental

In our experiments, we used a system with a vacuum chamber of dimensions 600×600×600 mm (figure 1) preliminary pumped by a turbomolecular pump to a pressure of no worse than 5×10^{-2} Pa at a rate of 500 l/s. In the chamber, an gas arc discharge was ignited by a PINK plasma generator with a thermionic and hollow tungsten cathodes [7]. At the outlet of its cathode cavity, a diaphragm of diameter 20 mm was located; the gas supplied to the cathode cavity was argon. The diaphragm allowed us to reduce the arrival of acetylene from the working chamber to the region of the hot cathode, precluding its fast chemical erosion. At the same time, the operation of the main discharge was impeded, and conditions were provided for electron acceleration in the diaphragm region. The acceleration of electrons at the cathode cavity outlet enhanced the ionization efficiency of acetylene in the vacuum chamber, and it was required to increase the discharge voltage to 120–130 V (against standard 35–60 V). The discharge was powered by a specially designed (homemade) power supply with an output current of 120–50 A and operating voltage of 60–200 V, respectively. The power supply was equipped with an arc quenching system. The water-cooled inner walls of the vacuum chamber and the heat-insulated emission grid electrode of the ion source were at the potential of the main discharge anode.

During the operation with plasma-forming hydrocarbon gases, the water-cooled walls of the vacuum chamber were covered with a non-conducting hydrocarbon film within several minutes. Under such conditions, the main discharge current took the path exclusively to the emission grid electrode, as the hydrocarbon films on its hot surface were graphitized and became conducting. The ion source represented a water-cooled casing which was electrically insulated from the chamber walls and contained a cylindrical cathode cup of inner diameter 100 mm and height 50 mm on the inside. The emission electrode was fixed on the side flange of the vacuum chamber such that its hole was fully overlapped. In the emission electrode, there was a window of diameter 100 mm covered with a braided fine grid (0.5 × 0.5 mm) having a geometric transparency of ≈ 45 %. The cathode cup and the emission electrode were coaxially fixed at 25 mm from each other and were placed at the potential of the vacuum chamber, forming a hollow cathode with a ring gap which opposed an axially located heat-insulated ring anode of dimensions ∅150×30 mm. The system operated as follows. The electrons from the main discharge plasma penetrated through the grid of the emission electrode to the hollow cathode of the ion source and oscillated there, ionizing the gas supplied through the gas inlet of the ion source casing. The ions from the auxiliary discharge plasma were accelerated in the cathode layer near the emission electrode and injected back to the vacuum chamber through the emission grid. The design has the advantage that all electrodes are heated to temperatures at which hydrocarbon films on their surface are graphitized and become conducting, and this allows long-term operation (for more than 2 h) with no cathode and anode spots. The non-self-sustained glow discharge was powered by a pulsed source with an output voltage of up to 1 kV, average current of up to 10 A, and frequency of up to 50 kHz at a pulse duty factor of 10–80 %.

The ion beam energy was measured with a single-grid energy analyzer. The analyzer is a collector to which a variable bias was applied, precluding the arrival of ions with an energy lower than its potential to its surface. At a distance of 1 mm upstream of the collector, there was a grid with a mesh size of 0.4×0.4 mm, diameter of 16 mm, and geometric transparency of 40 % fixed on the stainless steel casing. No electrons from the main discharge plasma arrived at the collector because a negative bias of 250 V was applied to the grid of the energy analyzer, and its value was sufficient for the formation of a cathode layer in the main discharge plasma. The collector was placed coaxially to the window of the emission electrode at a distance of 50 mm.

IOP Conf. Series: Materials Science and Engineering **387** (2018) 012047 doi:10.1088/1757-899X/387/1/012047

The instantaneous values of the potential and collector current were measured using an automated digital recording system. The beam energy characteristics were measured at an Ar pressure in the vacuum chamber of 0.3 Pa with no gas supply to the ion source and with the main discharge current chosen to provide an auxiliary discharge current of 5 A.

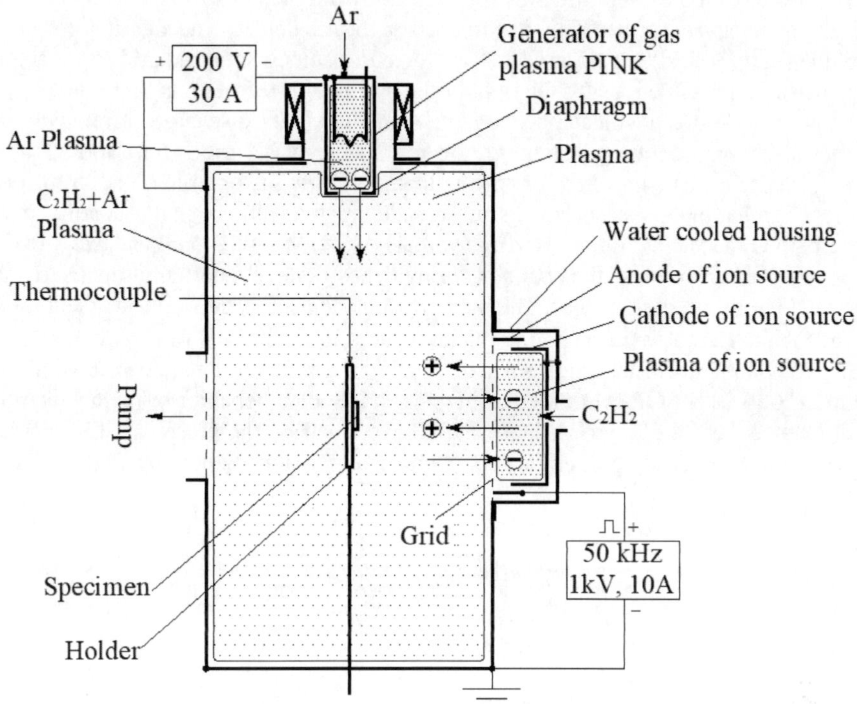

Figure 1. Schematic of experimental arrangement.

The substrate material for depositing a-C:H films was tungsten carbide (VK8 grade). The substrates were polished and were fixed on a holder kept at floating potential in the center of the chamber. The temperature of the holder was measured with a thermocouple. Directly before deposition, the substrates were exposed to ion cleaning and heating to 130 °C with an Ar ion beam. The deposition time was 40 min. The temperature of the holder at the end of the process was no greater than 170 °C.

The operating pressure in the chamber was measured using an ionization gauge. The system operates such that during the discharge operation in an acetylene-containing atmosphere, the pressure readings decrease 5–6 times, depending on the parameters of the main and auxiliary discharges. This is likely because acetylene is dissociated with the release of a large amount of hydrogen, and the relative sensitivity of the used pressure gauge to hydrogen differs greatly from its sensitivity to argon and acetylene. To avoid measurement errors, the acetylene flow during deposition was controlled with an RRG-10 flow meter, and the pressure in the chamber was measured with the discharges off. All coatings were deposited at a partial acetylene pressure of 1.2 Pa, which corresponds to a flow rate of 18 l/min, and at a partial argon pressure of 6×10^{-2} Pa. The partial pressure ratio was $Ar/C_2H_2 = 1/20$.

The coating microhardness was measured with a PMT-3M device, and the nanohardness with a Nanotest 600 device (Micro Materials).

IOP Conf. Series: Materials Science and Engineering **387** (2018) 012047 doi:10.1088/1757-899X/387/1/012047

3. Results and discussion

Our measurements of the ion energy (figure 2) show that at a pressure of 0.3 Pa, the maximum ion energy corresponds to the auxiliary discharge operating voltage with an accuracy of several volts. By estimates, the maximum range of ion energies is 60–70 V. Here, it should be taken into account that the range of ion beam energies depends strongly on the operating pressure, distance from the emission grid, and form factor of the emission and energy analyzer grids. The maximum ion beam density (accounting for the geometric transparency of the analyzer grid) is 1.25 mA/cm^2 and increases somewhat if the grid form factor is allowed for. The current density of plasma ions (to the right of the ordinate) is 0.5 mA/cm^2.

The coatings were deposited at constant average main and auxiliary discharge currents of 10 and 3.5 A, respectively. The discharge operating voltage in the ion source was varied from 300 V to 700 V. The auxiliary discharge current was controlled by varying the pulse duty factor from 65 to 40 %.

According to microhardness measurements at an indenter load of 50 g, the coating microhardness increases from 25–35 GPa at an auxiliary discharge voltage of 300 V to 40–45 GPa at 625–700 V. The wide microhardness spread occurs because the coatings during deposition were partially separated and were thus discontinuous and nonuniform in thickness. The poor coating adhesion is due to insufficient ion cleaning of the surface from cobalt. The maximum coating thickness is 5.7 μm, and the film growth rate is thus ≈ 8.5 μm/h. According to nanohardness measurements at a load of 2–20 mN, the nanohardness of all specimens increases from 12–16 to 19–24 GPa with increasing the load. Such a behavior of the nanohardness and its considerable difference from the microhardness can be explained by the presence of a thin surface layer (tens of nanometers) with modified properties due to chemical reactions with atmospheric air.

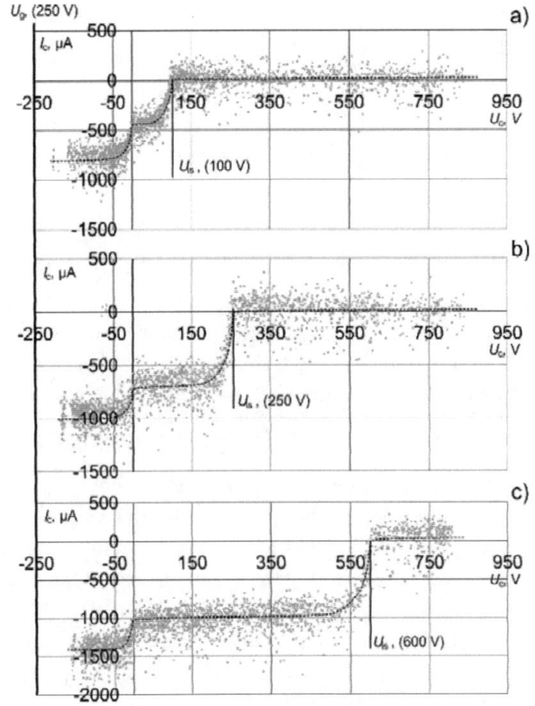

Figure 2. Collector current versus its voltage at auxiliary discharge voltages equal to 100 V (**a**), 250 V (**b**), and 600 V (**c**).

4. Conclusion

Thus, the ion-plasma system based on a non-self-sustained glow discharge provides the generation of a plasma and an ion beam of energy gradually variable in the range of hundred electron-volts in a carbon-containing gas atmosphere. The system allows the deposition of diamond-like *a*-C:H films with a thickness of several micrometers and microhardness of up to 40 GPa. The preliminary ion cleaning of VK8 alloy substrates in the system is insufficient for satisfactory adhesion of the diamond-like coatings, and additional experiments are needed for its improvement.

Acknowledgment

The work was supported by the Russian Science Foundation (project No. 14-29-00091).

References

[1] Trakhtenberg I S, Bakunin O M, Korneyev I N, Plotnikov S A, Rubshtein A P and Uemura K 2000 *Diam. Relat. Mater.* **9** 711–4
[2] Paik N 2005 *Surf. Coat. Technol.* **200** 2170–74
[3] Lackner J M, Major R, Major L and Schoberl T 2009 *Surf. Coat. Technol.* **203** 2243–8
[4] Gavrilov N V, Mamaev A S and Kaigorodov A S 2009 *Technical Physics Letters* **1** 33–5
[5] Fedotov S A, Efichik A A and Bueli A A 1997 *Diamond and Related materials* **6** 1638–42
[6] Martens V Yu, Tarala V A and Shevchenko E F 2011 *Perspect. Mater.* **3** 40–5
[7] Lopatin I V, Akhmadeev Yu H and Koval N N 2015 *Rev. Sci. Instrum.* **10** 103301

IOP Conf. Series: Materials Science and Engineering **387** (2018) 012048 doi:10.1088/1757-899X/387/1/012048

Structure and properties of solid lubricating coatings based on the TiN-Pb system

A A Lozovan, L N Lesnevsky, S Ya Betsofen, M A Liakhovetsky and A M Ushakov

MAI – Moscow Aviation Institute (National Research University), Orshanskaya 3, 121552, Moscow, Russia

E-mail: loz-plasma@yandex.ru

Abstract. This paper presents the results of comparative investigations of solid lubricating coatings of the TiN-Pb system including multi-layer coatings with layers thickness of the nanoscale range. Morphology, phase and chemical compositions and textures of the coatings are investigated, and the results of tribological tests are presented.

In the problems of tribology, the use of solid lubricating coatings is an effective and relatively inexpensive way to reduce friction and protect the surface of products against wear. The choice of a suitable coating to solve specific tribological problem is very difficult since the effect of many factors must be taken into account such as coating properties, substrate, interface, operating conditions and capabilities of deposition methods [1–4].

It is generally believed that alloying with soft metals, such as Pb, can reduce the friction coefficients of the coatings due to the effects of lubrication although they have poor wear resistance. The joint application of soft and solid metals, such as Ti [5], can reduce friction coefficient and improve mechanical properties of coatings. In this paper, the results of investigations on the capabilities of this method for the TiN-Pb system are presented.

Coatings were formed on specimens with the sizes of 20×15×10 mm made of VT9 titanium alloy. Before deposition, specimens were polished with diamond wheel over the surface with a larger area to the roughness of Ra ≤ 1.2 μm. The specimens were then cleaned in gasoline and isopropyl alcohol before deposition and placing into a vacuum chamber.

The specimen was placed into the chamber at a distance of 100 mm to the target of magnetron. After opening the shutter of diffusion pump, specimens were heated up to the temperature of 160 °C which then grew up to about 180 °C with the magnetron being turned on. The surface of the grounded substrate in the chamber was purified with Ar ions using an ion source for 20 minutes with the following process parameters: $P_{Ar} = 5 \times 10^{-2}$ Pa, current of $I = 0.5$ A, ion energy of the order of E ≈ 1.5 keV. After purification the ion source was not turned off when all types of coatings being formed, which thus provided ion-stimulated sputtering during the whole process. Sputtering was carried out in the pulsed medium-frequency mode of magnetron operation (50 kHz). To spray Ti + Pb, a mosaic target was used, i.e. titanium with lead inserts.

Comparative investigations of coatings with the following compositions were made: TiN (№ 1), TiPbN single-layer (№ 2) and TiPb / TiPbN multi-layer (№ 3). The TiN coating (№ 1) with 3 μm thickness was obtained at a chamber pressure of $P_{Ar+N_2} = 5\times10^{-2}$ Pa, a discharge current of $I_d = 3.5$ A and a discharge voltage of $U_d = 585$ V, and a bias voltage on the substrate of 75 V. The TiPbN single-layer coating (№ 2) with a thin (50 nm) TiPb sublayer were obtained using the following deposition mode. First, at a chamber pressure of $P_{Ar} = 5 \times 10^{-2}$ Pa, Ti + Pb layer with 50 nm thickness was deposited for 3 minutes at a discharge current of $I_d = 4$ A and a discharge voltage of $U_d = 525$ V, and a bias voltage on the substrate of 75 V. Nitrogen was then introduced into the chamber, and the second TiPbN layer with 1.5 μm thickness (immobile substrate) and 1 μm thickness (rotating substrate) was deposited at a pressure of $P_{Ar+N_2} = 2\times10^{-1}$ Pa for 85 minutes under the same discharge parameters.

The TiPb / TiPbN multi-layer coating (№ 3) with 3 μm thickness was obtained using the following deposition mode. At a chamber pressure of $P_{Ar} = 5 \times 10^{-2}$ Pa the TiPb layer with the thickness of about 150 nm was deposited for 6 minutes at a discharge current of $I_d = 4$ A and a discharge voltage of $U_d =$ 629 V, and a bias voltage on the substrate of 75 V. Then nitrogen was introduced into the chamber, and at the pressure of $P_{Ar+N_2} = 2\times10^{-1}$ Pa the second coating layer, TiNPb, was deposited also for 6 minutes under the same discharge parameters. Thus, 20 layers each of TiPb and TiPbN alternate one after another for 240 minutes (the last layer is TiPbN) forming a multi-layer coating. Morphology of the coatings and elemental composition were studied using the EVO-40 CarlZeiss scanning electron microscope with the INCAOxfordInstruments attachment for the energy-dispersive analysis; phase compositions and textures of the coatings were determined using the DRON-4 diffractometer in filtered CuKα radiation. To estimate the values of friction coefficients and compare tribological properties of various coatings compositions, the friction test machine was used shown in figure 1.

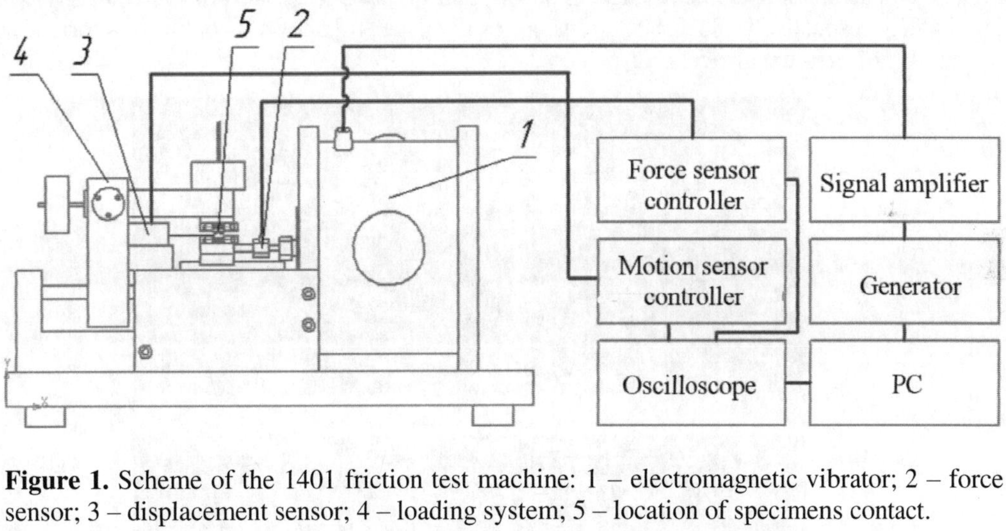

Figure 1. Scheme of the 1401 friction test machine: 1 – electromagnetic vibrator; 2 – force sensor; 3 – displacement sensor; 4 – loading system; 5 – location of specimens contact.

The 1401 friction test machine developed in MAI allows the reciprocating relative sliding of rubbing bodies to be excited by means of the electromagnetic vibrator (1). The normal force in the contact of $F_n = 1$ N was set by the lever system (4) with a plumb line; moving of rubbing bodies of $D = 100$ μm, frequency equaled $f = 20$ Hz. Movement control was carried out by the LC-2420 laser sensor (3) with the resolution of $\Delta = 0.01$ μm and the LC-2400 controller made by the Keyence (Japan). The number of friction cycles n was limited to 5×10^4 in all tests. The frictional force in the contact was recorded using the 208C02 piezoelectric force sensor (2) with the resolution of $\Delta = 4$ mN, and the 482A22 controller made by the PCB (USA). The rigidity of the friction test machine system with the device for fixing the samples was estimated by direct measurement and equaled $k_s = 31$ MN/m. Balls of ShH-15 material with the diameter of 12.7 mm were used as counter-bodies.

Figure 2 shows the surfaces of three coatings samples with the following composition: № 1 – TiN, № 2 – TiPbN (single-layer) and № 3 – TiPbN (multi-layered). As can be seen, the surface of sample

IOP Conf. Series: Materials Science and Engineering **387** (2018) 012048 doi:10.1088/1757-899X/387/1/012048

№ 2 (figure 2(b)) has a columnar structure, which appears in the form of spherical formations on the surface with the diameter of less than 2 μm closely adjacent to each other. The surface of sample № 3 has morphology similar to sample № 1 with TiN (figure 2(a)), but places with lens-shaped formations are observed (figure 2(c)). Considering the multi-layer construction of TiPb/TiPbN coating where the thickness of each layer does not exceed 150 nm, we can assume that the presented image (figure 2(c)) is a reflection of beginning for the columnar structure growth. At the same time the thickness of each coating layer limited to 150 nm results in an intermediate state of the surface when the columns formation is still at the initial stage.

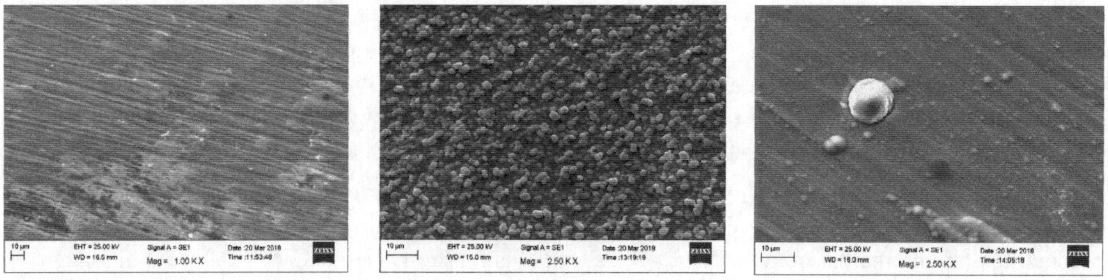

(a) **(b)** **(c)**

Figure 2. SEM images for the surfaces of three coatings: (**a**) – TiN, (**b**) – TiPbN (single-layer) and (**c**) – TiPbN (multi-layer).

The elemental analysis of specimens surfaces shows that the atomic % ratio of the coating elements for TiN is 45:55, respectively. Such a relationship, as the paper [6] shows, is optimal to form a wear-resistant coating of titanium nitride.

(a)

(b)

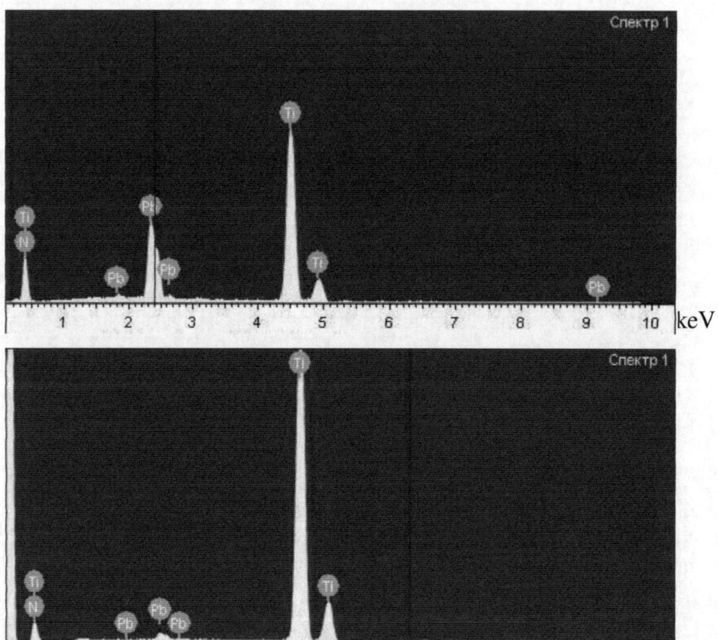

Figure 3. Spectra of energy-dispersive analysis for the coatings: (**a**) TiPbN – single-layer (full scale 2929 imp., cursor 2.410 (538 imp.)) and (**b**) TiPbN – multi-layer (full scale 3576 imp., cursor 6.180 (18 imp.)).

Diffraction pattern analysis of the coatings showed that the presence of α-Ti with the HCP lattice and TiN with NaCl-type lattice was detected in coating № 1, and strong (110) tex-ture of the substrate plane was observed. The phase compositions of coatings № 2 and № 3 are characterized by the presence of three phase reflections: α-Ti with hexagonal close-packed crystal lattice, δ-TiN titanium nitride with NaCl-type lattice, and Pb with FCC lattice. Reflections of α-Ti are the total reflections of the coating and the titanium substrate since the thickness (1–1.5 μm) of coatings is less than the penetration depth of X-rays.

Elemental analysis of the samples № 2 and № 3 (figure 3(a), (b), respectively) showed that the coatings consisted of titanium, nitrogen and lead, but had different ratios. So, the atomic % Ti:Pb:N ratio is 35:6:59 for the sample № 2, and that for the sample № 3 equals 79:0.5:20.5. Increased titanium content for the sample № 3 can be attributed to the small thickness of separate layers, their alternation of TiPb/TiPbN, and the sensitivity of the analysis method to the depth: the analyzed region is a drop-shaped volume with a characteristic size of 1 μm. Analysis of the spherical formations in the coatings of the samples № 2 and № 3 showed the elemental composition identical to that indicated above.

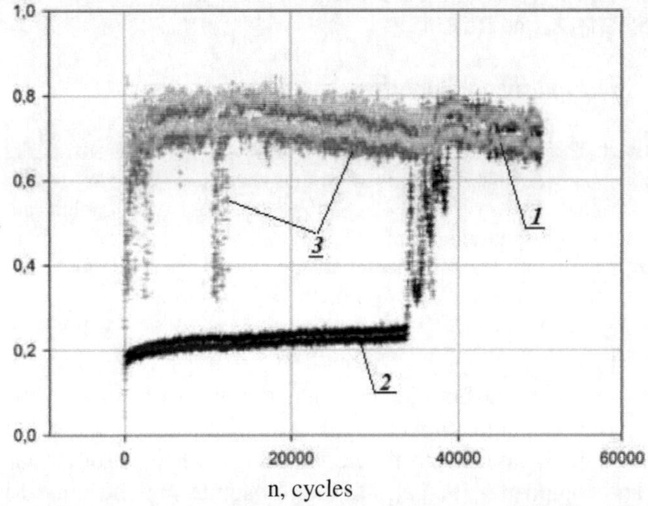

Figure 4. Comparative graph for friction coefficients of obtained coatings: 1 – TiPbN – single-layer; 2 – TiPbN – multi-layered; 3 – TiN.

Figure 4 shows a comparative graph for friction coefficients of obtained coatings. Friction coefficient of pure TiN with ShH-15 is approximately 0.7. The same friction coefficient is shown by the TiPb / TiPbN multi-layer coating, but irregular peaks of friction coefficient decrease to the values of 0.3 are observed. The best friction coefficient is shown by the TiPbN monolayer coating – 0.2. This value of friction coefficient was practically unchanged up to 33 thousand cycles, after which the destruction of tribological layer followed, and friction coefficient increased up to 0.7 as for the other samples.

References

[1] Cao M, Zhao L, Wu L and Wang W 2018 Coatings **8** 134
[2] Wang H, Xu B and Liu J 2012 *Micro and Nano Sulfide Solid Lubrication* (Springer-Verlag: Berlin/Heidelberg, Germany) 1–60
[3] Yang J F, Jiang Y, Hardell J et al. 2013 *Frontiers of Materials Science* **7** 28–39
[4] Zhao X, Zhang G, Wang L and Xue Q 2017 *Tribology Letters* **65** 64–71
[5] Zhou H, Zheng J, Wen Q P et al. 2011 *Physics Procedia* **18** 234–9
[6] Kondratev V A, Lesnevsky L N, Tyurin V N and Ushakov A M 2004 *Problems of machine building and machine reliability* **2** 49–54

Study of a variable deposition angle effect on the structure of Ti-Pb composite coatings

A A Lozovan, S Ya Betsofen, S S Alexandrova, M A Lyakhovetskiy and E P Kubatina

MAI – Moscow Aviation Institute (National Research University), Orshanskaya 3, 121552, Moscow, Russia

E-mail: loz-plasma@yandex.ru

Abstract. The results of a study on a variable deposition angle effect on the structure of TiPb / TiPbN composite coatings sprayed on titanium alloy (VT6) substrates by magnetron sputtering are presented. The structures of coatings deposited on immobile and rotating substrates and obtained by co-sputtering a mosaic target of Ti and Pb are compared. It is shown that the differences in the incidence angles of the flow particles deposited on the substrates result in different mechanisms for coating structures formation.

In spite of certain environmental difficulties in production use of lead, thin Pb films are of increasing interest due to a wide range of applications including tribology [1, 2], aerospace industry [3], thin-film metallic photocathodes [4, 5] and electron-beam guns [6, 7], electrode contacts [7, 8] as well as various compounds and composites [9, 10]. Magnetron sputtering and pulsed laser deposition (PLD) have been effectively used to fabricate thin Pb films. With the help of PLD high-quality thin films are obtained at room temperature [11]. However, for this purpose the problem of droplets in the deposited flow must be solved when spraying, which substantially interferes with the solution of various problems set by the researcher [12–14]. Therefore, in this work the priority was given to magnetron sputtering.

To co-precipitate several components of thin films by magnetron sputtering, several magnetrons with targets of different compositions [15] or so-called "mosaic" targets are used which consist of a base metal matrix with inserts of other metals [16], which allows using only one magnetron whose target design is produced according to a special program [17].

In this work, the second variant was used. Pb inserts (2 mm diameter bars) in the Ti target were placed in holes located along the axis of the target sputtering zone at the intervals of 18 mm.

Thus, the thin-film metal composition was obtained by magnetron sputtering of Ti and Pb in a single process cycle; the chamber had been previously evacuated using a diffusion pump (with cooling the trap in liquid nitrogen) to a pressure of $P = 3 \cdot 10^{-4}$ Pa. The coatings were deposited on specimens with the sizes of 20×15×10 mm made of VT6 titanium alloy. Once the shutter was opened, the specimens was heated to the temperature of 160–170 °C (20–30 minutes) which grew up to about 180 °C with the magnetron being turned on.

The substrates were immobile and located opposite the magnetron or rotated in front of the magnetron during deposition. The distance between the magnetron and the targets equalled 100 mm. Before being installed into the chamber the specimens were washed in an ultrasonic bath with gasoline and alcohol and then dried. The surface of the grounded substrate in the chamber was purified with Ar ions using an ion source for 20 minutes with the following process parameters: $P_{Ar} = 5 \times 10^{-2}$ Pa, current of $I = 0.5$ A, ion energy of the order of $E \approx 1.5$ keV. Argon was used as the atomizing gas for the magnetron.

Sputtering on immobile and rotating substrates was applied in one mode. The ion source was not turned off after purification, which thus provided ion-stimulated deposition of coatings during the whole process. First, at a pressure in the chamber of $P_{Ar} = 5 \times 10^{-2}$ Pa a Ti + Pb layer with 50 nm thickness was deposited for 3 minutes using a pulse mode (50 kHz) of the magnetron operation at a discharge current of $I_d = 4$ A and a discharge voltage of $U_d = 525$ V and a bias voltage on the substrate of 75 V. Nitrogen was then introduced into the chamber, and the second TiN + Pb layer with 1,5 μm thickness (immobile substrate) and 1 μm thickness (the substrate rotates) (figure 2) was deposited for 85 minutes under the same discharge parameters at a pressure of $P_{Ar+N_2} = 2 \times 10^{-1}$ Pa.

The quantitative phase composition and the crystallographic texture of multicomponent coatings on a titanium alloy substrate were studied by X-ray diffraction analysis using a DRON-4 diffractometer in filtered CuKα radiation [18–22]. The texture was evaluated by the method of inverse pole figures, and the pole density of the reflections was determined by the ratio of the reflections intensities in textured samples to the intensities of the corresponding reflections in the textureless standard which was estimated by calculation of all the intensity factors [23]. The fractions of the phases presented in the coating were evaluated on the basis of the mean pole densities of the reflections for each phase, i.e. taking into account the texture.

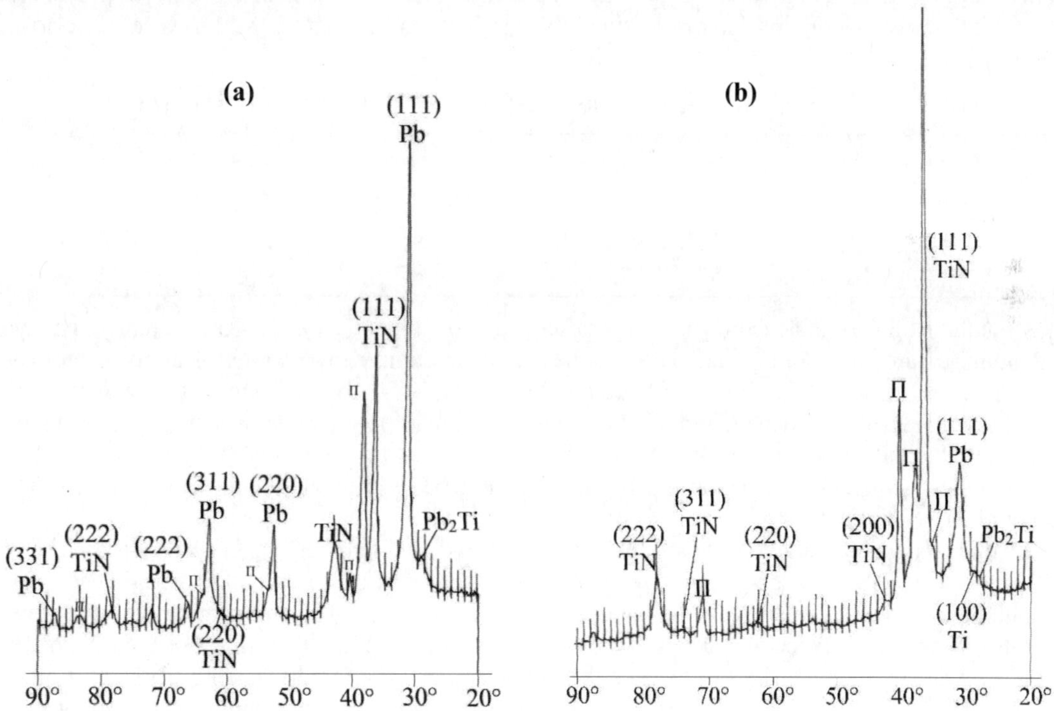

Figure 1. Diffraction patterns for coatings of samples № 1 – (**a**) and № 2 – (**b**); П – substrate.

Figures 1(a), (b) show the diffraction patterns of two coatings obtained by deposition on an immobile substrate (№ 1) and a rotating substrate (№ 2). The phase compositions of both coatings are the same and are characterized by the presence of three phase reflections: α-Ti with hexagonal close-packed crystal lattice, δ-TiN titanium nitride with NaCl-type lattice, and Pb with FCC lattice. Reflections of α-Ti are the total reflections of the coating and the titanium substrate since the thickness (1–1.5 μm) of coatings is less than the penetration depth of X-rays. Therefore, to evaluate the texture of titanium in the coating and to estimate its volume fraction in the coating, the coatings were also deposited on a nickel alloy substrate simultaneously with the coating deposition on titanium substrate. These coatings showed that titanium in the coating had a pronounced basal texture, and its volume fraction was ~ 10% of titanium nitride fraction.

Table 1. Pole densities of Pb reflections.

		№ 1	№ 2
	hkl	P_{hkl}	P_{hkl}
1	111	0.15	0.075
2	200	0	0.075
3	220	0.13	0.075
4	311	0.13	0.075
5	331	0.04	0.075
$P_{average}$		0.113	0.075

Table 2. Pole densities of TiN reflections.

		№ 1	№ 2
	hkl	P_{hkl}	P_{hkl}
1	111	2.50	5.9
2	200	1.25	0.23
3	220	0.32	0.24
4	311	0.52	0.33
$P_{average}$		1.147	1.673

Tables 1 and 2 show the values of the pole densities for TiN and Pb reflections for the two coatings, and table 3 gives the phases ratios in coatings which differ in a variable deposition angle. The differences are found in the volume ratio of the phases as well as in the texture. The texture of Pb in coating № 2 could not be estimated due to the small amount of Pb in the coating (3.9 %, see table 3); to calculate the quantitative ratio of the phases in this coating, we assumed that lead in this coating is textureless (table 1), therefore, the values of the pole density for all the reflections were assumed to be identical and equal to the pole density of the single reflection (111) which we were able to detect (figure 1(b)). This assumption is correct since the lead texture in coating № 1 is also close to the textureless state.

Table 3. Average values of pole densities and volume fractions of phases in coatings.

№№	α-Ti	TiN	Pb	$V_{\alpha\text{-Ti}}$	V_{TiN}	V_{Pb}
		$P_{average}$		%		
1	0.115	1.147	0.113	8.4	83.4	8.2
2	0.167	1.673	0.075	8.7	87.4	3.9

It was found that the variable deposition angle affects the texture of TiN and α-Ti differently. The type of titanium texture for both coatings is the same and characterized by parallel location of the basal plane to the substrate. Conditions for titanium nitride (table 2) are different – the texture of the immobile sample is much weaker, and in addition to the typical for ion-vacuum coatings {111} texture being parallel to the substrate a weak component of the cubic texture is present in the sample (see table 2). This indicates different formation mechanisms of the (α-Ti) solid solution structure and the (TiN) interstitial phase in the coating.

The obtained results show that varying geometric and, undoubtedly, energy conditions of the multicomponent coatings deposition one can obtain a controlled phase composition and phase orientations to which a certain set of service properties corresponds. Thus, in coating № 1 (immobile substrate) Pb is almost twice as much and TiN is correspondingly less than in coating № 2 (rotating substrate), which makes it possible to vary the phase composition of the coating depending on the required functional properties, e.g. tribological characteristics or wear resistance. An important result is the possibility of a quantitative evaluation of the phase composition on the basis of X-ray diffraction data, which makes it possible to search for quantitative correlations with properties. Quantitative textural results can also be used to optimize functional properties.

The fact that a two-component {111} + {001} nitride texture is found in coating № 1, and one-component {111} intensive texture is present in coating № 2 can be also used for practical purposes since a strong one-component {111} texture are generally associated with the columnar structure formation in ion-vacuum coatings that is optimal for providing high wear resistance but characterized by higher permeability compared to coatings with an equiaxed microstructure which multicomponent less pronounced texture usually corresponds to.

It is also been found that the mechanisms of texture formation in a solid solution (α-Ti) and interstitial phase (TiN) differ. This gives additional possibilities for creating a controlled texture in Ti / TiN / Ti /…multilayer coatings, since in this case it is possible to realize different variants of orientation of the components of the multilayer coating and thereby change the structure of interphase boundaries that largely determine the properties of multilayer coatings.

Acknowledgment
The work was carried out within the framework of the project part of the state task No. 11.1989.2017/ PC of the Ministry of Education and Science of the Russian Federation.

References

[1] Ren S, Li H, Cui M, Wang L and Pu J 2017 *Applied Surface Science* **401** 362–72
[2] Scharf T W and Prasad S V 2013 *Journal of Materials Science* **48** 511
[3] Miyoshi K 2007 *Solid Lubricants and Coatings for Extreme Environments: State-of-the-Art Survey* (NASA Glenn Research Center, Cleveland, OH)
[4] Gontad F, Lorusso A, Di Giulio M et al. 2017 *Journal of Vacuum Science & Technology A: Vacuum, Surfaces, and Films* **35** 031502
[5] Mistry S, Cropper M, Valizadeh R et al. 2016 *Proceedings of IPAC2016 (Busan, Korea)* pp 3691–4
[6] Power J G 2010 *AIP Conference Proceedings* **1299** 20
[7] Barday R et al. 2013 *Proceedings of IPAC2013 (Shanghai, China)* MOPFI001
[8] Lee S H, Bae J, Lee S W and Jang J W 2014 *Bulletin of the Korean Chemical Society* **35** 2415
[9] Sadovnikov S and Rempel A 2009 *Phys. Solid State* **51** 2375–83
[10] Sadovnikov S I, Kozhevnikova N S and Rempel A A *Semiconductors* **44** 7349–56
[11] Dewang S 2009 *Jurnal ILMU DASAR* **10** 186–9
[12] Lozovan A A, Prishepov S V, Frangulov S V et al. 2016 *Journal of Physics: Conference Series* **729** 012024
[13] Lozovan A A, Alexandrova S S, Mishnev M A and Prishepov S V 2014 *Journal of Alloys and Compounds* **586** 387–90
[14] Lozovan A A and Alexandrova S S 2014 *Journal of Surface Investigation. X-ray, Synchrotron and Neutron Techniques* **8** 607–11
[15] De La Roche J, Gonzalez J M, Parra E R and Sequeda F 2016 *Matéria* (Rio J.) **21**
[16] Nakano J, Miyazaki H, Kimura T et al. 2004 *Journal of the Ceramic Society of Japan* **112** S908–11
[17] Golosov D A and Melnikov S N 2011 *XVIXth Symposium on Physics of Switching Arc (Czech Republic)* pp 197–200
[18] Lozovan A A, Ivanova S V, Betsofen S Ya and Molostov D E 2013 *Technology of light alloys* **3** 34–9
[19] Betsofen S Ya, Petrov L M, Plikhunov V V, Nochnovaya N A and Sivakov D V 2008 *Aviation industry* **2** 44–7
[20] Betsofen S Ya, Spector V S and Bannikh I O 2008 *Technology of light alloys* **3** 111–6
[21] Petrov L M, Betsofen S Ya, Spector V S, Sarychev S M and Aleksandrov A A 2007 *Technology of light alloys* **3** 113–8
[22] Petrov L M, Betsofen S Ya, Tarasov Yu M, Chernyavsky A I, Sarychev S M and Ivanchuk S V 2003 *Aviation industry* **4** 53–7
[23] Dzunovich D A, Betsofen S Ya and Panin P V 2016 *Deformation and destruction of materials* **11** 8–16

IOP Conf. Series: Materials Science and Engineering **387** (2018) 012050 doi:10.1088/1757-899X/387/1/012050

Design of the evaporation source for Au-Ge alloy

M V Makarova[1], K M Moiseev[1], A A Nazarenko[2] and P A Luchnikov[3]

[1] Bauman Moscow State University, 105005, Moscow, Russia
[2] OOO RMT Ltd., 115230, Moscow, Russia
[3] Moscow Technological University (MIREA), 119454, Moscow, Russia

E-mail: m.v.makarova@list.ru, k.moiseev@bmstu.ru, alex.nazarenko@rmtltd.ru

Abstract. The new design of the evaporator is developed in which the material is heated by heat transfer from a plate heated in glow discharge plasma above the surface of a magnetron sputtering source. The evaporator makes it possible to obtain pure films with deposition rates around 1 μm / min, 100% material productions, and from 1.5 to 3 times higher material consumption efficiency. The design of the source is very simple and the new one could be made without great cost. This solution can be useful in prototype and small-scale production for the implementation of evaporation process in vacuum coating machines equipped with only magnetron sputtering sources without the use of additional power supplies and without any changes to the control system.

1. Introduction

For devices of microelectronics, such as thermoelectric coolers (TEC), silicon crystals, etc., which after their assembly are exposed to high temperatures, or are subject to further installation at high temperatures because of the features of the technological process, it is necessary to use solders with a higher temperature melting in comparison with solders on the basis of tin. One of these high-temperature solders used in modern microelectronics is an Au-Ge alloy with a melting point of 356 °C [1, 2, 4]. This temperature is comfortable for the assembly process on the described solder, and is also sufficient for further temperature effects, which are usually carried out with further installation [3–9]. This solder is used in the assembly of thermoelectric coolers, which are subject to further installation in devices with high heat dissipation. In order for the TEC to withstand such a mounting, the temperature of its soldering should not be less, so they are also mounted on the Au-Ge solder [3, 4].

It was experimentally determined that the thickness of the Au-Ge layer for soldering the TEC should be from 15 to 20 μm. The deposition of such layers can be realized by the following methods: galvanic deposition, resistive, electron beam or arc evaporation, magnetron sputtering with solid target as well as with liquid-phase target. The solder paste cannot be used in this case, because when soldering the branches from the thermoelectric material to the ceramic base, menisci appear, and the solder material diffuses into the thermoelectric material, which negatively affects the quality of the thermoelectric material and the characteristics of the TEC as a whole.

In prototype or small-scale production, the number of equipment is usually limited. In our case, it is possible to obtain thick-film coatings either galvanically or by magnetron sputtering from solid or liquid-phase target.

There is no proven technology of Au-Ge alloy deposition by galvanic method to date. In addition, the method itself has a number of unacceptable shortcomings. The non-uniformity of the galvanic coating during the good-quality process is not less than 15 %. The deposition time of the layer of such thickness is about 3 hours. The deposited films contain impurities of materials released from complexes added to the electrolyte. Pollution of the solder layer can lead to a shift in its melting point, as well as an increase in brittleness after soldering due to the formation of intermetallic inclusions in it [5, 6].

When a layer is formed by classical magnetron sputtering from a magnetron source with a target diameter of 100 mm, it takes about 8 hours to produce a film with a thickness of around 20 μm. A good solution successfully used in the technological process for the deposition of thick-film coatings would be the use of the ion sputtering method in magnetron systems in target vapors (liquid-phase magnetron sputtering) [10, 11]. This method provides all the requirements in this case: film unevenness is not more than 5 %, high productivity (deposition rates on a standing substrate are tens of micrometers per minute), the purity of the coatings obtained (the possibility of working in a deep vacuum). However, this method has a drawback – a very low effective material consumption of the order of 10 %, which in the case of deposition of an extremely expensive Au-Ge alloy is unacceptable.

Equipping already available magnetron sputtering vacuum coating machine with other technological sources – thermal, electron beam or arc evaporators is not possible due to its design features, as well as economic inexpediency.

As a result, it became necessary to develop a new device for depositing Au-Ge alloy films with a high effective material flow rate and compatibility with the design of the existing coating machine. In this paper we present a design of a new evaporation source based on the existing 4" magnetron source, as well as the results of evaporation regimes during the deposition of a copper coating which has similar properties to Au-Ge alloy.

2. Design of the evaporation source

As it was said above, on the existing vacuum coating machine the method of liquid-phase magnetron sputtering (LPMS) is realized which will allow forming films of various materials with high speed and uniformity, but at the same time possesses a low material consumption coefficient.

To increase the efficiency of material consumption, the only way is to reduce the diameter of the crucible. But this will lead to the fact that the discharge zone will be located outside the crucible, which will not allow the material to warm up in the crucible and will also lead to the sputtering of the crucible material. Therefore, we proposed the following construction (figure 1), which is an additional evaporation module placed directly on the existing 4" magnetron source. The distance between the cover plate 1 and the surface of the magnetron 2 is 50 mm, which is sufficient for the anomalously glow discharge between them.

The surface of the magnetron is a cathode, and the special cover is an anode. The special cover is made of molybdenum, since it has a low sputtering coefficient and is a refractory material. On the surface of the magnetron source molybdenum spacers 3 are installed. A molybdenum disk 4 with a groove is placed on the spacers. A crucible 5 with a diameter of 15 mm is placed in the groove on the disc. Between the molybdenum disk and the magnetron surface, there is a guaranteed gap of 0.5 mm in width, which insulates the disk from cooling. The protective screen 6 is mounted on the special cover. The special cover itself is mounted on a stainless-steel support 7, which in turn is mounted on an external screen 8.

The construction is assembled in such a way that the discharge burns exclusively between the special cover 1 and the disc 4. By bombardment with ions, the disc is heated to a high temperature. By thermal conductivity and radiation, the crucible having direct contact with the surface of the disk is heated to a temperature sufficient for the melting and evaporation of the material inside the crucible. The system of protective screens prevents the material of the disk and covers from getting onto the substrate. Since the discharge burns between the screens, there is a constant re-sputtering of the material of the disk and the cover, which is confirmed by the almost complete absence of the material of the disk in the erosion zone of the disk 4.

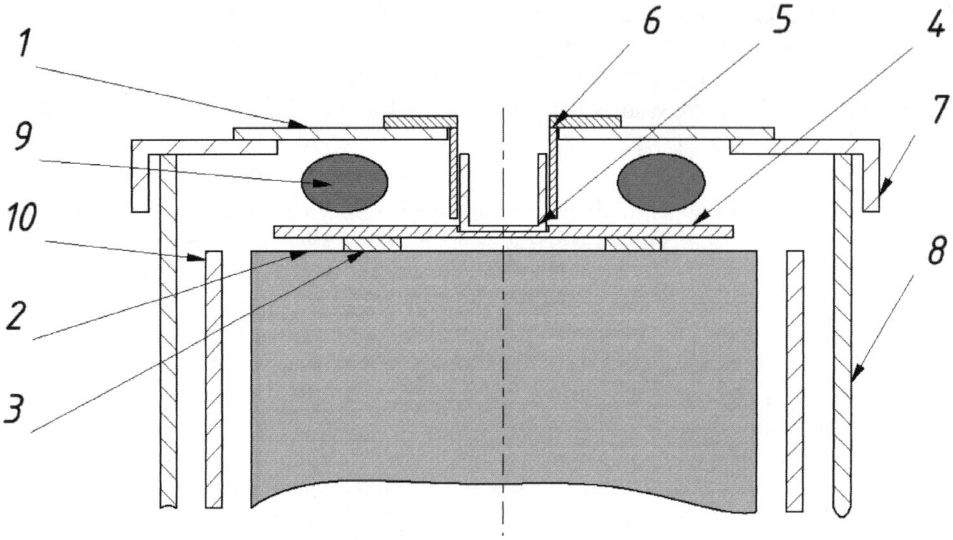

Figure 1. Design of the source: 1 – special cover, 2 – existing magnetron source, 3 – spacers, 4 – disc with groove, 5 – crucible, 6 – protective screen, 7 – support, 8 – external screen, 9 – plasma zone, 10 –anode of the magnetron source.

3. Pilot process of the evaporation and its results

To carry out the experiments we used 60×48 mm AlN substrates with a thickness of 0.25 mm and average roughness Ra 0.009. The distance from the substrates to the source was 115 mm. The regimes were tested using copper, since this material has similar values for saturated vapor pressure, but a higher melting point and a significantly lower cost than the Au-Ge alloy. The discharge power was 1.5 kW in direct current mode at an argon flow rate of 4 l/min. The deposition time was 12 min. The deposition was carried out on a substrate without rotation.

The weight of copper in the crucible was 2 grams and the weight of the deposited film, measured on UG Unigram microbalances, was 0.6 g. The efficiency of material consumption at these parameters was about 30 %. With similar parameters, the efficiency of material consumption for a liquid-phase magnetron is about 10 %, for a classical magnetron – about 20 %. The thickness of the copper film measured on the MarSurf PS1 profilometer was 12 μm and the calculated deposition rate was around 1 μm/min.

The resulting sample (figure 2) was studied using a Thermo Scientific X-ray microanalyzer for the presence of crucible materials or screens. The results of the study (figure 3, table 1) showed that these materials were not found. On the surface of the film, elements O and C are found, which indicates contamination of the surface in the atmosphere, and Al and N are found under the film, which are the main elements of the ceramic substrate.

Figure 2. Image of the substrate with a copper film.

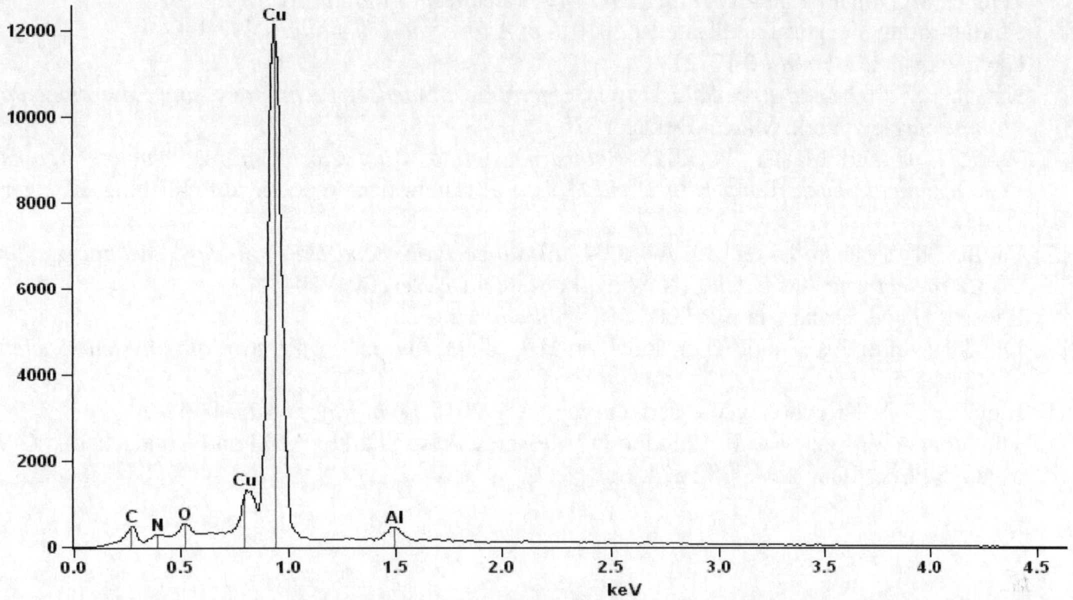

Figure 3. Auger spectrograph of the elemental composition of the sample by volume.

Table 1. Percentage of elements by volume of the film.

Element	Mass ratio, %	Atomic ratio, %
C	0,97	4,63
N	0,36	1,49
O	0,82	2,95
Al	2,14	4,55
Cu	95,70	86,38
Total	100,00	100,00

4. Conclusions

Developed new design of the evaporator allows to obtain clean films with high speed and the efficiency of material consumption, which exceeds by 1.5 times the classical magnetron sputtering and 3 times the liquid-phase magnetron sputtering. The presented solution may be useful if there is no evaporation system using existing magnetron sputtering coating machines, because it can be implemented very simply and cheaply. This circumstance is especially important for small production and research laboratories. An additional advantage is full integration with the existing control system of the vacuum coating machines, which is impossible with the installation of power supplies for additional standard evaporation sources.

Acknowledgment

This work was supported by the Ministry of Education and Science of the Russian Federation (№ 3.8074.2017/8.9 basic part of the State task).

References

[1] Liu L, Cui J, Wang J, Zhou Z, Johnson R W and Liu C 2017 *Electronic Components and Technology Conference (ECTC) IEEE 67th* pp 417–22

[2] Zeng G, McDonald S and Nogita K 2012 *Microelectronics Reliability* **52** 1306–22
[3] Chidambaram V, Hald J and Hattel J 2010 *J. of Alloys and Compounds* **490** 170–9
[4] Gayle F et al. 2001 *JOM* **6** 17–21
[5] Katsuaki S 2003 *Lead-Free Soldering in Electronics: Science, Technology, and Environmental Impact* (New York: Marcel Dekker) 376
[6] Wong E H and Mai Y W 2015 *Robust Design of Microelectronics Assemblies Against Mechanical Shock, Temperature and Moisture* (Cambridge: Woodhead Publishing, Elsevier) 458
[7] Puttlitz K J and Stalter K A 2004 *Handbook of Lead-Free Solder Technology for Microelectronic Assemblies* (New York: Marcel Dekker, Inc.) 1048
[8] Reuss R H and Chalamala B R 2003 *MRS Bulletin* **1** 11–20
[9] Lee J H, Shim J Y, Yoon H S and Kim H C 2003 *Journal of the Korean Physical Society* **42** 662–5
[10] Bleykher G A, Krivobokov V P and Yuryeva A V 2015 *Tech. Phys.* **60** 1790–5
[11] Tumarkin A V, Kaziev A V, Kolodko D V, Pisarev A A, Kharkov M M and Khodachenko G V 2015 *Phys. Atom. Nucl.* **78** 1674–6

IOP Conf. Series: Materials Science and Engineering **387** (2018) 012051 doi:10.1088/1757-899X/387/1/012051

Distribution of coating thickness applied by magnetron sputtering

S A Martsinukov, D K Kostrin, A A Lisenkov and V A Simon

Department of electronic instruments and devices, Saint Petersburg Electrotechnical University "LETI", 197376, Saint Petersburg, Russia

E-mail: sergm2006@mail.ru

Abstract. The Danilin model for flat ring evaporator was studied and used to calculate the thickness distribution of the coating. The calculations for different types of the cathode current density variations (uniform, triangle and third-degree polynomial) were made. Relation between the magnetron sputtering system parameters providing the most uniform coating was analysed. Research of the target sputtering profile with nonzero cone angle was performed. It was established that in case of the cone angle increasing in the areas of substrate situated under the cathode the sputtered coating becomes thicker. Also at the center and on the edges of the substrate the coating thickness is decreased due to the redistribution of thickness.

1. Introduction

In the magnetron sputtering systems (MSS) [1, 2] can be used targets of different shapes, but the flat and conical targets are used most often. At the initial moment of sputtering the ions bombard the smooth surface, and over time the target become V-shaped. Herewith the cavity inside the target is constantly increasing and the deposition rate of the sputtering material is reduced. The deposition rate depends on the design of MSS and the discharge current value [3, 4]. Longer deposition process compensates the decreasing in deposition rate. The accuracy of the deposition mode correction can be increased if the deposition rate dependence on the target operation time at a fixed discharge current is known [5].

2. The Danilin model

In case of the MSS with a flat cathode the Danilin model can be used to calculate the thickness distribution of the film or coating [6]. The Danilin model for a flat ring evaporator is a special case of the model for a cone evaporator (the angle of the cone target in this case will be zero). The free path of the sputtered atoms shouldn't be less than the distance from the cathode to the substrate (straight-span mode); the distribution of material in space occurs according to cosine law; the distribution rate is proportional to the density of the ion current.

When the target of MSS is an axially symmetrical body, on its surface there is a narrow annular sputtering zone with following dimensions: width l, internal radius R_1, external radius R_2 and average radius R_{avg}. The drawing of the MSS with annular-shaped target is presented in figure 1. From the drawing follows:

$$H_1 = H + (R_2 - R)\mathrm{tg}\gamma\,; \quad \cos\varphi(R,\alpha) = \frac{1}{r}\left[H_1 \cos\gamma + (R - R_{\mathrm{cond}} \cos\alpha)\sin\gamma\right];$$

$$\cos\theta(R,\alpha) = H_1 / r\,; \quad r(R,\alpha) = \left(H_1^2 + R_{\mathrm{cond}}^2 + R^2 - 2R_{\mathrm{cond}} R \cos\alpha\right)^{0.5}.$$

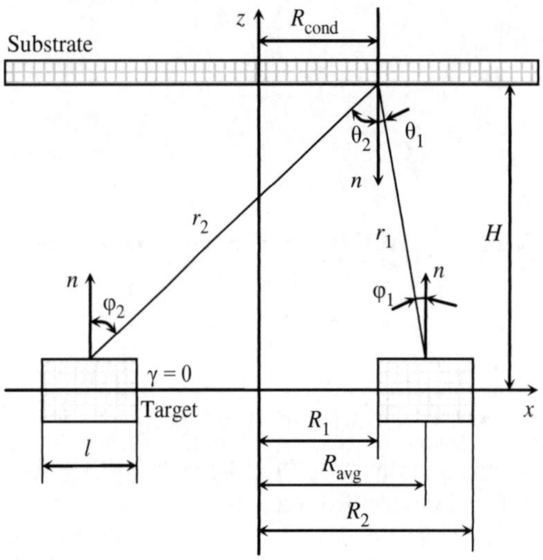

Figure 1. Geometrical representation of the MSS with annular-shaped target: R_{cond} is the distance from the point of condensation to the center of the substrate, l – the width of the annular zone of sputtering, H – the average distance from the substrate to the target, γ – the cone angle of the target. Other radial (r_1, r_2) and angular (θ_1, θ_2, φ_1, φ_2) dimensions are auxiliary.

The general formula for calculating the film thickness $h(R_{cond})$ in this case is

$$h(R_{cond}) = \int_{R_1}^{R_2}\int_0^{2\pi} \frac{\cos\varphi(R,\alpha)\cos\theta(R,\alpha)\,R\,\Pi(R)}{r^2(R,\alpha)\cos\gamma} d\alpha dR,$$

where $\Pi(R)$ is the Lagrange polynomial. If the target cone angle is zero, these expressions take the following form:

$$H_1 = H\,;\ \cos\theta(R,\alpha) = \cos\varphi(R,\alpha)\,;\ \cos\varphi(R,\alpha) = \frac{H}{r}\,;\ r(R,\alpha) = (H^2 + R_{cond}^2 + R^2 - 2R_{cond}R\cos\alpha)^{0.5}.$$

Thus the formula for the film thickness calculation is

$$h(R_{cond}) = \int_{R_1}^{R_2}\int_0^{2\pi} \frac{\cos\varphi(R,\alpha)\cos\theta(R,\alpha)R\,\Pi(R)}{r^2(R,\alpha)\cos\gamma} d\alpha dR = \int_{R_1}^{R_2}\int_0^{2\pi} \frac{H^2 R\,\Pi(R)}{(H^2 + R_{cond}^2 + R^2 - 2R_{cond}R\cos\alpha)^2} d\alpha dR.$$

To simplify the calculation, take typical values as the size of the sputtering system. The diameter of the target can vary within 0.005...5 m [7]. However, the MSS is most effective when the target has a relatively small diameter – 50...75 mm. The ratio R_{cond}/R_{avg} varies in the range of 0.2…1; the ratio H/R_{avg} varies in the range of 0.6…2. Therefore it can be assumed that l = 36 mm, R_1 = 32 mm, R_{avg} = 50 mm, R_2 = 68 mm, H = 20…100 mm, R_{cond} = 100 mm [6].

The current density distribution was studied by two methods: using a target consisting of seven isolated rings, each of which was removed from the share of the total current on the target, as well as by measuring the profile of the target spraying zone after a long time of the MSS operation at constant parameters. The obtained data were approximated by the Lagrange polynomial $\Pi(R)$, which takes into account the real radial distribution of the current density. There are three types of the radial current density distribution for consideration.

Uniform current density distribution. The polynomial describing the uniform current density distribution over the target radius is

$$\Pi_1(R) = \begin{vmatrix} 0, \text{ when } R \le R_1; \\ 1, \text{ when } R_1 \le R \le R_2; \\ 0, \text{ when } R \ge R_2. \end{vmatrix}$$

In this case, the general formula for calculating the film thickness become

$$h(R_{cond}) = \int_{R_1}^{R_2}\int_0^{2\pi} \frac{H^2 R}{(H^2 + R_{cond}^2 + R^2 - 2R_{cond}R\cos\alpha)^2} d\alpha dR.$$

The dependence of the thickness distribution of the inflicted film on the substrate $h(R_{cond})$ at different values of the distance H from the substrate to the target, if the current distribution over the radius of the target is uniform, is presented in figure 2(a).

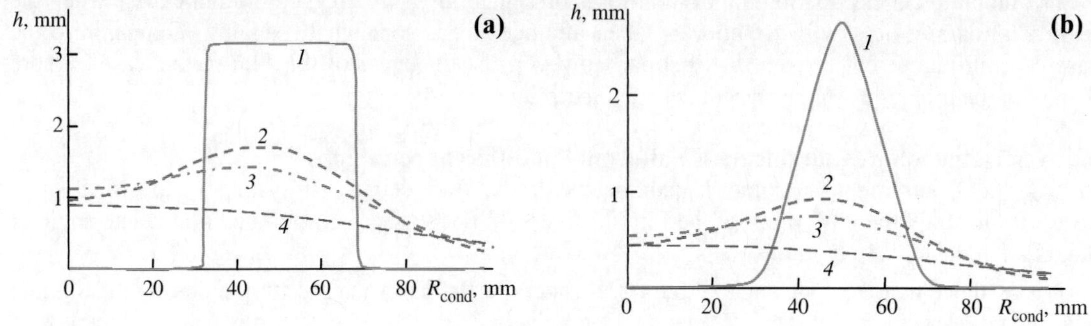

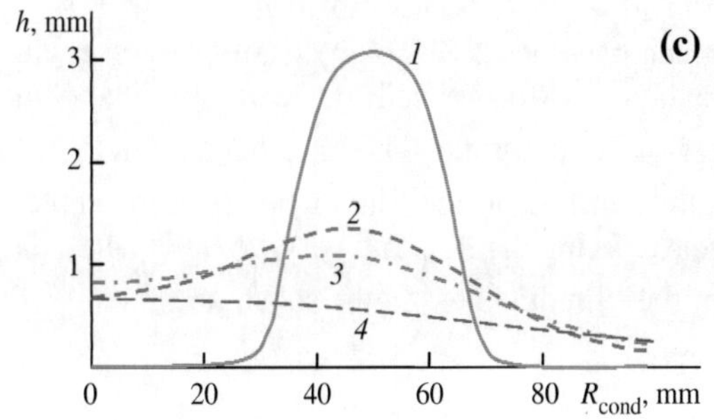

Figure 2. Film thickness distribution in case of: **(a)** – uniform current density over the target radius; **(b)** – triangle-shaped current density; **(c)** – current density approximated by third-order polynomial. The curve *1* is the spray profile directly from the target (excluding *H*), *2* – $H = 30$ mm, *3* – 40 mm, *4* – 80 mm.

Triangle-shaped current density distribution. In this case, the polynomial describing the current density distribution is given by the expression

$$\Pi_2(R) = \begin{cases} \dfrac{R - R_1}{R_{avg} - R_1}, \text{ when } R_1 \le R \le R_{avg}; \\ \dfrac{R_2 - R}{R_2 - R_{avg}}, \text{ when } R_{avg} \le R \le R_2. \end{cases}$$

The general formula for the film thickness calculation is

$$h(R_{cond}) = \int_{R_1}^{R_2} \int_0^{2\pi} \frac{H^2 R \Pi_2(R)}{(H^2 + R_{cond}^2 + R^2 - 2R_{cond} R \cos\alpha)^2} d\alpha dR .$$

The radial thickness distribution of the inflicted film is presented in figure 2(b), in comparison with the case of uniform current density distribution (figure 2(a)) the film thickness decreases slightly when $H = 0$ (ceteris paribus) and then the decreasing become stronger as the distance from the target to the substrate increases. This is due to the fact that now the cathode is sprayed unevenly across its surface – the nature of the distribution has become such that the current density is maximal at $R = R_{avg}$, and decreases linearly as R to R_1 on the one hand, and to R_2 on the other hand relative to R_{avg}.

Current density distribution is approximated by third-order polynomial. This case describes the current density distribution over the target radius most close to the real situation. The Lagrange polynomial for this type of current density distribution is

$$\Pi_3(R) = \begin{cases} 0, \text{ when } R \le R_1 \\ -0.24\left(|R - R_{avg}|10^{-1}\right)^3 + 0.44\left(|R - R_{avg}|10^{-1}\right)^2 - 0.64|R - R_{avg}|10^{-1} + 1, \text{ when } R_1 \le R \le R_2; \\ 0, \text{ when } R \ge R_2. \end{cases}$$

The general formula for the film thickness calculation is

$$h(R_{cond}) = \int_{R_1}^{R_2}\int_{0}^{2\pi} \frac{H^2 R \Pi_3(R)}{(H^2 + R_{cond}^2 + R^2 - 2R_{cond} R\cos\alpha)^2} d\alpha dR .$$

The radial thickness distribution of the film is presented in figure 2(c), the film thickness when the real radial current density distribution is taken into account is somewhat "sagging" compared to the case of uniform current density distribution. However, the thickness of the film in this case is more than with the triangular-shape current density distribution.

3. Investigation of the film thickness uniformity in different conditions

In order to determine the geometric parameters of the film deposition system in the MSS, it is necessary to study the uniformity of the film thickness on flat surfaces depending on the cone angle of the target and the H/R_{cond} ratio.

It is assumed that the cone angle of the target is zero (a flat disk) and the current density distribution over the target radius is constant ($\Pi(R) = 1$). Determine the value of H/R_{cond}, in which the thickness of the coating on the maximum area will not change with an accuracy of ±2 %. To do this, it is necessary to find the film thickness in the center of the substrate (at $R_{cond} = 0$), as well as the ratio of the film thickness at $R_{cond} \neq 0$ to the film thickness in the center of the substrate. The ratio of R_{avg} and H in which the quotient of the film thickness distributed over the substrate to the film thickness in the center of the substrate will be close to 100 % can be found further by fixing the R_{avg} and varying H.

The following expression is used for determination of the film thickness in the substrate center:

$$h_0|_{R_{cond}=0} = H^2 \int_{R_1}^{R_2} \frac{R\Pi(R)}{\left(H^2 + R^2\right)^2} dR.$$

The general formula for the film thickness calculation is

$$h(R_{cond}) = \int_{R_1}^{R_2}\int_{0}^{2\pi} \frac{H^2 R \Pi(R)}{(H^2 + R_{cond}^2 + R^2 - 2R_{cond} R\cos\alpha)^2} d\alpha dR .$$

Thus it is necessary to provide the validity of expression $h/h_0 \to 100\%$. The film thickness distribution for the annular evaporator is shown in figure 3.

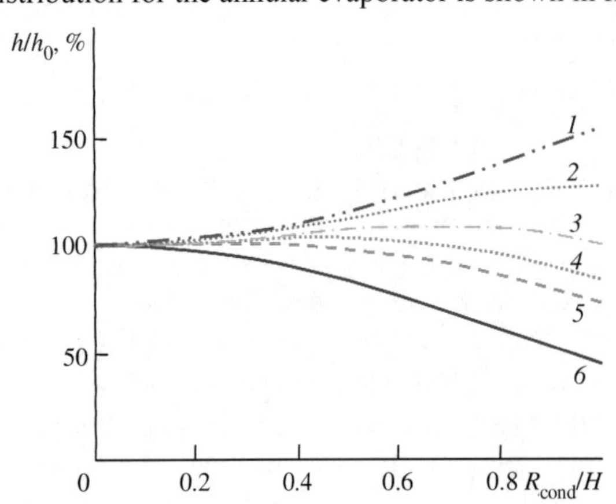

Figure 3. Film thickness uniformity vs. distance from the substrate center. Curve *1* is for the conditions $H = 30$ mm, $R_{avg}/H = 1.6$; *2* – $H = 40$ mm, $R_{avg}/H = 1.25$; curve *3* – $H = 50$ mm, $R_{avg}/H = 1.0$; curve *4* – $H = 60$ mm, $R_{avg}/H = 0.83$; curve *5* – $H = 65$ mm, $R_{avg}/H = 0.77$; curve *6* – $H = 100$ mm, $R_{avg}/H = 0.5$.

The graph in figure 3 shows that the maximum average uniformity of the film thickness distribution obtained in the range $H = 60\ldots65$ mm under condition of $R_{avg} = 50$ mm. Thus, R_{avg}/H is in the range of (50/65)…(50/60), R_{avg} – in the range of $0.77H\ldots0.83H$. Consequently $R_{avg} \approx 0.8H$, $H = 1.25R_{avg}$ – the distance from the target to substrate should be equal to $1.25R_{avg}$ for the maximum uniformity of the film thickness distribution. As mentioned earlier, only at the initial moment of sputtering the target surface is flat, and later the surface profile become V-shaped at that the profile depth is constantly increasing. In this case, the cone angle of the sputtered target $\gamma \neq 0$; this kind of MSS is shown in figure 4.

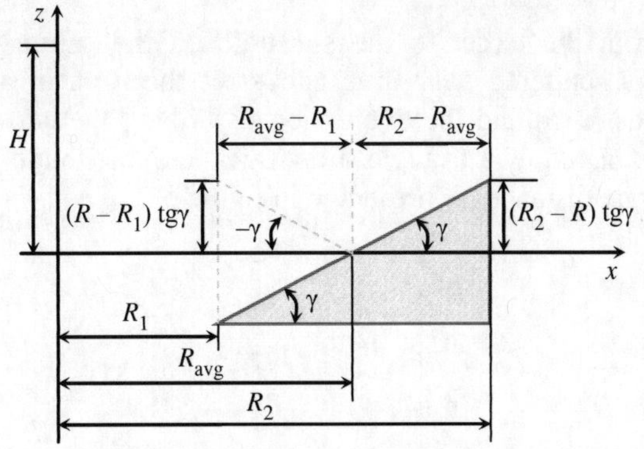

Figure 4. Geometrical drawing of the sputtering system with nonzero cone angle.

Two areas of sputtering are arising during the MSS functioning. The overall film thickness will be equal to the sum of the thicknesses resulting from the combined evaporation of these areas. It is assumed that the current density distribution over the target radius is uniform. The formulas characterizing the obtained areas are written below. The expressions for the first sputtering area are

$$H_1 = H + (R - R_1)\mathrm{tg}(-\gamma);$$

$$r(R,\alpha) = \sqrt{[H + (R - R_1)\mathrm{tg}(-\gamma)]^2 + R_{cond}^2 + R^2 - 2R_{cond}R\cos\alpha};$$

$$\cos\varphi(R,\alpha) = \frac{[H + (R - R_1)\mathrm{tg}(-\gamma)]\cos(-\gamma) + (R - R_{cond}\cos\alpha)\sin(-\gamma)}{\sqrt{[H + (R - R_1)\mathrm{tg}(-\gamma)]^2 + R_{cond}^2 + R^2 - 2R_{cond}R\cos\alpha}};$$

$$\cos\theta(R,\alpha) = \frac{H + (R - R_1)\mathrm{tg}(-\gamma)}{\sqrt{[H + (R - R_1)\mathrm{tg}(-\gamma)]^2 + R_{cond}^2 + R^2 - 2R_{cond}R\cos\alpha}}.$$

The thickness of the film deposited from the first sputtering area is

$$h_1(R_{cond}) = \int_{R_1}^{R_{avg}} \int_0^{2\pi} [H + (R - R_1)\mathrm{tg}(-\gamma)] R\Pi(R) \times$$

$$\times \frac{[H + (R - R_1)\mathrm{tg}(-\gamma)]\cos(-\gamma) + (R - R_{cond}\cos\alpha)\sin(-\gamma)}{\left\{[H + (R - R_1)\mathrm{tg}(-\gamma)]^2 + R_{cond}^2 + R^2 - 2R_{cond}R\cos\alpha\right\}^2 \cos(-\gamma)} d\alpha dR.$$

The expressions for the second sputtering area are

$$H_1 = H + (R_2 - R)\mathrm{tg}(\gamma);$$

$$r(R,\alpha) = \sqrt{[H + (R_2 - R)\mathrm{tg}(\gamma)]^2 + R_{cond}^2 + R^2 - 2R_{cond}R\cos\alpha};$$

$$\cos\varphi(R,\alpha) = \frac{[H + (R_2 - R)\mathrm{tg}(\gamma)]\cos(\gamma) + (R - R_{cond}\cos\alpha)\sin(\gamma)}{\sqrt{[H + (R_2 - R)\mathrm{tg}(\gamma)]^2 + R_{cond}^2 + R^2 - 2R_{cond}R\cos\alpha}};$$

$$\cos\theta(R,\alpha) = \frac{H + (R_2 - R)\mathrm{tg}(\gamma)}{\sqrt{[H + (R_2 - R)\mathrm{tg}(\gamma)]^2 + R_{cond}^2 + R^2 - 2R_{cond}R\cos\alpha}};$$

$$h_2(R_{cond}) = \int_{R_1}^{R_{avg}} \int_0^{2\pi} [H + (R_2 - R)\mathrm{tg}(\gamma)] R\Pi(R) \times$$

$$\times \frac{[H + (R_2 - R)\mathrm{tg}(\gamma)]\cos(\gamma) + (R - R_{cond}\cos\alpha)\sin(\gamma)}{\left\{[H + (R_2 - R)\mathrm{tg}(\gamma)]^2 + R_{cond}^2 + R^2 - 2R_{cond}R\cos\alpha\right\}^2 \cos(\gamma)} d\alpha dR.$$

The thickness of the coating deposited on the substrate will be determined by the total impact on it of both areas of the target:

$$h(R_{cond}) = h_1(R_{cond}) + h_2(R_{cond}) .$$

Consider the effect of the distance from the target to the substrate on the nature of coating distribution in the case where the target is sputtered over time and takes the form of a V-shaped groove. Let the cone angle γ be as small as π/15, and the expression describing the current density distribution over the target radius is approximated by a triangle. Film thickness distribution of for the annular evaporator at different target-substrate distances is presented in figure 5.

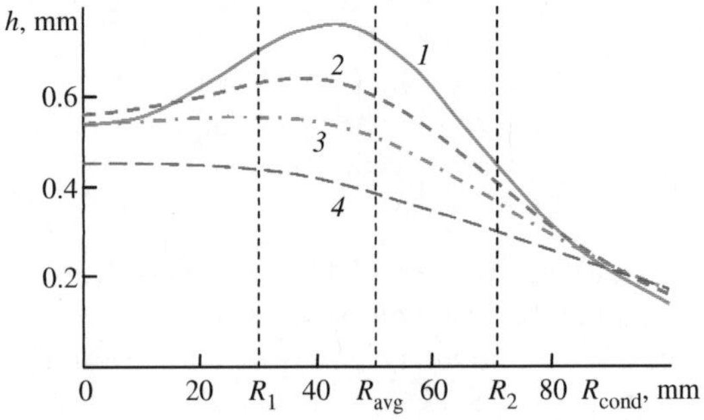

Figure 5. Film thickness distribution of for the annular evaporator at different distances to target. Curve *1* is for the condition H = 40 mm, *2* – H = 50 mm; *3* – H = 60 mm; *4* – H = 80 mm.

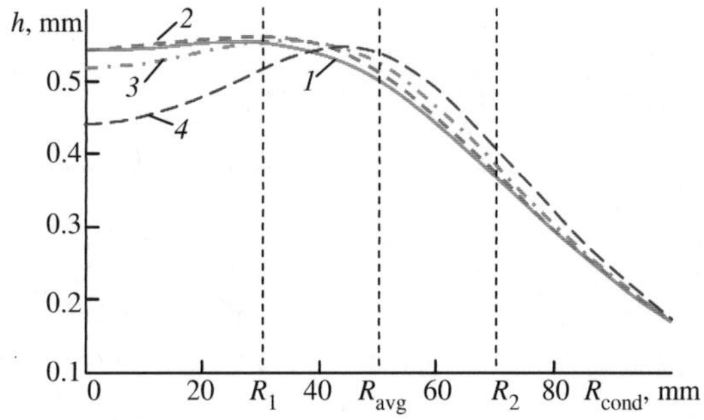

Figure 6. Cone angle impact on the deposited coating profile. Curve *1* is for condition γ = 0°; *2* – γ = 15°; *3* – γ = 30°; *4* – γ = 45°.

It was taken in the calculations that R_{avg} = 50 mm. The maximum uniformity of the film thickness distribution is observed at H = 1.25, R_{avg} ≈ 60 mm. The influence of the cone angle on the profile of the deposited coating at R_{avg} = 50 mm, $H = 1.25 \cdot R_{avg}$ ≈ 60 mm is shown in figure 6.

It can be seen from figure 6 that the achievement of the maximum uniformity of the film thickness distribution occurs at small values of the cone angle γ < 30°. With increasing γ in the substrate areas directly above the cathode, there is a strong thickening of the deposited film, but the thickness in the center and at the edges of the substrate is significantly reduced (thickness redistribution occurs). The MSS is already inoperative when γ > 30°. Therefore, to obtain the maximum area with a high uniformity of the film thickness distribution, it is necessary to avoid large values of γ and adhere to the ratio $R_{avg} \approx 0.8H$.

4. Conclusion

The study of changes in the profile of the sputtered cathode-target surface on the coating thickness distribution [8] allows to determine a number of parameters of the MSS requiring experimental determination, which have a significant impact on the deposited coating profile.

References

[1] Udovichenko S Yu, Kostrin D K et al. 2017 *Journal of Physics: Conference Series* **872** 012013
[2] Udovichenko S Yu, Lisenkov A A et al. 2018 *Journal of Physics: Conference Series* **967** 012018
[3] Gerasimov V A, Kostrin D K et al. 2016 *Journal of Physics: Conference Series* **729** 012019
[4] Kostrin D K, Lisenkov A A et al. 2016 *Journal of Physics: Conference Series* **735** 012055
[5] Barchenko V T, Egorov S K et al. I P 2009 *Vacuum Technique and Technology* **19** 173–9
[6] Danilin B S 1989 *Low-temperature plasma application for the thin film deposition* (Moscow: Energoatomizdat)
[7] Barchenko V T, Bystrov Yu A and Kolgin E A 2001 *Ion-plasma technologies in the manufacturing of electronics* (Saint Petersburg: Energoatomizdat)
[8] Lisenkov A A, Martsinukov S A et al. 2017 *Journal of Physics: Conference Series* **808** 012005

IOP Conf. Series: Materials Science and Engineering **387** (2018) 012052 doi:10.1088/1757-899X/387/1/012052

Investigation of electrochemical sensors with thin film nanocomposite Pt/C electrodes obtained by magnetron deposition

E A Medvedeva[1] and A M Baranov[2]

[1] FSUE SPA 'Analitpribor', Smolensk, Russia
[2] Moscow Aviation Institute (National Research University), Moscow, Russia

E-mail: elenamedvedeva78@mail.ru, baranov_64@mail.ru

Abstract. The electrochemical gas sensors with the film nanocomposite electrodes are produced and tested in several gas atmospheres. The experimental results demonstrate that a-C/Pt has high sensitivity and stability for continuous monitoring of carbon monoxide (CO) and hydrogen sulfide (H_2S) gases compare to with main counterparts are mass-produced.

1. Introduction

In connection with environmental pollution, the need for real-time air monitoring is constantly growing. For these purposes the different monitoring systems are used which include electrochemical [1], semiconductor [2], catalytic [3] and IR sensors [4].

The world trend aimed at solving problems of air monitoring over large areas associated with leaks toxic gases has been the development of sensory networks of various sizes that collect, analyze and make decisions on the management of executive devices to prevent emergencies and alert appropriate services when exceeding the maximum permissible concentration (MPC) of toxic gases.

Depending on the way data is transferred from the sensors to the data reception devices, wire and wireless sensor networks are allocated [5]. The drawback of wired systems is their dependence on power supply systems, which limits the use of wire-based monitoring systems for air composition, in particular, in places with no network power supply. Overcoming of existing limitations is possible due to the creation of autonomous wireless systems for continuous monitoring of toxic gases that provide air composition measurements on the territory of industrial enterprises.

To create autonomous wireless monitoring systems, in addition to low-power electronic components and energy efficient algorithms, it is required to use gas sensors with minimal power consumption. From the whole variety of gas sensors, minimal energy consumption and, at the same time, sufficient sensitivity and selectivity have only electrochemical sensors. In the electrochemical sensor, the detectable component reacts with the sensitive layer directly on the working electrode. Therefore, the working electrode must simultaneously perform several functions that ensure the operation of sensors: to have the ability to pass gas, play the role of a catalyst, have high electrical conductivity, and prevent the liquid electrolyte from flowing out.

The electrochemical sensor electrode in the general case is a gas permeable membrane with a catalytically active layer deposited on it (figure 1). The catalyst layer can be applied by using well-known deposition techniques such as screen printing [6], chemical and electrochemical deposition [7], ultrasonic spraying [8] and laser technics [9].

In recent years, much attention has been paid to the composite materials based on amorphous carbon and metals obtained by magnetron deposition. This is due to the interesting physical properties and possible wide application of amorphous carbon-metal nanocomposites [10, 11].

Recently, the technology of magnetron deposition of thin-film nanocomposite catalysts based on platinum and amorphous carbon (graphite) has been developed to create working electrode of electrochemical sensor, which improves the reproducibility of electrode parameters [12]. Amorphous carbon is a highly inert material with good stability in many acids, solvents and electrolytes and has a high electrical conductivity. The co-sputtering of graphite and platinum provides the increasing platinum surface and its porosity. Note that platinum and graphite do not participate in the oxidation-reduction process, but are only carriers of atomic oxygen.

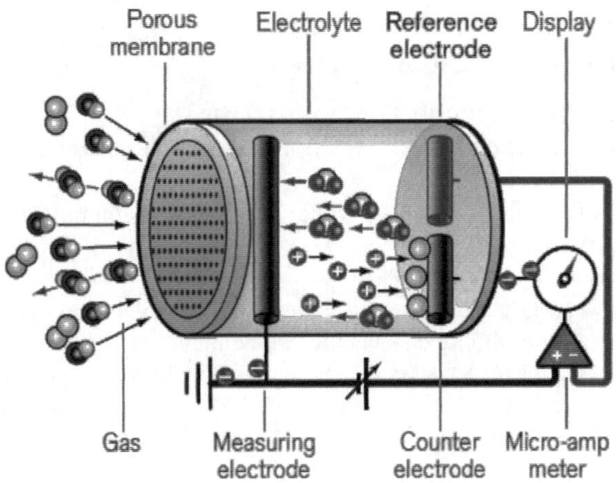

Figure 1. Scheme of the electrochemical sensor.

2. Experimental

Thin a-C/Pt films of the composite were produced on a "Elato 600-R" (Izovak) unit equipped with a DC magnetron and a rotation system for the substrate holder. The deposition of the films was carried out in an argon atmosphere at a pressure of ~10^{-1} Pa. Spray mode: U = 500V, I = 200mA, t = 20 min. Substrates for the application of the catalyst were Teflon FM-400 (PTFE) with a porosity of 50 % and a pore diameter of 1 μm without any additional preparation (figure 2(a)). Photos of the working and auxiliary electrodes are shown in figure 2(b), (c). The composition of a-C/Pt catalytic layers was investigated by EDX spectra (figure 3). The developed thin film electrodes are embedded in the sensor nodes with a liquid electrolyte and tested in the presence of CO and H_2S. We evaluate the sensor performance in terms of sensitivity, selectivity and response.

(a) (b) (c)

Figure 2. Teflon surface (**a**) and magnetron sputtering electrodes: (**b**) working, (**c**) reference and auxiliary.

IOP Conf. Series: Materials Science and Engineering **387** (2018) 012052 doi:10.1088/1757-899X/387/1/012052

3. Results and discussion

Table 1 demonstrates the sensor sensitivity (response) in the presence of two gases CO and H_2S in the environment. The sensors improve the results of commercial sensors by Nemoto (NAP-505 for CO and Ne4 for H_2S) while guarantying low background current which is lower than 1 %.

The evaluation of the electrochemical sensor cross sensitivity to different acid gases is highly important when considering the long-term operation of the sensor. The results of this evaluation are presented in table 2. As a matter of fact, H_2S sensor has no sensitivity to other gases listed in the column 'Tested Gas' in table 2. In contrast, Nemoto's NE4 sensor for H has significant response to SO_2 and NO_2. As for the CO sensor fabricated in this work, it has negligible sensitivity (at the level of an error rate) to H_2S and NO_2.

Comparison of the characteristics – sensitivity (K), lower background currents (I_0), response times (t_{09}), temperature dependence of background current (T_w) and detection limits for gases (C_{min}) are presented in figure 4. It is seen that developed electrochemical sensors have better characteristics.

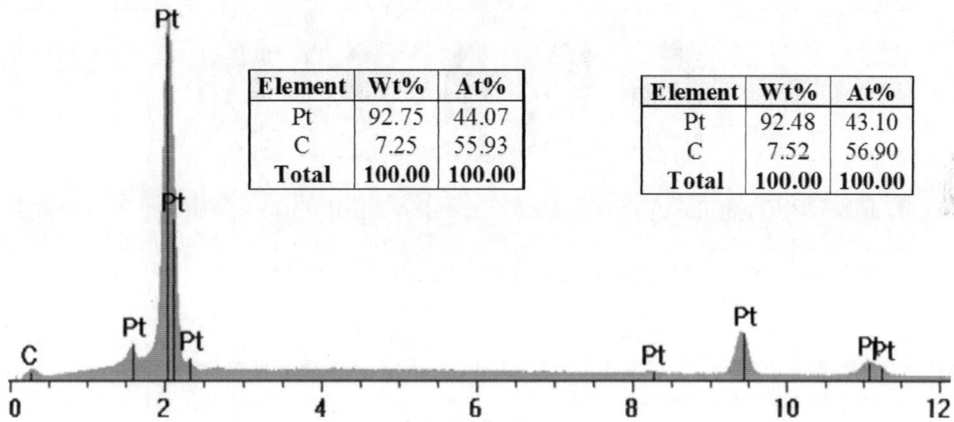

Figure 3. EDX spectra and calculated values of Pt and C.

Table 1. Sensitivity of electrochemical sensors.

Gas	Sensitivity of electrochemical cell	
	Background current, μA	Sensitivity, μA
CO	0,05	48,6 (100 ppm)
H_2S	0,01	30,1 (30 ppm)

Table 2. Cross sensitivity of carbon monoxide and hydrogen sulfide sensors.

Tested Gas	Concentration (ppm)	Sensors' readings (ppm)	
		CO	H_2S
Chlorine (Cl)	100	0	0
Carbon monoxide (CO)	100	100	0
Hydrogen sulfide (H_2S)	30	0.5	30
Nitrogen dioxide (NO_2)	30	0.3	0
Hydrogen chloride (HCl)	100	0	0
Sulfur dioxide (SO_2)	25	0	0
Formaldehyde (CH_2O)	10	0	0
Ethylene oxide (C_2H_4O)	100	0	0
Ammonia (NH_3)	200	0	0

4. Conclusions

It was shown that the developed electrochemical sensors with Pt|/C electrodes have greater sensitivity, lower background currents, comparable response times and detection limits for gases.

To better understand the relationship between the structure of a-C/Pt catalyst layers and their electrochemical properties, it is planned to use the in-situ X-ray monitoring method [13].

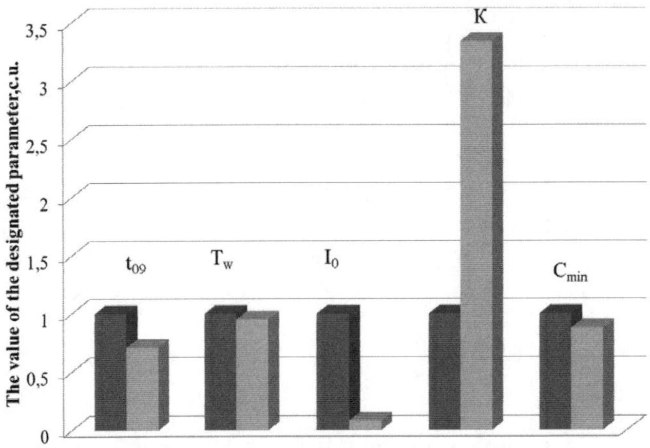

Figure 4. Comparison of the characteristics of the developed sensors (columns on the right) in comparison with the sensors of the main manufacturers: Alphasense (Great Britain), City Technology (Great Britain), Drager (Germany), Nemoto (Japan); MST (Germany).

References

[1] Spilker B, Randhahn J, Grabow H, Beikirch H and Jeroschewski P 2008 *Journal of Electroanalytical Chemistry* **612(1)** 121–30

[2] Vasiliev A A, Pisliakov A V, Sokolov AV, Samotaev N N et al. 2016 *Sensors and Actuators, B: Chemical* **224** 700–13

[3] Karpov E E, Karpov E F, Suchkov A, Mironov S, Baranov A M, Sleptsov V V and Calliary L 2013 *Sensors and Actuators A: Physical* **194** 176–80

[4] Makeenkov A, Lapitskiy I, Somov A and Baranov A 2015 *Sensors and Actuators B: Chemical* **209** 1102–7

[5] Samotaev N, Ivanova A, Oblov K, Soloviev S and Vasiliev A 2014 *Procedia Engineering* **87** 1294–7

[6] Eggins Brian R 2002 *Chemical Sensors and Biosensors* (John Wiley & Sons Ltd., London)

[7] Hart J, Crew A, Crouch E, Honeychurch K and Pemberton R M 2004 *Anal. Lett.* **37** 789–830

[8] Vork F, Janssen L, Barendrecht E 1986 *Electrochim. Acta* **31** 1569–75

[9] Litvinov A V, Nikolaev I N and Samotaev N N 2010 *Procedia Engineering* **5** 1216–9

[10] Speranza G, Calliari L, Ferrari M, Chiasera A, Tran Ngoc K, Baranov A M, Sleptsov V V, Nefedov A A, Varfolomeev A E and Fanchenko S S 2004 *Applied surface science* **238** 117–20

[11] Nechitaĭlov A A, Zvonareva T K, Remenyuk A D, Goryachev D N et al. 2008 *Semiconductors* **42** 1249–54

[12] Medvedeva E, Baranov A and Somov A 2016 *Sensors and Actuators B* **236** 858–64

[13] Baranov A M 1998 *Surface and Coatings Technology* **102** 154–8

IOP Conf. Series: Materials Science and Engineering **387** (2018) 012053 doi:10.1088/1757-899X/387/1/012053

Composite nanostructured wear-resistant coatings for high-speed cutting processing

M Sh Migranov, S R Shekhtman, N A Suhova and A M Migranov

Ufa State Aviation Technical University, 450000, Ufa, Russia

E-mail: MigMars@mail.ru

Abstract. The wear resistance of coatings, chip formation conditions, microstructure and chip shrinkage, the angle of the conventional shear plane were studied. The data on the characteristics of the surface structure and the adhesion of such coatings, etc., were studied and analyzed. It is shown that the main reason for high wear resistance of these coatings during high-speed processing is the formation of thin protective oxide films on the surface of a tool made from aluminum oxide with a complex amorphous-crystalline structure. Such protective films significantly improve the friction and wear characteristics of coated tools: the adhesion of the tool surface to the workpiece decreases and the thermal intensity of the cutting zone decreases due to the intense heat removal into the chips.

1. Introduction

Increasing the wear resistance of the cutting tool can now be achieved through the use of coatings made of titanium and aluminum nitrides (Ti, Al) N. These coatings are characterized by unique combinations of properties: high hardness at elevated temperatures in the contact zone together with thermal and chemical stability and low thermal conductivity. An important advantage (TiAl) N coatings is their tendency to oxidation and to the formation of relatively stable surface oxide films [1–6]. Further increase in wear resistance (TiAl) N coatings during cutting can be obtained by adjusting the grinding of the grain to the nanoscale level (grain size less than 100 nm). This can be done by doping (TiAl) N coatings onto the tool material with advanced advanced methods (for example, magnetic arc filtration (MAF), etc.) [3, 5, 6].

Obviously, a study of the mechanism of tribo-oxidation coating during cutting will most fully utilize the reserves to increase the wear resistance of such coatings. It can be assumed that the actual task in this case is the formation in the cutting zone of a thermodynamically stable and constant aluminum oxide film with high tribological properties [3, 5–8].

The purpose of this work is to study the friction and wear processes of cutting tools with multilayer composite coatings and to determine the influence of nanocrystalline structure in MAF coatings on these processes under high-speed processing conditions.

2. Experimental research

Various coatings were investigated, including: conventional monolayer, multilayer and nanostructured (improved).

Improved coatings were applied by filtration with a magnetic arc on an installation of the type NNV 6.6 - I1. These installations for the physical deposition of coatings (FOP) have up to three removable targets, the so-called modules. The temperature regime on the surface of the sample during coating was monitored with an optical pyrometer. The substrate was heated to the coating deposition temperature by means of argon ions during their deposition at a bias voltage of the substrate at 1 kV. When the desired substrate temperature was reached, the amount of bias voltage was lowered, and nitrogen was applied to the chamber for coating (TiAl) N.

Such a scheme makes it possible to influence the physicochemical and plasmochemical processes when applying heat-resistant coatings by increasing the ionization rate of both metals and chemically active gases. The effectiveness of MDF technology to prevent coating during the "droplet" phase is well known. But there is another advantage of MDF, which is important for the application of hard coatings. Due to the high rate of plasma ionization in the coating chamber chamber and the relatively low coating speed, the temperature at the beginning of the crystallization of MAF coatings is also low.

The coating conditions are close to the heating of the nanoscale, when the kinetic energy of the bombarding ions is transferred to very small regions of actively growing thin layers of coating, which then quickly move it deep into the substrate material. Thus, the cooling rate during a sufficiently nonequilibrium MDF precipitation process is high. Ion bombardment with low energy of growing films can limit the growth of grains and promote the formation of nanocrystalline layers. So nanocrystalline structures of wear-resistant coatings on the cutting tool are formed (especially for high-speed processing applications).

In the paper, an experimental method was used to estimate the tribotechnical parameters ($\tau_{nn,}$ $p_{rn,}$ τ_{nn}/p_{rn}) [4]. This method is based on the physical model (figure 1), which in the first approximation reflects the real conditions of friction and wear on the local contact. According to this model, the spherical indenter 2 (simulating a single roughness of the contact spot of rubbing solids), squeezed by two plane-parallel samples 1 (with high accuracy and cleanliness of the contacting surfaces) rotates under load around its own axis. The force F expended on the rotation of the indenter and applied to the cable 3 laid in the groove of the disc 4 is mainly related to the shear strength τnn of the adhesive bonds.

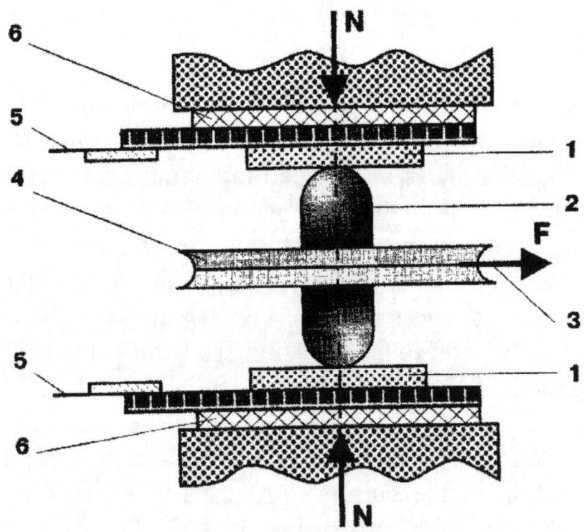

Figure 1. Frictional contact model.

To use this method, special equipment was developed and created [3], which makes it possible to produce an electric contact heating (via bus bars 5, insulated from the body with gaskets 6, figure 1), in order to use this method, and to provide a characteristic temperature distribution for the tribocontact in the near-surface layer. Tare methods have also been developed for controlling the temperature of the contact, as well as obtaining the dependence of the value of τnn on the normal stresses prn at different contact temperatures.

Indenters are made of tool material (TT8K6) without coating and with coatings (TiN, (TiCr) N, (TiAl) N, (AlTi) N, (TiAlCr) N, (AlTiCr) N with different percentages of each of the coating elements up to 15 μm). Samples of the counterbody (processed material) were made of steel 40X (HB 280–290). The studies were carried out using a 5% water-based emulsion and without coolant.

Full-scale tests of the cutting tool were made during milling and turning. The milling was carried out on the vertical milling machine HECKERT with end mills (d = 12 mm, z = 4) and tool milling cutters (d = 90 mm, z = 1) with interchangeable tetrahedral carbide plates TT8K6. The investigation was carried out under different milling conditions (n = 500–900 rev/min, S = 60-100 mm / min, t = 1–3 mm, b = 4–10 mm); longitudinal turning was performed at V = 150–450 m / min, S = 0.11 mm / rev, t = 0.5 mm. The coatings for the tools were applied by various companies (Balzers, Caromant, Carbide, Rimet) using factory technologies. The turning of the blanks was carried out on the lathe-cutting machine 16K20 with non-re-sharpening carbide plates TT8K6 $\gamma = 0°$ $\alpha = 10°$; $\varphi = \varphi_1 = 45°$; r = 0,5 mm) with all the above coatings.

In actual experiments, during milling and turning, tool wear on the back surface, cutting temperature, cutting forces and roughness of the treated surface were investigated. Some of the results obtained are shown in figures 2–6.

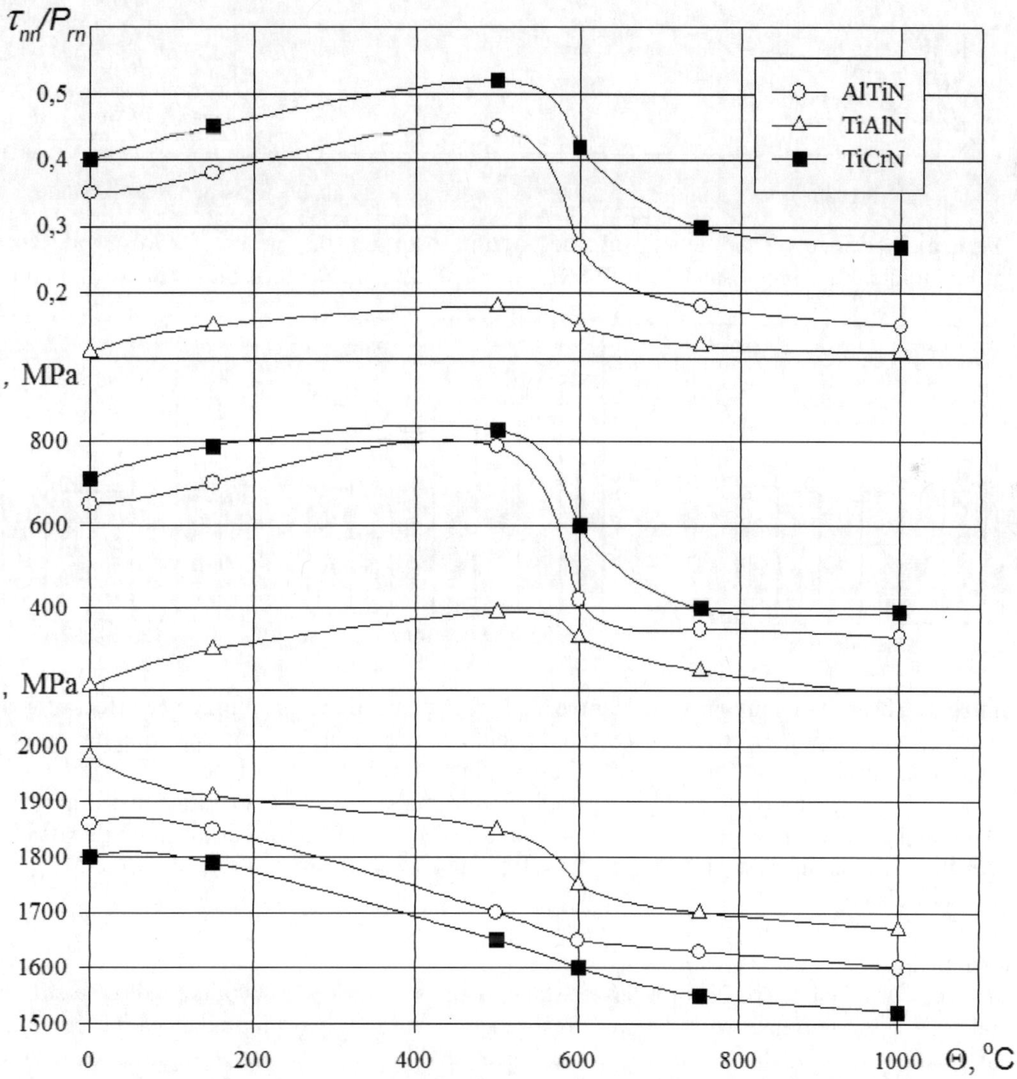

Figure 2. The effect of temperature on the characteristics of the adhesive interaction of the pair steel 40X - indenter TT8K6 + coating ((TiCr) N, (TiAl) N, (AlTi) N).

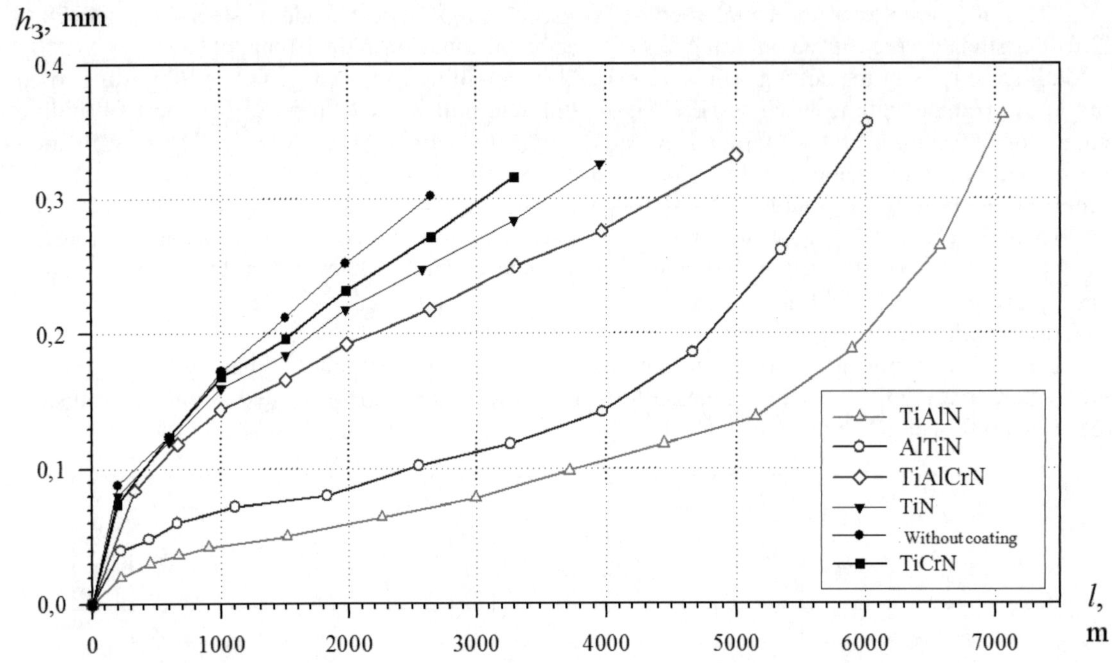

Figure 3. Effect of the length of the cutting path on the wear of tools with coatings at longitudinal turning of steel 40X-TT8K6 (V = 450 m / min, S = 0.11 mm / rev, t = 0.5 mm).

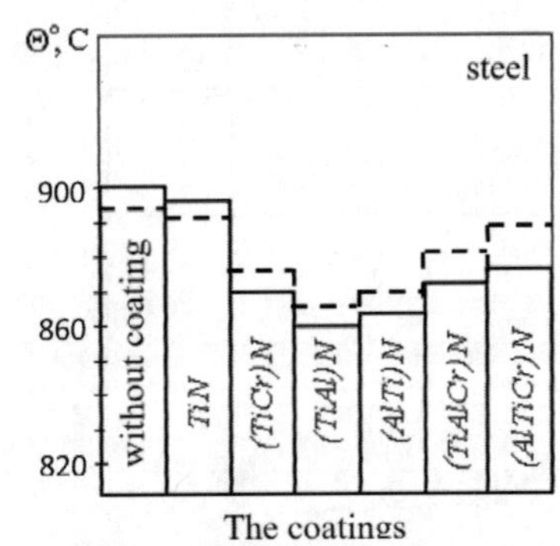

Figure 4. Effect of coating on the cutting temperature for longitudinal turning V = 450 m / min, S = 0.11 mm / rev, t = 0.5 mm, hz = 0.1 mm; —— experimentally obtained steel, – – – calculated).

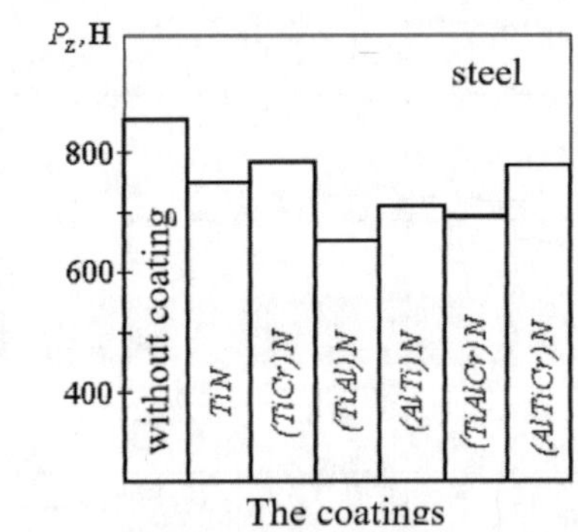

Figure 5. Effect of coating on the tangential component of the cutting force at longitudinal turning (V = 450 m / min, S = 0.11 mm / rev, t = 0.5 mm, hz = 0.1 mm).

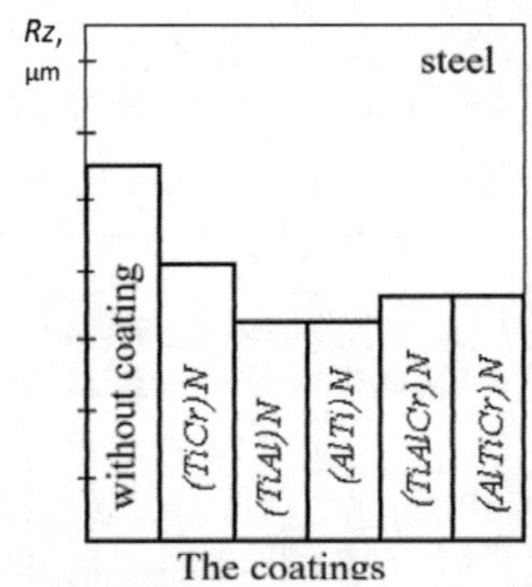

Figure 6. Effect of coating on the roughness of the treated surface at longitudinal turning (V = 450 m / min, S = 0.11 mm / rev, t = 0.5 mm).

3. Conclusions

1. Tribotechnical characteristics of wear-resistant coatings, investigated with the aid of an adhesiomer in full-scale tests, correlate with the wear resistance of the tool and this indicates the possibility of using inexpensive and labor-intensive tribotechnical studies for a qualitative rapid assessment of the wear resistance of coatings (figure 2);

2. The use of wear-resistant coatings ((TiAl) N, (AlTi) N) leads to a decrease in the intensity and magnitude of run-in and normal wear and tear, and increases the wear resistance of the cutters and the

period of their durability by an average of 2 to 2.5 times reduces the level of temperature-force loading of the cutting zone (figures 4, 5) and the roughness of the machined part surface (figure 6).

3. Two major improvements in the characteristics of the coating surfaces can be attributed to the magnetic filtering method. The first improvement is due to the complete or partial filtration of the "droplet" phase. As a result, a surface with a lower roughness is formed, which affects the adhesion (adhesion) of the material being processed to the tool surface. The second improvement is associated with a decrease in the frictional forces and wear of the filtered coatings.

4. The main advantage of MDF coatings for use in high-speed processing is a significant grinding of grains (TiAl) N coatings, which leads to the formation of a surface layer of the nanoscale (grain size about 60–80 nm). It is shown that the main cause of high wear resistance of filtered coatings under high-speed processing is the formation of thin protective oxide films on the surface of the cutting tool (as a result of self-organization phenomenon).

5. The formation of alumina films on the surface of the tool significantly changes the heat fluxes and the heat removal into the chips. The chip formation conditions (chip shrinkage ratio and the angle of the conventional shear plane), as well as the coefficient of friction along the front surface of the cutting tool, measured during the cutting, also show a significant improvement in the tribotechnical parameters for instruments with filtered coatings.

References

[1] Vereshchaka A S 1993 *Efficiency of the cutting tool with wear-resistant coatings* (M.: Mechanical Engineering) 336
[2] Fox-Rabinovich G S et al. 1997 *Wear* **206** 214
[3] Fox-Rabinovich G S, Shuster L Sh and Migranov M Sh 2003 *Metallurgical coatings and thin films* **51**
[4] Shuster L S and Migranov M Sh 2003 *Device for investigating adhesion interaction. Patent for utility model No. 34249*
[5] Shuster L S, Krioni N K, Sholom V Yu and Migranov M Sh 2008 *Coatings and greases in high-temperature moving interfaces and metalworking* (M.: Mechanical Engineering) 318
[6] Cryoni N K and Migranov M Sh 2016 *Coatings and greases for high-temperature tribocoupling* (Moscow: Innovative mechanical engineering) 377
[7] Bershadsky L I 1981 *Self-organization and reliability of tribosys*tems (Kiev: Knowledge) 35
[8] Polzer G, Abileng V and Firkovskii A 1988 *Friction and wear* **1** 12

IOP Conf. Series: Materials Science and Engineering **387** (2018) 012054 doi:10.1088/1757-899X/387/1/012054

Formation of the thin film heterostructures CuO/ZnO by RF magnetron sputtering powder targets

N V Mukhin, D A Chigirev, L D Bakhchova, G A Konoplev, K V Kochunov and I N Kashkul

Saint-Petersburg State Electrotechnical University "LETI", 197376, St. Petersburg, Russia

E-mail: gakonoplev@mail.ru, mukhin.nikolay.v@gmail.com

Abstract. Method for forming thin film heterostructures ZnO/CuO in a single technological cycle with vacuum circuit with use RF magnetron reactive sputtering of copper powder and zinc oxide was offered. The resulting film CuO had band gap of 1.6…1.7 eV and possessed stability up to 350°C. ZnO film had a transparency of 85% in the wavelength range of 400…1100 nm. Selection of medium temperatures for forming the heterostructure (200…250°C) allowed to reduce its series resistance, and the addition of the thin heterojunction structure of the intermediate layer i-ZnO allowed to improve nonlinearity voltage characteristics and improve the photoresponse.

1. Introduction

Existing limitations of modern photovoltaic cells leads to the needs to find more efficient materials for solar cells, as well as the development of new technologies for their production [1–3]. As photoactive layer for such heterostructures considerable interest has copper oxide, as it's absorption spectrum is consistent with the solar radiation spectrum [4]. Furthermore, copper oxide has p-type conductivity, which in combination with other oxides, such as zinc oxide films (n-type conductivity) [5–7], allows the formation of heterostructure solar cells [1]. In addition to the n-ZnO/p-CuO heterojunctions themselves, the heterostructures of zinc and copper oxides with PZT are also of interest for photovoltaic applications [8–10] due to the ferroelectric photovoltaic effect at the interface of the PZT and semiconductors [2, 10, 11].

The main methods of producing films of copper oxide or zinc oxide are reactive ion-plasma sputtering of metal targets, CVD or sol-gel methods. In this work was chosen the RF magnetron sputtering from a target of powder to produce oxide films of zinc and copper. It should be noted that the developed method allows forming heterostructure n-ZnO/p-CuO in a single technological cycle.

This paper is a continuation study of works [3, 12, 13] as development of thin film oxide layers and heterostructures for photovoltaic applications.

2. Obtaining of thin film heterostructures CuO/ZnO by RF magnetron sputtering

Heterostructures ZnO/CuO prepared layering films of zinc oxide and copper by reactive RF magnetron sputtering on a specialized installation scheme of which and its inner casing shown in figure 1. Inside the vacuum chamber are two magnetrons 100 mm (figure 1(b)), connected to the RF generator operating at 13.56 MHz and allowing to vary the power input to the magnetron from 10 W to 1000 W.

As the target for sputtering films of zinc oxide and copper were used CuO and ZnO chemically pure powders distributed evenly across the surface of the magnetrons.

Complex conducted experiments showed that the investigated oxides characterized by low sputtering rates. Dependence of growth rate of zinc oxide and copper on the RF discharge are shown in figure 2. It should be noted that a further increase in the RF power discharge to increase the film growth rates is undesirable because it will lead to overheating and eventual magnetrons failure. The thicknesses of the obtained ZnO layers were 120 nm and CuO 400 nm; optimal parameters of the films (table 1) were selected.

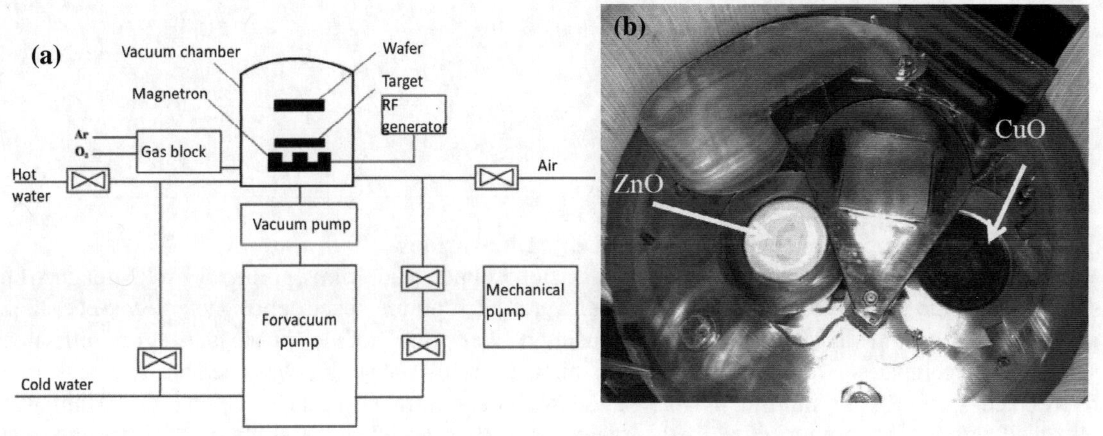

Figure 1. Flowchart of reactive RF magnetron sputtering (**a**) and the inner casing of the vacuum chamber (**b**).

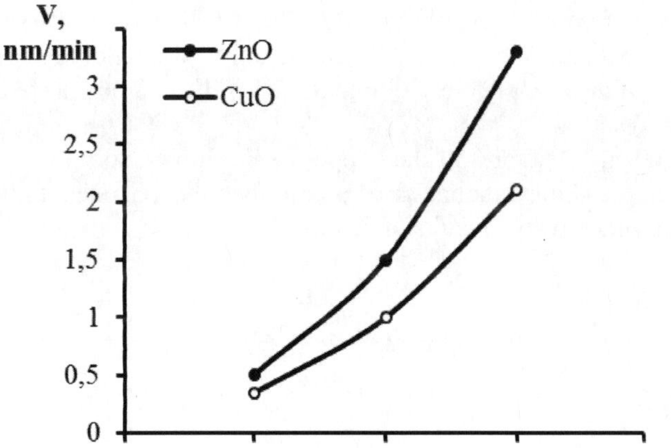

Figure 2. Depending velocity ZnO and CuO film growth on the power RF discharge.

Thus, in this work it was first used by RF magnetron sputtering zinc oxide powder target and copper oxide to form heterostructures ZnO/CuO in one vacuum cycle.

In this work as a transparent contact substrate was used a glass with the conductive layer of zinc oxide doped with boron. As the opaque contact to heterostructures ZnO/CuO used thin films Pt, Ag, Ni, Al. Top electrodes were deposited on the already formed heterostructure. A distinctive feature of forming the upper electrodes are decreased deposition temperature. Modes of obtaining top electrodes are basing on previous studies. The configuration of upper electrodes created by the use of a shadow mask. Electrode area was 3,14 and 6,75 mm^2.

Table 1. Major parameters of the deposition process, oxide films.

Process parameters	ZnO	CuO
RF discharge power, W	100	150
The composition of the gas mixture	76% Ar + 24% O_2	76% Ar + 24% O_2
The operating pressure of the gas mixture, mm Hg	0.01	0.01
Substrate temperature (Self-heating), °C	260...270	260...270
Film growth rate, nm/min	1,3	2,1
Deposition time, min	90	180

3. Study optical properties of the copper and zinc films oxides

Were measured spectra of reflection and transmission copper oxide films prepared by RF magnetron sputtering on the glass substrate. Measurements were taken on the spectrophotometer AvaSpec-2048 in the visible spectral range of 400...1000 nm using a fiber-optic module for measuring the reflection spectra. As a comparison reference for the transmission spectra measuring was used air, for measuring reflectance spectra – aluminized mirror. In the part of spectra near the wavelength 500 nm was observed intrinsic absorption edge in the region of longer wavelengths – alternating maxima and minima due to interference of the radiation reflected from the upper and lower sides of the film.

In accordance with the Lambert-Bouguer-Beer law from the reflection and transmission spectra was calculated absorption spectrum of the copper oxide. The calculation results showed that used in this work method of optical characteristics of thin films, including a parallel registration transmission and reflection spectra allows almost completely eliminate the influence of interference in the analysis of absorption spectra.

Figure 3 shows the results of studies of the copper oxide films absorption spectra. The value of the band gap width 1,6…1,7 eV corresponds to the fact that the obtained thin films are CuO. This conclusion was also confirmed by the results of X-ray analysis.

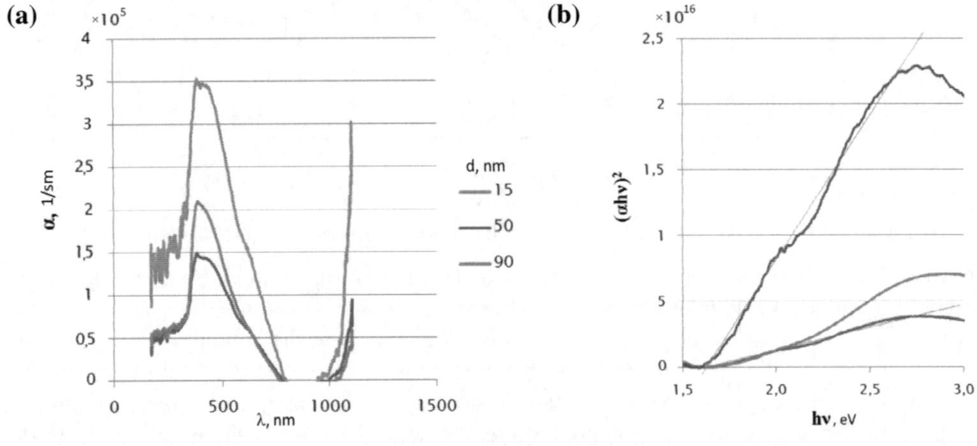

Figure 3. The absorption spectra of copper oxide (**a**) and determining of the copper oxide band gap (**b**).

Reflection (R) and absorption (T) spectra of films CuO showed no thermal effect to 350 °C inclusive that indicates the stability of the formed oxide, that showed in figure 4.

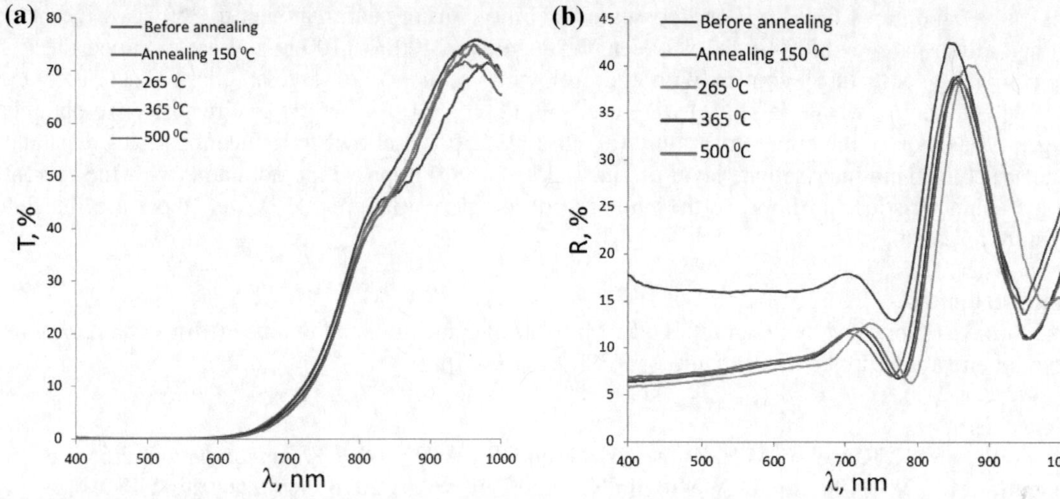

Figure 4. The spectra of transmission (**a**) and reflection (**b**) copper oxide without annealing and with at 150, 265, 365 and 500 °C.

Study of transmission spectra zinc oxide films have shown that they have a transparency 85 % in the wavelength range from 400 to 1100 nm.

4. Study electrical properties of thin film heterostructures based on oxides of copper and zinc

Investigation of the current-voltage characteristics of thin film heterostructures ZnO/CuO was carried out in the dark mode and under illumination using a light source with a spectral characteristic close to the solar spectrum. Figure 5 shows the results of research.

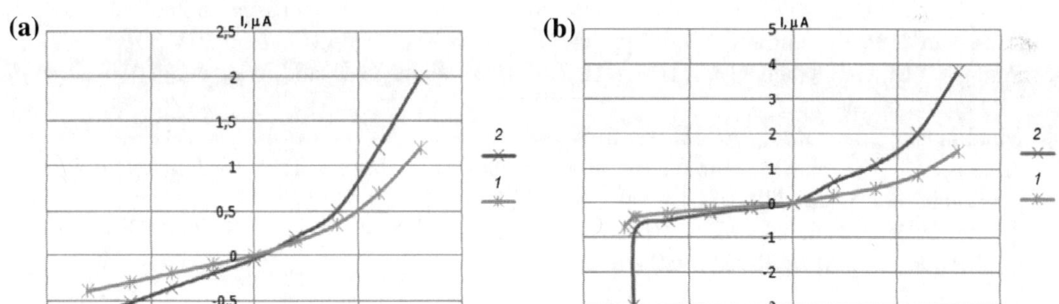

Figure 5. The current-voltage characteristics of heterostructures Glass/n-ZnO/p-CuO/Pt (**a**) and Glass/n-ZnO/i-ZnO/p-CuO/Pt (**b**) by dark mode (1) and under illumination (2).

Has been found that in the samples where the copper oxide applied at higher temperatures up to 300 °C, the characteristics were weak nonlinearity and had large series resistance, that was presumably a consequence of the diffusion of boron in the structure. Samples which were formed at low temperatures exhibited weak photoresponse and high series resistance, presumably due to imperfectly formed crystalline structure of the films. Selection of medium temperature (about 200...250 °C) allows to reduce the series resistance of the structure. Adding a thin intermediate layer of zinc oxide (i-ZnO) in heterostructure n-ZnO/p-CuO improved the nonlinearity of the current-voltage characteristics and the photoresponse of heterostructures.

5. Conclusions

A new technology of forming heterostructures n-ZnO/p-CuO in a single process cycle with using RF magnetron reactive sputtering of copper and zinc oxide powder targets. The obtained copper oxide films had a band gap 1.6...1.7 eV and possessed stability during heat treatment to 350 °C. Zinc oxide film had a transparency 85 % in the wavelength range from 400 to 1100 nm. Were obtained the first thin film oxide structure samples with the following sequence of layers: Glass/n-ZnO/i-ZnO/p-CuO/Me. As electrode materials (Me) were used Cu, Pt, Ag, Ni, Al, but the best results were obtained using Pt. Selection of medium temperature (about 200...250 °C) allows to reduce the series resistance of the structure. Thin intermediate layer of zinc oxide (i-ZnO) improve the nonlinearity of the current-voltage characteristics and improve the photoresponse heterostructures. However, there are the field for future research.

Acknowledgment

This work was supported by the grant of The Ministry of Education and Science (Minobrnauka) of the Russian Federation (project Goszadanie № 3.3990.2017/4.6).

References

[1] Adilov Sh R, Kumekov M E, Kumekov S E and Terukov E I 2013 *Semiconductors* **5** 655–6
[2] Afanasjev V P, Chigirev D A, Mukhin N V and Petrov A A 2016 *Ferroelectrics* **1** 170–6
[3] Agafonova D S, Kozyrev A B, Mukhin N V, Red'ka D N, Elanskaya K G, Rudenko M V, Oseev A and Hirsch S 2018 *Glass Phys. Chem.* **1** 15–20
[4] Adilov Sh R, Afanasiev V P, Kashkul I N, Kumekov S E, Mukhin N V and Terukov E I 2017 *Glass Phys. Chem.* **3** 272–5
[5] Redka D N, Mukhin N V and Zakharov I G 2016 *Technical Physics* **61** 1744–6
[6] Iakovlev A, Redka D, Afanasjev V, Mukhin N, Grishkanich A, Terukov E and Hirsch S 2017 *Proc. of SPIE* **10091** 100911B
[7] Afanasjev V P, Mukhin N V, Redka D N, Rudenko M V, Terukov E I, Oseev A and Hirsch S 2017 *Ferroelectrics* **508** 124–9
[8] Cao D, Wang C, Zheng F, Dong W, Fang L and Shen M 2012 *Nano Letters* **6** 2803–9
[9] Afanas'ev V P, Vorotilov K A and Mukhin N V 2016 *Glass Phys. Chem.* **3** 295–301
[10] Pan D-F, Bi G-F, Chen G-Y, Zhang H, Liu J-M, Wang G-H and Wan J-G 2016 *Scientific Reports* **6** 22948
[11] Mukhin N V 2016 *Glass Phys. Chem.* **1** 64–9
[12] Afanasjev V, Bazhan M, Klimenkov B, Mukhin N and Chigirev D 2016 *Journal of Physics: Conference Series* **729** 012013
[13] Afanasjev V P, Konoplev G A, Chigirev D A, Mukhin N V, Terukov E I and Terukova E E 2016 *Proc. of SPIE* **9896** 98980V

IOP Conf. Series: Materials Science and Engineering **387** (2018) 012055 doi:10.1088/1757-899X/387/1/012055

Vacuum system "MAGNA TM 150-01K" for magnetron sputtering deposition of multilayer functional coatings

V V Odinokov, R A Karakulov, V V Panin and A V Shubnikov

JSC Research Institute of Precision Machine Manufacturing (NIITM), 124460, Moscow, Zelenograd, Russia

E-mail: vodinokov@niitm.ru

Abstract. The new vacuum system "MAGNA TM 150-01K" developed at Research Institute of Precision Machine Manufacturing is described. Its structure and principle of operation are considered. This system was used for magnetron sputtering deposition of multilayer coatings on 100 – 150 mm circular silicon wafers. Coatings were deposited in common vacuum cycle from three targets with use of cassette holder in load-lock built-in "clean room". Titanium (Ti) and titanium nitride (TiN) layers were deposited consequently from common titanium target, aluminum (Al) layer was deposited from two aluminum targets.

1. Introduction

All layers must be deposited on the substrates in a common vacuum cycle to obtain high-quality multilayer coatings when creating various semiconductor devices based on thin-film technologies. This technology provides not only a growth of productivity, but also an increase of the adhesion between the substrate and the adhesive layer, as well as between individual layers of coating [1].

An important task is creation of the domestic special process equipment (SPE) with cassette holder in load-lock built-in "clean room". This could provide good quality of coatings and allow to operate in automatic mode with control of all parameters at all stages of the deposition process [2].

2. Description and principle of operation of the system

Figure 1 shows the appearance of the system. The system consists of load-lock with cassette, transport chamber with vacuum robot, process chamber and control rack.

The system operates as follows. Loading and unloading of cassette *12* with wafers occurs manually by means of load-lock *1* (figure 2). Then high-vacuum gate valve *2* is opened and vacuum robot *10* conveys wafers individually from cassette to substrate holder *8* which heats substrate *6* up to 300 ^{O}C when the required ultimate pressure is achieved. The individual gas channel with a mass-flow controller of argon under the substrate provides an even warming up of its surface.

After that, the wafer is undergone ion cleaning in the plasma of the high-frequency (HF) gas discharge due to the supply of the HF potential to the substrate holder. Multicathode sputtering device (MSD) with three 100 mm circular targets *3* and *4* is used to produce material of coatings (third target is not shown on figure 2). Shutter *5* rotates and opens the targets sequentially when the process technology is executed. The substrate holder ensures a smooth rotation of the wafers around the vertical axis at a frequency of up to 20 rpm. At the end of the process, the manipulator conveys the processed wafer into its cell in the cassette and takes the next untreated wafer.

IOP Conf. Series: Materials Science and Engineering **387** (2018) 012055 doi:10.1088/1757-899X/387/1/012055

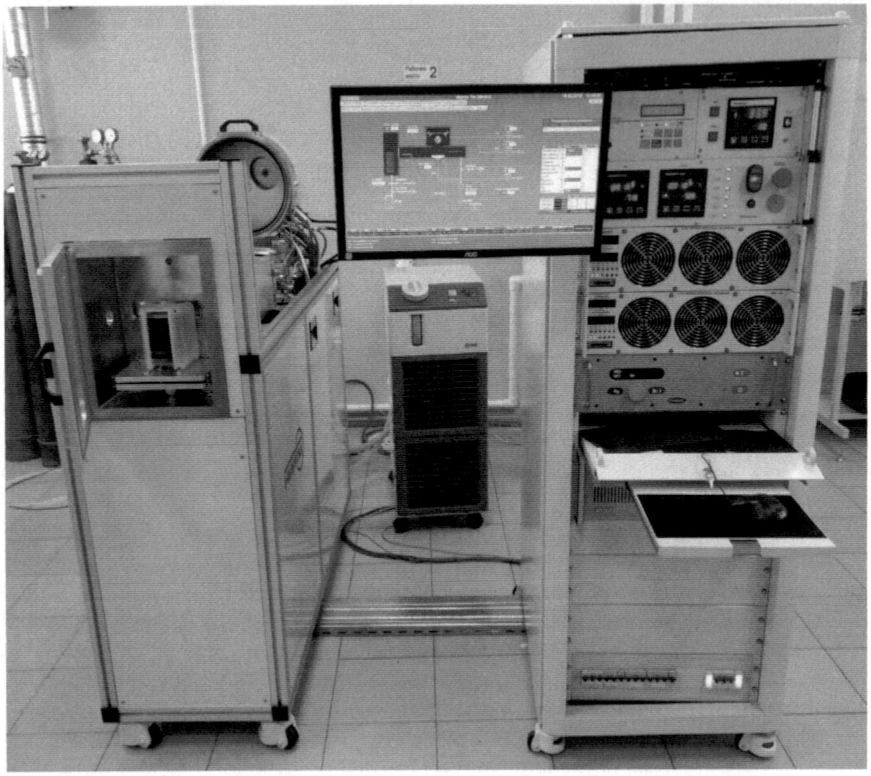

Figure 1. The appearance of the «MAGNA TM 150-01K» system.

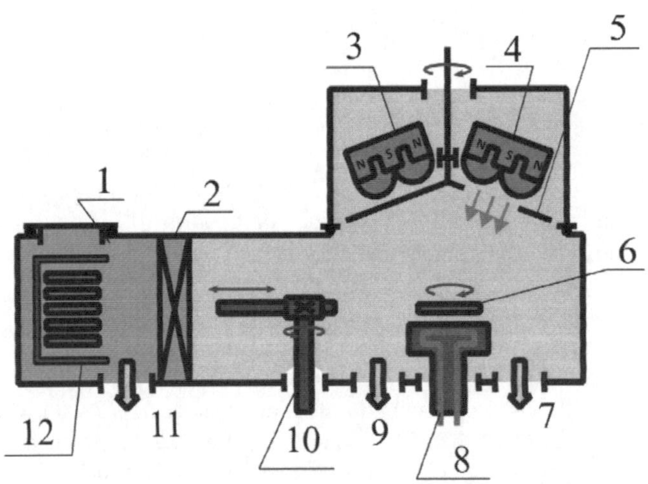

Figure 2. The principal scheme of the «MAGNA TM 150-01K» system.

3. Results and discussion

The «MAGNA TM 150-01K» system was used for deposition of multilayer metallization on 100 mm circular silicon wafers. Titanium (Ti), titanium nitride (TiN) and aluminum (Al) layers were deposited consequently in common vacuum cycle in a co-sputtering mode. Such multilayer functional structures are used to create thin-film circuits in microwave devices.

The modes of HF cleaning of the wafers, presented in table 1, were selected. The 600 W power HF generator was used.

Table 1. The modes of HF cleaning.

Argon flow, l/h	Process pressure, Pa	Power, W	Time, min
1,4	1,8	100	10

The efficiency of this operation was evaluated by measuring the thickness of the silicon dioxide layer (SiO_2) before and after purification. Thickness was controlled on the wafer at 5 points in its various sections (figure 3). Thus, information about the uneven etching of the material was received. The average cleaning rate was 1.511 nm / min. The results of the measurements are shown in table 2.

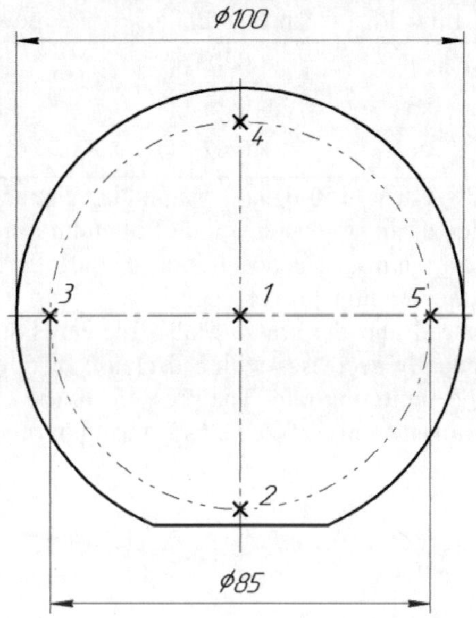

Figure 3. Location of control points on the wafer.

Table 2. Efficiency of HF cleaning

Point number	SiO_2 before etching, nm	SiO_2 after etching, nm	Thickness changing (Δ), nm	Δ_{av}, nm
1	318	300,93	17,07	
2	325,98	313,14	12,84	
3	316,79	302,93	13,86	15,11
4	316,53	302,24	14,29	
5	320,25	302,76	17,49	

Optimum modes of deposition of Ti, TiN and Al layers, shown in table 3, were chosen on the basis of the following criteria: maximum productivity, magnetron discharge stability, composition and structure of the coating. Titanium and titanium nitride coatings were deposited using a single titanium target (Ti) and aluminum coating was deposited using two aluminum (Al) targets. Two 3 kW power direct current (DC) sources was used. A switching device was used to switch first power source between the titanium target and one aluminum target. The second power source has always been connected to another aluminum target.

A series of experiments was performed to obtain a stoichiometric coating of titanium nitride (Ti_1N_1). It is known that the best characteristics (specific electrical resistance) is the titanium compound with nitrogen in the ratio of atoms 1:1 [3]. At the initial stage, the primary evaluation of the

IOP Conf. Series: Materials Science and Engineering **387** (2018) 012055 doi:10.1088/1757-899X/387/1/012055

coating structure was its color: stoichiometric titanium nitride has a golden sheen, a purple hue indicates excess nitrogen and gray color indicates its lack. After this, the final optimization of the modes for the formation of the required structure was carried out after an analysis of the elemental composition. The result of the investigation of the coating obtained after the final adjustment of the deposition parameters showed the content (quantitative) of nitrogen at the level of 53.65 %, and the specific electrical resistance of the film was 216 μΩ-cm.

Table 3. Optimal modes of deposition

Coating	Argon flow, l/h	Nitrogen flow, l/h	Process pressure, Pa	One target power, W	Temperature, ^0C
Ti	1,0	-	0,45	1500	200
TiN	0,4	0,34	0,33	1500	200
Al	1,0	-	0,52	1500	150

In addition, for the «MAGNA TM 150-01K» system the character of the pressure recovery depending on the flow of nitrogen during the application of titanium nitride in the working chamber is established (figure 4). Because the nitrogen concentration in the plasma increases on path 1-2, the process of active nitriding of the titanium target begins. As a result, the titanium content in the working volume is sharply reduced and the pressure also increases spasmodically as titanium is a getter. The rate of nitriding gradually decreases with a decrease of nitrogen consumption (path 3-4) and becomes equal to the nitride sputtering rate. The pressure in the chamber becomes the same as soon as nitride is completely removed from surface and sputtering of pure titanium begins.

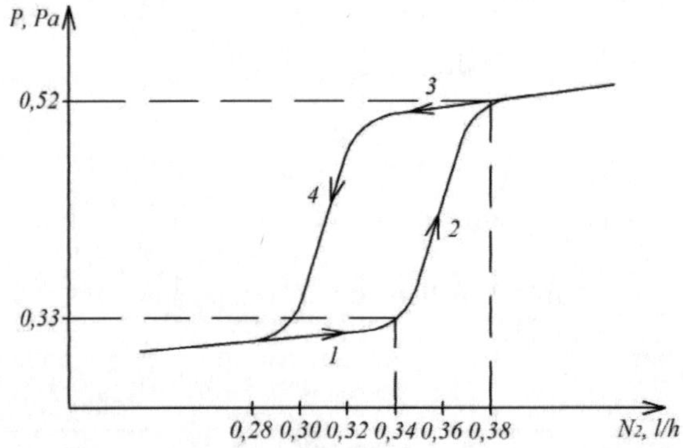

Figure 4. Hysteresis of working pressure during TiN deposition.

The performance of the "MAGNA TM 150-01K" system for metallizing of wafers with such metals can be judged by knowing the rates of their deposition. The results of measurements of the thickness of the coatings of each material are shown in table 4.

Table 4. Thickness of coatings.

Coating	Time of deposition, min	Thickness, nm
Ti	1	70
TiN	2	120
Al	5	1270

Thus, the average deposition rate of titanium and titanium nitride using a single magnetron is 70 nm/min and 60 nm/min respectively. The average deposition rate of aluminum using two magnetrons is 254 nm/min.

4. Conclusions

The «MAGNA TM 150-01K» system allows deposit multilayer functional thin film coatings in common technological cycle by means of magnetron sputtering. In this case, high-quality multilayer thin films are formed having good adhesion strength between the substrate and the adhesive layer, as well as between the individual layers. The automatic control system monitors all process parameters and maintains their stability according to the set values. This ensures good reproducibility of the properties of the coatings. The placement of equipment of this type in clean industrial premises will significantly reduce the introduced defectiveness on the wafers and lead to an increase in the yield of suitable products.

References

[1] Odinokov V V, Pavlov G Ya, Panin V V, Putyrskiy V V, Rashchinskiy V P, Shpakov A N and Shubnikov A V 2014 *Nanoindustry* **9** 8–11
[2] Karakulov R A, Odinokov V V, Panin V V, Shubnikov A V, Vladimirov D S, Vladimirov S V and Golubtsov A A 2017 *Nanoindustry* **2** 80–6
[3] Grigoriev S N, Melnik Yu A, Metel A S, Panin V V and Prudnikov V V 2011 *Japan. J. Appl. Phys.* **50** 08JG04

IOP Conf. Series: Materials Science and Engineering **387** (2018) 012056 doi:10.1088/1757-899X/387/1/012056

Non-self-sustained low-pressure glow discharge for nitriding steels and alloys

E V Ostroverkhov, V V Denisov, Yu A Denisova, N N Koval and I V Lopatin

Institute of High Current Electronics SB RAS, 634055, Tomsk, Russia

E-mail: evgeniy86evgeniy@mail.ru

Abstract. Structural steel 40Cr and die steels 8Cr4W2MoVSi2 and Cr12Mo were nitrided in the plasma of a non-self-sustained hollow-cathode glow discharge at a pressure of 1 Pa, negative bias of 300 V, and temperature of 520 °C and 400 °C. The method has been shown promising for low-temperature nitriding of steels to a depth of several tens of micrometers with minimum tempering.

1. Introduction

The most efficient and ecologically clean methods of increasing the wear resistance of metal products are those based on electron-ion plasma surface treatment [1]. In a hollow-cathode glow discharge at a pressure of ≈1 Pa, the electron free path measures several to tens of centimeters, allowing efficient gas ionization by electrons oscillating in the hollow cathode. In self-sustained mode, the discharge operates due to electrons produced through secondary ion-electron emission at the cathode surface and accelerated in the region of near-cathode potential fall. Such self-sustained operation is disadvantageous in the necessity of increasing the voltage to provide discharge currents acceptable for practical use, and this causes substantial etching of the cathode walls and contamination of the plasma by erosion products. In non-self-sustained mode, additional electrons are injected into the plasma, making it possible to decrease the operating pressure several times and the discharge voltage to several tens of volts, and to greatly increase the glow discharge current [2]. Thus, one can produce a plasma of density higher than 10^{18} m^{-3} in vacuum volumes larger than 0.1 m^3 [3]. In such a discharge, the average ion current density to a target reaches 15 mA/cm^2, which is promising for practical applications. The high ion current density in a non-self-sustained glow discharge greatly improves the efficiency of ion cleaning and shortens the nitriding time compared to anomalous glow discharges operating at 50–1000 Pa. Another important factor is that the positive ions produced in the nitrogen-containing discharge plasma and delivered to the surface of a target under negative bias are a source of atomic nitrogen which is absorbed by the surface and diffused deep into the target. Thus, it is valid to say that the plasma of a non-self-sustained glow discharge provides a high saturation efficiency, and this should assist in decreasing the nitriding time and temperature. The decrease in temperature is particularly urgent for hardened steels which are tempered during long-term nitriding at about 520 °C.

The aim of our study was to assess the possibility to obtain hardened layers several μm thick in steels through their low-temperature nitriding in the plasma of a non-self-sustained hollow-cathode glow discharge.

IOP Conf. Series: Materials Science and Engineering **387** (2018) 012056 doi:10.1088/1757-899X/387/1/012056

2. Material and research techniques

Figure 1 shows a schematic of the experimental test bench used in the study and described in detail elsewhere [3]. The glow discharge was ignited in its vacuum chamber of dimensions 650×650×650 mm^3 between hollow cathode *1* (inner chamber surface) and plane anode *2*. The glow discharge was powered by a stabilized source with an output voltage of up to U_d = 300 V and average output current of up to I_d = 200 A. For providing stable low-pressure operation of the glow discharge in high-purity nitrogen (99.999 %), we used an electron source based on an arc discharge with an integrally cold hollow cathode[4]. The test materials were specimens of structural steel 40Cr and die steels 8Cr4W2MoVSi2 and Cr12Mo with a diameter of 20 mm and thickness of 4–6 mm. The specimens were fixed on a holder kept at the hollow cathode potential in the center of the chamber and were nitrided for 3 h at a temperature of 520 and 400 °C, pressure of 1 Pa, discharge voltage of 100 V,negative bias of 300 V, and pulse duty factor of 50 % (table 1). Their heating and cleaning were provided via bombardment by nitrogen ions accelerated in the cathode layer of the glow discharge.

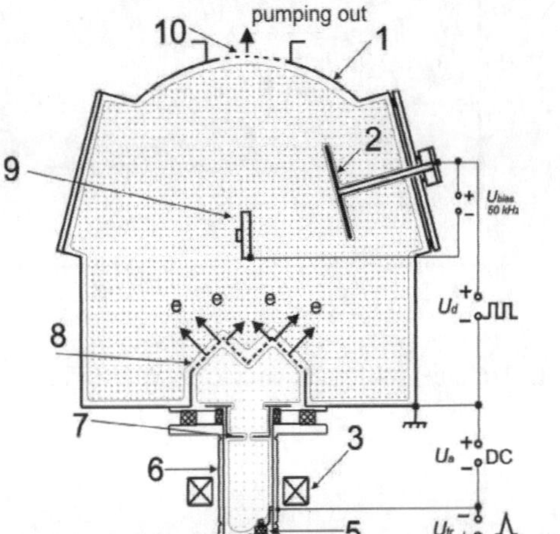

Figure 1. Schematic of experimental arrangement: *1* – cathode of glow discharge, *2* – anode of glow discharge, *3* – magnetic coil, *4* – gas inlet, *5* – igniter electrode, *6* – hollow cathode of auxiliary arc discharge, *7* – arc arrester, *8* – conical grid anode of auxiliary arc discharge, *9* – specimen holder, *10* – grid.

The nitriding efficiency was assessed by measuring the Vickers microhardness of lateral specimen sections in depth from the surface with a step of 10 μm at an indenter load of 0.5 N. Before nitriding, the surface microhardness of steels40Cr, 8Cr4W2MoVSi2, and Cr12Mo was 5.4, 7.1 and 7.3 GPa, respectively.

Table 1. Nitriding modes and surface microhardness of steels.

T, °C	U_d, V	I_d, A	j_i, mA/cm^2	W, kW	U_{bias}, V	$HV_{0.5}$, GPa 40Cr / 8Cr4W2MoVSi2
400	100	29	1	2.9	–300	7.39 / 9.85
520	100	66	1.9	6.6	–300	7.2 / 9.85

T – nitriding temperature, U_d – discharge voltage, I_d – discharge current, j_i – ion current density, W – discharge power, U_{bias} – bias, $HV_{0.5}$ – surface microhardness after nitriding

3. Results and discussion

The microhardness distributions in depth of steels 40Cr, 8Cr4W2MoVSi2, and Cr12Mo are presented in Figures 2(a), (b) and 3, respectively. As can be seen from figures 2, 3, all steels after nitriding reveal a substantial increase in their surface microhardness. The nitrided layer thickness depends on the chemical composition of the steels, and its maximum value of 300 μm is found in low-alloy steel 40Cr nitrided at 520 °C. Decreasing the nitriding temperature to 400 °C decreases the nitrided layer thickness to 140 μm. As it follows from figure 2(a), the bulk microhardness of steel 40Cr after nitriding at 400 °C is 40 % higher than that after nitriding at 520 °C. For steel 8Cr4W2MoVSi2, there is no need to decrease the nitriding temperature because the thickness of its nitrided layer at 400 °C is small, and its heating and aging at 520 °C fails to provide a substantial increase in the bulk microhardness. When nitrided both at 520 and 400 °C, die steel Cr12Mo and steel 40Cr are involved in tempering which decreases their bulk microhardness by 25 % and by 8−10 %, respectively, compared to the initial state. Decreasing their nitriding temperature decreases the nitrided layer thickness to 50 μm, i.e., almost two times.

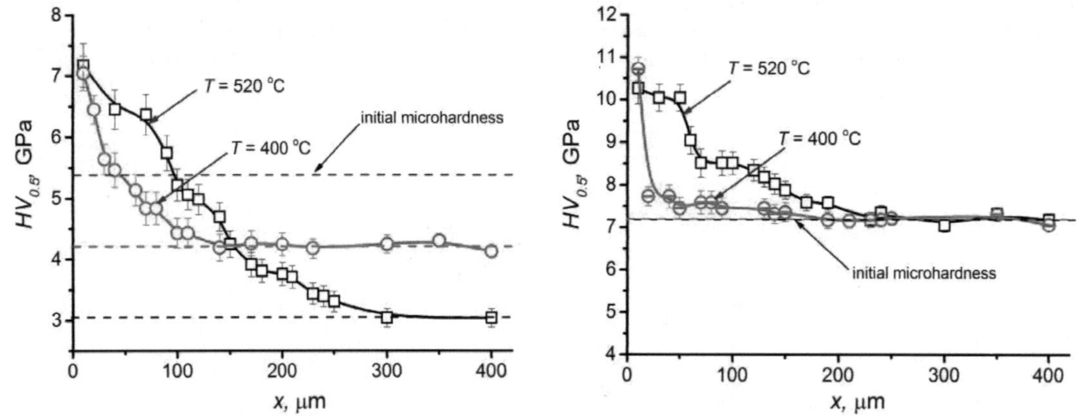

Figure 2. In-depth microhardness distributions in nitrided steels 40Cr (**a**) and 8Cr4W2MoVSi2 (**b**).

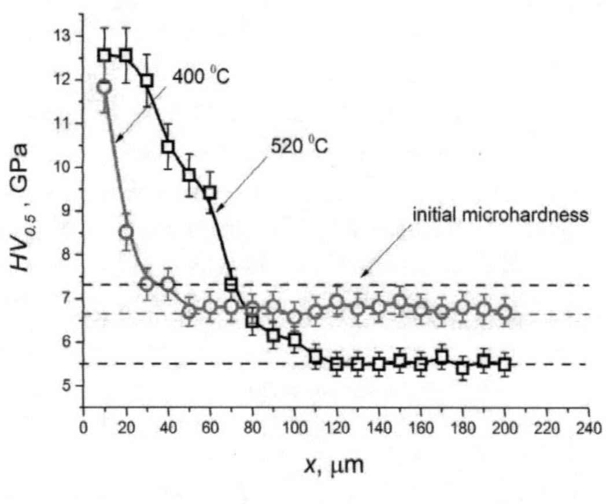

Figure 3. In-depth microhardness in nitrided steel Cr12Mo.

4. Conclusion

Thus, our study demonstrates the efficiency of using a non-self-sustained low-pressure glow discharge with a hollow cathode for nitriding structural and tool steels. The nitrided layer thickness after treatment for 3 h at 400 and 520 °C is respectively 140 and 300 μm in steel 40Cr,≈30 μm and 150 μm in steel 8Cr4W2MoVSi2, and ≈50 μm and 110 μm in steel Cr12Mo.Decreasing the nitriding temperature decreases the nitriding rate and the tempering, and this greatly increases the bulk hardness of steels 40Cr and Cr12Mo. From the data obtained, one can select appropriate temperatures and times of nitriding to provide a desired thickness of nitrided layers and desired bulk hardness of steels.

Acknowledgements

The work was supported by the Russian Science Foundation (grant No. 14-29-00091).

References

[1] Budilov V V, Koval N N, Kireev R M and Ramazanov K N 2013 *Intergrated Methods of Processing of Constructional and Instrumental Materials by Glow and Vacuum-Arc Discharges* (Mashinostroenie, Moscow) 320
[2] Vizir A V, Oks E M, Schanin P M and Yushkov G Yu 1997 *Tech. Phys.* **6** 611–4
[3] Akhmadeev Yu Kh, Denisov V V, Koval N N, Kovalsky S S, Lopatin I V, Schanin P M and Yakovlev V V 2017 *Plasma Phys. Rep.* **1** 67–74
[4] Schanin P M, Koval N N, Akhmadeev Yu Kh and Grigoriev S V 2004 *Tech. Phys.* **5** 545–50

IOP Conf. Series: Materials Science and Engineering **387** (2018) 012057 doi:10.1088/1757-899X/387/1/012057

Informatics support of thin film deposition with specified structure development

Y V Panfilov

"Electronic technology in machinery" Chair of MSTU named after Bauman, 105005 Moscow, RUSSIA

Abstract. Virtual experiment method which consists of acquisition, systematization, and analysis of large experimental results of thin film deposition technology is described. The data base examples and thin film deposition of any materials result analysis are adduced. Dependence of thin films structure from thin film deposition rate and method and substrate temperature is shown. DLC thin films with amorphous and crystalline structures deposition regimes are represented. WSi thin films for using as thin film resistor in microcircuit and as a superconductor coating in one photon detectors are described. It is shown that the elaborated method of informatics support of research and development in the field of thin film deposition technology dives opportunity to analysis and choose of thin film deposition method and regimes for manufacturing of thin films with specified structure and characteristics.

"Doing anything well, take some information" – thus sounds the motto of a number of successful researchers and businessmen and others. So the item of this article – the reasoning of relevance and practicability of creation of a methodic providing of a research in the field of thin film deposition with some information is actual task. This methodic is intended for science beginners studying, and for specialist for deciding new tasks, as well.

The suggesting methodic consists of acquisition, systematization, and analysis of large number of publications and experimental results of thin film deposition technology by means of a special algorithm, and it's representing as the relation date base [1].

The results of information analysis, which may be carried out by means of the data base control systems, one can represent in a view of curves and histograms and other methods of information representation. These are a virtual experiment results, in substance.

As an example, the information about materials, deposition methods and regimes of thin film technology is shown in table 1. The parameters of the data base were chosen on the base of work experience with data base control systems and on the reasoning of relevance to research of thin film deposition method and regimes influence to thin films structure.

So, the results of publication analysis for thin film deposition of silica, aluminum oxide, ITO, titanium nitride, bismuth telluride, and DLC are represented in the table 1. Atomic Layer Deposition (ALD) and Chemical Vapor Deposition (CVD) and Plasma Enhance Chemical Vapor Deposition (PECVD) and Electron Beam Evaporation (EBE) and Pulsed Laser Deposition (PLD) and Magnetron Sputtering (MS) and Arc Evaporation (Arc) and Ion Beam Deposition (IBD) methods are represented in the table 1, as well. There are analyzing some parameters of thin film deposition such as particle energy (E), deposition rate (V_d), substrate temperature (T_s), gas pressure into vacuum chamber (p), thin film structure in the table 1.

IOP Conf. Series: Materials Science and Engineering **387** (2018) 012057 doi:10.1088/1757-899X/387/1/012057

To deposit of thin film with crystalline structure we have to know thin film structure influence from thin film deposition regimes. Well known models of thin films grow do not give opportunity to have common dependence of thin film structure from thin film deposition regimes. It is necessary to have thin film structure dependence from variated process parameters. Thin film deposition rate, atoms and molecules energy, substrate temperature, gas pressure into vacuum chamber are such parameters.

Table 1. Thin film deposition methods and regimes data base example.

Thin film material	Method	E, eV	V_d, nm/s	T_s, K	p, Pa	Thin film structure
Al_2O_3	ALD	0,05	0,01	873	100	Crystalline
Si	CVD	0,06	10	1075	150	Crystalline
Bi_2Te_3	PLD	400	0,4	523	10^{-3}	Crystalline
InSnO (ITO)	EBE	0,2	0,24	373	$2{\cdot}10^{-2}$	Amorphous
AlN	MS	4	0,1	573	10^{-2}	Crystalline
ZnO	PLD	1000	0,01	573	10^{-5}	Crystalline
DLC	PE CVD	10	1	350	2	Amorphous with crystals
DLC	PE CVD	300	80	1200	2	Crystalline
DLC	Arc	10	8	293	10^{-3}	Amorphous
DLC	Arc	10	100	373	$4{\cdot}10^{-6}$	Amorphous with crystals
DLC	IBD	70	0,02	473	10^{-2}	Amorphous with crystals
DLC	IBD	100	0,05	900	10^{-2}	Monocrystalline
DLC	EBE	0,2	0,8	293	10^{-3}	Amorphous
DLC	EBE	0,2	5	573	10^{-3}	Crystalline
DLC	PLD	0,5	0,1	453	10^{-4}	Amorphous with crystals
DLC	PLD	400	10	773	10^{-4}	Monocrystalline

For ease of representation of thin film deposition regimes influence to thin film structure it is marked, in two dimension coordinate system “thin film deposition rate – substrate temperature”, the areas of thin film deposition methods and thin film materials and structures from literature and data base in table 1 (figure 1). Lines of Ge thin film structure junction from amorphous to crystalline and monocrystalline condition are marked by solid lines in figure 1 [2]. We have an opportunity to compare and analyze of results of new research by means of this plot – so we have opportunity to carry out of virtual experiments.

As it follow from figure 1, the results of virtual experiments with thin films from different materials and deposited with different regimes are coincided with the results of Ge thin film deposition real experimentsas a rule (solid lines in figure 1).

More detailed analysis of literature [3–6] and other tables of elaborated data bases give opportunity to resume the next:

- Thin films with amorphous structure can be manufactured without substrate heating even and with any thin film deposition rate. So, we may resume that all thin films deposited with room temperature of substrate have amorphous structure. But thin film epitaxy can be happened with room temperature of substrate and with very small thin film deposition rate.
- Thin films with crystalline structure can be deposited to any substrate materials with melting temperature more than 600–700 K by means of any thin film deposition method and with any thin film deposition rate.

- Besides this, thin films with crystalline structure can be manufactured by means of substrate heating to 600 K and more. In addition, thin film deposition rate must be more than 0,2–0,25 nm/s, and there are not limit of thin film deposition rate maximum.
- Thin films with mono crystalline structure can be manufactured with very small thin film deposition rate (less than 0,2–0,25 nm/s) and with high substrate temperature (more than 600 K).
- Besides this, thin films with mono crystalline structure and with diminution minimum can be manufactured by means of ion beam deposition method with particle energy E=4 keV, substrate temperature T_s=800 K, thin film deposition rate V_d=0,5 nm/sand ionization degree k=0,1 %.

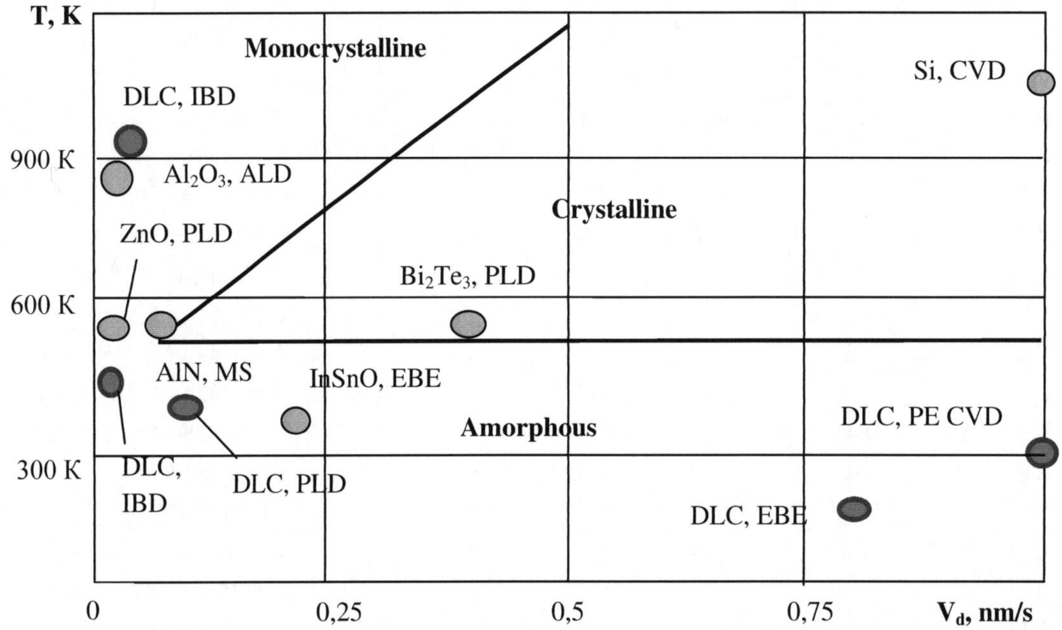

Figure 1. Dependence of thin films structure from thin film deposition rate V_d and method and substrate temperature T.

The experimental research results of DLC thin film deposition with plasma discharge voltage U_p and substrate temperature T are represented in figure 2 (the rate of DLC thin film deposition is corresponded with the circle diameter). This results were provided with the help of dates in table 1. In the left down corner of figure 2, regimes of DLC thin film deposition on laboratory coater is shown and some variants of DLC thin film deposition rate increase by means of plasma discharge voltage U_p or substrate temperature T increasing are represented.

It is impossible to obtain the same results by means of real experiments because the real experimental equipment has, as a rule, only one or two source of thin film deposition and it has not opportunity to diversify in a large range of thin film deposition regimes such as particle's stream intensity and energy, gas pressure into vacuum chamber, substrate temperature and so on.

WSi thin films are used as a thin film resistor in microcircuit and as a superconductor coating in one photon detectors [7–10]. Thickness of WSi superconductor coating must be less than 10 nm, and the main methods of WSi thin film deposition are PLD and ion plasma sputtering and CVD, as well. Thin film deposition rate is depended from thin film deposition method and has amount variation from 0,032 to 2,5 nm/s (table 2). Thin film deposition method with productivity maximum is CVD method, whereas PVD thin film deposition method has limit about 0,7 nm/s. Thin film deposition method with particle energy maximum (10–1000 eV) is PLD method. WSi thin films can be manufactured by means of ion plasma sputtering of multicomponent WSi target or of two targets made of W and Si.

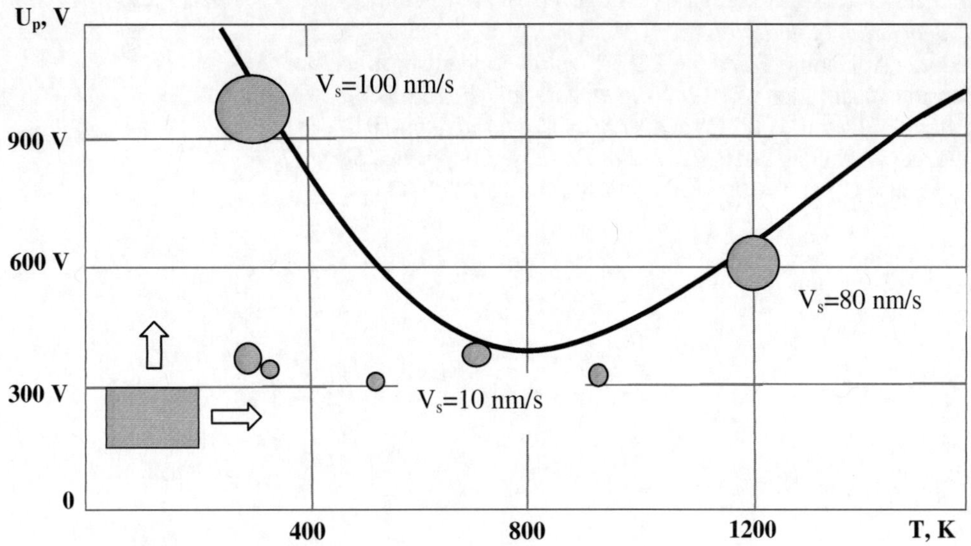

Figure 2. The influence of DLC thin film deposition rate maximum from substrate temperature T and plasma discharge voltage U_p (rectangle in the left corner is the working regimes of the laboratory coater).

Table 2. WSi thin film deposition rate and particle energy with some thin film deposition methods.

Thin film deposition method	Thin film deposition rate, nm/s	Particle energy, eV
PLD	0,032 – 0,7	10 – 1000
Ion plasma sputtering	0,39 – 0,71	1 – 10
CVD	1,33 – 2,5	0,05 – 0,1

Comparison of any materials thin film deposition regimes shown that thin film nanostructured condition is realized in a restricted diapasons of particle mass and energy streams.

So, the elaborated method of informatics support of research and development in the field of thin film deposition technology gives opportunity to analysis and choose of thin film deposition method and regimes for manufacturing of thin films with specified structure and characteristics.

The virtual experiment method is recommended for using in the automatic expert systems for decide of development modern equipment begin, of resource assignment to buy of available equipment or to correct of thin film deposition regimes of operated equipment for manufacturing of thin films with specified structure. By means of automatic expert systems one can evaluate the prospect of a process and equipment stock or newly creating and analysis of modern process and equipment progress. Predict of thin film with specified structure deposition opportunity, and carry out of design expertise, practicability of a very expensive equipment procurement, and of research and development and engineering operations financial support request expertise, as well.

References

[1] Panfilov Y V and Ivanchikov I V 2000 *Handbook. Engineer journal* **7** 21–3
[2] Ivanov R D 1972 *Cathode method of microcircuit thin film elements creation* (Moscow, “Energy”) 112
[3] *Nanotechnology in the nearest decade. Prediction of a development trends* 2002 ed. M K Roko, R C Wiliams and P Avilisatos (Moscow: Mir) 292
[4] Zettlemoyer C 1960 *J. Am. Chem. Soc.* **16** 4440

[5] Michael N R, Claeyssens A F, Fuge G M and Henley S J 2003 *Pulsed laser ablation and deposition of thin films*
[6] Fedotov A, Uchino T and Ou J 2012 *Optical Society of America*
[7] Henrion W and Lange H 1989 *Phys. Stat. Sol. (b)* **151** 375–82
[8] Radulescu F et al. 2006 *CS MANTECH Conference* pp 101–3
[9] Seleznev V A et al. 2016 *Journal of Physics: Conference Series* **737**
[10] Backhouse C J et al. 1997 *Thin Solid Films* **311** 299–303

IOP Conf. Series: Materials Science and Engineering **387** (2018) 012058 doi:10.1088/1757-899X/387/1/012058

Investigation of the influence of technological growth parameters on the characteristics of organic photosensitive structures based on the ZnPc:C_{60} system

M D Pavlova, I A Lamkin and S A Tarasov

Saint-Petersburg State Electrotechnical University "LETI", 197022, Saint-Petersburg, Russia

E-mail: minaepanae@mail.ru

Abstract. The influence of technological growth parameters, namely, the mass of the evaporated substance, on the spectral characteristics of organic photosensitive structures based on the ZnPc:C_{60} system is considered. The absorption, transmission, and photosensitivity spectra of the structures were studied.

1. Introduction

The production of thin films is one of the main tasks of modern electronics technology, which is caused by the tendency to reduce the linear dimensions of devices and chip elements. Over the past decades, vacuum deposition has become the main method of creating thin films for use in science, in the fields of research and development, industry. For application in the field of research, the main manufacturing method is high-vacuum deposition of films.

Thermal deposition is one of the most common methods for making films of simple composition, and also allows you to flexibly and quickly change the conditions in which sputtering occurs, to change the spray materials, etc. This makes it easy to vary the growth parameters, which is especially important in the studies. For example, when creating solar cells based on organic materials. Since organics are easily oxidized in the atmosphere, and, consequently, the operating characteristics of the devices deteriorate, it is necessary to produce structures in a vacuum to create highly efficient solar cells [1].

2. Experiment, measurements and results

In this work, the influence of the thickness of the deposited film of organic materials on the photosensitivity of structures was considered. As a substrate, glass was used, coated with a layer of FTO (Fluorine-doped Tin Oxide). The FTO layer served as an optically transparent contact. On the surface of FTO, ZnPc (Zinc phthalocyanine):C_{60} (Fullerene)/C_{60}/Al [2] layers were successively deposited by vacuum thermal evaporation. Layers of organic materials and aluminum contact were deposited at a residual gas pressure of $2 \cdot 10^{-5}$ torr. During the sputtering process of organic layers, the substrate was heated to 60 °C, and when the metal contact was applied, the substrate temperature was maintained at 70 °C.

Evaporation of the substance occurs when it is heated. In operation, the heating of the evaporated substance is carried out by the flow of the heating current through the boat into which the organic material or metal of the required mass is preloaded. It was found that at a filament current of 80 A, which roughly corresponds to a temperature of 680 °C, ZnPc and C60 evaporate in the best way. Because when the substance is heated, the kinetic energy of its atoms and molecules increases and becomes sufficient to break away from the surface of the boat and spread into the surrounding space.

Heating of the substrate, during the deposition of organic materials, up to 60 °C, improves adhesion to the substrate and diffusion over the surface, which improves the structural quality of the layers. The thickness of the deposited film is directly proportional to the mass of the evaporated matter and back from the distance between the substrate and the boat. In the experiment, the height from the substrate to the evaporated substance remained unchanged at 15 cm. A series of photosensitive structures with various thicknesses of active layers was manufactured. During the experiment, 3 types of structures were created, in which the mass of the evaporated substance varies. Below are the types of structure and mass of evaporated substances:

- 1 sample: ZnPc (15mg): C60 (15mg) / C60 (25mg) / Al (120mg);
- 2 sample: ZnPc (15mg): C60 (20mg) / C60 (30mg) / Al (120mg);
- 3 sample: ZnPc (15mg): C60 (10mg) / C60 (20mg) / Al (120mg);
- 4 sample: ZnPc (15mg): C60 (10mg) / C60 (20mg) / Bphen (7mg) / Al (120mg).

Sample 4 was also created, in which the additional transport layer Bphen (Bathophenanthroline), used to create an additional energy barrier for holes, suppress recombination and improve the collection of charge carriers, was introduced. Figure 1 shows the energy diagram of the structures under study.

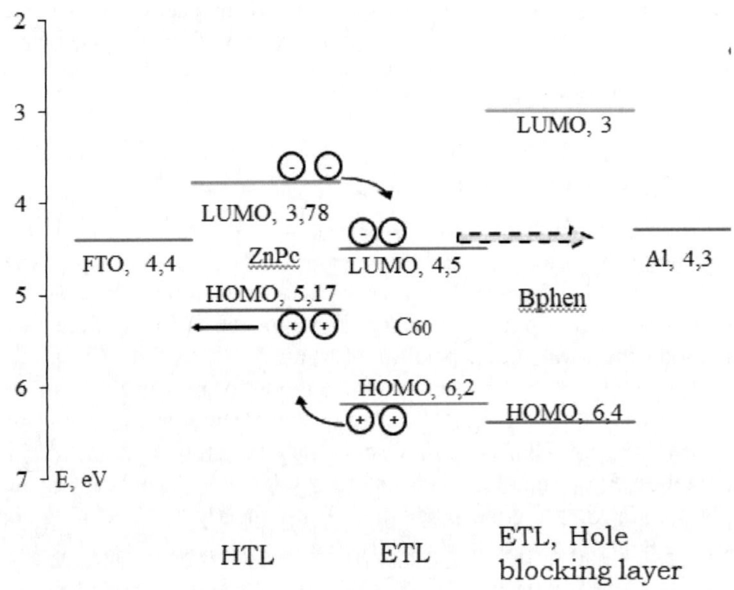

Figure 1. Position of the energy levels of the structures studied.

The absorption and transmission spectra of the ZnPc:C_{60}/C_{60} layers were measured, which were measured on a USB4000 fast scanning spectrometer using an incandescent lamp.

From the analysis of absorption and transmission spectra it is established that in order to optimize the characteristics of photosensitive structures, it is necessary to select the optimal thicknesses of active layers, which ensure high absorption and effective generation of charge carriers. This is due to the peculiarity of organic materials in which, under the influence of light, not free charge carriers are formed, and the associated ones are excitons. The dissociation of an exciton by free charge carriers can be carried out at the ZnPc/C_{60} heterojunction. The characteristic distance that the exciton travels during its lifetime in zinc phthalocyanine is about 10 nm [3], therefore, only photons absorbed near the heterojunction region can contribute to the photocurrent, which can effectively move to the ZnPc/C60 interface. The addition of a hole-blocking layer Bphen effectively increases the absorption in the green-red region of the spectrum.

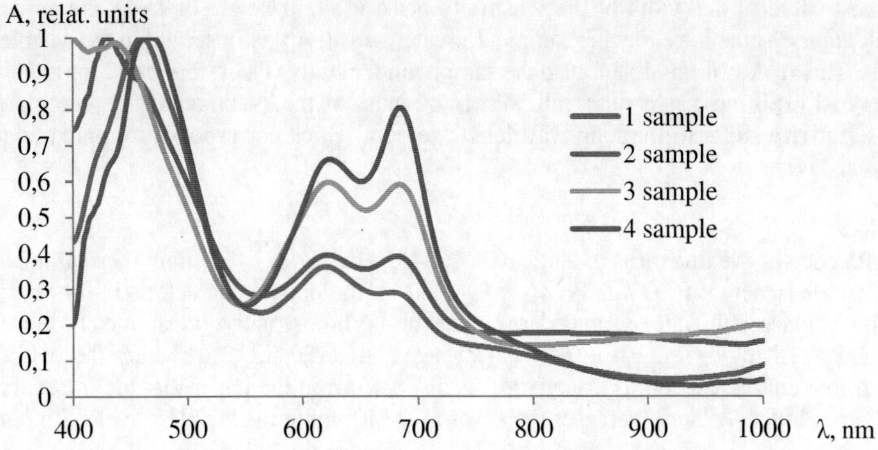

Figure 2. Absorption spectrum of structures with different masses of C_{60}.

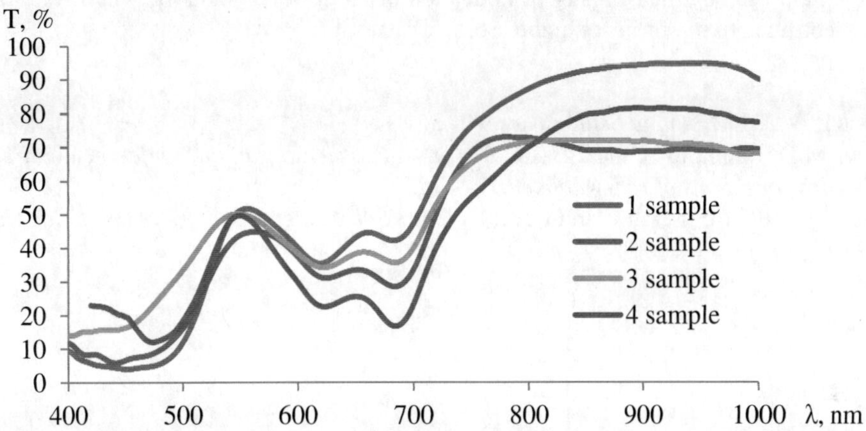

Figure 3. Transmission spectrum of structures at different masses of evaporated C_{60}.

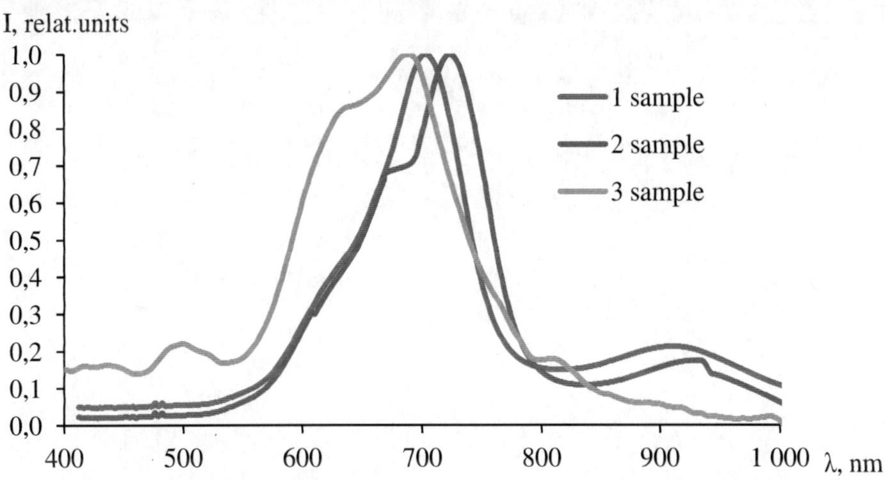

Figure 4. The spectrum of photosensitivity for images 1, 2, 3.

The photosensitivity spectrum were studied using a diffraction grating monochromator using a Keithley 6483 picoammeter, an incandescent lamp was used as a radiation source.

The highest value of the width of the sensitivity peak at a half-height of 158.8 nm corresponds to the 3 sample, in the same way for this sample a greater sensitivity is observed in the visible range of wavelengths. This makes it possible to use the sample under study (FTO / ZnPc (15 mg): C_{60} (10 mg) / C_{60} (20 mg) / Al (120 mg)) as a solar cell. When the mass of the evaporated fullerene increases, the maximum sensitivity shifts to the long-wavelength region, which corresponds to a greater absorption in the fullerene layer.

3. Conclusion

Thus, the influence of the thickness of ZnPc: C_{60} / C_{60} layers grown by vacuum thermal deposition for the photosensitive structure FTO / ZnPc: C_{60} / C_{60} / Al on the optical characteristics of photosensitive structures on a glass substrate is considered. The best photosensitivity parameters correspond to sample 3 (ZnPc (15 mg): C_{60} (10 mg) / C_{60} (20 mg) / Al (120 mg)), in which the dissociation of excitons into free charge carriers is effective. It is also shown in the paper that the introduction of an additional barrier layer (BPhen) increases the absorption of light in the visible wavelength range.

Acknowledgment

The work was supported by the Ministry of Education of the Russian under project of the state task in the field of scientific activity, project number 16.1750.2017/4.6.

References

[1] Gordon P W 2017 *Pomona Senior Theses* 1–36
[2] Pavlova M D, Lamkin I A and Tarasov S A 2018 *Proceedings of the conference "Physico-chemical problems of renewable energy"* 125
[3] Hazar D A, Dilber E Y, Ali C and Levent T 2013 *Solar Energy Materials and Solar Cells* **113** 100–5

IOP Conf. Series: Materials Science and Engineering **387** (2018) 012059 doi:10.1088/1757-899X/387/1/012059

A combined surface layer strengthening of aluminum alloys by plastic deformation and vacuum ion plasma processing methods

V V Plikhunov, L M Petrov, Yu S Rumyantsev and A N Smirnova

Joint Stock Company "National Institute of Aviation Technologies"
Moscow, Russia

E-mail: info@niat.ru

Abstract. The influence of tool effects of the processing media on formation of corrosion characteristics (reactive capacity and corrosion resistance) in the surface layer of aluminum alloys during strengthening by surface plastic deformation and vacuum ion-plasma processing has been investigated. It has been shown, by example of V95pchT2 alloy, that formation of the surface layer corrosion characteristics under the integrated effect of processing media (shot, plasma) is determined by the interaction modes of the tool/work surface contact pair. The influence of reactive capacity, which has been formed during strengthening, on changing corrosion resistance of the strengthened surface layer under the action of 5 % solution of NaCl (120 hours test base) has been considered.

1. Introduction

Formation of the surface layer reactive capacity during generation of modified surface structures using surface plastic deformation and vacuum ion-plasma processing is the prime task implemented in surface strengthening processes. The given parameter of the surface layer characterizing its readiness to interact with the environment is responsible for origination of new internal (subsurface) and external (supersurface) structural compositions, which provide development of the required service properties. Reactive capacity of the surface layer is determined by its structural condition formed by preceding process effects as well as the presence of different types of barrier layers on it that appear under these effects or when they are finished. The study of dependence of the surface reactive capacity on operating modes with surface strengthening processes is very important both for the use of this knowledge at justification of processing modes and for development of additional protective barrier layers during surface forming with the aim to protect them against in-service effects of corrosive outdoor environment.

2. The change in energy characteristics of the surface layer under tool effects of the surface plastic deformation and vacuum ion-plasma processing

The complex physicochemical processes occurring at the surface of the processed part under tool effects as well as effects of high-energy flows of gas and metal plasma and other types of process effects generate the structural condition and appropriate surface energy state. The quantitative and qualitative characteristics of the surface energy state are represented by a reduced surface potential Vp

IOP Conf. Series: Materials Science and Engineering **387** (2018) 012059 doi:10.1088/1757-899X/387/1/012059

[1–3]. That surface layer parameter is a structural-dependent value as well as reactive capacity reactive capacity of the surface layer; therefore according to the variation of Vp values we can also judge changes in reactive capacity.

During the progress of surface strengthening the process effects primarily influence the change in physicochemical, mechanical and energy properties of the surface layer of metal materials being processed including those of aluminum alloys. Investigation of the interaction of the surface energy state with its reactive capacity at different stages of surface strengthening is highly topical for performance evaluation of the processing media effects in treatment, and also for quality assessment of created surface structures.

Investigations conducted have shown that the interaction of the surface energy state with its reactive capacity is traced from the initial stage of the process tool effect performed during part manufacture. As a rule, a blank of the future part has no design surfaces and possesses initial surface structure inherent in the "inner" structure of the basic material. When the processing tool influences the blank, the structurally changed "defective" surface with a very high reactive capacity is formed. The time of existence of the surface in such a state is extremely small and measurements of the energy state and reactive capacity are not practical. The subsequent exposure of highly activated surface to the process atmosphere gives rise to formation of "contaminated" multicomponent surface layer consisting of various adsorbents, simple and composite oxides. The change in the energy properties of the surface layer takes place over the whole processing time (especially active at the beginning); and with processing completed, we can measure the value of the energy state of the surface and evaluate its reactive capacity. The in-process cleaning of the surface after processing leads to change of the energy state of the surface, and measuring of the reduced surface potential Vp makes it possible to draw the conclusion that its reactive capacity changes. In storage the surface of the finished part is under the action of time and atmospheric factors with the resulting formation of the final structure of the part surface layer. The Vp value can be measured when parts are in storage and in doing so to keep track of stability of the finished part surface, figure 1 (modes 1 and 2).

The values of Vp are determined by material grade. It is shown by the example of V95pchT2 alloy that aluminum based alloys the values of Vp are, as a rule, in the negative region through formation of considerable oxide layer at the surface. Being non-mechanically stable and under the effect of tangential stresses arising during machining, surface oxide is easily removed as design surfaces of the part are being developed.

Recovery of the removed surface oxide at the newly formed surface occurs almost instantly because there remains its high reactive capacity. The value of Vp, which characterizes the presence of the oxide, is hardly changed, figure 1 (mode 2) despite the fact that defect structure of the surface layer has changed under machining (milling). The change in the defect structure of the surface layer generated under machining has more intensively influenced the change in resistance of the surface layer to the action of corrosive medium (5 % solution of NaCl). The conducted investigations have demonstrated that existence of the defective surface layer brought about an increase in intensity of the corrosive medium action and formation of a large amount of corrosion product, figure 2 (mode 2), which was favourable to reduction in mass of the test specimens (salt-spray cabinet, test base 120 hours).

Technological effects of the processing media during surface plastic deformation affect to a great extent the variation of the surface energy state since deformation and contact processes, while being their basis and facilitating the change of structural condition of the surface layer, change its surface chemical composition of elements and the surface energy state.

At the initial stage of treatment, when realizing surface plastic deformation methods we observe shattering of metal (alloy) grains into blocks (subgrains) and production of mosaic structure with subsequent formation of the texturized structure with anisotropic mechanical properties. As a rule, for every structural condition there is a definite value of surface energy state. Therefore every stage of the technological effect during surface plastic deformation has its value of the surface energy state that determines its reactive capacity.

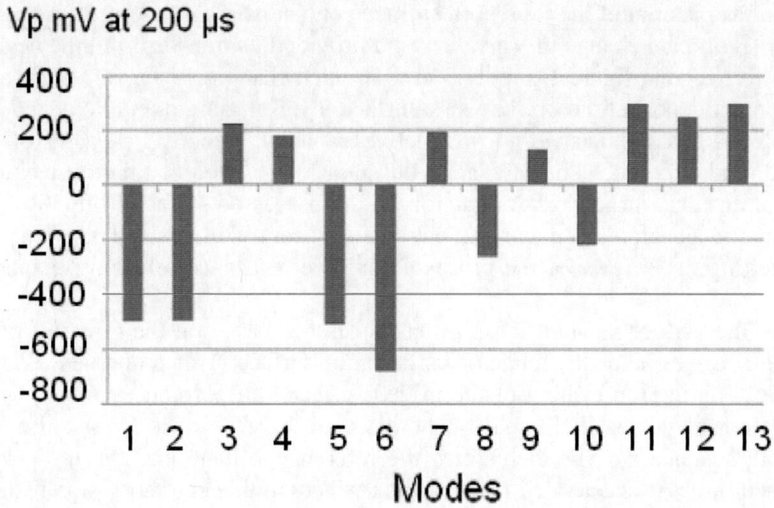

Figure 1. Change in values of Vp of the surface layer of V95pchT2 aluminum alloy against surface condition: 1 - pure Al-99,9%; 2 - milling; 3 - milling + peen forming (30 s); 4 - milling + peen forming (120 s); 5 - milling + ion etching Uref. = 0V; 6 - milling + ion etching Uref. = -100V; 7 - milling + plasma flow Ar-N-Al (20 min); 8 - milling + plasma flow Ar-N-Al (60 min); 9 - milling + plasma flow Ar-N-Al (40 min); 10 - milling + vibro strengthening with chips (Si) 60 s; 11 - milling + abrasive vibro strengthening + plasma Ar-N-Al (20 min); 12 - milling + abrasive vibro strengthening + plasma Ar-N-Al (40 min); 13 - milling + abrasive vibro strengthening + plasma Ar-N-Al (60 min).

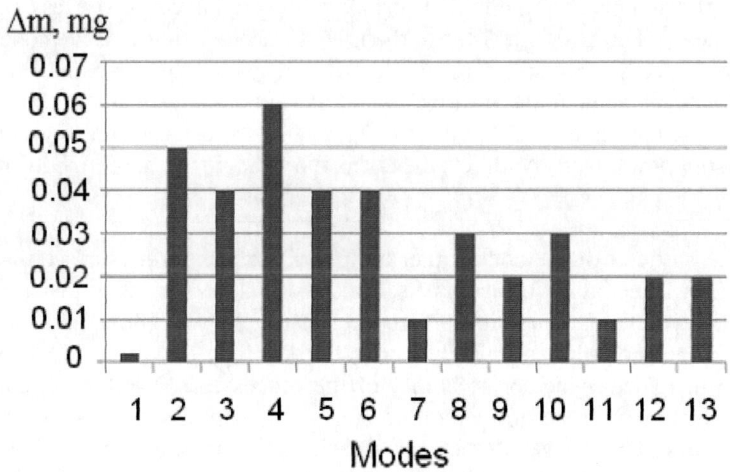

Figure 2. Mass variation of test specimens under the action of corrosive medium (120 hours, salt-spray cabinet – 5% NaCl) of the surface layer of V95pchT2 aluminum alloy formed by different processing modes: 1 - pure Al-99,9%; 2 - milling; 3 - milling + peen forming (30 s); 4 - milling + peen forming (120 s); 5 - milling + ion etching Uref = 0V; 6 - milling + ion etching Uref = -100V; 7 - milling + plasma flow Ar-N-Al (20 min); 8 - milling + plasma flow Ar-N-Al (60 min); 9 - milling + plasma flow Ar-N-Al (40 min); 10 - milling + abrasive vibro strengthening; 11 - milling + abrasive vibro strengthening + plasma Ar-N-Al (20 min); 12 - milling + abrasive vibro strengthening + plasma Ar-N-Al (40 min); 13 - milling + abrasive vibro strengthening + plasma Ar-N-Al (60 min).

The intensity of variation in surface energy state values depends on the energy and time of work tool action on the surface layer under processing, its geometric parameters and scheme of its contact with the surface as well as material of the tool/work surface contact pair. The use of shot peening with steel spheres made for significant change in Vp values and provided its transfer from the negative region of values to the positive region, figure 1 (modes 3 and 4). Such changes of Vp values are associated with appearance of contact deposited Fe on the work surface layer of aluminum alloys and forming on its base of a complex alloyed oxide with low reactive capacity. Reactive capacity of such oxide is structurally stable and Vp values do not change during a long period of time (test base is more than 2 years). Variation in time of the work tool action on the processed surface from 30s to 120s actually had no effect on the generated Vp values, but it significantly influenced the amount of contact deposited Fe resulting in decrease of the processed surface resistance to corrosive medium, figure 2 (modes 3 and 4).

Application of the surface vacuum ion-plasma processing based on the effect of high-energy gas and metal plasma flows expands the range of variation of surface layer properties. Thus the action of Ar gas plasma flow during ion etching of the milled surface with reference voltage Uref=0V retains initial value of Vp and its reactive capacity, here we observe corrosion resistance of the formed surface, figures 1, 2 (mode 5). The increase in the reference voltage has raised Vp value to -100V through impact action of Ar accelerated flow on the surface, in this case corrosion resistance remains at the previous level, figures 1, 2 (mode 6).

Exposure of the milled surface to gas –metal plasma flow (Ar-N-Al) considerably influenced the change in Vp values and corrosion resistance of the surface being formed, figures 1, 2 (modes 7, 8, 9). Processing with plasma flow within 20 minutes has facilitated the transition of Vp values to the positive region and significant growth in corrosion resistance of the processed surface (mode 7). The increase in processing time to 60 minutes has led to recovery of the negative value of Vp and decreasing of corrosion resistance (mode 8). Reduction in processing time to 40 minutes favoured the recovery of surface layer parameters up to values obtained at 20 minutes mode of processing (mode 9).

The mode of vibro strengthening the milled surface with ceramic chips to a less extent affected the variation in Vp values and corrosion resistance than peen forming, figure 1, 2 (mode 10). Additional treatment of the generated surface structure (mode 10) with gas – metal plasma flow (Ar-N-Al) has caused the transition of Vp value to the positive region of values and growth of corrosion resistance of the surface under formation, figure 1, 2 (modes 11, 12, 13). As a result, increasing time of exposure to the gas – metal plasma flow adversely affects the corrosion behavior of the formed surface layer.

3. Conclusions

Consequently, process effects on the contact pair tool/work surface during surface plastic deformation lead to interaction between contact surface layers, which results in changes of the structural and phase states of the surface layer, variations in its surface microhardness, roughness, and energy state as well as its chemical composition that ultimately provides development of residual stresses, reactive capacity, which, in turn, determine serviceability of the processed surface. The generated structures and implementable variety of properties determine objectively controlled characteristics of the surface layer including control of the changes in the chemical composition of the surface layer, its roughness, energy state, surface microhardness, residual stress levels, corrosion resistance, which evaluates the subsequent operating capacity of created surfaces.

Post processes of the interaction of argon and nitrogen plasma flows as well as aluminum plasma flow with the processed surface of V95pchT2 aluminum alloy at the stages of vacuum ion-plasma processing are associated with the change in chemical composition of the processed surface during removal of adsorbed elements and oxides. The realizable interaction of the cleaned surface with elements of gas and metal plasma at modification stage and deposition of coatings based on the oxynitride aluminum composition changes the relation of alloying elements in the surface layer, which promotes plasma chemical reaction behavior. V95pchT2 alloy belongs to Al-Zn-Mg-Cu system and oxide film at the surface is formed involving all elements found on the surface. Processing with argon-

IOP Conf. Series: Materials Science and Engineering **387** (2018) 012059 doi:10.1088/1757-899X/387/1/012059

nitrogen-aluminum gas metal plasma provides nitrogen enrichment in the surface layer and change in its structure depending on the value of reference voltage.

References

[1] Sulima A I, Shulov V A and Yagodkin Yu D 1988 *Surface Layer and Operational Properties of Machine Parts* 240
[2] Petrov L M and Plikhunov V V 2012 *Aviation Industry* **1** 22–6
[3] Iliyin A A, Plikhunov V V and Petrov L M 2011 *Aviation Industry* **2** 28–32

Mass transfer of niobium and copper in binary systems obtained by magnetron sputtering

D V Postnikov[1], A I Blesman[1,2], D A Polonyankin[1,2], S L Ibnoyaminova[1] and A A Teplouhov[1]

[1] Physics Department, Omsk State Technical University (OmSTU), 11 Mira Avenue, Omsk, 644050, Russia

[2] Scientific–educational Resource Centre «Nanotechnology», Omsk State Technical University (OmSTU), 11 Mira Avenue, Omsk, 644050, Russia

e–mail: nano@omgtu.ru

Abstract. This paper provides the mutual low–temperature mass transfer in binary systems obtained by magnetron sputtering of niobium onto a heated copper substrate. Based on the kinetic equation of diffusion, the distribution of the niobium through the substrate's depth at temperatures of 100, 200 and 300 °C is calculated. The method of calculating the concentration profiles of niobium, proposed in the manuscript, was verified using experimental data of energy–dispersive analysis.

1. Introduction

Niobium has a high corrosion resistance and performance characteristics under conditions of temperature influences and is widely used in the automotive, chemical and aerospace industries [1]. Application of niobium in bipolar plates manufacture for hydrogen fuel cells is also explained by its high corrosion resistance in sulfuric acid solutions compared to other transition metals, herewith niobium deposition into plates of austenitic AISI 316L steel can be implemented by magnetron sputtering [2].

At present, superconducting niobium thin films are used in a large number of microelectronic devices, such as Josephson junctions, quantum dot, superconducting quantum interference devices (SQUIDs), coplanar waveguide resonators [3, 4], and superconducting qubits [5]. Thin niobium films formed by spraying–deposition methods are used for manufacture of ultra–large integrated circuits, magnetometers, elementary particle detectors [6], high–frequency cryogenic switches [7] and gyroscopes [8]. Due to the high electrical conductivity, niobium is used for manufacture of such elements of micro electromechanical systems (MEMS) as wave filters, nanomechanical oscillators of microwave superconducting resonators [9].

In comparison with other superconductors, niobium has the highest critical temperature (T_C=9.23 K) and lower critical magnetic field ($H_{C1}(0)$=143.24 kA/m) and can be deposited to substrates of various materials [10]. The magnetron sputtering method is used for thin niobium coatings deposition on the internal cavities (cells) of superconducting radio–frequency resonators (SRF) of elementary particle accelerators.

Magnetron sputtering of niobium on the cavities of oxygen–free copper, which has a high conductivity, provides technological and economic advantages in comparison with the production of resonators from the bulk niobium [11]. One of the ways to improve the functional characteristics of SRF (including higher magnetic field gradients in comparison with macroscopic superconductors) is to form two–layer structures by magnetron sputtering. Such structures consist of superconducting and insulating layers (for example, Nb/MgO) and provide the best shielding of the internal surfaces of SRF resonators [12]. Sputtering of niobium is applied for covering the cavity's inner walls, made of oxygen–free copper with high conductivity (OFHC copper), to increase the quality factor and reduce the surface resistance of the cavities. The advantages of this approach are the better thermal stability of OFHC copper, its insensitivity to the magnetic field of the earth and the possibility of coating from a material with a high critical temperature (niobium, coating thickness of about 2 μm) [13].

Niobium magnetron sputtering onto a graphite substrate followed by high–temperature annealing of films with a thickness of 8 to 12 μm in the temperature range from 800 to 1800 °C is carried out for niobium carbides' formation (Nb_2C and NbC) and application in aggressive (corrosive, erosive) environments [14]. Heating leads to an increase in the hardness of the niobium coating and its adhesion to graphite, and also to a decrease in the porosity of the coating [15]. Investigation of thermal diffusion transfer in binary systems «niobium–zirconium dioxide stabilized with yttrium oxide (Y_2O_3–ZrO_2)» in the temperature range from 780 to 1000 °C allows the authors [16] to obtain an empirical relationship between the diffusion coefficient of niobium, the energy of its activation, and the pre–exponential factor.

Based on the literature review, it could be concluded that the method of magnetron sputtering is widely applied to the formation of thin niobium coatings of various functional designation. In a number of works, the increase in the physicochemical and mechanical characteristics of niobium films is achieved both during their deposition by heating the substrate and during subsequent annealing. For example, during the condensation of high–energy niobium ions with simultaneous heating of the substrate, a synergistic effect occurs, consisting in the epitaxial growth of niobium films with a low defect concentration [17].

A significant influence on the properties of binary systems is also rendered by the mutual thermal diffusion, which leads to a transition layer formation from the coating's and the substrate's atoms. The theoretical and experimental investigation of the mutual mass transfer of niobium and copper is carried out in the paper when a niobium coating is sputtered to a heated copper substrate.

2. Mass transfer model in a binary system

To calculate the niobium distribution through the depth of the «niobium–copper» binary system during the substrate heating, it was used the kinetic equation of diffusion:

$$\frac{\partial c_B}{\partial t} = -\frac{\partial}{\partial x}\tilde{D}\left[1+\left(\frac{P}{E}\right)^2\right]\frac{\partial c_B}{\partial x} + \frac{\partial}{\partial x}\frac{c_A c_B}{kT^2}\left(E_B D_B - E_A D_A\right)\left[2+\left(\frac{P}{E}\right)^2\right]\frac{\partial T}{\partial x} \quad (1)$$

here c_A, c_B – are the atoms' concentration of A and B type, E_A, E_B – activation energy of atoms' diffusion of A and B type, $D_A = D_{OA}e^{-\frac{E_A}{kT}}$, $D_B = D_{OB}e^{-\frac{E_B}{kT}}$ – the diffusion coefficients of A and B atoms' type, $\tilde{D} = D_B c_A + D_A c_B$ – is the mutual diffusion coefficient, P – distribution of internal stresses, T – is the temperature in absolute scale, k – Boltzmann's constant.

The first summand in equation (1) reflects the mutual diffusion in the binary system by the concentration mechanism; the second summand corresponds to the thermal diffusion under the influence of the temperature gradient. In addition, this model takes into account internal stresses in the crystal lattice, caused by an external impact or an uneven temperature distribution. The diffusion coefficients (D_0) and activation energy (E_a) of niobium and copper used for calculations are presented in table 1 [18].

In accordance with the proposed model [19], the atoms' flow of the coating into the inner layers of the substrate occurs both by the concentration mechanism and by the thermal diffusion mechanism.

Table 1. Diffusion coefficients and activation energy.

Element	D_0 (m^2·c^{-1})	E_a (kJ/mol)
Nb	0.376·10^{-2}	381.2
Cu	1.5·10^{-2}	77.7

3. Experimental results

A niobium coating with a controlled thickness of 2 μm [20] was obtained by the magnetron sputtering method in the ADVAVAC VSM–200 installation at a copper substrate temperature of 200 °C. Analytical part of the study was carried out via scanning electron microscopy by using Jeol JCM–5700 equipped with an X–ray energy dispersive spectrometer JED–2300.

Figure 1 shows the experimental and calculated niobium concentration profiles (through the depth) depending on the heating temperature. Figure 2 shows a micrograph of the cross–section of «niobium–copper» system with the points in which the energy–dispersive spectroscopy (EDS) analysis was performed. To the left of LG10000 point (figure 2), a compound is seen, into which the binary Nb–Cu system was mounted.

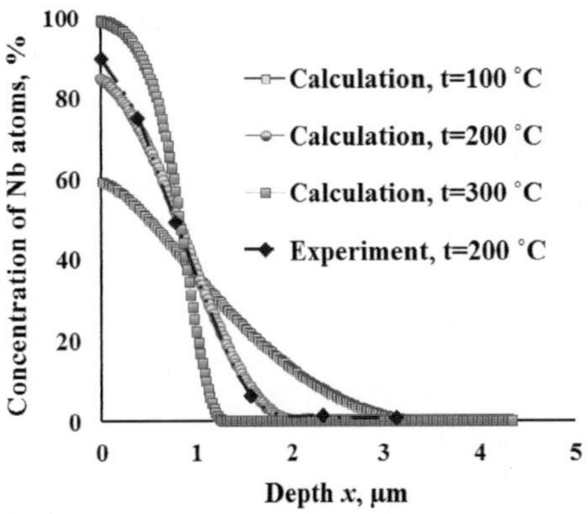

Figure 1. Niobium concentration profiles (through the depth) depending on the heating temperature.

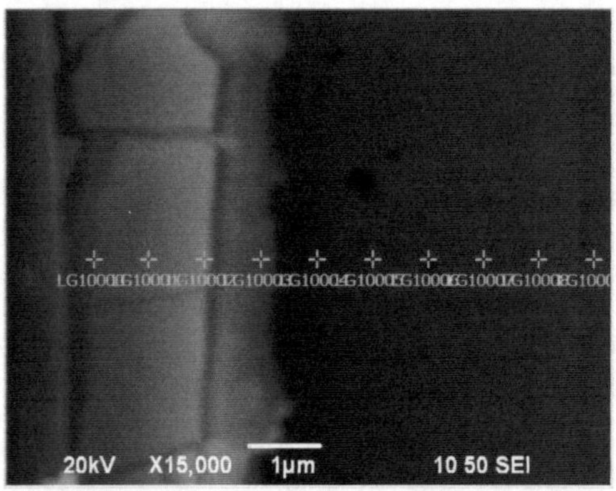

Figure 2. A micrograph of the cross–section of Nb–Cu system with the EDS points.

As can be seen from figure 1, niobium concentration profile obtained by the experimental method coincides with the calculated data within the margin of error.

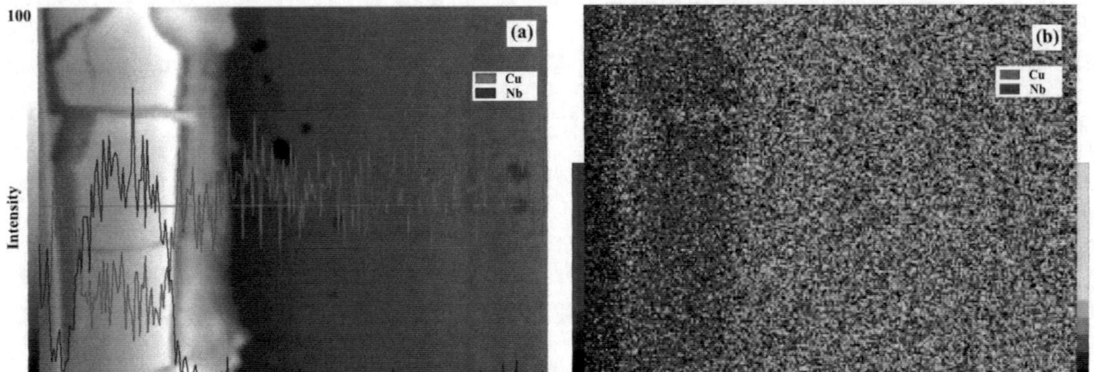

Figure 3. EDS analysis «along the line» (**a**) and Nb–Cu distribution map (**b**).

The thickness of the transition diffusion layer is determined, which is of the order of 1 μm (figure 3(a)). Based on the analysis of the Nb–Cu distribution map (figure 3(b)), it can be concluded that the diffusion of copper (the element of substrate) into the coating occurs predominantly by the grain boundary mechanism.

4. Conclusions

The kinetic equation of diffusion makes it possible to obtain concentration profiles of niobium through the depth of the copper substrate, which allows to estimate the thickness of the transition layer and to carry out mass transfer simulation without experimental studies.

Niobium concentration profile obtained by the experimental method coincides with the calculated data within the margin of error. In the course of the work it was established that the diffusion of copper (the element of the substrate) into the niobium coating in the temperature range from 100 to 300 °C mainly occurs by the grain boundary mechanism. The flux of niobium ions incident on the substrate with the energy significantly higher than the energy of the lattice thermal vibrations leads to an increase in the mass transfer intensity in the binary system.

Acknowledgment
This work was supported by the Omsk State Technical University [grant number 18041B].

References

[1] Liu E B, Cui X F, Jin G, Li Q F and Shao T M 2012 *Key Eng. Mater.* **525–6** 9–12
[2] Kim J–H, Jung D–W, Kim S, Hong S, You Y and Kim D 2012 *Vacuum* **12** 1789–94
[3] Bothner D, Clauss C, Koroknay E, Kemmler M, Gaber T, Jetter M, Scheffler M, Michler P, Dressel M, Koelle D and Kleiner R 2012 *Supercond. Sci. Technol.* **6** 065020
[4] Broussard P R 2017 *J. Low Temp. Phys.* **1–2** 108–19
[5] Bruno A, Mengucci P, Mercaldo L V and Lisitskiy M P 2013 *Supercond. Sci. Technol.* **3** 035004
[6] Esposito A, Nakagawa H, Akoh H and Takada S 1999 *J. Vac. Sci. Technol., A* **6** 3525–28
[7] De Freitas T C, Gonzalez J L, Nascimento V P and Passamani E C 2016 *Thin Solid Films* **611** 33–8
[8] Kolosov V N and Shevyrev A A 2014 *Phys. Met. Metallogr+* **8** 786–92
[9] Kim Y W, Lee S–G and Choi J–H 2011 *Physica C* **21–22** 1193–5
[10] Beebe M R, Valente–Feliciano A–M, Beringer D B, Creeden J A, Madaras S E, Li Z, Yang K, Phillips L, Reece C E and Lukaszew R A 2017 *IEEE T. Appl. Supercon.* **4** 1–4
[11] Cattarin S, Musiani M, Palmieri V and Tonini D 2006 *Electrochim. Acta* **8–9** 1745–51
[12] Beringer D B, Roach W M, Clavero C, Reece C E and Lukaszew R A 2013 *J. Appl. Phys.* **22** 223502
[13] Bemporad E, Carassiti F, Sebastiani M, Lanza G, Palmieri V and Padamsee H 2008 *Supercond. Sci. Technol.* **12** 125026
[14] Barzilai S, Raveh A and Frage N 2005 *Vacuum* **3–4** 171–7
[15] Barzilai S, Frage N and Raveh A 2006 *Surf. Coat. Technol.* **14–15** 4646–53
[16] Kuri G, Gupta M, Schelldorfer R and Gavillet D 2006 *Appl. Surf. Sci.* **3** 1071–80
[17] Krishnan M, Valderrama E, James C, Zhao X, Spradlin J, Feliciano A–M V, Phillips L, Reece C E, Seo K and Sung Z H 2012 *Phys. Rev. ST Accel. Beams* **3** 032001
[18] Heumann T 2007 *Diffusion in Solids* (Berlin–Heidelberg: Springer–Verlag)
[19] Postnikov D V, Blesman A I, Logachev I A, Logacheva A I, Tkachenko E A and Polonyankin D A 2016 *Proc. Eng.* **152** 576–81
[20] Polonyankin D A, Blesman A I, Postnikov D V, Logacheva A I, Logachev I A, Teplouhov A A and Fedorov A A 2017 *IOP Conf. Ser.: Mater. Sci. Eng.* **168** 012069

IOP Conf. Series: Materials Science and Engineering **387** (2018) 012061 doi:10.1088/1757-899X/387/1/012061

Modern technologies of X-ray systems for control of electronic components

N N Potrakhov, V B Bessonov, A V Obodovskiy, A Y Gryaznov, K K Guk and N E Staroverov

Saint Petersburg Electrotechnical University "LETI", Saint Petersburg 197376, Russia

E-mail: nn@eltech-med.com

Abstract. Some issues of technology and design of modern X-ray devices for quality control of microelectronic devices are considered. The functional capabilities of both types of technologies are compared. The design of the Russian microfocus X-ray tube – a key element of the devices for control, is described.

1. Introduction

In recent years, the range of tasks in the electronics industry with the help of X-ray systems significantly expanded. X-ray systems for the control of possible defects in the various components of microelectronics and quality of PCB for new technologies intensively developed [1]. The main element of such a system is a microfocus X-ray tube. It defines its scope of application. Two classes of microfocus tubes are used: opened tube with constant maintenance of vacuum and closed tube.

2. Materials and methods

The functional diagram of the X-ray system is shown in figure 1.

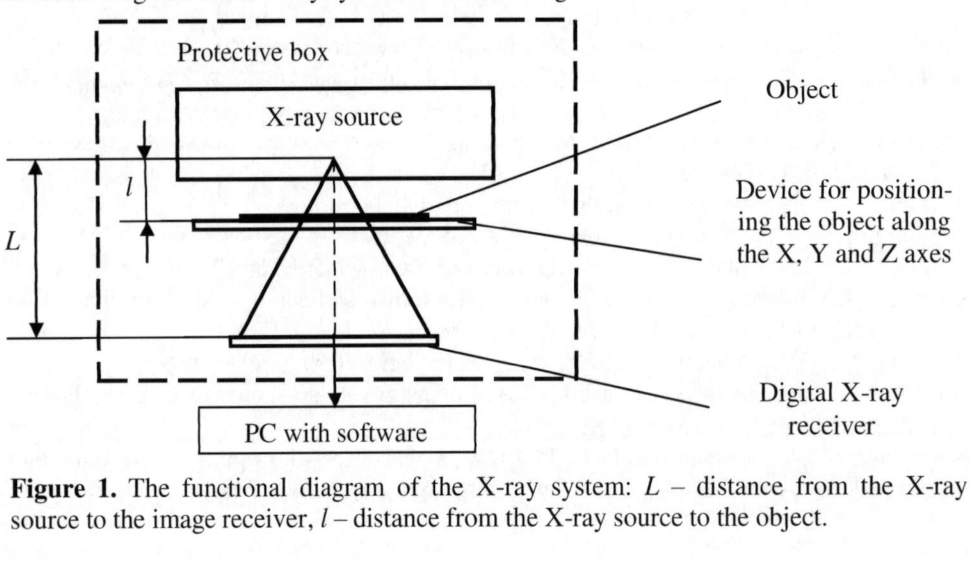

Figure 1. The functional diagram of the X-ray system: *L* – distance from the X-ray source to the image receiver, *l* – distance from the X-ray source to the object.

IOP Conf. Series: Materials Science and Engineering **387** (2018) 012061 doi:10.1088/1757-899X/387/1/012061

The source, object, object positioning device and receiver are located in a special box with leaded walls (X-ray protection box), which does not allow x-ray radiation to go beyond it.

In foreign X-ray systems, three types of X-ray tubes are used. Open tubes – with a rod anode and closed tubes - of two modifications: with massive and rod anodes (figure 2).

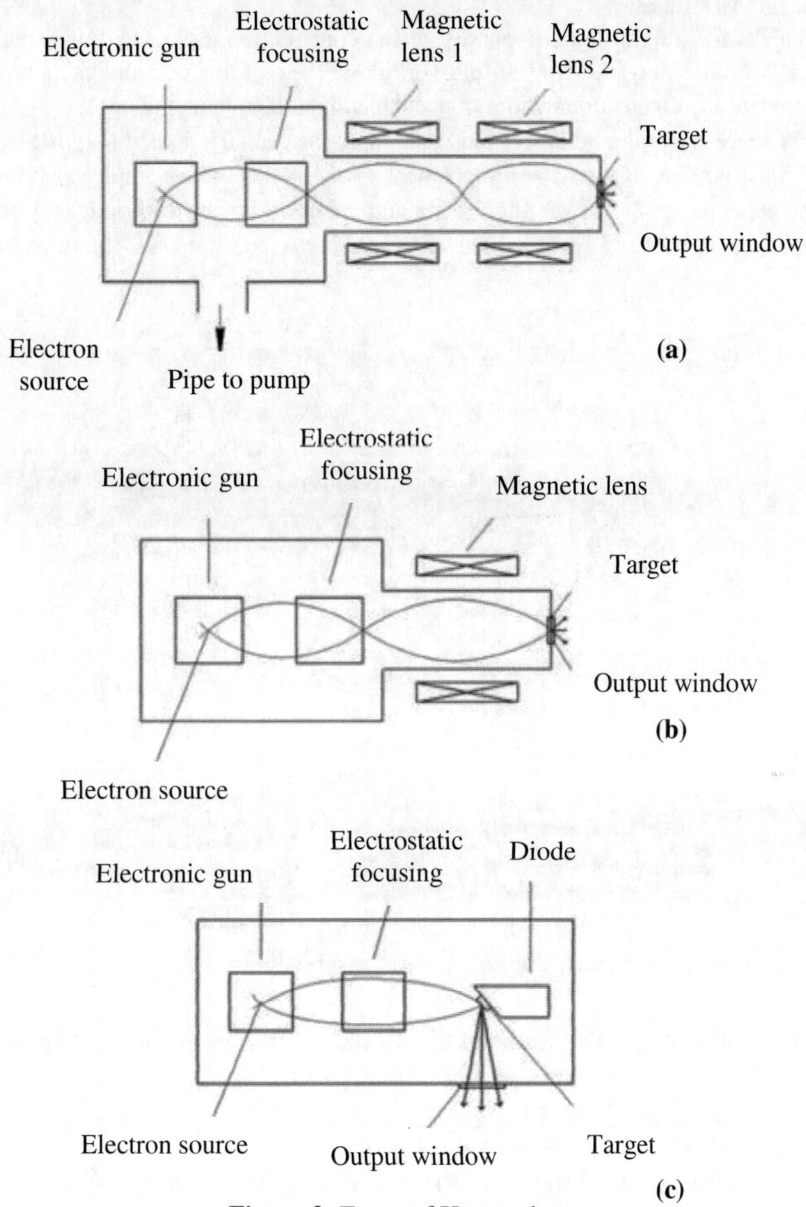

Figure 2. Types of X-ray tubes.

In Russia, such X-ray systems are not commercially available, but X-ray tubes for the development of the discussed systems have already been developed. Figure 3 shows a closed microfocus X-ray tube of the family BS for voltage of 150 kV with rog anode.

3. Analysis of technology

Each of the two technologies for developing X-ray systems has its advantages and disadvantages.

X-ray systems on open tubes allow for thin objects to detect defects less than 1 µm due to a more perfect focusing system and the choice of anode design. Operation of the X-ray tube at maximum modes is possible, since we can replace the cathode and the target of the anode.

However, X-ray systems on open tubes have a whole set of significant disadvantages compared to X-ray systems on closed tubes:

- large dimensions, weight and complexity of the construction, associated with the use of a cermet insulator, which is designed for the full working voltage of the tube and vacuum-tight mechanical collapsible tubing connections with replaceable units (cathode and anode);
- in the tube there is an integral constructive element - a specialized vacuum post;
- the tube must be trained after the replacement of the nodes, which imposes additional requirements on the generator of the power supply and alignment of the electron-optical system;
- increase in the size of the focal spot is possible due to mechanical vibrations of the vacuum post;
- high price.

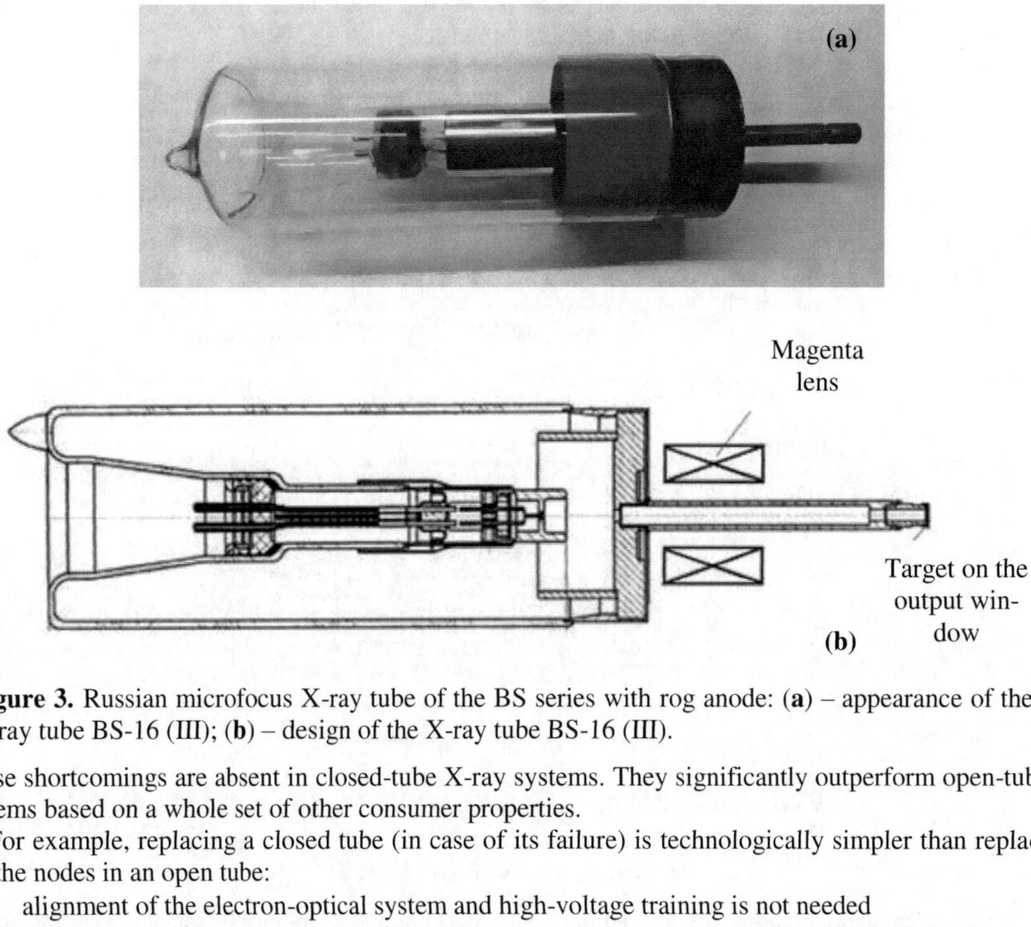

Figure 3. Russian microfocus X-ray tube of the BS series with rog anode: (**a**) – appearance of the X-ray tube BS-16 (III); (**b**) – design of the X-ray tube BS-16 (III).

These shortcomings are absent in closed-tube X-ray systems. They significantly outperform open-tube systems based on a whole set of other consumer properties.

For example, replacing a closed tube (in case of its failure) is technologically simpler than replacing the nodes in an open tube:

- alignment of the electron-optical system and high-voltage training is not needed
- replacement of the tube takes less time than the replacement of the anode or cathode in the open tube;
- the power consumption is much lower.

Particularly effective tubes with rog anode. They are stable in work and durable. The negative effect on their electrical strength of the electrons scattered on the target is eliminated by the absence of an electric field in the rog anode.

IOP Conf. Series: Materials Science and Engineering **387** (2018) 012061 doi:10.1088/1757-899X/387/1/012061

4. Conclusion

X-ray inspection systems built on both technologies are in demand in the electronics industry. Systems on closed tubes with massive anodes are effective for controlling large and dense objects, where it is sufficient to detect defects of 5 μm or more in size

Systems based on closed tubes with rog anode should be chosen to detect defects up to 2.5 μm in size. The most advanced systems based on open tubes with anode allow detecting defects less than 1 μm. However, the higher the resolution of the system, the lower such consumer properties as price, dimensions, weight, and also the translucent capability due to the reduction of the maximum value of the voltage on the X-ray tube.

Multifocus open X-ray tubes are promising in the development of X-ray systems. They have modes of different power to control dense objects or medium-density objects, as well as a high-resolution mode for controlling nanostructures. The mode switches in the process control objects. However, in many cases, a closed tube with an anode is more practical. The size of the focal spot can be adjusted by means of a magnetic or electromagnetic lens.

Acknowledgment

This work was supported by the Russian Science Foundation in the framework of the project entitled "The robotic system of multi-view microfocus X-ray control aircraft parts and assemblies made of polymer composite materials with complex action". The project number is 16-19-00155.

References

[1] Pirogova E V 2005 *Design and technology of printed circuit boards* (M.: Forum Intra-M) 560

[2] Podemskiy A A and Potrakhov N N 2017 *Control. Diagnostics* **4** 4–8

A model of the vacuum level effect on the sensor response of vacuum sensors based on nanostructured semiconductor oxide materials

I A Pronin[1], N D Yakushova[1], I A Averin[1], A A Karmanov[1] and V A Moshnikov[1,2]

[1] Department of Nano- and Microelectronics, Penza State University, 440026, Penza, Russia
[2] Department of Micro- and Nanoelectronics, Saint-Petersburg Electrotechnical University "LETI", 197376, Saint-Petersburg, Russia

E-mail: pronin_i90@mail.ru

Abstract. A mathematical model of the sensor response of vacuum sensors based on wide-band gap semiconductor nanostructured oxides, using a small number of parameters, is proposed. The results of the research can be used to develop a methodology for the synthesis of sensitive elements with a given response for a selected range of vacuum.

1. Introduction

An analysis of recent publications has shown an increasing interest in a new type of vacuum sensors, the sensitive element of which is made on the basis of nanostructured wide-band gap oxides [1, 2]. The principle of operation of such devices is based on the reversible process of chemical adsorption of oxidizing gases and reducing agents, leading to a change in the concentration of free charge carriers in the material [3]. For example, oxygen, being an oxidizing gas, is desorbed from the surface of a sensitive element with an increase in the vacuum level, and returns electrons to the conduction band of the semiconductor, increasing its electrical resistance. The advantage of this type of sensor is low power consumption, the possibility of microminiaturization, and compatibility with planar silicon technology.

At present, there are no model representations describing the change in electrical resistance of a given type of sensitive elements with a change in the vacuum level. Therefore, the aim of the article is to develop a model of the vacuum level effect on the sensor response of vacuum sensors based on nanostructured semiconductor oxide materials.

2. Model development and discussion

To calculate the response of sensory elements during adsorption-desorption of the gases under investigation (the investigated atmosphere can contain both oxidizing gases (usually, oxygen) and reducing gases (for example, hydrogen), it is necessary to analyze the dependence of the carrier concentration in the material on their partial pressure p', i.e., analytically determine the dependence $n = f(p')$. We will consider the region of low gas concentrations, assuming that there is no direct reaction between them, the chemisorption of the oxygen molecule is accompanied by the capture of the electron, and the chemisorption of the reducing agent is by transfer of the electron to the conduction band of the semiconductor.

The condition of adsorption equilibrium in the case when adsorption is not accompanied by dissociation has the form [4]:

$$\gamma P(N^* - N) = \nu^0 N^0 \exp\left(-\frac{q^0}{k_b T}\right) + \nu^+ N^+ \exp\left(-\frac{q^+}{k_b T}\right) + \nu^- N^- \exp\left(-\frac{q^-}{k_b T}\right) \tag{1}$$

where γ is Langmuir kinetic coefficient; P is pressure; N^* is surface concentration of adsorption centers; N is surface concentration of all varieties of adsorbed particles; ν^0, $\nu^\pm$ are probabilities of particle desorption in neutral, positive and negatively charged forms; N^0, $N^\pm$ are surface concentrations of the corresponding forms of adsorbed molecules; q^0, $q^\pm$ is differential heat of adsorption of the corresponding forms. However, the equilibrium with the gas phase in the case of a nondegenerate semiconductor is maintained exclusively due to the "weak" neutral form of chemisorbed particles, i.e. "firmly" bound particles do not participate in interaction with the environment. Then equation (1) can be rewritten in the form:

$$\gamma P(N^* - N) = \nu^0 N^0 \exp\left(-\frac{q^0}{k_b T}\right). \tag{2}$$

Introducing the notation $p_i = \dfrac{\gamma P_i}{\nu_i^0 \exp\left(-\dfrac{q_i^0}{k_b T}\right)}$ for partial normalized pressures of i-th gases, we rewrite the conditions of adsorption equilibrium (2) for oxygen and a reducing agent:

$$\begin{cases} p_{gr}(N^* - N_{gr} - N_{go}) = N_{gr} - N_{gr}^+ = N_{gr}^0, \\ p_{go}(N^* - N_{gr} - N_{go}) = N_{go} - N_{go}^+ = N_{go}^0, \end{cases} \tag{3}$$

where N_{gr}, N_{go} is the surface concentration of the adsorbed (charged and neutral) forms of the reducing agent and oxygen, respectively; N_{gr}^+, N_{gr}^- is the surface concentration of the charged forms of the reducing agent and oxygen, respectively; N_{gr}^0, N_{gr}^0 is the surface concentration of the neutral forms of the reducing agent and oxygen, respectively. The probability that the particle remains neutral on the surface of the semiconductor is described by the Fermi-Dirac statistics:

$$N_{gr}^0 = N_{gr}\left(\frac{1}{1 + \frac{1}{2}\exp\left(\frac{E_D - E_F}{k_b T}\right)}\right), \tag{4}$$

$$N_{go}^0 = N_{go}\left(\frac{1}{1 + \frac{1}{2}\exp\left(\frac{E_F - E_A}{k_b T}\right)}\right), \tag{5}$$

where E_D, E_A are ionization energies of the donor and acceptor levels, respectively; E_F is Fermi energy of a semiconductor.

For further analysis, let us pass to the counting of energy from the bottom of the conduction band, and add a subexponential expression (E_C – E_C) to each numerator, where E_C is energy of the conduction band bottom. Then, taking into account equality $n = N_C \exp\left(\frac{E_F - E_C}{k_b T}\right)$, where N_C is the effective density of states in the conduction band, equations (4), (5) can be rewritten in the form:

$$N_{gr}^0 = \frac{2nN_{gr}}{2n + N_C \exp\left(\frac{E_D - E_C}{k_b T}\right)}, \quad (6)$$

$$N_{go}^0 = \frac{2N_{go} N_C \exp\left(\frac{E_A - E_C}{k_b T}\right)}{2N_C \exp\left(\frac{E_A - E_C}{k_b T}\right) + n}. \quad (7)$$

Introducing notations $\Lambda_1 = N_C \exp\left(\frac{E_D - E_C}{k_b T}\right)$, $\Lambda_2 = N_C \exp\left(\frac{E_A - E_C}{k_b T}\right)$ we rewrite the system of equations (3) in the form:

$$\begin{cases} p_{gr}(N^* - N_{gr} - N_{go}) = \dfrac{2nN_{gr}}{2n + \Lambda_1} \\ p_{go}(N^* - N_{gr} - N_{go}) = \dfrac{2N_{go}\Lambda_2}{2\Lambda_2 + n} \end{cases} \quad (8)$$

To calculate the dependence $n = f(p_{gr})$ at a constant oxygen pressure p_{go} in addition to system (8), one more equation is needed to exclude the terms N_{go} and N_{gr} [5]. For this we introduce the electroneutrality equation:

$$nV_f + N_{go}^- S_f = N_D V_f + N_{gr}^+ S_f \quad (9)$$

where V_f, S_f are volume and area of the film, respectively; N_D is concentration of donor impurity in a semiconductor. Dividing both parts of equation (9) by S_f, we get its final version:

$$nh_f + N_{go}^- = N_D h_f + N_{gr}^+ \quad (10)$$

Rewriting equation (10) taking into account the previously adopted notation, as well as the degree of filling of impurity levels, we arrive at the expression:

$$\frac{nN_{go}}{2\Lambda_2 + n} - \frac{\Lambda_1 N_{gr}}{\Lambda_1 + 2n} = h_f (N_D - n) \quad (11)$$

To express the desired dependence in the analytic form, we find the parameter N_{gr} from the first equation of the system (8):

$$N_{gr} = (N^* - N_{go})\left(1 - \frac{2n}{2np_{gr} + \Lambda_1 p_{gr} + 2n}\right) \quad (12)$$

Then, we divide the first and second equations of the system (8) by each other, and from the obtained identity $\frac{N_{gr}}{N_{go}} = \frac{p_{gr}\Lambda_2(2n + \Lambda_1)}{p_{go} n(2\Lambda_2 + n)}$ we substitute N_{go} to equation (12), as a result of which we express the dependence of the parameter N_{gr} on quantities that do not depend on N_{go}:

$$N_{gr} = \frac{N^* p_{gr}\Lambda_2(2n + \Lambda_1)}{p_{go} n(2\Lambda_2 + n) + 2np_{gr}\Lambda_2 + \Lambda_1\Lambda_2 p_{gr} + 2n\Lambda_2}. \quad (13)$$

Substituting equation (13) into the identity obtained earlier by dividing the equations of the system (8), we express the parameter N_{go}:

$$N_{gr} = \frac{N^* p_{go} n(2\Lambda_2 + n)}{p_{go} n(2\Lambda_2 + n) + 2np_{gr}\Lambda_2 + \Lambda_1\Lambda_2 p_{gr} + 2n\Lambda_2}. \quad (14)$$

Substituting (13) and (14) into the electroneutrality equation (11), we arrive at the final expression in which there are no terms N_{gr}, N_{go}:

$$\frac{N^*(p_{go}n^2 - \Lambda_1 p_{gr}\Lambda_2)}{p_{go} n(2\Lambda_2 + n) + 2np_{gr}\Lambda_2 + \Lambda_1\Lambda_2 p_{gr} + 2n\Lambda_2} = h_f(N_D - n). \quad (15)$$

An analysis of (15) shows that expression $n = f(p_{gr})$ in the analytical form is very difficult, and involves the solution of a cubic equation with the use of the Cardano formula. Therefore, we express the inverse relationship $p_{gr} = f(n)$, and further analysis will be carried out with the help of its numerical solution:

$$p_{gr} = \frac{N^* p_{go} n^2 - h_f(N_D - n)(p_{go} n(2\Lambda_2 + n) + 2n\Lambda_2)}{N^*\Lambda_1\Lambda_2 - h_f(N_D - n)(2n\Lambda_2 + \Lambda_1\Lambda_2)}. \quad (16)$$

Within the framework of the developed model, let us analyze the dependence of the concentration of charge carriers in the zinc oxide film on the partial pressure of the reducing agent. We will assume that hydrogen acts as a reducing gas. Let us translate the partial normalized pressures of oxygen and hydrogen into Pa and *ppm*, respectively. Then the partial pressure of atmospheric oxygen will be $1.167 \cdot 10^{34}$ rel. units, and the concentration of hydrogen in 1 *ppm* will be $5.83 \cdot 10^{28}$ rel. units.

Figure 1 shows the dependence of the electron concentration in the ZnO film on the hydrogen concentration in a gas mixture containing some air. The calculation was carried out for a film thickness of 200 nm. An analysis of the graphs shows that two regions can be distinguished in the gas concentration ranges under study: the first one in which the electron concentration does not depend on the ethanol concentration, and the second one in which the growth of the electron concentration occurs according to a power law.

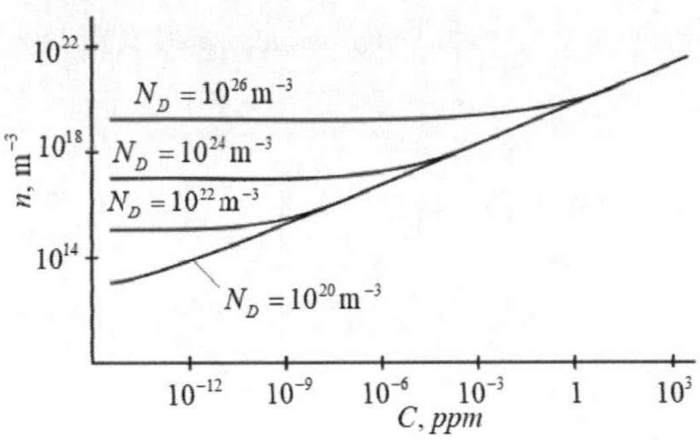

Figure 1. Dependences of the electron concentration in the ZnO film on the concentration of ethanol in air at various amount values of the donor impurity.

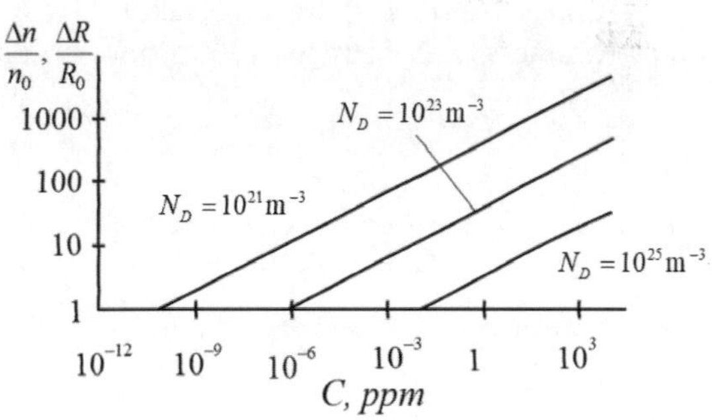

Figure 2. Dependence of the relative change in carrier concentration and film resistance on the concentration of ethanol at various concentrations of the donor impurity.

The research of the relative change in the concentration of charge carriers in a semiconductor film under the action of a gas analyzer is of particular interest. Figure 2 shows the dependence $\frac{\Delta n}{n_0}$. If we neglect the dependence of the mobility of the charge carriers on their concentration, then $\frac{\Delta n}{\Delta n_0} = \frac{\Delta R}{R_0}$,

where $\frac{\Delta R}{R_0}$ is the relative change in resistance.

An analysis of the dependence shows that there is a certain threshold value of the ethanol concentration, below which the relative change in the electron concentration is negligible. When this threshold concentration is exceeded, power-law dependence grows proportional to the square root of the analyzer gas concentration.

3. Conclusions

Thus, in the article a mathematical model of sensor response of vacuum sensors based on nanostructured semiconductor oxide materials is developed. The obtained model can be used for a technical vacuum containing both oxidizing and reducing gases. These results can be used in the design of sensing elements based on ZnO, SnO_2, In_2O_3 and other wide-band gap oxides obtained in the framework of various synthesis methods

4. Acknowledgments

The work was carried out with the financial support of the Ministry of Education and Science of the Russian Federation (project No. 16.897.2017/4.6), as well as in the framework of the Presidential Grant and the Presidential Scholarship (project No. MK-1882.2018.8 and SP-84.2018.1).

References

[1] Wu L, Song F, Fang X, Guo Zh and Liang S 2010 *Nanotechnology* **21** 475502

[2] Karmanov A A, Averin I A, Pronin I A, Yakushova N D, Igoshina S E, Moshnikov V A and Vishnevskaya G V 2017 *Journal of Physics: Conference Series* **872** 012005

[3] Averin I A, Igoshina S E, Karmanov A A, Pronin I A, Moshnikov V A and Terukov E I 2017 *Technical Physics* **62** 799–806

[4] Vol'kenshteyn F F 1987 *Elektronnyye protsessy na poverkhnosti poluprovodnikov pri khemosorbtsii* (Moskva: Nauka)

[5] Kissine V V, Sysoev V V, Voroshilov S A and Simakov V V 2000 *Semiconductors* **34** 308–11

IOP Conf. Series: Materials Science and Engineering **387** (2018) 012063 doi:10.1088/1757-899X/387/1/012063

The change in the phase state of thin PZT layers in the region of the morphotropic phase boundary obtained by the RF magnetron sputtering with varying target-substrate distance

V P Pronin[1], D M Dolgintsev[1], V V Osipov[1,2], I P Pronin[2], S V Senkevich[2] and E Yu Kaptelov[2]

[1]Herzen State Pedagogical University, 191186, St.-Petersburg, Russia
[2]Ioffe Institute, 194021, St.-Petersburg, Russia

E-mail: pronin.v.p@yandex.ru

Abstract. The relationship between the microstructure, composition, phase state, and dielectric properties with a change in the distance from the target to the substrate (in the range 30–70 mm) was studied for high-frequency magnetron deposition of thin layers of lead zirconate-titanate (PZT). The composition of the sputtered target of the solid solution of PZT (PbZr0.54Ti0.46O3) corresponded to the region of the morphotropic phase boundary. After depositing the film, the perovskite structure was formed at 580 °C. It is shown that the variation of the distance leads to a change in the composition (elemental ratio of Zr/Ti atoms) within 2%, the character of the spherulite radiant microstructure of the layers, the volumes of monoclinic and tetragonal modifications of the ferroelectric phase coexisting in this concentration region, and also the dielectric properties.

1. Introduction

The solid solutions of lead zirconate-titanate (PZT), corresponding to a morphotropic phase boundary (MPB) between rhombohedral and tetragonal modification of the ferroelectric phase, is characterized by anomalously high electromechanical and piezoelectric moduli [1]. Ceramic materials, corresponding to these compositions, are currently the basic materials of modern piezo-technic. In a number of recent studies, these anomalies have been attributed to the existence of an intermediate monoclinic phase in the MPB region [2–3]. In ceramic solid solutions of PZT, the MPB region, as a rule, occupies an interval of concentrations not exceeding 1–2 percent. However, segregation of zirconium (or titanium) atoms in perovskite microregions can occur due to the use of insufficiently finely dispersed initial metal oxide powders, and also due to the difference in the crystallization temperatures of the perovskite phase in lead titanate ($PbTiO_3$) and lead zirconate ($PbZrO_3$). This leads to elemental microinhomogeneity and, as a result, to the expansion of the region of the MPB and to the decrease of the abnormally high physical characteristics, both in bulk solid solutions and in thin layers of the PZT [4–5]. The aim of the present work was to study the phase state and dielectric properties of thin PZT layers in the MPB region, obtained by high-frequency sputtering of a ceramic target with a variation of the distance from the target to the substrate.

2. Sample preparation and research

A ceramic target of a solid solution (1-x) $PbTiO_3$-x $PbZrO_3$ of composition equal to x = 0.54 was chosen for the studies. Thin PZT layers with a thickness of ≈500 nm were formed by a two-step procedure [6]. In the first stage, amorphous thin PZT layers were deposited by the RF magnetron sputtering of a ceramic target in an argon-oxygen atmosphere at ~150 °C on silicon substrates, and also on silicon substrates with silicon dioxide sublayers (500 nm thick), titanium dioxide (20–30 nm) and platinum (80 nm) ($Si/SiO_2/TiO_2/Pt$). In the second stage, the structures were annealed in air at 580 °C for 1 hour to obtain single-phase perovskite PZT layers deposited on the $Si/SiO_2/TiO_2/Pt$ substrate. The composition, microstructure, and dielectric properties of thin PZT layers were studied by varying the distance from the target to the substrate (d) in the interval 30-70 mm in steps of 10 mm.

The microstructure of the films was determined using a scanning electron microscope Carl Zeiss EVO-40. To determine the elemental composition, we used an electron-dispersive attachment INCA, the energy of the probing electrons was 12 keV. The phase state was determined using X-ray diffraction analysis (DRON-7), as well as the method of electron backscatter diffraction (EBSD). To determine the dielectric constant, the immittance meter E7-20 was used. The measurement frequency was 10 kHz, the measuring voltage was 40 mV. C-V characteristics were measured by applying a constant bias voltage in the range ± 0–8 V.

3. Results and discussion

The results of the change in the composition of the deposited films are shown in figure 1. To obtain them, amorphous thin films of PZT deposited on a silicon substrate (in the absence of a platinum sublayer) were used to avoid distortion in the composition due to the proximity of the L_α- line of zirconium (2.042 keV) and M_α- the platinum line (2.048 keV) [7] The change in the distance (d) between the target and the substrate, as seen in figure 1, a, led to a change in the composition of the solid solution of thin amorphous films.

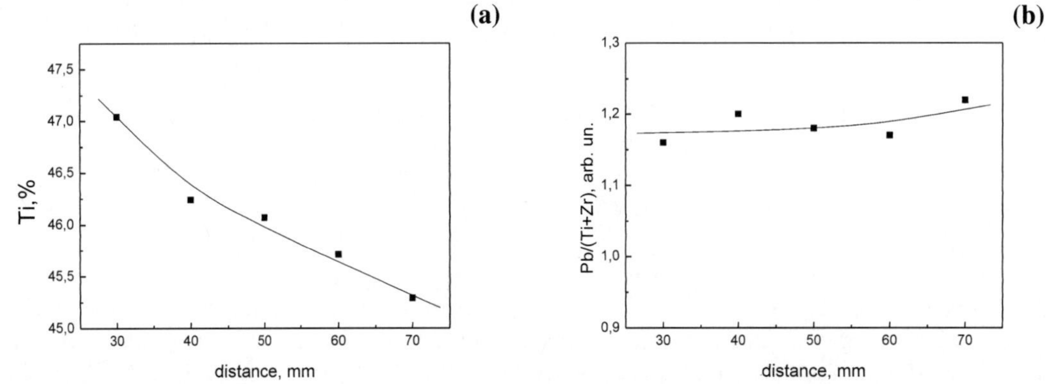

Figure 1. Changes in the content of Ti (**a**) and the ratio of Pb / (Ti + Zr) atoms (**b**) in PZT films with varying target-substrate distance.

The elemental ratio of atoms varied from Ti/Zr ≈ 47/53 in films deposited at d = 30 mm to ≈ 45.3/54.7 in films deposited at d = 70 mm. Thus, the composition change in the region of the MPB was about 1.7% with a change in d from 30 to 70 mm. Somewhat more, about 5%, the content of lead atoms changed as d increases, expressed as the elemental ratio Pb/(Ti + Zr), figure 1(b).

The reasons for such changes are related to the processes of scattering of sputtered target atoms in a gas plasma under conditions of a specific geometry of the mutual location of the circular zone of erosion of the ceramic target and the substrate placed in the center. Analysis of the literature data shows that in the investigated target-substrate distance interval at the working gas pressure of 8 Pa, the lead atoms become thermalized when the directed motion of the atomized atoms is replaced by a diffusion motion characterized by an isotropic form of atom propagation [8].

The SEM images in figures 2(a), (c), (e) reflect the change in the spherolitic microstructure of the perovskite phase in thin PZT layers formed on $Si/SiO_2/TiO_2/Pt$ substrates with increasing d obtained in the mode of selection of backward reflected electrons. The change in the spherulite microstructure consisted in a decrease in the average size of the spherulite blocks with increasing d (i.e., with an increase in the concentration of the nucleation centers of the perovskite phase), as well as in the density of radial rays in spherulite blocks.

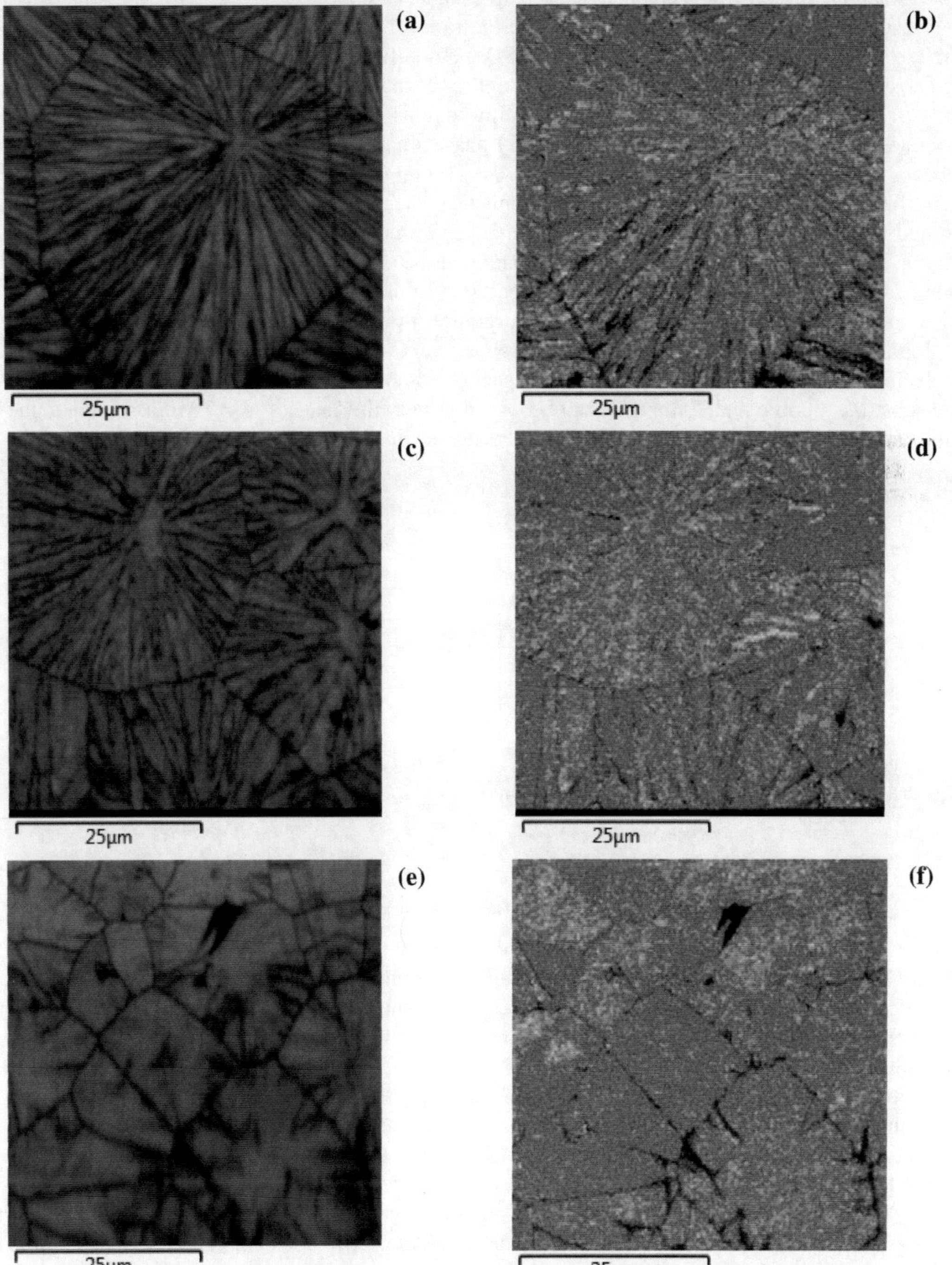

Figure 2. SEM image in the mode of selection of backward reflected electrons (**a**), (**c**), (**e**) and phase images (**b**), (**d**), (**e**) of PZT films formed at different distances from the target to the substrate: 30 mm (**a**), (**b**), 50 mm (**c**), (**d**) and 70 mm (**e**), (**f**).

It can be seen that in films deposited at d = 70 mm, the radial character of the structure is close to disappearing (figure 2(d)). It can be assumed that such changes are associated with a change in the microinhomogeneity of the atoms that make up the composition of the film. However, in order to understand the reasons for such changes, separate studies are required.

Figures 2(b), (d), (f) shows the changes in the ratio of monoclinic and tetragonal modifications of the ferroelectric phase determined by the EBSD method [9]. Variations in the distance from the target to the substrate led to a slight change in the stability of the monoclinic and tetragonal modifications of the ferroelectric phase. Averaging over scanning regions equal to 50*50 μm showed that the ratio of volumes of monoclinic (M) and tetragonal (T) phases increased from M/T ≈ 52/42 in films deposited at d = 30 mm to ≈ 55/40 in films deposited at d = 70 mm (part of the volume of the films could not be identified). The reasons for such minor changes in the stability of the M and T phases when the composition of the solid solution is changed by 2 % may indicate a significant width of the MPB region. However, it should be taken into account that the obtained data give information only about the thin near-surface layer of the PZT (not exceeding several nanometers). Changes in the volume of the film can be more significant, which, however, requires the use of other methods of investigation.

Figure 3 shows the characteristic volt-capacitive C-V curves (or reverse dependences of the dielectric constant, ε-V) for the Pt/PZT/Pt capacitor structures at different distances from the target to the substrate (30 mm and 70 mm). It can be seen that in a film made at d = 70 mm, a large value of the dielectric constant is observed, and the C-V dependence itself takes a more symmetrical form.

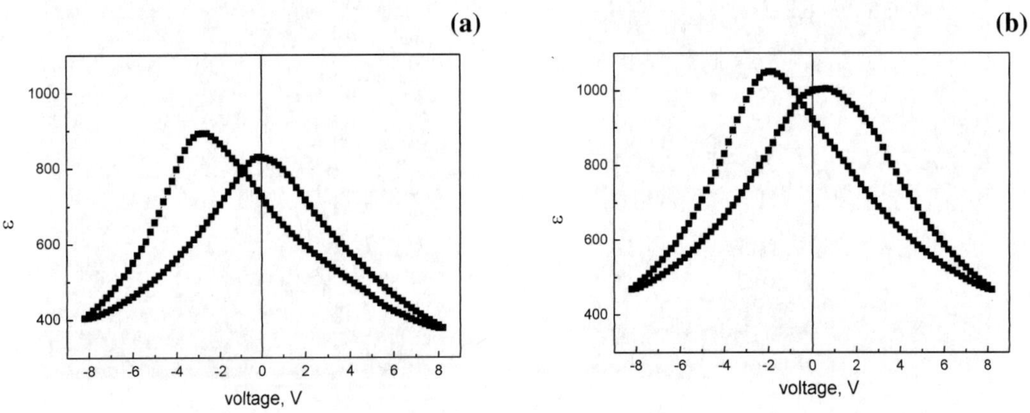

Figure 3. C-V(ε-V) characteristics for d = 30 mm (**a**) and d = 70 (**b**).

Figure 4 shows the dependence of the value of the internal field E_{int} (a) and the dielectric permittivity (b) of the PZT films with a change in d, measured at room temperature. The value of E_{int} (figure 4(a)) was determined from the shift of the C-V characteristics along the abscissa axis toward negative voltages (see figure 3), which corresponded to the orientation of the field vector directed from the free surface to the lower electrode.

With increasing target-substrate distance, the E_{int} fell monotonically. This behavior can be explained by the increase in the amount of lead oxide, but this contradicts its role in the formation of the internal field and the naturally unipolar state in thin PZT films [10]. However, the nature of the dependence can be explained if to assume that the observed changes are associated with a change in the location of excess lead oxide between the lower and upper interfaces.

Figure 4(b) reflects the anomalous behavior of the dielectric constant caused by the change in the composition of the PZT solid solution, the microstructure, and the phase state. It was expected that in the region of the MPB the dielectric constant will pass through a maximum, but the experimental data go against expectations. The reasons for this behavior are not yet clear and require further research.

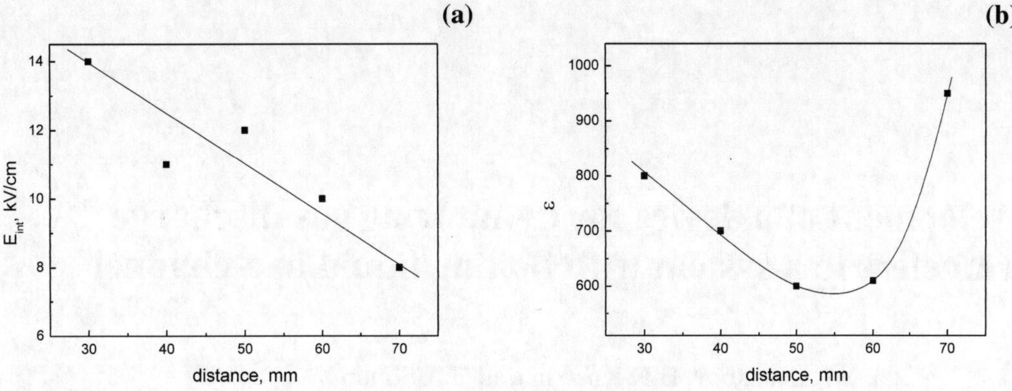

Figure 4. The experimental dependence of the permittivity (**a**) and the internal field (**b**) of thin PZT films with a change in the target-substrate distance.

4. Conclusions

The carried out researches have shown that by changing the distance from the target to the substrate, it is possible to vary (within 2 %) the composition of thin films of PZT in the region of MPB, which provides a tool for searching for optimal electromechanical parameters. It is shown that this leads to a strong change in the character of the spherolitic microstructure of the films, to the values of the internal field and the dielectric permittivity. However, additional research is required to find out the reasons for such changes.

Acknowledgment

The work was carried out in the framework of the project part of the state assignments, project No. 16.2811.2017 / 4.6 and partially supported by RFBR grants No. 16-02-00632 and 16-08-00808.

References

[1] Jaffe B, Cook W R and Jaffe H 1971 *Piezoelectric Ceramics* **3**
[2] Noheda B, Cox D E, Shirane G, Gonzalo J A, Cross L E and Park S-E 1999 *Appl. Phys. Lett.* **74** 2059
[3] Noheda B, Cox D E, Shirane G, Guo R, Jones B and Cross L E 2001 *Phys. Rev.B* **63** 014103
[4] Ledermann N, Muralt P, Baborovski J, Gentil S, Mukati K, Cantoni M, Scifert and Setter N 2003 *Sens. Actuators A* **105** 162–70
[5] Calame F and Muralt P 2007 *Appl. Phys.Lett.* **90** 062907
[6] Pronin I P, Kaptelov E Yu, Senkevich S V, Klimov V A, Zaytseva N V, Shapligina T A, Pronin V P and Kukushkin S A 2010 *Physics of the Solid State* **52** 124–8
[7] Senkevich S V, Flegontova E Yu, Kaptelov E Yu, Kanareykin A G, Pronin V P, Sergeeva O N and Pronin I P 2015 *Ferroelectrics* **477** 103–11
[8] Volpyas V A and Kozirev A B 2011 *Journal of Experimental and Theoretical Physics* **140** 196–204
[9] Pronin V P, Kanareykin A G, Dolginstev D M, Kaptelov E Yu, Senkevich S V and Pronin I P 2017 *Surface* **2** 40–6
[10] Osipov V V, Kislelev D A, Kaptelov E Yu, Senkevich S V and Pronin I P 2015 *Physics of the Solid State* **57** 1748–54

IOP Conf. Series: Materials Science and Engineering **387** (2018) 012064 doi:10.1088/1757-899X/387/1/012064

Development of a device for monitoring gas discharge parameters in a system with boiling liquid in a channel

A N Ramazanov, D K Kostrin and V A Simon

Department of electronic instruments and devices, Saint Petersburg Electrotechnical University "LETI", 197376, Saint Petersburg, Russia

E-mail: ramazanow.alexander@yandex.ru

Abstract. Experimental prototypes of the high-voltage power sources for initialization of a gas discharge in the channel with boiling liquid are designed. The development principle of the device for control of the discharge system parameters is presented. The emission characteristics of a gas discharge at different frequencies are given.

1. Introduction

Today there is an intensive development of devices for water composition analysis [1]. Designed devices vary in the studied parameters, dimensions and price. Special attention is paid to a compact device capable of analyzing an aqueous solution sample in the field conditions. The gas discharge at atmospheric pressure is often used for this purpose [2, 3]. This type of discharge has high temperature and pressure plasma, which serves as a high brightness light source. If liquid is placed between the electrodes and the discharge is initiated, the spectral lines of chemical elements which salts are dissolved in the liquid will arise in the spectrogram, saying about the qualitative composition of the analyte under study [4, 5].

In this regard, when creating devices for gas discharge initialization in the field conditions, a number of problems should be solved, related not only to the choice of the discharge cell, but also to the development of an electrical circuit for excitation and control of discharge parameters. Currently there are a number of systems that are intensively used to maintain the gas discharge at atmospheric pressure that do not require high voltages and frequencies of the power source. There are three types of these devices of frequent occurrence: a system with a liquid electrode, a system with a liquid supply and a system with boiling liquid in a channel. Each of them has a number of advantages and disadvantages. The system with liquid supply is used in industrial enterprises, has a complex structure and is inefficient in the analysis for small samples. The system with a liquid electrode has simple structure, but it is quite difficult to ensure the safety of the spectral lines registration in this case, because the optical fiber should be placed quite close to the discharge arc. The most safe and easy to use is a system with boiling in the channel which is shown in figure 1. The electrodes *2* are located in the vessels *1*, interconnected by a quartz glass capillary *3*.

To excite and maintain the gas discharge in this system the high-voltage pulse power sources capable of providing the output voltage up to 5 kV and pulse repetition rate in the range from 15 to 50 kHz should be used. These settings should be maintained in the specified ranges to ensure optimal initialization conditions in a portable device. In order to simplify the actions of the intended user, the system should be automated. In this regard, it is necessary not only to develop a high-voltage power source, but also a device for monitoring its parameters.

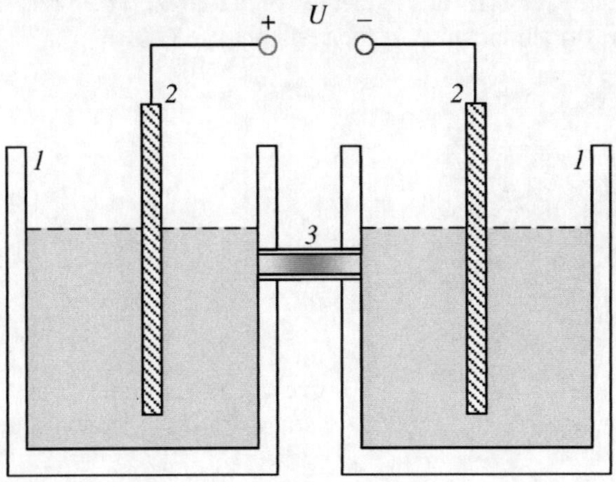

Figure 1. Gas discharge system with boiling liquid in a channel.

2. Development of a high-voltage power source

During the development of a high-voltage pulse power source, two electrical circuits were created: one based on a half-bridge self-clocked driver IR2153, the second was based on the commonly used self-clocked driver NE555N.

The circuit of the power source based on the IR2153 driver is shown in figure 2. The circuit is powered by a battery to ensure portability of the device. The driver controls the transistors Q1 and Q2 in the booster mode, which leads to the formation of rectangular pulses of a given amplitude and frequency. To obtain high output voltage, the step-up transformer and a voltage multiplier consisting of four amplifying stages is used.

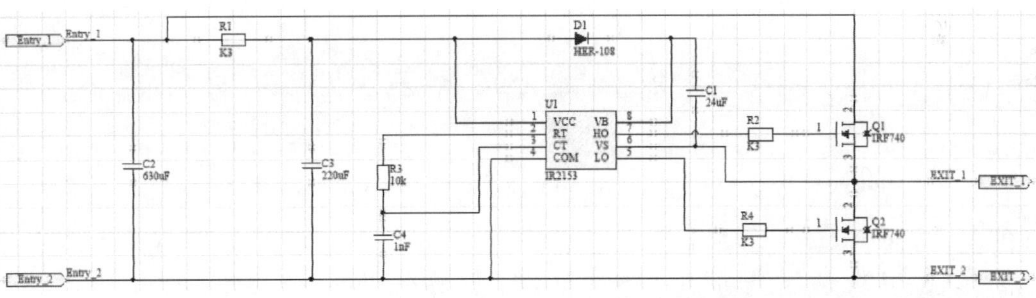

Figure 2. Power source circuit based on the IR2153 driver.

In order to ensure the efficiency of the developed circuit for use in portable applications, its experimental prototype was created. The three-dimensional (3D) printed circuit board (PCB) model of this prototype is shown in figure 3. Changes were made to adjust the value of the output pulse frequency: instead of a fixed value resistor R3 a potentiometer was used. The waveform of the circuit output signal is demonstrated in figure 4. The presence of a rectangular pulse source at the output indicates the suitability of this circuit for use in the discharge parameters control device.

The electrical circuit diagram based on the NE555N timer is shown in figure 5. Due to further use in portable applications, power was also provided by a battery. By pressing the S1 button, the optocoupler sensor U1 is activated, which has an output resistance of about 50 Ohms. Then the transistor Q1 is opened, and the supply voltage is feeding the timer. Transistor Q2 is used to amplify the output pulses from the current timer. Pulse duty cycle is regulated by a potentiometer R4. The pulses are fed from the circuit output to the primary winding of the step-up transformer. The voltage

IOP Conf. Series: Materials Science and Engineering **387** (2018) 012064 doi:10.1088/1757-899X/387/1/012064

formed by the secondary winding is increased by several times and supplied to the high voltage multiplier. Figure 6 shows the topology of the developed power source based on the NE555N timer. Figure 7 shows the output waveform at frequency of 22 kHz. The presence of emissions on the waveform is due to the parasitic inductance of the resistors used.

Figure 3. 3D PCB model of the IR2153-based power source.

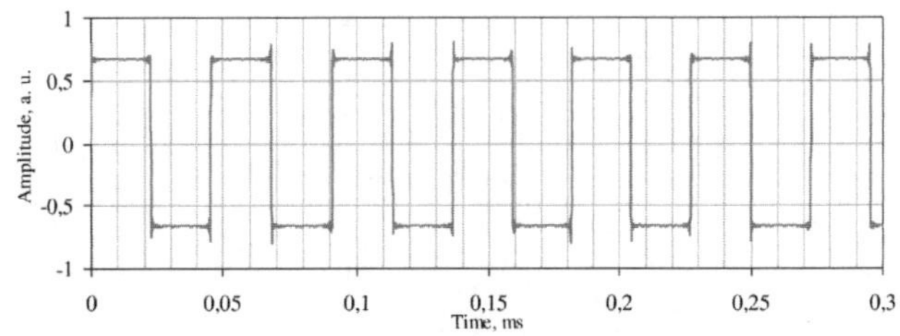

Figure 4. Output signal of the IR2153-based power source circuit (pulse frequency is 22 kHz).

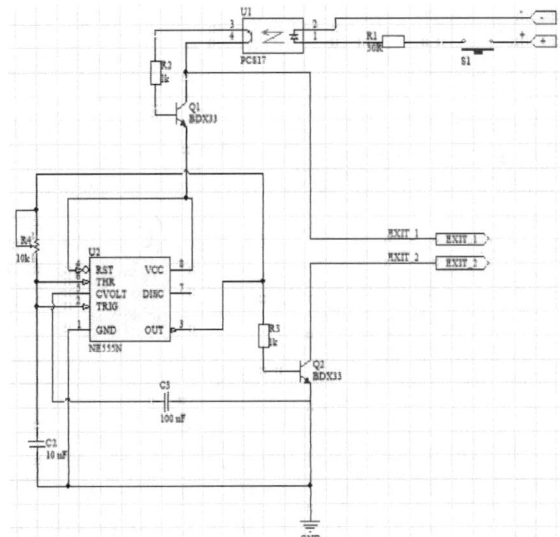

Figure 5. High voltage power source circuit based on the NE555 timer.

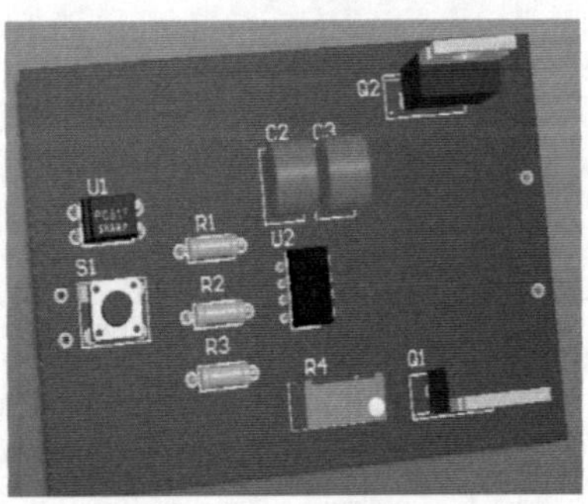

Figure 6. 3D PCB model of the NE555-based power source.

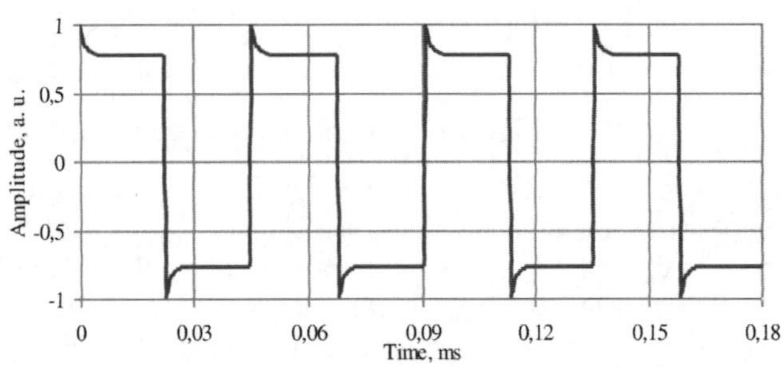

Figure 7. Output signal of the NE555-based power source circuit (pulse frequency is 22 kHz).

Both developed circuits are applicable to the device for gas discharge parameters monitoring, but it is preferable to use a power source based on the IR2153 driver, since the output waveforms have virtually no parasitic pulses, providing the device performance stability.

IOP Conf. Series: Materials Science and Engineering **387** (2018) 012064 doi:10.1088/1757-899X/387/1/012064

3. Glow discharge control device

The developed control device must have high functionality to minimize the work of the intended user, so the device must be able to independently control the discharge parameters. It is possible to implement this using a variety of peripherals operated by microcontroller. The device must have the following components: stepper motor driver, which will be used to monitor the position of the tip-shaped electrode, ultrasonic sensor to determine the water level in the discharge cell, display indicating the current status of the system, matrix keyboard for the data input. It is also necessary to add programming connectors to be able to make adjustments to the program code of the device not only by connecting a special programmer, but also by connecting to a laptop computer with specialized software.

The 3D PCB model of the developed control device is shown in figure 8. All terminals are located relative to the convenience of connecting the battery, and the terminals for connecting the stepper motor, ultrasonic sensor and step-up transformer are at a distance from the microcontroller and connectors for programming and output information. Matrix keyboard and display aren't located on the PCB and connected as input-output peripherals.

The electronic components used in the circuit were selected under the condition of minimum power consumption to increase the system operation time. Some circuit elements, such as DC-DC converters and drivers, are heated during long-term tests. To improve the stability of the circuit, the necessary calculations were made to determine those chips that need an additional heat sink. Based on the thermal calculations the heat dissipating pads for voltage regulators were increased and the heat sinks for transistors cooling were added.

Figure 8. 3D PCB model of the glow dischage control device.

4. Application of the glow discharge control device in the experiment

The spectrograms of discharge plasma radiation were obtained under various operating modes of the device in the course of further research. In this case, the influence of pulse frequency on the spectrum curves, as well as the presence of obvious impurities (potassium permanganate in particular) was considered. The spectrometer ISM3600 [6, 7] with 2 s accumulation time setting was used. The software processing of the obtained spectra was carried out with the ASpect2010 software [8].

Figure 9 shows a gas discharge spectrogram of the system with boiling liquid in the channel at different frequencies. Tap water was used as the analyte under study.

The spectral curves are arranged as the intensity increases, which corresponds to an increase in the pulse repetition rate: 20, 25 and 30 kHz. In the range 200–300 nm low-intensity spectral lines of nitrogen and hydroxyl group are observed. The lines with the highest intensity are located in the wave range from 300 to 400 nm and correspond to the hydroxyl group OH^-, single-charge nitrogen N^+ and oxygen O^+ ions, molecular nitrogen and oxygen. The wave range from 600 to 800 nm is of greatest interest, because it contains various impurities containing in the sample e. g. potassium, sodium, aluminum, calcium.

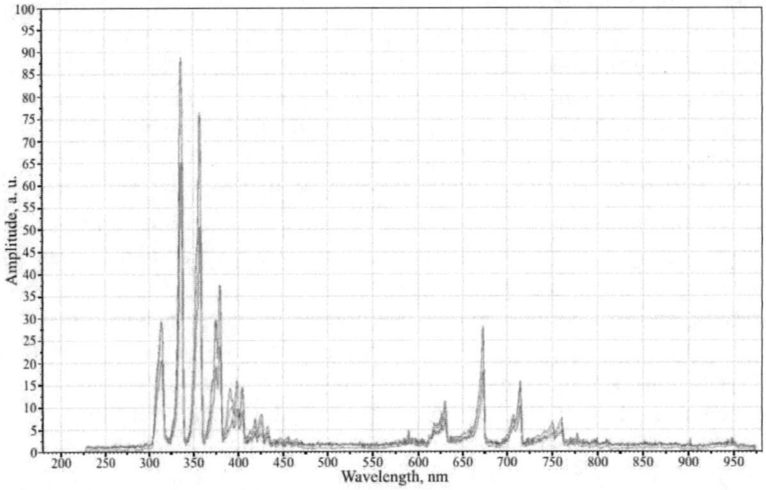

Figure 9. Gas discharge emission characteristic for different pulse frequencies

The presented results suggest the possibility of using small-sized circuits and PCBs in the creation of a portable device for environmental monitoring of water resources using glow discharge spectra. For further minimization, the spectrometric equipment should be significantly simplified, leaving the accuracy of the analyte composition determination unchanged.

5. Conclusion

The actual research proves the implementation possibility of the gas discharge parameters monitoring device in the setup with analyte boiling in the narrow channel. The circuit based on a half-bridge self-clocked driver IR2153 as a high-voltage pulse power source will be used in the future, since it has no output voltage surges, compared to the similar circuit utilizing the NE555N timer.

The developed gas discharge parameters control device will be further used in a portable setup for water resources quality monitoring due to its high functionality. The device obtains high-accuracy spectra, allowing the determination of the analyte qualitative composition.

Acknowledgment

The research was partially supported by FASIE (project № 11764GU/2016).

References

[1] Vasilevskii A M et al. 2014 *Biomedical Engineering* **1** 13–6
[2] Ramazanov A N et al. 2016 *Journal of Physics: Conference Series* **729** 012004
[3] Ramazanov A N et al. 2017 *Journal of Physics: Conference Series* **929** 012062
[4] Ramazanov A N et al. 2017 *Proceedings of the 2017 IEEE Russia Section Young Researchers in Electrical and Electronic Engineering Conference* 58–62
[5] Ramazanov A N et al. 2018 *Proceedings of the 2018 IEEE Conference of Russian Young Researchers in Electrical and Electronic Engineering* 754–7
[6] Uhov A A et al. 2014 *Journal of Physics: Conference Series* **567** 012039
[7] Kostrin D K et al. 2016 *Vakuum in Forschung und Praxis* **3** 34–7
[8] Kostrin D K et al. 2016 *Journal of Physics: Conference Series* **729** 012030

Development of physico-technological basics of formation of structural organized ferroelectric films by the method of magnetron sputtering followed by laser processing

D M Redka[1], T V Gordinskaya[1], N V Mukhin[1] and A B Kozyrev[1,2]

[1]St Petersburg Electrotechnical University "LETI", 197376, St. Petersburg, Russia
[2]Dagestan University of National Economy, 367008, Makhachkala, Russia

E-mail: dnredka@etu.ru, muhinnv_leti@mail.ru, gordinskaya_tatyana@mail.ru

Abstract. The current research considers physico-technological aspects of obtaining structural organized ferroelectric films on the basis of the oxide system of lead zirconate titanate with use of the HF methods of magnetron sputtering with the subsequent laser processing. The approach allows to obtain three types of structural organized films by local laser annealing on the basis of one oxide system: a ferroelectric-semiconductor composite, structure with periodically changing index of refraction and a ferroelectric material with periodically changing residual polarization.

1. Introduction

The development of laser technology is increasingly attracting researchers due to possibilities of creation of metamaterials and structurally-organized environments for various applications [1]. Thus, at present, world scientific groups are providing research on the use of laser technologies in the production of plasmon and acoustic metamaterials consisting of artificial micro- or nanostructures with unique optical and acoustic properties, such as negative refraction, inverse Doppler effect, etc. [2, 3]. One of the most promising and rapidly developing technologies is LDW (laser direct writing) and as its version – LIFT (Laser-induced forward transfer) [4], with which metal structures [5], carbon nanotubes [6], chalcogenide glasses [7], various metamaterials [8, 9] and plasmon devices were produced. Recent works in this area [1, 10] shows perspective of femtosecond laser technologies in the creation of new nanomaterials and structures.

Among the existing variety of materials, perovskites, "active" materials are of high interest due to possibilities to create "smart" composite systems based on these materials. Laser processing of materials that require crystallization in the perovskite structure has its own specifics. Laser technologies have been used in [11–15] for structuring and crystallization of perovskite ferroelectric systems.

In the majority of these researches, laser annealing is used to locally crystallize amorphous films into the ferroelectric phase in order to minimize the heating of the elements surrounding the functional region. When annealing with excimer lasers from photon energies substantially exceeding the width of the forbidden band, in this case absorption, and consequently annealing, takes place in a thin near-surface layer [12]. In addition, the mode structure of the spot of excimer lasers does not allow localization of annealing in the submicrometer region.

It was proposed in [13] to use a single-mode femtosecond laser with a wavelength that falls in the transparency region of the film and at the same time into the absorption region of platinum for annealing a ferroelectric film on a platinized substrate. This method allows to simultaneously solve three problems. First, to approximate the annealing conditions to thermal annealing in the furnace, since the film is heated from the side of platinum. Secondly, heating is performed locally with a Gaussian temperature distribution along the radius of the laser spot. Thirdly, the use of a femtosecond laser makes it possible to diagnose the formation of the ferroelectric phase during annealing. This diagnostic is based on the method of generation of the second optical harmonic, which is an effective method for studying phase transitions, including crystallization into the perovskite phase.

The methods of local crystallization of perovskite microstructures in an amorphous film by a femtosecond laser greatly increase the controllability of the laser annealing process and can be used not only for ferroelectrics but also for a wide range of materials that crystallize into a noncentrosymmetric phase.

2. Methods and approaches

The research is aimed the development of physical basics and technological method of obtaining new functional materials using the interaction of femtosecond laser radiation with multicomponent oxide systems of certain compositions, leading to local phase transformations of the matrix and providing the necessary structural organization. In this work a lead zirconate titanate oxide system is planned to use as a matrix material due to its phase diversity and unique electrophysical and mechanical properties that are of interest in the development of various converters [16] and memory devices [17].

The use of femtosecond laser technology allows solving two problems at once:

1) expansion of the functionality of the perovskite-like oxide system by its spatial and phase structuring; 2) solving the problem of high crystallization temperature by local annealing, which is critical for integrating with other layers and structures. The development of this method requires a deep understanding of the mechanisms of local phase transformations and structural modification of the matrix under the action of laser radiation. It requires fundamental theoretical and experimental studies. In the long term, developed methods and approaches can be used for a wide range of materials in the selection of appropriate modes of laser microprocessing.

The approach consists in a special methodology, which includes:

– the choice of a unique oxide system of lead zirconate titanate, which allows to obtain various phase states of the substance (dielectric and semiconductor phases, pyrochlore and perovskite phases, ferroelectric and paraelectric phases), including possibilities of initiating necessary phase transformations and structural changes under the action of laser radiation;

– special preparation of samples allowing one to obtain, on the basis of one oxide system, three types of structurally-organized media by means of local laser annealing: a ferroelectric-semiconductor composite, a structure with periodically varying refractive index and a ferroelectric film with periodically changing residual polarization.

The methods of obtaining samples in this work consisted of the following stages:

1) Deposition of a structure-forming platinum electrode on a silicon or sapphire substrate with an underlayer of a titanium by method of ion-plasma sputtering.

2) Deposition of an amorphous film of PZT with various excess of lead by the HF method of magnetron sputtering.

3) A heat treatment of a film of PZT to obtain required structure and phase of a matrix.

4) Local laser processing of PZT system to form the structural organized ferroelectric film.

3. Technological points of formation of the structure and discussion

Lead oxide and PZT films were obtained by the method of reactive HF magnetron sputtering on special equipment. In a vacuum chamber there are two magnetrons with a diameter of 100 mm connected to the high-frequency generator working at frequency of 13,56 MHz and allowing to vary power input to a magnetron from 10 W to 1000 W.

IOP Conf. Series: Materials Science and Engineering **387** (2018) 012065 doi:10.1088/1757-899X/387/1/012065

Powder PbO_2 and ceramics $Pb_{0,81}Sr_{0,04}\cdot Na_{0,075}Bi_{0,075}(Zr_{0,58}Ti_{0,42})O_3$ were used as the targets. This PZT ceramics was chosen because it is characterized by good ferroelectric properties with values of dielectric constant and loss of angle 1630 and 0,021 respectively (passport data). As a rule, when the oxygen-containing materials are sputtered, the deviation from stoichiometry in oxygen is observed in the resulting film. Therefore, to compensate the loss of oxygen in the growing films, target sputtering was carried out in a argon-oxygen gas mixture. The thickness of the PZT films were varied from 250 to 600 nm (Fig. 1). To choose the optimal parameters for the deposition of PbO_x film, samples were obtained with a power variation. As a result, the optimal working power of 14W was choosen, which provides a reproducible result in thickness. Table 1 shows the deposition regimes of the layers.

Table 1. Parameters of deposition processes of PbO_x and PZT films.

Process parameters	PbO_x	PZT
Power of HF-charge, W	14	100
Diameter of a target, mm	100	100
Composition of gas mixture	76%Ar+24%O_2	76%Ar+24%O_2
Pressure of gas mixture, mmHg	10^{-2}	10^{-2}
Substrate temperature, °C	130	130
Growth rate of films, nm/min	0,7	5,8

The next stage in the formation of the structure was temperature annealing in air in a muffle furnace. Depending on the problem being solved, the annealing temperature was varied over a wide range. The final stage of the treatment was local annealing with femtosecond laser radiation from the near-IR range.

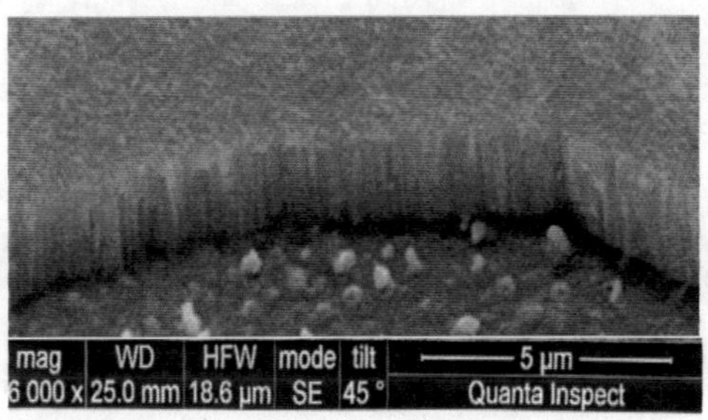

Figure 1. The SEM image of a PZT film.

Figure 2 shows the results of the absorption spectra investigation of PZT films using a Nicolet 6700 FT-IR spectrometer. It can be seen that the absorption maximum lies in the wavelength range from 0.3 µm to 0.4 µm. The laser processing modes were chosen on the basis of the spectra.

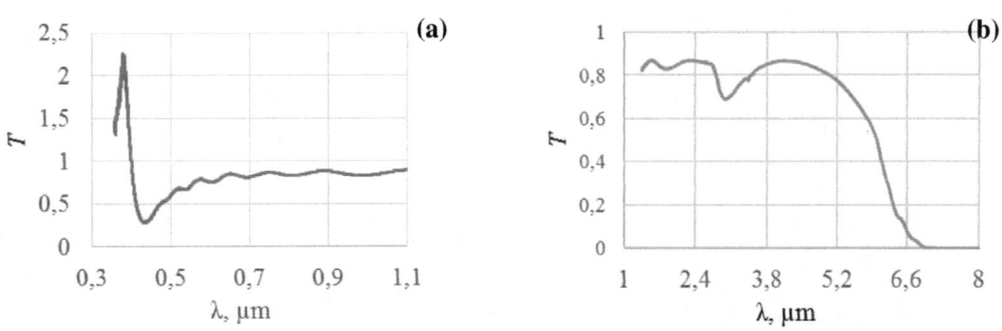

Figure 2. PZT absorption spectra in wavelength range from 0,3 to 1,1 µm (**a**) and from 1 to 8 µm (**b**).

IOP Conf. Series: Materials Science and Engineering **387** (2018) 012065 doi:10.1088/1757-899X/387/1/012065

Table 2 shows three types of structures formed.

Table 2. Types of structural organized ferroelectric films on the basis of the PZT system.

Type of structure	Ferroelectric-semiconductor composite	Structure with periodically varying index of refraction	Ferroelectric with variable residual polarization
Phase composition	Heterophase film consisting of ferroelectric PZT grains and lead oxide inclusions	PZT film, locally crystallized in perovskite structure in pyrochlore matrix	Self-Polarized PZT film (perovskite phase) with locally depolarized regions
The layers obtained by the high-frequency magnetron sputtering method (at 130 °C)	Multilayer system of of PZT and PbO_x layers	PZT film of stoichiometric composition	PZT film with a nanosized PbO_x sublayer
Heat treatment in furnace	–	450 °C, 60 min	600 °C, 15 min
Laser processing parameters: power input; pulse width	0,3…0,35 J/cm^2; 500 ms	0,3...0,35 J/cm^2; 100…500 ms	0,18…0,25 J/cm^2; 100…500 ms

Composite ferroelectric-semiconductor films, which are of interest due to their unique photoelectric properties, were formed by crystallization of PZT-PbO_x multilayer structure. The formation mechanism of such heterophase system is associated with the moving of the excess lead oxide on the periphery of growing PZT crystallites from the centers of nucleation at the interfaces of the structure at temperatures of 580...620 °C [18, 19]. By varying the thicknesses and the number of layers of PZT and PbO_x, as well as the laser processing modes, it is possible to control the ratio and size of the phase inclusions in the system.

Creation of structures with periodically changing index of refraction on the basis of the PZT is possible due to the significant optical contrast of the low-temperature and high-temperature PZT phases. PZT film crystallizes in the perchloric phase at temperatures around 450 °C. When the temperature is raised to 570 °C and higher, a growth of crystallites with a perovskite structure occurs in the pyrochlore matrix.

Obtaining of a ferroelectric film with periodically varying residual polarization is achieved by the formation of a self-polarized film and subsequent local destruction of the self-polarized state by the action of laser radiation. The preparation of self-polarized PZT films in the present work is carried out by forming a two-layer system from a PZT with a nanosized (10 nm) PbO_x sublayer and its further crystallization in air at a temperature of 600 °C.

The study of the samples phase composition was carried out by X-ray diffraction on Shimadzu XRD 6000 diffractometer. The phase identification was performed using radiometric file cabinets Powder Diffraction File (PDF-2). The study of layers microstructure was carried out by the atomic force microscope (AFM) "Integra Thermo" and by raster electron microscopy (REM) with the help of the technological complex Helios Nanolab D449 FEI Company. The self-polarization and depolarization states of PZT films were determined by the method of volt-faradic characteristics. For this case, the upper platinum electrodes with an area of 0.18 mm^2 were precipitated by ion-plasma sputtering method through a free mask.

It should be noted that the main problem in the preparation of structures based on the PZT is the high volatility of lead oxide [20], which leads to difficulties in controlling the composition of the films and the formation of pores during laser processing. This requires a serious optimization of technological regimes.

4. Conclusions

The general physico-technological aspects of obtaining structural organized ferroelectric films on the basis of PZT oxide system were studied with use of the method of HF magnetron sputtering with the subsequent laser processing. The approach allows to obtain three types of structural organized films by local laser annealing on the basis of one oxide system: ferroelectric-semiconductor composite, structure with periodically changing index of refraction and ferroelectric material with periodically changing residual polarization. The main problem in the preparation of such structures is the high volatility of lead oxide, which leads to difficulties in controlling the composition of the films and the formation of pores during laser processing.

Acknowledgment

The key part of the work associated with studying the structural and phase features of ferroelectric films and laser processing was supported by the Ministry of Education and Science of the Russian Federation within the framework of "Research and development in priority areas of advancement of the Russian scientific and technological complex for 2014–2020", agreement No. 14.608.21.0002 of 27.10.2015 (unique number of agreement RFMEFI60815X0002) (A.B. Kozyrev). Study of formation conditions of self-polarized PZT films and ferroelectric-semiconductor structures was supported by the Russian Foundation of Basic Research (project no. 17-52-04127 Bel_mol_a) (N.V. Mukhin).

References

[1] He S, Yu J and Hu M 2016 *Current Nanoscience* **12** 676–84
[2] Klimov V, Baudon J and Ducloy M 2011 *Europhys. Lett.* **94** 20006
[3] Garcia-Meca C, Hurtado J, Marti J, Martinez A, Dickson W and Zayats A V 2011 *Phys. Rev. Lett.* **106** 067402
[4] Tseng M L, Wu P C, Sun S, Chang C M, Chen W T, Chu C H, Chen P-L, Zhou L, Huang D-W, Yen T-J and Tsai D P 2012 *Laser Photonics Rev.* **5** 702–7
[5] Banks D P, Grivas C, Mills J D, Eason R W and Zergioti I 2006 *Appl. Phys. Lett.* **89** 193107
[6] Chang-Jian S-K, Ho J-R, Cheng J-W J and Sung C-K 2006 *Nanotechnology* **17** 1184–7
[7] Tseng M L, Chen B H, Chu C H, Chang C M, Lin W C, Chu N-N, Mansuripur M, Liu A Q and Tsai D P 2011 *Opt. Express* **19** 16975–84
[8] Gansel J K, Thiel M, Rill M S, Decker M, Bade K, Saile V, Freymann G, Linden S and Wegener M 2009 *Science* **325** 1513–15
[9] Chen Z, Rahmani M, Gong Y D, Chong C T and Hong M 2012 *Adv. Mater.* **24** OP143–7
[10] Nakata Y, Matsuba Y, Miyanaga N and Murakawa K 2015 *Photonics and Nanostructures – Fundamentals and Applications* **17** 10–4
[11] Kazusa J, Koga K, Umeyama N, Imura F, Noda D, Khumpuang S and Hara S 2016 *IEEJ Transactions on Sensors and Micromachines* **12** 493–8
[12] Khmelenin D N, Zhigalina O M, Vorotilov K A and Lebo I G 2012 *Physics of the Solid State* **5** 939–41
[13] Firsova N Yu, Mishina E D, Sigov A S, Senkevich S V, Pronin I P, Kholkin A, Bdikin I and Yuzyuk Yu I 2012 *Ferroelectrics* **1** 164–9
[14] Pronin I P, Kaptelov E Yu, Senkevich S V, Klimov V A, Zaitseva N V, Shaplygina T A, Pronin V P, Kukushkin S A 2010 *Physics of the Solid State* **1** 124–8
[15] Firsova N Yu, Elshin A S, Marchenkova M A, Bolotov A K, Ivanov M S, Pronin I P, Senkevich S V, Kiselev D A and Mishina E 2014 *Nano- and microsystem technology* **7** 43–6
[16] Afanasjev V P, ChigirevD A, Mukhin N V and Petrov A A 2016 *Ferroelectrics* **1** 170–6
[17] Afanas'ev V P, Vorotilov K A and Mukhin N V 2016 *Glass Phys. Chem.* **3** 295–301
[18] Mukhin N V 2016 *Glass Phys. Chem.* **1** 64–9
[19] Agafonova D S, Kozyrev A B, Mukhin N V, Red'ka D N, Elanskaya K G, Rudenko M V, Oseev A, Hirsch S. 2018 *Glass Phys. Chem.* **1** 15–20
[20] Mukhin N V 2014 *Glass Phys. Chem.* **2** 238–42

IOP Conf. Series: Materials Science and Engineering **387** (2018) 012066 doi:10.1088/1757-899X/387/1/012066

Formation of coatings from the flow of metal plasma of a vacuum-arc discharge

A A Rikov, M I Yurchenkov, A A Lisenkov and D K Kostrin

Department of electronic instruments and devices, Saint Petersburg Electrotechnical University "LETI", 197376, Saint Petersburg, Russia

E-mail: rikov247@gmail.com

Abstract. Modification of the surface properties of a material or application of a protective coating are the most effective ways to improve the service life and reliability of parts. For the production of coatings vacuum-arc discharge with integrally cold cathode is widely used. The advantages of this method are high degree of purity of the coating, the possibility of changing the structure of the coating and obtain complex multicomponent coatings. In this paper, we consider the production of anti-emission coating based on titanium; the results of metallographic studies of this coating are presented.

1. Introduction

In many cases, the performance properties of parts are determined by the state of their surface layer, as it is exposed to external active mechanical, thermal, chemical or other types of effects. Therefore, modification of the surface properties of a material or application of a protective coating by ion-plasma method are one of the most effective ways to improve the service life and reliability of parts [1, 2].

Method of surface treatment by forming the coating allows precision dimensional surface treatment, significantly increase wear resistance, mechanical strength and thermal protection of parts, improve anti-corrosion properties, change the hardness and friction coefficients.

Of great scientific and practical interest is the development of technological processes based on the formation of film structures from metal plasma of a vacuum-arc discharge, allowing the most effective control of the state of the surface layer of a solid body [3–8]. Operational properties of parts are determined by all stages of the process: evaporation of the material, its transportation and deposition on the treated surface [9, 10]. Comprehensive study of the conditions of coatings formation and study of the surface layer taking into account operating conditions is a progressive trend in improving the quality of structural materials, in solving the problem of achieving high structural strength and compatibility with the environment. Analysis of the behavior of the surface layer of a solid body in the process of ion-plasma treatment, and in the process of operation, with the goal to study the physical nature of the relationship of structure, composition and construction with the physical, mechanical and chemical properties of materials and the development of scientific foundations for the management of these properties is an urgent scientific problem.

2. Results and discussion

One of the important applications of surface layer modification is the deposition of anti-emission coatings on grid electrodes of electrovacuum devices to protect them from thermoelectronic and secondary electron emission [11, 12]. The vast majority of modern radio transmitting devices operating at frequencies up to 30 MHz, with an output power level of more than 40 kW, as the amplifying devices of the end stages use powerful generator lamps, which are structurally complex products. The main directions of development of this type of devices are to increase the specific power while increasing their durability. The maximizing efficacy parameters required the solution of problems of thermal resistance of the shielding grids of the tetrode, withstanding in the conditions of pulsed gain significant electronic loads. This problem becomes very relevant when creating powerful generator lamps using overstressed modes and high specific power dissipation on the grids up to 20 W/cm^2 or more, with a corresponding increase in their operating temperature. Therefore, the anti-emission properties of the grid coatings will determine the quality and reliability of the devices.

The most effective is the use in the manufacture of grid electrodes of materials and coatings with high values of the integral coefficient of radiation and the work function, and in some cases representing a complex system of multilayer structures consisting of alloys or chemical compounds. To reduce the thermoelectronic emission, the grid electrode is coated with metal of group VIII of the periodic system, in particular with platinum. To reduce the diffusion of platinum into the core of the grid and increase the radiating capacity between the base metal and platinum, an intermediate layer consisting of Zr–Pt or Ti–Pt compounds is applied. The most effective compound is Pt_3Zr [13].

Titanium is distinguished from the group of metals used in the production of electronic vacuum devices. It allows creating protective coatings serving to prevent evaporation and also blackening and anti-emission coatings having a sufficiently high adhesion with a substrate. The properties of titanium include high mechanical strength, lightness, high melting point, low coefficient of secondary electron emission, good getter properties and stable characteristics to oxidation and interaction with alkali metal vapors. The mechanism of operation of titanium as an anti-emission coating consists in the desorption of atoms of the active cathode substance from the mesh surface under sufficient thermal load. The estimated temperature of the thorium desorption from the surface of titanium is 900 K. The work function from the titanium coating, under the conditions of spraying the active substance from the cathode, at a temperature of 1000 K is 3.0 eV, at 1200 K – 3.16 eV and at 1400 K – 3.35 eV. The emission factor for these temperatures is 0.53, 0.56 and 0.58, respectively.

In addition to these properties of titanium, it should be noted that in the working modes of operation, titanium coatings present a gettering surface, actively contributing to the maintenance of vacuum conditions in the device. Titanium absorbs active gases and keeps them in a wide temperature range starting from 473 K. Sorption capacity of titanium coatings depends on the thickness, roughness and porosity of the coating, and, to a large extent, is determined by the method of its production.

During the combustion of a vacuum-arc discharge with an integrally cold cathode, droplets exist in the erosion products of the cathode material, which lead to a rough coating with a developed surface (figure 1). This condition has a favorable effect on the gas-absorbing properties of the titanium coating, which most effectively absorbs the residual gases of the active group: hydrogen, oxygen, nitrogen prevailing in vacuum systems.

Figure 1. The appearance of the mesh fabric with anti-emission titanium coating (**a**); morphology of a gas-absorbing titanium coating (**b**) (average particle size ~2.0 µm, the coating is loose (porous) and agglomerated).

IOP Conf. Series: Materials Science and Engineering **387** (2018) 012066 doi:10.1088/1757-899X/387/1/012066

The generation of droplets is due to the formation of erosion craters on the cathode surface. The localization of the electric field in the area of microscopic inhomogeneities on the cathode surface leads to the flow of an autoemission current, the density of which can exceed the critical value. In this case, an explosion of a microscopic protrusion occurs, leading to the melting of the surrounding surface and the formation of a plasma cloud over the cathode spot.

The quantitative composition of droplets is associated with changes in the operation of the integral cathode temperature. With the growth of the cathode operating temperature, the percentage of neutral steam and droplets in the flow increases. With the decrease of the cathode temperature, the growth rate of the coating is determined by the positively charged component of the plasma flow. It should be noted that the percentage of the droplet fraction in the generated plasma flow depends on the melting temperature of the cathode material. So for refractory metals such as molybdenum and tungsten, these values are at the level of units of percent, while for copper this value is about 50 %. The results obtained in the experiments suggest that as the vacuum-arc device works, the angular distribution of the droplets in space and the number of generated droplets with dimensions up to 1 μm practically do not change, while the quantitative composition of larger particles undergoes significant changes.

The pulse operation mode of the evaporator significantly reduces the portion of the droplet fraction in the generated plasma flow, which is due to the thermal and mechanical inertia of their formation. The flow rate of the plasma-forming material is easily controlled by changing the pulse repetition frequency. During the arc burning for no more than 5 s, drops not exceeding 1 μm were mainly recorded. At the same time, the total number of them was 4–5 times less than in the ten-second interval of the evaporator functioning. The decrease in the arc burning delay time led to an increase in the number of droplets having larger sizes, since with a decrease in the pause, the initial temperature of the cathode for the subsequent combustion stage increases.

The intensity of generation of droplets increases with increasing discharge current. However, only by reducing it, it is impossible to decrease the content of droplets to an arbitrarily small value, since stationary arc combustion is possible only with a discharge current of not less than some critical value for this material, which significantly depends on the thermal properties of the cathode material. The maximum number of droplets evaporates from the cathode working surface at an angle of 20...30° to its plane, and for the formed flow a pattern of radial symmetry is observed.

Metallographic studies of coatings in the process of operation of electrovacuum devices showed that for titanium at a cathode temperature below 1000 K, the obtained coatings had a complex structure. The growth of the negative bias (in absolute value) on the treated substrate leads to the creation of a coating with strongly pronounced components, while at a cathode temperature of more than 1000 K the orientation effect was weak and the coating had a loose rough character. As it is known, the reduction of the secondary emission coefficient is achieved by increasing the roughness of the coating, which is explained by the difficult output of the secondary electron in a layer with a labyrinth structure than from a smooth layer. Figure 2 shows the section of a grid coating with thickness up to 10 μm and more.

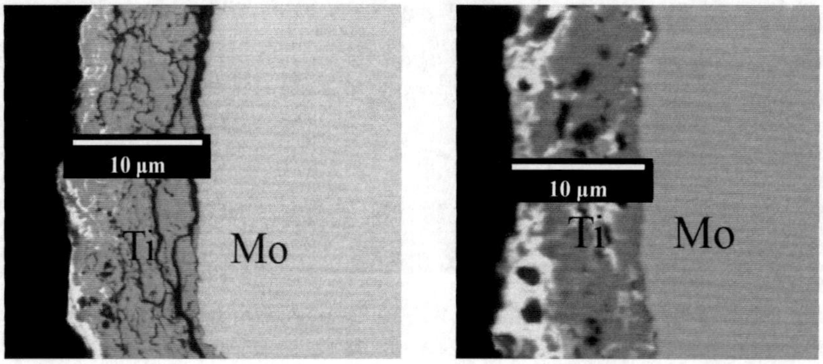

Figure 2. Metallographic section of an anti-emission titanium coating.

During operation at high temperature of the grid mesh, there is a counter diffusion of the mesh material and titanium coating (figure 3), leading to deterioration of the operational properties of the grid electrode. However, a characteristic feature of the diffusion of molybdenum and titanium is that they do not enter into a chemical reaction and do not form intermetallic compounds, while solid solutions are observed.

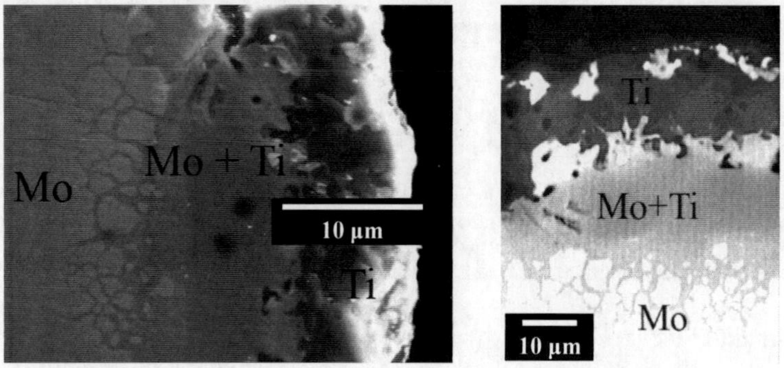

Figure 3. Counter diffusion of titanium and molybdenum.

3. Conclusion

With the growth of the cathode operating temperature, the percentage of neutral steam and droplets in the flow increases. With the decrease of the cathode temperature, the growth rate of the coating is determined by the positively charged component of the plasma flow. The results obtained in the experiments suggest that as the vacuum-arc device works, the angular distribution of the droplets in space and the number of generated droplets with dimensions up to 1 µm practically do not change, while the quantitative composition of larger particles undergoes significant changes. The pulse operation mode of the evaporator significantly reduces the fraction of the droplet fraction in the generated plasma flow, which is due to the thermal and mechanical inertia of their formation.

References

[1] Kostrin D K and Lisenkov A A 2016 *Materials Science Forum* **843** 278–83
[2] Kostrin D K, Lisenkov A A and Potrakhov N N 2017 *Biomedical Engineering* **4** 262–6
[3] Kostrin D K, Lisenkov A A, Ramazanov A N and Semenova A N 2016 *Proceedings of the higher educational institutions. Physics* **9/2** 240–3
[4] Bystrov Yu A, Kostrin D K, Lisenkov A A 2014 *Vacuum technique and technology* **1** 164–6
[5] Bystrov Yu A, Vetrov N Z, Lisenkov A A and Kostrin D K 2014 *Vakuum in Forschung und Praxis* **5** 19–23
[6] Bystrov Yu A, Kostrin D K, Lisenkov A A and Vetrov N Z 2015 *Vakuum in Forschung und Praxis* **2** 22–5
[7] Kostrin D K and Lisenkov A A 2017 *Vakuum in Forschung und Praxis* **3** 35–9
[8] Kostrin D K, Pikus M I, Smirnov E A and Lisenkov A A 2017 *Journal of Physics: Conference Series* **789** 012030
[9] Vinogradov M L, Kostrin D K, Smelova V V, Trifonov S A and Lisenkov A A 2016 *Proceedings of the 2016 IEEE North West Russia Section Young Researchers in Electrical and Electronic Engineering Conference* 729–30
[10] Kostrin D K, Vinogradov M L, Pikus M I, Mosolova Y M and Lisenkov A A 2017 *Proceedings of the 2017 IEEE Russia Section Young Researchers in Electrical and Electronic Engineering Conference* 1169–72
[11] Vetrov N Z, Kostrin D K, Lisenkov A A and Popova M S 2015 *Journal of Physics: Conference Series* **652** 012032
[12] Kostrin D K and Lisenkov A A 2016 *Materials Science Forum* **870** 371–6
[13] Lisenkov A A, Vetrov N Z and Kostrin D K 2017 *Technical Physics Letters* **4** 390–2

Pump-down model of vacuum chambers with adsorbing walls

L N Rozanov

Peter the Great St. Petersburg Polytechnic University
Politechnicheskaya Str. 29, St. Petersburg, 195251, Russia

E-mail: roz@vacuum.ru

Abstract. The mathematical model is considered for the calculation of the pump-down characteristics of vacuum chambers with adsorbing walls. This model uses the dependence of the absorption heat on the logarithm of the coverage degree, which is obtained from the Freundlich adsorption isotherm. The dependences of pressure, gas release, adsorption heat and the amount of adsorbed gas on the pump-down time have been calculated. It is shown that there is a steady pump-down mode for water at 298 K, which does not depend on the chamber volume and the initial pressure, but is determined only by the area of the internal surface of the chamber.

1. Introduction

Vacuum chambers can contain a lot of gas adsorbed on their internal surfaces. The adsorbed gas is energetically connected to the surface and pumped down more slowly than the gas which is in the chamber volume. At high vacuum the pump-down time of the vacuum chamber is virtually completely determined by the time within which the adsorbed gases are pumped down [1–4]. It takes a lot of time to heat the vacuum chamber and remove the adsorbed gas, which reduces the productivity of vacuum plants. The adsorption of gases on the walls of vacuum chambers is largely dependent on the adsorption of water vapor when they are decompressed.

In literature there are two types of mathematical models which determine the amount of adsorbed gas on the internal surfaces of vacuum systems: a diffusion one [1] and an adsorption one [2]. The diffusion model implies that there are surface pores where gas is adsorbed, while the coefficient of gas diffusion in pores limits the rate at which it gets removed during pumping-down. The adsorption model considers a smooth surface, with the rate of gas removal being determined by condensation and evaporation processes.

This paper is dedicated to the development of the adsorption model. The paper [2] considers the capabilities for applying various adsorption equations in the adsorption pump-down model of vacuum systems and makes a conclusion about the viability of utilizing the Temkin isotherm for these purposes. The weak point of the Temkin equation is the assumption of the linear dependence of the adsorption heat on the coverage degree, which is not executed for water vapor adsorption on the walls of vacuum chambers. The experimental data on water adsorption on a stainless steel surface is well described by the Freundlich equation [3].

In our model we used the logarithmic dependence of adsorption heat on the coverage degree. The model makes it possible to calculate the dependences of pressure, the amount of absorbed gas, gas release, the adsorption heat on the pump-down time, and to determine the coefficients of the Freundlich adsorption isotherm by an inverse method from the pump-down curve.

2. Pump-down adsorption model of vacuum chambers

The mathematical pump-down model of vacuum chambers with adsorbing walls can be made if the processes of evaporation q_e, condensation q_c and pump-down q_p are considered simultaneously (figure 1).

The differential equation of mass transfer on the internal surfaces of vacuum chambers at the pump-down productivity q_p=0 can be represented as:

$$\frac{da}{dt}+\frac{a}{\tau_a(Q_a)}-Kp=0, \qquad (1)$$

where a is the specific amount of adsorbed gas, m^3 Pa/m^2; t is the time, s; p is the pressure, Pa;

$K=\frac{s\cdot N_A}{2.65\cdot 10^{20}\sqrt{2\pi MRT}}=\frac{s\cdot 1.08\cdot 10^4}{\sqrt{MT}}$, s is the probability of condensation, $\tau_a(Q_a)=\tau_0\exp(\frac{Q_a}{RT})$,

$\tau_0 = 1.6\cdot 10^{-13}\cdot 298/T$, $R = 8.31\cdot 10^3$J/kmol, $N_A = 6.02\cdot 10^{26}$ kmol^{-1}.

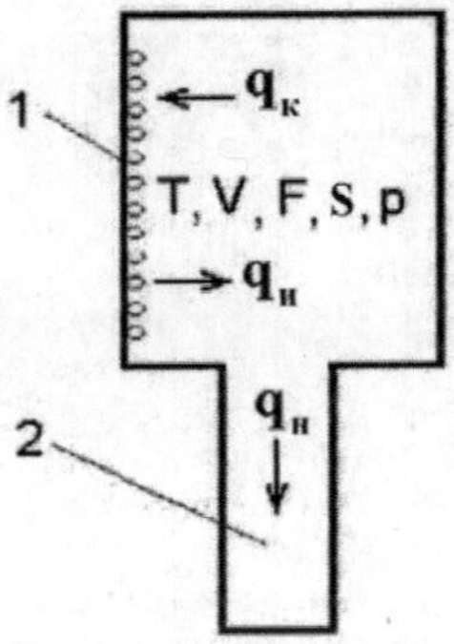

Figure 1. Conceptual pump-down model of vacuum chambers with adsorbing walls: 1 – vacuum chamber, 2 – vacuum pump.

We use the linear dependence of absorption heat on the logarithm of the amount of adsorbed gas in the form [4]:

$$Q_a = Q_1 - K_Q \lg(\theta), \qquad (2)$$

where

$$Q_1 = RT\ln\left(\frac{d^{\frac{1}{m}}}{K\tau_0 a_m^{\frac{1-m}{m}}}\right) \qquad K_Q=\frac{RT(1-m)}{0{,}43\cdot m}. \qquad (3)$$

Here d and m are the coefficients of the Freundlich adsorption isotherm:

$$a = d\cdot p^m, \qquad (4)$$

where a_m is the amount of gas in the monolayer per unit of the real surface, m^3Pa/m^2.

The equation (2) can be represented as:

$$Q_a = E - K_Q \lg(\theta), \qquad (5)$$

where θ is the coverage degree, $\theta = a/a_m$., E is the condensation heat. For water on non-processed stainless steel at 298K $E = 4.4\cdot 10^7$ J/kmol, a_m=0.116 m^3 Pa/m^2, d= 0.0185, m = 0.34 [6]. At τ_0 = 1,6.10^{-13}c, s= 0.4 and K =59 it follows from (3) that Q_1 = 3,34.10^{7}J/kmol, K_Q = 1,12.10^{7}J/kmol.

The differential equation of pumping down vacuum chambers with adsorbing walls at $q_н = S\cdot p$:

$$\frac{dp}{dt}+\frac{S\cdot p}{V}-\frac{da}{dt}\cdot\frac{F}{V}=0, \qquad (6)$$

where V, F and S are the volume, the area of internal surface and effective pump-down rate of the vacuum chamber:

$$S = K_u \cdot S_m \left(1 - \frac{p_m + p_в / K_c}{K_u p}\right), \tag{7}$$

where S_m is the maximum pump-down rate, p_m is the ultimate pressure of the pump, K_u is the coefficient of the pump utilization (ratio of the effective and real pump-down rate of the pump), K_c –is the coefficient of the pump compression, $p_в$ is the outlet pressure.

Specific gas release q_a is determined as:

$$q_a = \frac{da}{dt}. \tag{8}$$

The mathematical pump-down model of vacuum chambers with absorbing walls can be represented by the following set of equations:

$$\begin{cases} \dfrac{da}{dt} + \dfrac{a}{\tau_a} - Kp = 0, \\ \dfrac{dp}{dt} + \dfrac{p}{V/S} + \dfrac{da}{dt} \cdot \dfrac{F}{V} = 0, \\ \tau_a = 1{,}6.10^{-13} \cdot \dfrac{298}{T} \exp\left(\dfrac{Q_a}{RT}\right), \\ Q_a = E - K_Q \cdot \lg\left(\dfrac{a}{a_m}\right), \\ q_a = \dfrac{da}{dt}, \\ S = K_u \cdot S_m \left(1 - \dfrac{p_m + p_в / K_c}{K_u p}\right). \end{cases} \tag{9}$$

The specific features of this model include utilizing: 1) the logarithmic dependence (5), which allows determining the value of absorption heat at any moment of the pump-down time; 2) the monomolecular coverage of the real rough surface; 3) the approximating function (7) for rapid pump-down of the chamber.

3. Calculations

The model (9) allows us to obtain a numerical solution, which determines the following, depending on the pump-down time: pressure, the amount of absorbed gas, the rate of adsorption gas release, adsorption heat and effective pump-down rate of the vacuum chamber. The parameters of the pump-down curve of water vapors (figures 2, 3) are calculated at T=298 K, $v=V/S$, $f=F/S$, $p_m=10^{-8}$ Pa, $p_в$ = = 10^0 Pa, $K_c=10^{11}$, K_u=1, a_m=0.116 m^{3}Pa/m^2, E =4.4· 10^7 J/kmol, K_Q = 1,12·10^7J/kmol.

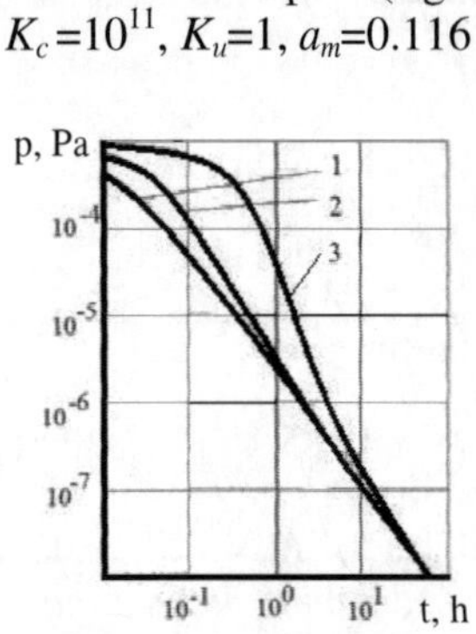

Figure 2. The pump-down curves of water vapors at f=25s/m, p_1= 10^{-3} Pa for various relative volumes of the chamber v: 1 – 1 s, 2 – 100 s, 3 – 1000 s.

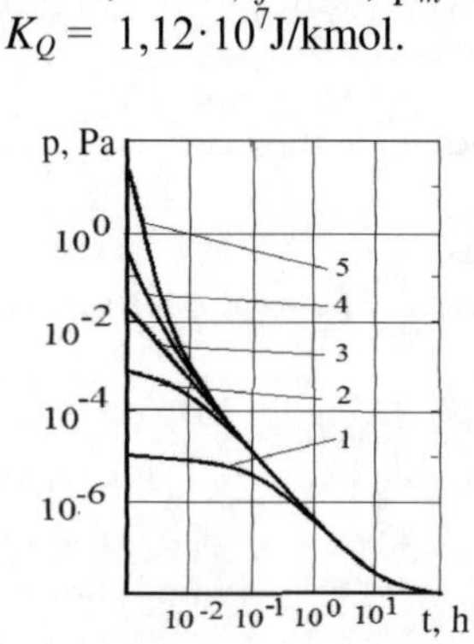

Figure 3. The pump-down curves of water vapors at f=25s/m, v=1s for various initial pressures p_1: 1 – 10^{-5} Pa; 2 – 10^{-3} Pa; 3 – 10^{-2} Pa, 4 – 1 Pa; 5 – 10 Pa.

The effect of the chamber volume on the pump-down curve (figure 2) is observed only in the initial phase. For *V/F*< 100 the duration of the initial phase does not exceed one hour.

The effect of the initial pressure on the pump-down curve in shown in figure 3. One hour after the pump-down starts, the curve is no longer dependent on the initial pressure.

4. Regular mode

When water vapors are pumped out at 298 K, the differential equation (1) in the model (9) can be substituted without noticeable loss of accuracy for the algebraic equation of the Freundlich adsorption isotherm (4). It means that the adsorption balance on the surface is achieved faster that the change of pressure occurs in the volume during pumping-down. The simplified model (9) after this substitution has an analytical solution. The pump-down time can be found from (4) and (6) after integrating dp/dt from t_1 to t and from p_1-p_m to p-p_m.

$$\Delta t = 2.3 \cdot v \cdot \lg \frac{p_1 - p_m}{p - p_m} + \frac{d \cdot m \cdot f}{1-m}\left[(p - p_m)^{m-1} - (p_1 - p_m)^{m-1}\right], \tag{10}$$

where d and m are the coefficients of the Freundlich adsorption isotherm for water at 298 K, p_m is the ultimate pressure of the pump, Pa; p_1 is the initial pressure, Pa; $v=V/S$,s; $f = F/S$, s·m^{-1}; V is the chamber volume, m^3; F is the area of the chamber internal surface, m^2; S is the effective rate of chamber pump-down in m^3/s by approximation (7). The first summand in (10) determines the pump-down time of water vapors in the chamber volume, while the second one – on the internal surface.

The regular pump-down mode is understood as a mode when the pump-down process does not depend on the relative chamber volume, ultimate pressure of the pump and initial pump-down pressure.

The effect of the chamber volume on the pump-down curve in shown in figure 2. Meeting the condition:

$$2.3 v \lg \frac{p_1 - p_m}{p - p_m} << \frac{d \cdot m \cdot f}{1-m}\left[(p - p_m)^{m-1} - (p_1 - p_m)^{m-1}\right] \tag{11}$$

the pump-down curve stops depending on the chamber volume. In standard vacuum systems the effect of the working chamber volume on the pump-down curve usually becomes insignificant one hour after the pump-down process starts.

Under the condition

$$(p_1 - p_m)/(p-p_m) > 10^2 \tag{12}$$

the pump-down curve does not depend on the initial pressure of the pump. At that Δt=t.

Under the condition

$$p>>p_m \tag{13}$$

the pump-down curve does not depend on the ultimate pressure of the pump.

Under meeting the conditions (11-13) the regular mode settles and the equation (10) takes the form:

$$p(t) = \left[\frac{d\, m\, f}{3600\,(1-m)\,t}\right]^{\frac{1}{1-m}}, \tag{14}$$

where t is the pump-down time, h;

The pressure in the regular mode depends on the relative area of the internal surface of the chamber (Fig.4), and has the form of a straight line.

The dependence of the rate of water vapor adsorption gas release on the pump-down time according to the simplified model (9) is determined by the following expression:

$$q_a(t) = \frac{da}{dt} = \frac{d\, m\, p^m}{v + f\, d\, m\, p^{m-1}} \tag{15}$$

In the regular mode, when

$$v << f\, d\, m\, p^{m-1} \tag{16}$$

the relative volume v in the denominator (15) can be neglected. Then the rate of adsorption gas release becomes approximately equal to the productivity of the pump, while the specific rate of gas release

$$q_a(t) = \frac{p(t)}{f} \quad . \tag{18}$$

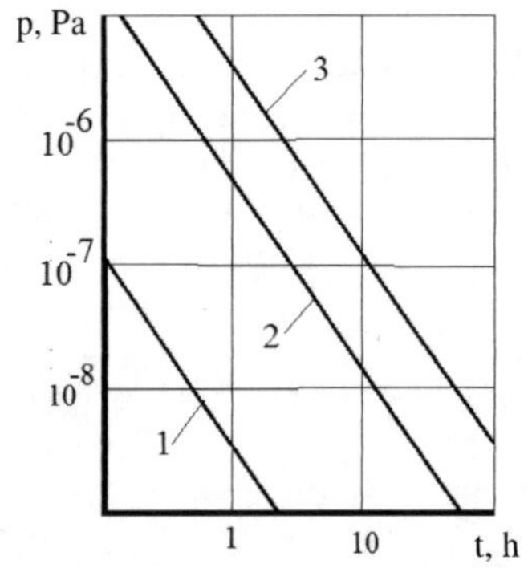

Figure 4. The pump-down curves in the regular mode (14) at d=0.0185: m =0.34: 1-f=1, 2 –f=25, 3 -f= 100.

The experimental data on the pressure in the regular pump-down mode can be be represented as:

$$p(t) = A_1 t^{-B}, \tag{19}$$

here t is the pump-down time, h. From the comparison of (14) and (19) it follows

$$A_1 = \left[\frac{d\,m\,f}{3600\,(1-m)}\right]^{\frac{1}{1-m}}, \qquad B = \frac{1}{1-m}. \tag{20}$$

The coefficients of the Freundlich adsorption isotherm can be determined from the pump-down curves (19), using (20):

$$d = 3600\frac{(1-m)\,A_1^{1-m}}{m\,f}, \qquad m = \frac{B-1}{B}. \tag{21}$$

In the Freundlich adsorption isotherm m>0. In order to apply this model based on the Freundlich equation, as it follows from (21) condition B>1 has to be met.

5. Conclusion

The specifics of this model imply utilizing: the logarithmic dependence of adsorption heat on the degree of surface coverage; the mono-molecular coating of the real rough surface; the expanded approximating dependence of the effective rate of pumping the pressure down from the chamber. The model for water at 298 K has been used to calculate the dependences of pressure, gas release, adsorption heat and the amount of adsorbed gas on the pump-down time for a stainless steel vacuum chamber. It is shown that there is a regular pump-down mode, which does not depend on the chamber volume and the initial pressure, but is determined only by the area of the internal surface of the chamber. The coefficients of the Freundlich adsorption isotherm have been determined by inverse calculation from the pump-down curve.

References

[1] Li M and Dylla H F 1994 *JVST* **A12** 1772
[2] Redhead P A 1995 *JVST* **A13** 467
[3] Rozanov L N and Akimov Y D 1985 *Journal of applied chemistry of the USSR* **58** 1381–4
[4] Rozanov L N 2018 *Vacuum* **147** 194–8

IOP Conf. Series: Materials Science and Engineering **387** (2018) 012068 doi:10.1088/1757-899X/387/1/012068

Features of the formation of non-vertical profiles on the surface of 4H-SiC by the reactive-ion etching

A V Serkov, V A Golubkov, V A Ilyin and D A Savchenko

St. Petersburg State Electrotechnical University «LETI», 5 Professora Popova str., St. Petersburg, Russia

E-mail: av.serkov@yandex.ru

Abstract. The features of the formation of non-vertical profiles on 4H-SiC by reactive-ion etching (RIE) using various masking coatings are studied. The formation of 4H-SiC mesa structures was carried out using automated airlock reactive-ion etching and plasma etching system "Caroline PE 15" with the ICP-source of plasma in a gas mixture of SF6, O_2 and Ar. Using photoresist AZ4533 as a mask, mesa structures with a wall inclination angle of more than 130° were obtained at the etching rate of 4H-SiC was ~0.5 μm/min. The developed technology of dry etching can be further used in the preparation of avalanche photodiodes or power electronics devices.

1. Introduction

The problem of detecting weak signals (up to individual photons) has been very relevant for many years. This problem has been solved by means of photoelectron multipliers in the form of a diode system or tubes with a very high secondary electron emission coefficient, which, as a result of multiplying the number of electrons as they move toward the anode, significantly increase the initial signal. However, the cost and dimensions of such devices are quite large.

In recent years, solid-state analogs of these devices, so-called avalanche photo detectors (APDs) based on diodes, have started to develop rapidly. Small size, compatibility with standard CMOS technology, low voltage, lack of sensitivity to magnetic fields and low cost make this class of photo electronic devices promising [1]. The advantage of avalanche photo detectors based on the wide-band silicon carbide (SiC) is that they are able to detect ultraviolet (UV) radiation up to gamma radiation [2].

When creating an APD, one of the important conditions is to ensure stable operation when high voltages are applied. The main danger in the application of which is the breakdown of the structure, and in the case of avalanche photodiodes - premature breakdown. Such a breakdown is most likely at the periphery of the structure, where the surface and its quality play a big role. To ensure a fully volumetric breakdown, it is necessary to reduce the intensity of the electric field at the boundary of the mesa structure.

There are many ways to provide a volumetric breakdown, such as security rings, additional metallization to reduce the field strength inside the structure, stepwise limitation of MJTE structures (figure 1(a)), as well as the formation of mesa structures with a large inclination angle (more than 100°) (figure 1(b)) and others.

This article is devoted to the formation of mesa structures on 4H-SiC with a large angle of inclination of the walls by ion-etching using fluorine, oxygen, and argon-containing plasma. The main technological parameters during the RIE process: the current of magnetic coils I_k = 1.2 A; gas consumption: argon - 2.5 l / h, oxygen - 1.5 l / h, sulfur hexafluoride - 3.5 l / h; the total pressure in the chamber is P = 1.7 Pa. Formation of mesa structures was carried out automated airlock reactive-ion etching and plasma etching system "Caroline PE 15" ("Zelenograd Development Corporation", Moscow, Russian Federation) with the ICP-source of plasma and a lower RF-electrode, to which RF power (13.56 MHz) can be separately applied. This design of the equipment provides the ability to control the chemical and physical components of the etching process.

(a)

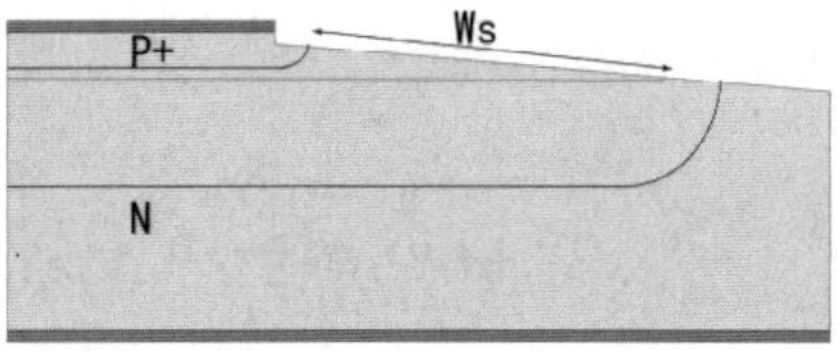

(b)

Figure 1. Methods for providing bulk breakdown – a step-by-step limitation of MJTE structures (**a**) and mesa structure with a large angle of inclination of the walls (**b**).

When forming mesa structures with a large angle of inclination of the walls by the RIE, it is necessary to choose a masking coating. The following materials were studied as a mask material: metallic (two-layer Ti-Ni composition), polymer (photo resist AZ4533).

At the preparatory stage, complex cleaning of the 4H-SiC surface was carried out by liquid treatment with a mixture of $H_2SO_4 + H_2O_2$ at a temperature of 60...70 °C.

2. Formation of mesa structures using the metallic mask

At the first stage, studies were carried out on formation non-vertical mesa structures in 4H-SiC using a masking coating based on metal. Two-layer metallization of Ti-Ni, with thicknesses of 20 and 200 nm, respectively, was deposited using modified equipment for vacuum magnetron sputtering "KONT" (Zelenograd, Russian Federation). Nickel was used as the main masking coating, and titanium – as under layer to improve adhesion of Ni. The photolithography was carried out *via* metallic mask and then the mesa structures were formed by RIE in the medium of gas mixture comprising SF_6, O_2, and Ar.

According to data published earlier [3, 4], an increase of power at the lower electrode (from 0 to 400 W) increases the anisotropy of the etching process. This effect is due to the increase in the energy of ions moving perpendicular to the working surface, which provides the almost vertical slope of the mesa structure (figure 3(a)).

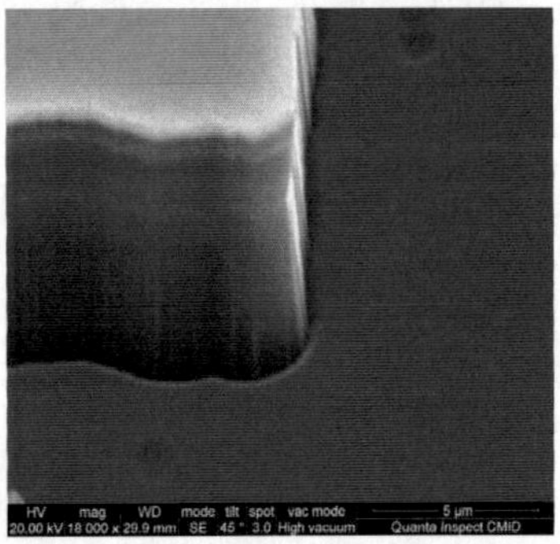
(a)

(b)

Figure 2. SEM image of mesa structures formed in 4H-SiC: vertical (**a**) and inclined (**b**).

On the other hand, by varying the power on the ICP-source of plasma (from 100 to 700 W), it is possible to control the chemical component of the etching process, forming thus the non-vertical mesa structures (figure 3(b)). Figure 2(b) shows a SEM image of the mesa structure with a non-vertical slope of the walls formed using a metal masking coating and variation of ICP-plasma source power.

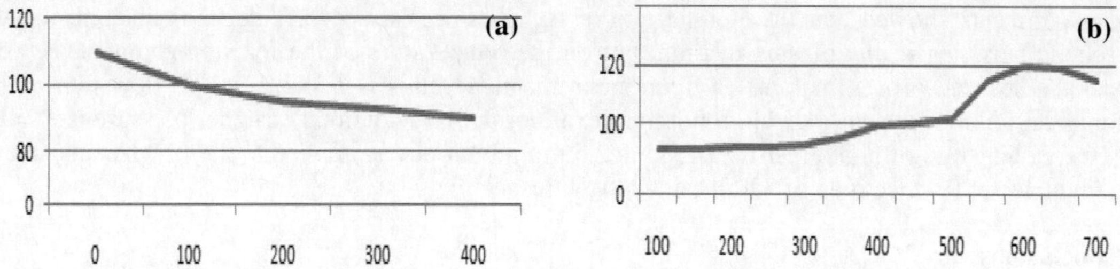

Figure 3. Relationship of mesa structure slope versus power at lower HF-electrode (**a**) and upper ICP-source of plasma (**b**).

As can be seen from figure 3(b), the mesa structure slope angle depends on the ICP-source of plasma power. When the ICP-source power increases, the amount of high energy and chemically active particles increases, which leads to increase of chemical component of the process and consequently the increase of horizontal etching speed.

3. Formation of mesa structures using photo resistive mask

Further studies were conducted towards the formation non-vertical mesa structures on 4H-SiC surface using a polymer masking coating based on photo resist AZ4533. A thick layer of viscous photo resist AZ4533 was deposited, followed by photolithography and annealing operations. The angle of the photo resist mask can be controlled by the temperature and time of annealing (table 1). The thickness of the photo resist is sensitive to the speed of rotation of the substrate during application. As a result, it is possible to obtain the edge of the applied photo resist in the form of a chamfer. The shape of the photo resist mask and the topological pattern formed after photolithography are transferred to 4H-SiC by RIE using the gas composition as above (figure 4).

Table 1. Dependence of the slope angle of photo resist AZ4533 versus temperature and time of annealing.

Slope angle (deg.)		Annealing temperature (°C)	
		135	150
Annealing time (min)	1	30	25
	20	35	25

The selection of a thick photo resist layer is due to the fact that this type of photo resist is not resistant enough (selectivity to 4H-SiC is 0.2) during RIE of 4H-SiC particularly if the plasma medium contains oxygen. On the other hand, without oxygen, the rate of 4H-SiC etching decreases several times, due to the lack of some volatiles (such as CO, CO_2 etc.) which are generated in the presence of oxygen. Therefore, the AZ4533 photo resist, as a masking coating, can only be used for the forming of shallow (up to 2 μm) mesa structures.

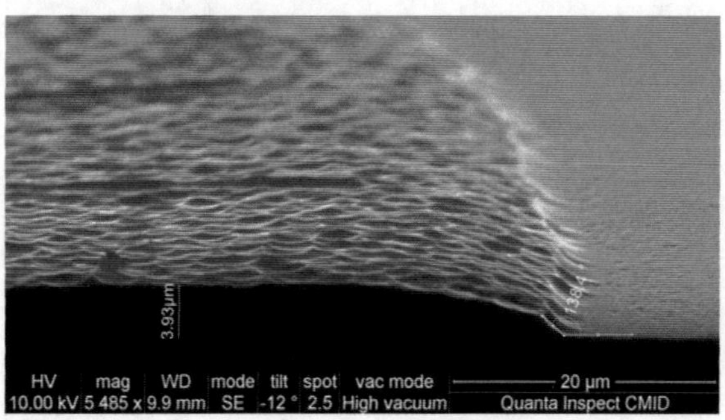

Figure 4. SEM image of non-vertical mesa structure formed on 4H-SiC surface using photoresist AZ4533 mask.

The experiments showed that the etching rate of the photo resist AZ4533 depends not only on the presence of oxygen in the plasma medium but on the temperature in the reaction chamber as well. Under the selected process parameters (current in magnetic coils is I_k = 1.2 A, gas flow rate is: for argon – 2.5 l/h, for oxygen – 1.5 l/h, for sulfur hexafluoride – 3.5 l/h, total chamber pressure is P = 1.7 Pa), the etching rate of photo resist AZ4533 is 2.5 μm/min at power applied to the ICP plasma source and to the lower RF electrode of 350 W and 200 W, respectively.

4. Conclusions

The possibility of formation of non-vertical mesa structures on the surface of 4H-SiC using various masking materials by the RIE in fluorine-, oxygen- and argon-containing plasma was studied using the technological installation "Caroline PE 15" with the ICP plasma source.

It has been shown that using a metal masking coating based on nickel with titanium sublayer, it is possible to obtain non-vertical mesa structures (with walls slope of more than 100°) by increasing the power applied to the ICP source of plasma (at 600 W the angle of inclination of the mesa structure is 120°). The effect is due to the increase of energy and number of chemically active particles and thus overall increase of the chemical component of the process, i.e. the etching speed is increased horizontally. If the power applied at the lower electrode is increased (from 0 to 400 W), the anisotropy of the etching process increases. This effect is due to the increase of energy of ions moving perpendicular to the working surface; as a result the slope angle of the mesa structure remains almost vertical.

Using photo resist AZ4533 as a masking coating, it is possible to form the non-vertical mesa structures of the slope angle more than 130°. After technological processing (the temperature and the annealing time are 150° and 20 min, respectively), the angle of the photo resist mask is 25°. The shape of the photo resist and the topological pattern formed after photolithography are transferred to 4H-SiC with RIE.

The technology developed can be applied in manufacturing of avalanche photodiodes and power electronics devices.

Referencies

[1] Klemin S A 2007 *ELECTRONICS: Science, Technology, Business* **8** 80–6
[2] Bai X, Liu H-D, Mcintosh D C and Campbell J C 2009 *IEEE J. Quantum Electron* **3** 300–3
[3] Serkov A V, Afanasiev A V, Ilyin V A, Romanov A A and Chigirev D A 2016 *Journal of Physics: Conference Series* **729** 012008
[4] Serkov A V, Afanisiev A V and Ilyin V A 2017 *Proc. of the 24th Scientific-technical conference with international participation «Vacuum technique and thechnology – 2017»* p 191

IOP Conf. Series: Materials Science and Engineering **387** (2018) 012069 doi:10.1088/1757-899X/387/1/012069

Current-voltage characteristics of magnetron with hot target

V I Shapovalov, E A Minzhulina and V V Smirnov

Saint-Petersburg State Electrotechnical University "LETI", St. Petersburg, Russia

E-mail: vishapovalov@mail.ru

Abstract. In this paper, the discharge characteristics of the magnetron with a hot titanium target in inert and reactive environments have been experimentally studied for the first time. It is established that the current-voltage characteristics of the magnetron operating in the argon environment correspond to the abnormal glow discharge mode with a low current. When the current increases, they reach a maximum and depend on the pressure in the same way as in the case of a cold target. When the magnetron operates in the $Ar+O_2$ environment, the current-voltage characteristics have three sections corresponding to the oxide, transition and metallic target operation modes. In the transition mode, an increase in current leads to a significant decrease in voltage (by about 80 V or by 17%). In the region of higher currents, the target operates in stationary metallic mode, and the discharge voltage decreases, when the current increases. In this case, the transition from the oxide mode starts at the current density, which is approximately four times smaller than for a magnetron with a cold target. The V-I characteristics of a magnetron with a hot target are changed in $Ar + N_2$ environment at any nitrogen flow similarly to the V-I characteristics in the $Ar+O_2$. However, in this case the transition from the nitride mode to the metallic mode is observed at a current density, which is about twice as less as for a magnetron with a cold target.

1. Introduction

Films of metals and their compounds are deposited by magnetron sputtering, laser ablation and other methods [1–3]. The use of DC sputtering systems with hot targets is considered promising. In these systems, the magnetron is designed so that the surface temperature of the target can be increased up to the melting point. The results of the research of this topic can be found in [4–7], which include the extended reference list. In addition, the operations of magnetrons with liquid targets are investigated [8].

However, such systems have not yet found industrial application, since physical processes during the sputtering of hot targets have not been completely investigated. In particular, the concept of the current-voltage characteristic is very ambiguous. Even for a magnetron with a cold target, there is no single point of view on this issue. Many specialists recognize the equivalence between a vacuum diode and a gas discharge in a diode system and believe that the so-called «low 3/2» is valid for a gas discharge [9]:

$$I \sim U^{3/2}. \tag{1}$$

It should be noted that the recording of the current-voltage characteristic of a gas-discharge device in the form (1) can not be accepted as correct, since the current in the gas discharge is an independent variable. Using here (1), the authors of this paper follow the notations of the cited publications.

Strictly speaking, the law (1) is indeed valid, but only for a collisionless layer [10]. In a more general form, the relationship between the current I and the voltage U is described as [11, 12]

$$I = kU^n. \qquad (2)$$

where k and n are constants that are adjustable parameters, when expression (2) is used to approximate the experimental results.

The influence of the target temperature on the magnetron V-I characteristics has not been discussed in the literature. In this paper, the distinctive features of the magnetron discharge with a hot titanium target in the inert and reactive environments have been experimentally studied for the first time.

2. Details of the experiment

The investigations were carried out in a high-vacuum system equipped with an oil-diffusion and mechanical pumps with a nominal operating speed of 0.3 m^3/s and 0.005 m^3/s, respectively. The volume of the vacuum chamber is 0.076 m^3, its residual pressure did not exceed $5 \cdot 10^{-2}$ mTorr. The experiments are performed on a balanced magnetron with a diameter of 130 mm, equipped with titanium targets:

- 6 mm thick with typical cooling by running water (cold target);
- 1 mm thick with cooling through a vacuum gap of 1 mm from a cooper water-cooled board and by fastening elements (hot target).

The processes of magnetron sputtering in three environements are studied:

- Ar at a partial pressure in the range 2, 3 and 7 mTorr;
- Ar + O_2 at a partial argon pressure of 2 mTorr and oxygen consumption of 4, 5 and 6 cm^3/min;
- Ar + N_2 at a partial argon pressure of 2 mTorr and a flow rate of nitrogen of 4, 5 and 6 cm^3/min.

Argon, oxygen and nitrogen of high purity were used to create working gas.

In all cases, the magnetron operated at a constant current from 0.5 to 4.0 A at a voltage of 200–500 V.

3. Results and discussion

Figure 1 shows the V-I characteristics of magnetrons operating in argon. For a magnetron with a cold target, the result is typical: as the pressure is increased, the current density on the target remains unchanged by decreasing the discharge voltage. A similar pressure effect is also observed for a magnetron with a hot target. But there is a significant difference between them: the V-I characteristics of a magnetron with a hot target in the region of ~ 90 mA/cm^2 have extremes, which can be caused by an increase in ion-electron emission from a target heated to a temperature corresponding to a discharge current density higher than 90 mA/cm^2.

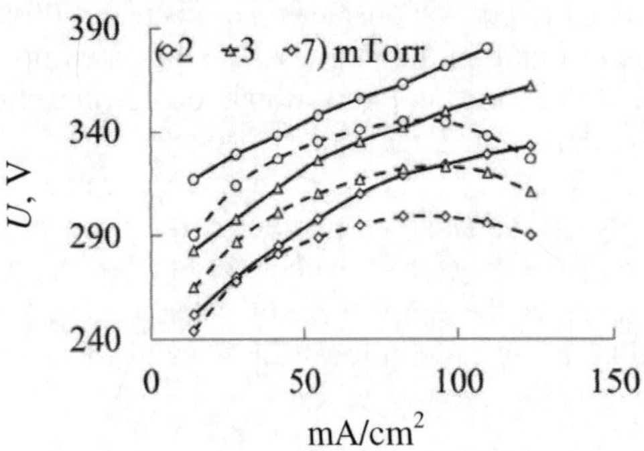

Figure 1. In Ar, discharge V-I characteristics of a magnetron with cold (solid lines) and hot (dashed lines) targets at argon pressure of 2, 3 and 7 mTorr.

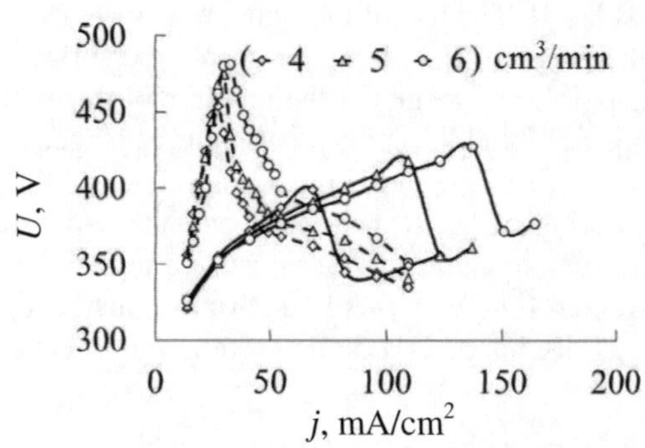

Figure 2. In Ar + O_2 discharge V-I characteristics of a magnetron with cold (solid lines) and hot (dashed lines) targets.

Figure 2 shows the V-I characteristics of a magnetron operating in the Ar + O_2 environment. The measurements were performed at an oxygen flow rate of 4, 5 and 6 cm^3/min. The curves for a magnetron with a cold target fit the well-known concepts of processes in a reactive sputtering system [12]. In the first section of the V-I characteristics, when the discharge current is increased, the voltage gradually increases up to the critical value of the current, at which it abruptly decreases by approximately 60 V or 15 %. This jump reflects the transition of the target from the oxide mode to the metallic mode.

The V-I characteristics of a magnetron with a hot target in figure 2 at any oxygen consumption have three sections:

- in the field of low currents, there is a similarity to the magnetron with a cold target. The target operates in the oxide mode, and the discharge voltage increases proportionally to the current density, but at a higher rate than for the cold target. This effect can be attributed to the known rarefaction of gas (a decrease in density) in the target region due to its heating during sputtering [13]. When the magnetron is operated with a hot target, the heating occurs not only due to the elastic interaction of ions with neutral particles, but also comes from a strongly heated target. The mode exists until the conditions for the transition of the target to the metallic mode of operation arise. The transition from the oxide regime begins at a current density approximately 4–5 times smaller than for a magnetron with a cold target (figure 3, curve 1). One of the reasons for this shift becomes more obvious if we consider the reactive sputtering model [4]. The mode changes when the oxide film growth rates on the target surface are equal to its sputtering with argon ions. The latter value is proportional to the sputtering coefficient, which, in turn, is approximately proportional to the ion energy or discharge voltage. Figure 2 shows that with a current density of 30 mA/cm2, the discharge voltage for a magnetron with cold and hot targets is approximately 350 and 480 V, respectively;
- in the region of higher currents, the target operates in a stationary metal mode, but as compared with the cold target, the voltage decreases with increasing current density;
- between these extreme states, there is a transient mode, in which an increase in the current density leads to a significant decrease in voltage (by about 80 V or by 17 %). Most likely, this is due to the equilibrium purification of the target surface from titanium oxide, which leads to an increase in the coefficient of ion-electron emission. For a titanium target, this value increases from 0.078 for an oxide to 0.113 for a pure metal [14]. The current density at which this process terminates depends on the consumption of oxygen (figure 3, curve 3).

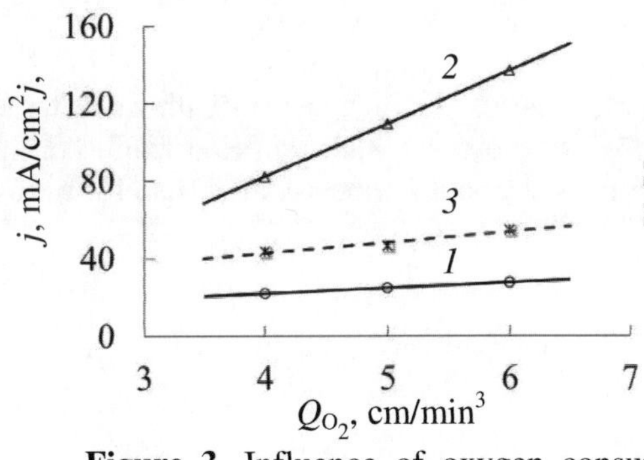

Figure 3. Influence of oxygen consumption on the V-I characteristics transition points from oxide to metallic modes for the target: 1 - cold; 2 – hot.

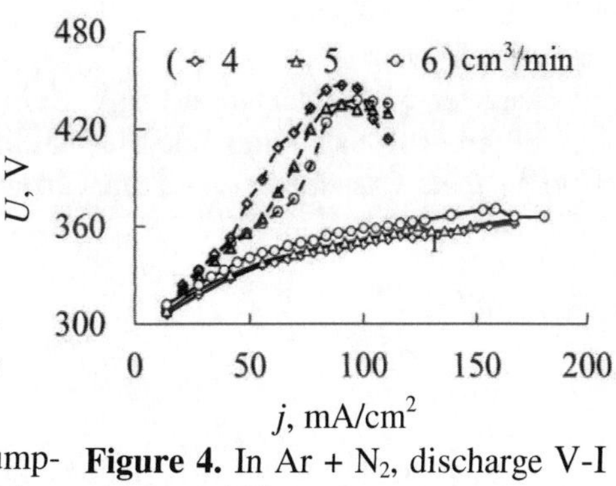

Figure 4. In Ar + N_2, discharge V-I characteristics of a magnetron with cold (solid lines) and hot (dashed lines) targets.

The results of the experiments carried out in the Ar + N_2 environment are given in figure 4. The measurements were fulfilled at a flow rate of nitrogen of 4, 5 and 6 cm^3/min. The character of the curves for a magnetron with a cold target differs from the analogous dependences in Fig. 2 only by the magnitude of the voltage jump, when the target operation mode changes, which does not exceed 5 V (1.5 %). Such a slight decrease in discharge voltage is associated with a slight difference in the coefficients of ion-electron emission of titanium nitride (0.110) and titanium (0.113).

The character of the change in the V-I characteristics of a magnetron with a hot target in figure 4 at any nitrogen flow is slightly different from what is reflected in figure 2. Only in the transition from the nitride mode to the metallic mode in this case begins at a current density about 2 times smaller than for a magnetron with a cold target.

4. Conclusions

The study of magnetrons with a cold and hot titanium target allowed to determine that:

- The V-I characteristics of a magnetron with a hot target operating in an argon environment in the pressure range of 2-7 mTorr at a low current density correspond to the abnormal glow discharge mode; have a maximum, when the current increases; depend on the pressure typically for the magnetron discharge: with increasing pressure, the current on the target is provided by a decrease in discharge voltage;
- in the Ar + O_2:
  ✓ in the V-I characteristics of the magnetron with a cold target, a characteristic jump-like decrease of about 60 V or 15% is observed, reflecting the transition of the target from the oxide mode to the metallic mode;
  ✓ the V-I characteristics of a magnetron with a hot target contains 3 sections corresponding to the oxide, transition, and metallic modes of operation of the target;
  ✓ in the oxide regime, the voltage on the discharge increases approximately in proportion to the current density, but at a higher rate than for the cold target;
  ✓ the transition from the oxide mode starts at a current density approximately four times smaller than for a magnetron with a cold target;
  ✓ in the transition mode, the current increase leads to a significant decrease in voltage (by about 80 V or by 17%);
  ✓ in the region of higher currents, the target operates in a stationary metallic mode, but the voltage decreases with increasing current;
- in the Ar+N_2:
  ✓ in the V-I characteristics of a magnetron with a cold target, the magnitude of the voltage jump when the mode of its operation changes does not exceed 5 V (1.5%); a characteristic jump-like decrease of about 60 V or 15% is observed, reflecting the transition of the target from the oxide mode to the metallic mode;
  ✓ the character of the change in the V-I characteristics of a magnetron with a hot target is slightly different from that obtained in the Ar + O_2. However, the transition from the nitride mode to the metallic mode occurs at the current density about 2 times smaller than for a magnetron with a cold target.

Acknowledgement

This work was supported by the Russian Foundation for Basic Research (grant 18-29-11043).

References

[1] Ferreira F, Sousa C, Cavaleiro A et al. 2017 *Sur. Coat. Technol.* **314** 97
[2] Gao X-H, Guo Z-M, Geng Q-F et al. 2016 *Sur. Engineer.* **32** 840
[3] Ashitkov S I, Komarov P S, Struleva E V et al. 2015 *J. Phys.: Confer. Ser.* **653** 012001
[4] Shapovalov V I, Karzin V V and Bondarenko A 2017 *Phys. Lett. A.* **381** 472
[5] Lapshin A E, Levitskii V S, Shapovalov V I et al. 2016 *Glass Phys. Chem.* **42** 359
[6] Bondarenko A S, Kolomiytsev A A and Shapovalov V I 2016 *J. Phys.: Confer. Ser.* **729** 012006
[7] Sidelev D V, Bleykher G A, Krivobokov V P et al. 2016 *Sur. Coat. Technol.* **308** 168
[8] Bleykher G A, Krivobokov V P, Yuryeva A V et al. 2016 *Vacuum* **124** 11
[9] Westwood W D, Maniv S and Scanlon P J 1983 *J. Appl. Phys.* **54** 6841
[10] Raizer Yu P 1991 *Gas Discharge Physics* (Springer, Berlin, New York) 449
[11] Saikia P, Kakati B and Saikia B K 2013 *Phys. Plasm.* **2** 103505
[12] Komlev A E, Shapovalov V I and Shutova N S 2012 *Techn. Phys.* **57** 1030
[13] Gudmundsson J T, Brenning N, Lundin D et al. 2012 *J. Vac. Sci. Technol. A.* **30** 030801
[14] Depla D, Li X Y, Mahieu S et al. 2008 *J. Phys. D: Appl. Phys.* **41** 202003

IOP Conf. Series: Materials Science and Engineering **387** (2018) 012070 doi:10.1088/1757-899X/387/1/012070

Magnetron with a sputtering unit for deposition of binary alloy films and solid solutions of two compounds

V I Shapovalov, A A Kozin, E A Minzhulina and V V Smirnov

Saint Petersburg Electrotechnical University "LETI", 197376, St. Petersburg, Russia

E-mail: vishapovalov@mail.ru

Abstract. In this work, for the first time we present a magnetron with a sputtering unit used to deposit films of two metal alloys and binary solid solutions of simple compounds (oxides and nitrides). This unit contains two metal targets located on the same axis. The inner (cold) target is effectively cooled by running water, the outer (hot) target is cooled through the fastening elements and by radiation. The inner target is sputtered through slots in the outer target. Current voltage characteristics of a new magnetron working on direct current in inert and reactive environments have been studied. In pure argon with a low current, the V-I characteristics correspond to the mode of abnormal glow discharge, when the current increases, a maximum appears. In the reactive Ar + N_2 environment, the V-I characteristics have a maximum and three sections corresponding to the nitride, transition and metal modes of operation of the sputtering unit.

1. Introduction

An increasing interest was observed for inorganic coatings in the form of binary solid solutions of transition metal nitrides in the world. In 2010-2017 over 5000 articles were devoted to this topic (according to the Scopus database).

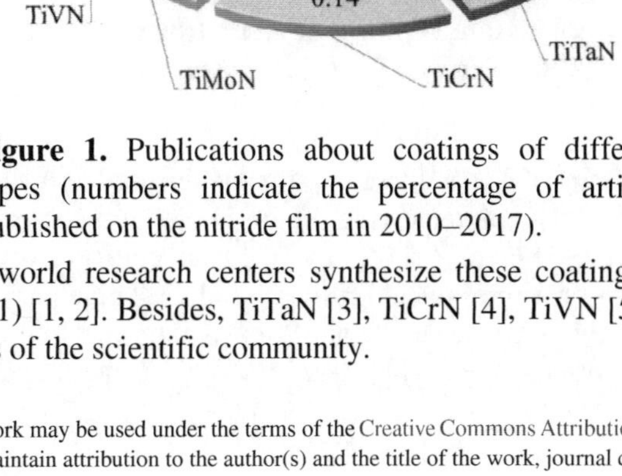

Figure 1. Publications about coatings of different types (numbers indicate the percentage of articles published on the nitride film in 2010–2017).

Presently, many leading world research centers synthesize these coatings. The greatest attention is paid to the TiAlN (figure 1) [1, 2]. Besides, TiTaN [3], TiCrN [4], TiVN [5, 6], TiMoN [7], TiZrN [8], AlCrN [9] are in the focus of the scientific community.

A variety of sputtering devices are used to deposit above films. They include, first of all, the devices containing two magnetrons with targets from different metals [10] or one magnetron with a compound target [11]. Usually the target is effectively cooled in these devices. At the same time, the magnetrons with targets that can be heated up to the melting temperature and higher are studied [12], [13]. The hot target mode is provided by removing heat from it through a gap of up to 1-mm-width and by fastening elements [14].

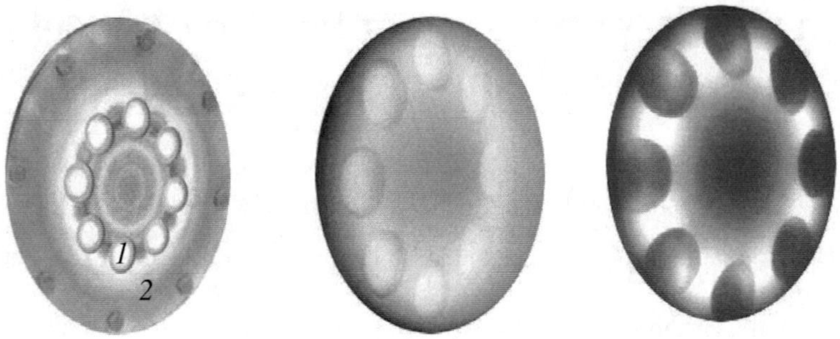

(a) **(b)** **(c)**

Figure 2. Magnetron with a sandwich–target: (**a**) – general view (1 – inner target, 2 – outer target); (**b**) and (**c**) are the photos of discharge with an external titanium target at low (heating is not noticeable) and medium (weak heating) power, respectively.

In this paper, we study a new type of magnetron with a sputtering unit, which was called a "sandwich–target". The unit contains two targets of different metals located on the same axis (figure 2(a)). The inner target 1 is sputtered through the gaps of the outer target 2 and operates in a cold mode. The external target is set at a distance of 1 mm from the internal target and operates in hot mode. Obviously, in order to deposit films by the new device, it is necessary to study it thoroughly. In this paper, we focused on the study of the magnetron V-I characteristics in different gas environment.

2. Experimental

The experiments were performed on a balanced dc magnetron with a diameter of 130 mm. The magnetron is installed in a high-vacuum system equipped with oil-diffusion and mechanical pumps having a nominal operating speed of 0.3 m^3/s and 0.005 m^3/s, respectively. The volume of the vacuum chamber is 0.076 m^3, its residual pressure did not exceed $5 \cdot 10^{-2}$ mTorr. We studied the V-I characteristics of a magnetron equipped with:

- a hot titanium target of 2-mm-thick (without gaps) cooled through an aluminum board;
- the sputtering unit with targets:
  - ✓ inner one with thickness of 4 mm, made of aluminum;
  - ✓ outer one with thickness of 2 mm, made of titanium. Its gaps occupied 50% of the area of the sputtered ring region.

The V-I characteristic of each magnetron is measured in two media:

- Ar at a partial pressure in the range 2-7 mTorr;
- Ar + N_2 at a partial argon pressure of 2 mTorr and nitrogen flow in the range of 4-6 cm^3/min.

Argon and nitrogen of high purity were used to create gas environment.

3. Results and discussion

Figures 2(b), (c) show the photos of the sputtered unit made during the operation of the magnetron in different modes. In figure 2(c), the sputtered ring region is clearly visible, which is heated up more strongly than other target surface. Figure 3 shows the V-I characteristics of a magnetron with a sandwich–target operating in an argon environment (curves 1 and 2). These curves , same as for a magnetron with a single hot target (figure 3, curves *3* and *4*), have extrema. And the discharge voltage of a magnetron with a sandwich–target is higher by 10–17 %. We assume that this result relates to an in-

IOP Conf. Series: Materials Science and Engineering **387** (2018) 012070 doi:10.1088/1757-899X/387/1/012070

ternal aluminum target. It is sputtered under 50 % discharge current. The lower coefficient of ion-electron emission of aluminum compared to the titanium one requires a higher voltage for providing a given current (see [15] for relevant references). It should be kept in mind that an additional increase in the discharge voltage can be caused by the known effect of pressure reduction in the target region caused by the heating of the gas layers adjacent to it and called a "sputtering wind" [15].

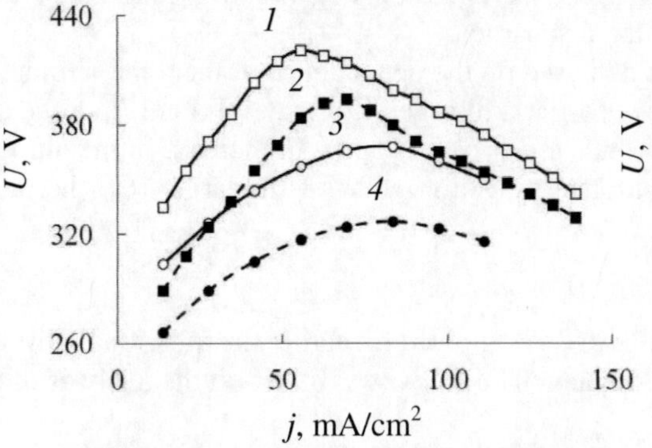

Figure 3. V-I characteristics of a magnetron with an Al + Ti (*1* and *2*) sandwich–target and a single hot titanium target (*3* and *4*) at argon pressure (mTorr): 2 – solid lines; 7 – dashed lines.

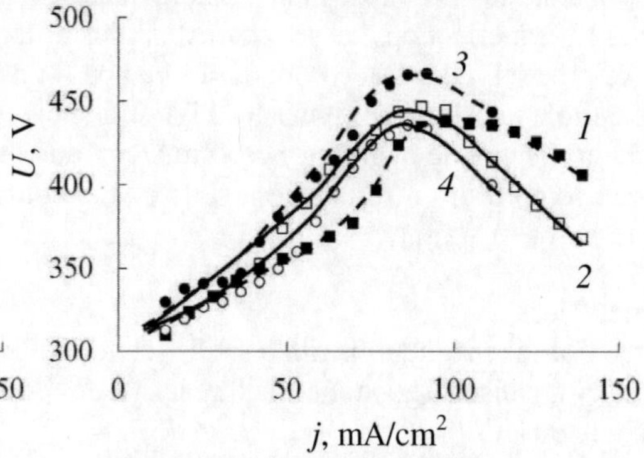

Figure 4. V-I characteristics of a magnetron with an Al + Ti (*1* and *2*) sandwich target and a single hot titanium target (*3* and *4*) in Ar + N_2 at an argon pressure of 2 mTorr and nitrogen flow rate (cm^3/min): 4 – solid lines; 6 – dashed lines.

In the initial section (up to about 50 mA/cm^2), the V-I characteristic in figure 3 corresponds to an abnormal glow discharge, which is typical of a magnetron with a cold target. But for the above reasons, they grow with a larger derivative than the ones of the magnetron with cold and hot targets. In figure 3 there is another feature characteristic of an abnormal discharge. When the pressure increases, the invariance of the current density on the target is due to a decrease in the discharge voltage.

At a current of more than 50–60 mA / cm^2, the current-voltage characteristic of the studied magnetron has a form of the dependence, which is usually observed in the transition to the arc discharge. It can result from an increase in ion-electron emission when an external titanium target heats up to a temperature corresponding to the discharge current densities to the right of the maxima.

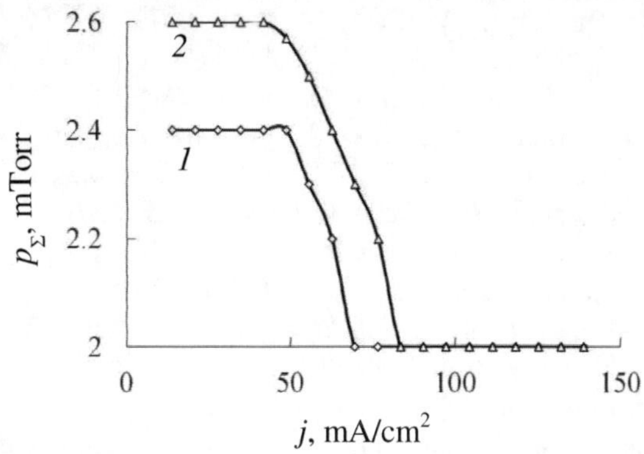

Figure 5. Dependence of the total pressure on the discharge current density for the Al + Ti sandwich–target in Ar + N_2 at a nitrogen flow rate (cm^3/min): *1*–4; *2*–6

Figure 4 shows the V-I characteristic of a magnetron with a sandwich–target operating in a mixture of argon and nitrogen (curves *1* and *2*). These dependencies with extrema are similar to the V-I characteristic of a magnetron with a single hot target (figure 4, curves *3* and *4*). Everything related to the emission of electrons in this case also has an effect, so the curves in figure 4 qualitatively do not differ from the dependencies in figure 3. The observed difference in the form of inflection points on the left sections, a larger discharge voltage and a shift to the right of the extremum position are determined by other physical processes. At low discharge current the sputtering unit operates in a reactive mode, when both targets are coated by a film of the corresponding nitride. In the region $j > 100$ mA/cm^2, the targets operate in the metal mode. These features affect the emission in other way, and hence they have a different effect on the V-I characteristics of the discharge.

More precisely, the mode boundaries can be found based on the dependence of the total pressure p_Σ on the current density (see figure 5). Figure 5 shows that at a nitrogen flow rate of 4 cm^3/min the exit from the nitride mode begins at $j = 50$ mA/cm^2 and ends at $j = 70$ mA/cm^2. An increase in the nitrogen flow rate to 6 cm^3/min results in a shift of the complete transition point of the target into the metal mode to $j = 85$ mA/cm^2.

4. Conclusions

The study of the magnetron with a sandwich–target (Al + Ti) allowed revealing the individual physical features of the discharge in inert and reactive environment. The study was based on the analysis of the V-I characteristics.

In the argon environment, the current-voltage characteristics in the low-current region are similar to the dependences observed in a magnetron with a cold target. When the current increases, the voltage of the discharge increases smoothly, but with a larger derivative than the one of the magnetron with a cold and hot target. The curves reach an extremum, and then behave similar to the arc discharge.

In a mixture of argon and nitrogen, the V-I characteristics qualitatively are similar to the dependences obtained in pure argon. They differ by inflection points in areas to the left of the extrema, by a large discharge voltage and by shifting the extreme positions to the right.

All the revealed features are likely to be resulted from variations in the coefficient of ion-electron emission under different conditions and for different materials. Additionally, the effect of pressure reduction in the target region may influence the discharge.

Acknowledgement

This work was supported by the Russian Foundation for Basic Research (grant 18-29-11043).

References

[1] Pervaiz S and Samad 2018 *Mech. Addit. Adv. Manufact.* **9** 101–7
[2] El-Hossary F, Abd El – Rahman A M, Raaif M et al. 2017 *Appl. Phys. A: Mater. Sci. Proces.* **122** 242
[3] Liu X, Ma G, Sun G et al. 2012 *Surface and Coatings Tech.* **212** 128–33
[4] Krishnan R M, Jeyanth M, Silambarasan R et al. 2016 *Intern. J. Sci. Eng. Res.* **7** 203–6
[5] Chang Y-Y, Weng S-Y, Chen C-H et al. 2017 *Sur. Coat. Technol.* **332** 494–503
[6] Baran Ö, Keleş A, Çiçek H et al. 2017 *Sur. Coat. Technol.* **332** 312–8
[7] Kislitsin S B, Uglov V V and Larionov A 2016 *AIP Conf. Proc.* **1783** 020091
[8] Lin Y, Chen H, Yu G et al. 2016 *Thin Solid Films* **618** 13–20
[9] Tatemizo N, Imada S, Miura Y et al. 2017 *J. Phys. Conden. Matter.* **29** 085502
[10] Stone D S, Gao H, Chantharangsi C et al. 2014 *Sur. Coat. Technol.* **244** 37–44
[11] Zaleska A 2008 *Recent Patents on Engineering* **2** 157–64
[12] Shapovalov V I, Karzin V V and Bondarenko A C 2017 *Phys. Lett. A.* **381** 472
[13] Bleykher G A, Borduleva AO, Krivobokov V P et al. 2016 *Vacuum* **132** 62–9
[14] Lapshin A E, Levitskii V S, Shapovalov V I et al. 2016 *Glass Phys. Chem.* **42** 359
[15] Shapovalov V I 2017 *J. Phys.: Confer. Series* **872** 012024

IOP Conf. Series: Materials Science and Engineering **387** (2018) 012071 doi:10.1088/1757-899X/387/1/012071

Coefficient of friction of titanium oxynitride films deposited by hot target reactive sputtering

D S Shestakov, A S Useinov, K S Kravchuk, E V Gladkikh, A V Zavyalov, A A Morozova and V V Smirnov

Saint-Petersburg State Electrotechnical University "LETI", St.Petersburg, Russia

E-mail: guitar.pro1994@mail.ru

Abstract. The influence of the oxygen flow rate on the friction coefficient of titanium oxynitride films deposited by reactive magnetron sputtering at a constant current of a hot target was investigated. The study used the plans of the active experiment and regression models of the first and second order. It is found that the increase of the oxygen flow rate in the gas environment leads to the increase of the film friction coefficient.

1. Introduction

TiO_xN_y films are materials drawing attention of many researchers [1]. They are widely applied to: produce solar [2] and lithium batteries [3], protect metals from the influence of aggressive media, increase the metal surface mechanical properties [4], etc. A wide variety of possible applications of oxynitride films stimulated the development of methods for their deposition. At present, both physical and chemical methods are used. Among the most popular methods are reactive magnetron sputtering with direct and alternating current, impulse and reactive high-power pulsed magnetron sputtering [5]. In these cases, magnetrons with both cold and hot targets are used [6].

The purpose of this paper is to study the effect of the independent parameters of the deposition process on the friction coefficient of titanium oxynitride films. The films were deposited by the method of reactive magnetron sputtering of a hot target with a direct current. The independent parameters include discharge current density, nitrogen consumption and oxygen consumption.

2. Experiment details

The films of titanium oxynitride were deposited by a magnetron with a hot titanium target operating on a direct current in argon, oxygen, and nitrogen. In order to deposit the films, a vacuum chamber with a volume of $7.8 \cdot 10^{-2}$ m^3 was equipped by a flat magnetron with a titanium target of 130 mm in diameter operating on a direct current. The hot target mode is provided by fixing a titanium disk of 1 mm thick with a gap of 1 mm on a copper plate of 4 mm thick cooled by running water. The residual pressure in the chamber did not exceed 10^{-2} mTorr. In the research, the theory of active experiment was used. Two batches of samples were produced:

- films of titanium nitride (a special case of TiO_xN_y films at $x = 0$) were deposited at a pressure of $p_{Ar} = 2$ mTorr, nitrogen consumption (2–6) sccm, and discharge current density (14–33) mA/cm^2. The batch contained four series of three samples, manufactured according to the plan of the first order, shown by round markers in figure 1(a). In this batch, an additional sample was made at the current density of 22 mA/cm^2 with the addition of oxygen to the gas environment at the rate of 1 sccm (the square marker in figure 1(a));

- titanium oxynitride films were deposited at the pressure of p_{Ar} = 2 mTorr, discharge current density of 22 mA/cm^2, nitrogen flow rate of (4–6) sccm, and oxygen consumption of (1–3) sccm. The batch contained nine samples made according to the second-order plan on the cube, shown by round markers in figure 1(b).

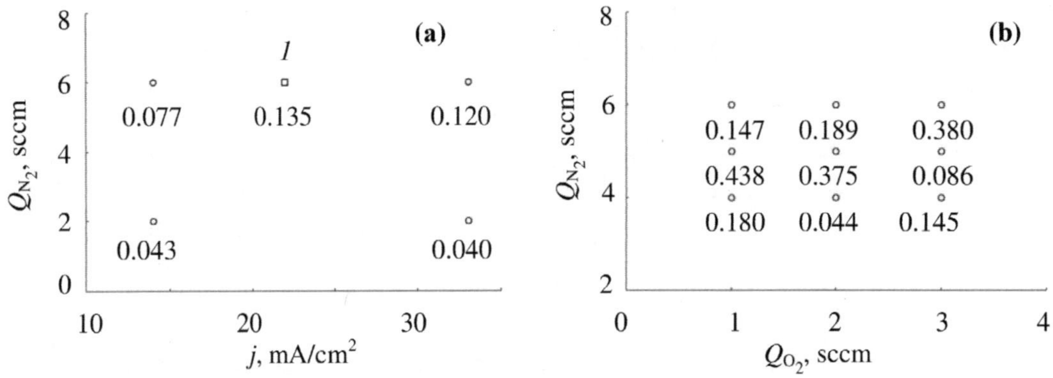

Figure 1. Experimental design: *a* – first order for TiN film deposition (*1* – plan point corresponding to TiO_xN_y deposition at Q_{O_2} = 1 sccm; *b* – second order on the cube for TiO_xN_y deposition at j = 22 mA/cm^2 value. The friction coefficient of each sample is indicated below the corresponding point in the plan.

Each sample was tribotested using a scanning hardness tester "NanoScan-4D" (manufacturer of FGBNU TISNUM). The measurement method is a multi-cycle friction by a spherical tip with the registration of the pressing force, lateral force and deepening of the tip into the sample. Tip: AISI 440C steel ball with a radius of 3 mm. Parameters of the measurement procedure:

- length of wear track 100 μm,
- number of cycles of reciprocating motion – 200,
- force of loading when tested — 1 mN,
- tip speed — 10 μm/s.

3. Results and discussion

The first batch of samples provided the following results. Figure 1 shows the average value of the friction coefficient below each point of the plan. The above value has a mean-square deviation of less than 0.01. The independent variables in this plan (the discharge current density j and the nitrogen flow rate Q_{N_2}) were changed at two levels, as shown in figure 1(a).

The experimental plan used for this batch allowed determining the influence of the selected parameters on the TiN friction coefficient in the form of a polynomial of the first order:

$$\mu \approx 0.001j + 0.014Q_{N_2}. \quad (1)$$

The coefficients in the model (1) were determined by the method of least squares with the variables in measurement units indicated in figure 1(a). In accordance with the theory of experimental planning, expression (1) can be used for interpolation. Using this possibility, the authors estimated the value of μ for the sample of point *1* shown in figure 1. Calculations resulted in μ = 0.1. While the measurement yielded a value of μ = 0.135. The observed difference obviously indicates an increase in the coefficient of friction when oxygen is introduced into the TiN film.

In the second batch, nine samples were produced (figure 1(b)). The experimental plan allowed to determine the influence of the selected parameters (Q_{N_2} nitrogen flow rate and Q_{O_2} oxygen) on the TiN friction coefficient in the form of a polynomial of the second order:

$$\mu \approx b_0 + b_1 Q_{N_2} + b_2 Q_{O_2} + b_{11} Q_{N_2}^2 + b_{22} Q_{O_2}^2 + b_{12} Q_{N_2} Q_{O_2}. \quad (2)$$

Just as in the model (1), the coefficients in the model (2) were determined by the method of least squares with the variables in units of measurement indicated in figure 1(b). The values of the model (2) parameters are given in the table 1.

Table 1. The values of the model (2) parameters.

Model (2) parameters	b_0	b_1	b_2	b_{11}	b_{22}	b_{12}
Values	–2.12	1.09	–0.52	0.12	0.06	0.07

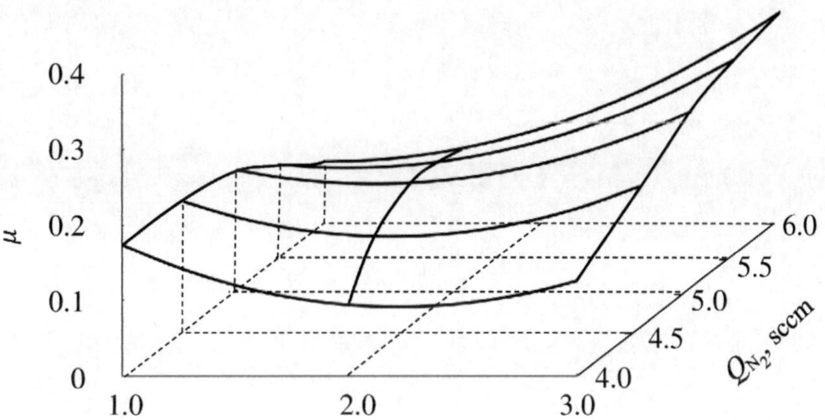

Figure 2. 3D model of the influence of nitrogen and oxygen consumption on the value of μ.

Figure 2 shows the surface modeled based on expression (2), where the dependence of $\mu = f(Q_{N_2}, Q_{O_2})$ reflects a tendency for the coefficient of friction to increase when oxygen fraction increases in the gas environment. The films with the smallest value of μ are deposited near the values of $Q_{N_2} = 4$ sccm and $Q_{O_2} = 2$ sccm.

4. Conclusions

The performed study revealed that:

- in the specified ranges of the discharge current density j and the nitrogen flow rate Q_{N_2}, their influence on the friction coefficient of TiO_xN_y films at $x = 0$ can be described by a polynomial of the first order;
- in the given ranges of the nitrogen flow rate Q_{N_2} and oxygen Q_{O_2}, their influence on the friction coefficient of TiO_xN_y films can be described by a second-order polynomial. The smallest value of μ can be obtained if the films are deposited at the current density of 22 mA/cm^2 near the values of $Q_{N_2} = 4$ sccm and $Q_{O_2} = 2$ sccm;
- an increase in the consumption of oxygen during deposition leads to an increase in the friction coefficient of TiO_xN_y films.

Acknowledgement
The authors thank the Russian Foundation for Basic Research for the support of the experimental research performed by Technological Institute for Superhard and Novel Carbon Materials, Troitsk, Moscow, Russia (grant 18-08-00558)

References

[1] Konischev M E, Kuzmin O S, Pustovalova A A et al. 2014 *Russ. Phys. J.* **56** 1144–9
[2] Zhang J, Chen T P, Liu Y C et al. 2017 *Solar Energy* **142** 33–8
[3] Chiu K-F, Su S-H, Leu H-J et al. 2015 *Thin Solid Films* **596** 29–33
[4] Lapshin A E, Shapovalov V I, Useinov A S et al. 2017 *Glass Phys. Chem.* **43** 477–9
[5] Shapovalov V I 2017 *J. Phys.: Confer. Series* **872** 012024
[6] Kozin A A, Shapovalov V I, Smirnov V V et al. 2017 *J. Phys.: Confer. Series* **872** 012019

Advanced development of chemical sources of current (CSC) and ultra high-volume capacitor structures (UCS) for energy storage devices

V V Sleptsov, S N Kulikov, D Yu Kukushkin and A O Diteleva

[1] Moscow Aviation Institute (National Research University), Volokolamskoe Shosse, 4, 125310, Moscow, Russia

E-mail: 08fraktal@inbox.ru, anna.diteleva@mail.ru

Abstract. This paper addresses advanced development of chemical sources of current (CSC) and ultra-high-volume capacitor structures (UCS) for energy accumulation. For estimated specific energy, CSC with the lithium anode ranks first. Recently, along with the CSC, the ultra-high-volume capacitor structures (UCS) are intensely developing. The most advanced area for a drastic improvement of batteries properties is the use of nanomaterials that acquire unique physical and chemical properties due to the nanoscale level structurization. To implement development of nanomaterials and nanostructures-based CSC and UCS, conventional microelectronics thin-film technologies are used.

CSC based on lithium materials is the champion in energy intensity. Lithium has the highest specific energy: 11760 W × h/kg [1]. First lithium electric cells (LEC) entered the world market in the early 1970s, and this was due to the development of non-aqueous electrolyte solutions, and the use of high-performance and low-cost thick-layer technology for their production only [2].

Regarding estimated specific energy (one of the key parameters for the electrochemical cells energy efficiency) CSC with the lithium anode hit the first place. Real energy advantages of current sources with lithium anodes are evident in the following comparison.Specific energy value of one of the widely spread in practice energy-intensive Clark cell systems is 80 to 100 W × h/kg, among other standard (silver-zinc and magnesium silver-chloride) systems it does not exceed 120 to 130 W × h/ kg, and with various lithium anode based current sources specific energy value amounts to 300 to 500 W × h/kg. Advanced CSC developments can provide obtaining specific energy about 1000 W × h/kg. Discharge voltage values of lithium anode based electro-chemical couples used amount to about 3 to 4.8 V.This is more than twice the voltage of components widely used in household appliances. (1.5 V).

The last steps in developing LIB was the discovery by Japanese scientists the possibility of using not lithium metal but chemically-inactive carbon capable of reversible intercalation of Li + ions as anode and subsequent combination with lithiated transition metals oxides in organic electrolytes. Specific characteristics of modern anode materials produced by using the thick layer technology are shown in table 2. Advanced developments in the area of CSC is energy intensity increase, operating safety improvement and production and operation costs reduction. The major carbon materials used in the LIB compositions are graphite and amorphous carbon [8].

The history of LIB development and improvement goes to show that emergence of new LIB potential uses coincided with the development of new active materials, while the innovations maturity

IOP Conf. Series: Materials Science and Engineering **387** (2018) 012072 doi:10.1088/1757-899X/387/1/012072

processes starting with the laboratory study via RTD to real applications took several years. *However, it should be noted that since the time of development and over more the last twenty years the lithium-ion batteries capacity underwent no massive improvement.* That is because both anode and cathode materials in fact remained the same many laboratories around the world are developing new materials for electrodes, batteries, and numerous breakthrough results are being reported, but commercial applications are hardly in evidence. This is due to the wide range of requirements for such materials - except that they should contain larger amount of lithium than compatibles used, their behavior in batteries should not differ from that of existing electrodes. The most advanced area of drastic batteries properties improvement is the use of nanomaterials that acquire unique physical and chemical properties due to the nanoscale level structurization.

Table 1. Parameters of commercial LIB cathode materials [3–7].

Material	Average potential (V)	Specific capacitance (mA × h/g)	Specific energy (W × h/kg)
LiCoO2	3.8-3.9	140-145	546
LiNi0.8Co0.15Al0.05O2	3.7-3.8	180-200	680-760
LiNi0.33Co0.33Mn0.33O2	3.7-3.8	160-170	610-650
LiMn2O4	4.1	100-120	410-492
LiFePO4	3.4-3.45	150-170	518-587

In more recent times, along with the CSC, the ultra-high-volume capacitor structures (UCS) are intensely developing. At the first stage of UCS development high capacitance of capacitor structures was ensured by developing electrode materials with highly developed surface impregnated with electrolytes. Such capacitor structures are called supercapacitors. The surface quantity of carbon-based electrode materials amounts to 2000÷2500 m^2/g. The thick-film technology for supercapacitors differs from the CSC technology solely in that instead of a chemically-active material a highly porous carbon mixture is applied onto the current collector in the form of a thin metal (15-20 μm) foil.

At the second stage of development with the aim of improving energy intensity by way of increasing operating voltage polymer electrolytes became to be used that allowed operating voltage (U) increase from 1 ÷ 1.2 V to 4 V. The next stage of development is the electrical capacitance C and operating voltage U increase by way of using dielectrics. Dielectrics with high dielectric permeability, which among nanostructured materials amounts to 107 and higher (Ɛ ~ 104 for the monolithic ferroelectrics), can enable to considerably increase not only capacitance (200F/g and more), but also UCS operating voltage to the level of 4 V and higher. In this case the layer, in which electrical energy is accumulated, has a thickness on nanometer scales and is called double-electric layer (DEL).

The most advanced areas for drastic improvement of battery properties are the use of nanomaterials that due to the nanoscale level structurization acquire unique physical and chemical properties. To development of nanomaterials and nanostructures based CSC and UCS traditional microelectronics thin-film technologies are used (layers thickness changes from micro- to nanometers) [2]. The UCS technical standards are primarily determined by electrode materials and electrolytes design.

Being inferior to the chemical sources of current as for specific energy the supercapacitors significantly outperform them for the specific capacitance and charge-discharge characteristics stability in a wide temperature range. In addition, the ultracapacitors today exceed chemical sources of current by the number of charge-discharge cycles by an order of magnitude [9].

The amount of energy stored in the supercapacitors is determined by the sum capacitances:

1. - the electrochemical capacitance of double-electric layer [9], which is generated at the electrode-electrolyte interface;

2. - the pseudocapacitance, caused by chemical reactions between the electrode and electrolyte [10] [10]. In this case, the electrons are formed by oxidative reactions and transferred across the border electrode-electrolyte interface region.

In addition to this, among the metals potentially usable for the capacitance increase there are RuO_2, MnO_2, NiO, Co_3O_4, SnO_2, ZnO, TiO_2, V_2O_5, CuO, Fe_2O_3, WO_3, etc. [11,12].

Capacitor structures-based power sources are mainly presented in the market by supercapacitors which capacitance under industrial conditions is 8 to 12 W-h/kg and 15 to 25 W-h/kg under laboratory ones, but they have significantly higher characteristics of power, charging time and the number of charge-discharge cycles (see table 2) [13]. Ceramic capacitors and hybrid power sources that integrate CSC and UCS properties are under laboratory studies. Hybrid power sources allow attaining energy intensity of about 40 to 200 W-h/kg [14]. Such power sources are also called pseudo capacitors, they store energy through reversible Faraday charges that assume rapid and reversible electrochemical redox reactions at the electrodes and electrolyte interface.

Table 2. Specific DEL (double-electric layer) capacitance in capacitor structures.

Electrode	Electrolyte	Specific capacitance (F/g)	Specific capacity (W/kg)	Specific energy (W × h/kg)	Retention ability
Single-wall nanotubes (SWNTs)	1 M propylene carbonate	160 F/g at 4 V	24 kW/kg	17 W * h/kg	96.4% after 1000 cycles
Peeling reduced graphene oxide (rGO)	1-ethyl-3-methyl imidazole tetrafluoroborate/ acetonitrile ($EMIMBF_4/\ AN$) 166 F/g at 5.7 A/g and 3.5 V		250 kW/kg	70 W * h/kg	97% after 10000 cycles
Nitrogen doped graphene	1 M tetraethanolammonium tetrafluoroborate	280 F/g at 20A/g in 6 M KOH	800 kW/kg	48 W * h/kg	99.8% after 230000 cycles
Graphene aerogel	0.5 M H_2SO_4	325 F/g at 1 A/g	7 kW/kg	45 W * h/kg	~98% after 5000 cycles
2D microscopic covalent triazine-based structure	Non-soluble ionic liquid $EMIMBF_4$	151/3 F/g at 0.1 A/g	10 kW/kg	~42 W * h/kg	85% after 10000 cycles at 10 A/g
Co - Polyaniline (PANI) (poly (aniline - Co - 3 - fluoroaniline))	6 M KOH	989 F/g 2 mV/s	1.581 kW/kg	352 W * h/kg	79% after 1000 cycles
$Co(OH)_2$/ graphene film layers	1 M KOH	622 F/g at 2 A/g	15.8 kW/kg	86.6 W * h/kg	80% after 10000 cycles at 10 A/g
$Ni(OH)_2$/ graphene	1 M KOH	1335 F/g at 2.8 A/g	10 kW/kg	37 W * h/kg	100% after 2000 cycles at 28.6 A/g
MnO_2/ reduced graphene oxide (rGO) (hydrogel and aerogel)	1 M Na_2SO_4	242 and 131 F/g at 1 A/g	0.82 kW/kg	212 W * h/kg	89.6% 1000 cycles
Polyaniline (PANI)-graphene/Carbon nanotube (CNT)	1 M KCl	271 F/g with 0.3 A/g	2.7 kW/kg	188.4 W * h/kg	82% after 1000 cycles at 2 A/g
Patronite (VS_4)/ single-wall nanotube (SWNT)/reduced graphene oxide (rGO)	0.5 M K_2SO_4	558.7 F/g at 1 A/g	13.85 kW/kg	174.6 W * h/kg	97% after 1000 cycles
Polyaniline (PANI)/Graphene oxide (GO)/graphene	2 M H_2SO_4	793.7 F/g at 1 A/g	2.14 kW/kg	50.2 W * h/kg	80% after 1000 cycles at 100 A/g

Graphene structures (figure 1) with the hierarchical porous structure, high ion-accessible surface area, effectiveness of the electron and Ion transfer routes allow developing electrode materials with a capacitance of 298 F/g or 212 F/cm^3 in organic electrolyte [13]

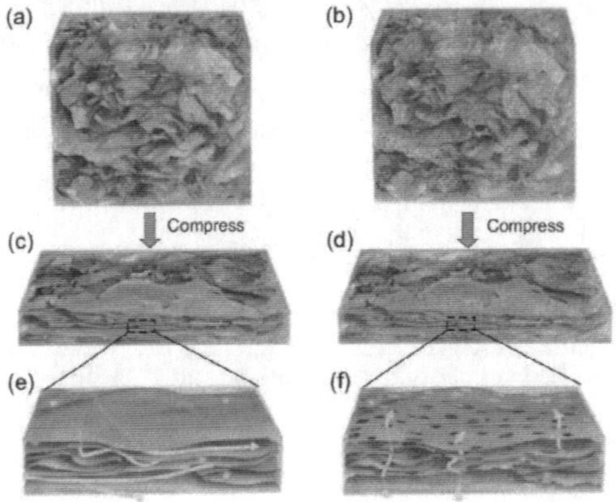

Figure 1. Schematic illustration of graphene and porous graphene foams: (**a**), (**b**) - initial 3D macro-porous Graphene foam (**a**) and (**b**) - porous Graphene foam; (**c**), (**d**) - thin films (**c**) Graphene foam and (**d**) - porous Graphene foam; (**e**), (**f**) - closed type (**e**) Graphene and (**f**) porous Graphene film. Arrows indicate ions transfer routs [13].

Lithium-ion capacitor entirely on the basis of graphene, formed by reduced graphene oxide with the oxide nanoparticles coating (SnO^2-rGO), and on the other hand, anode - capacitor electrode based on activated reduced graphene oxide (a-TEGO). The power density of this supercapacitor is about 200 Wh/kg. CNT-NiO composite synthesized by way of nickel oxide nanoparticles growth in suspension with carbon nanotubes during the hydrothermal process demonstrated capacitance increase from 122 F/g to 1600 F/g at the scan rate of 2 mV/sec in the alkaline electrolyte as compared to pure nickel oxide. The specific capacity increase was also observed with V_2O_5 composite nanoconductors and multi-walled carbon nanotubes, which demonstrated 450 F/g specific capacity at current density of 1 A/g, which is much higher (100 F/g) than the characteristics of pure V_2O_5 nanoconductors [12].

Thin-film technologies produce not only new and advanced structures for CSC and UCS but provide for significant improvement of characteristics of standard structures by thick-film technologies [14].

On the basis of the CSC and UCS development analysis it was shown that the development prospect in this area is the development of new nanomaterials and nanotechnologies. Therefore, the development of advanced CSC and UCS design should be carried out in the following areas:

- development of a uniform nanostructurized electrode material for the CSC and UCS;
- development of an active nanostructurized mass for the CSC;
- development of ceramic and polymer-ceramic materials of high inductive capacitance;
- development of uniform electrolytes and separators.

The breakthrough and foremost goals should be should be considered development of new electrochemical systems, structures and technologies that provide developing current sources with wider charge-discharge cycles range (over 2000-10 000) and higher specific energy (over 500-1000 W - h/kg).

Acknowledgments
The research has been financed by the State represented by the Ministry of Education and Science of the Russian Federation. Subsidies provision agreement No. RFMEFI57717X0275.

References

[1] Kozaderov O A 2017 *Modern chemical current sources* (Lan)

[2] Proleyko V M 2009 *Basic lectures on electronics* **2** (Tekhnosfera)
[3] Zhang Z and Zhang S S 2015 *Rechargeable Batteries: Materials*
[4] Martinet S 2016 *Nanomaterials for Sustainable Energy*
[5] Nitta N, Wu F, Lee J T and Yushin G 2015 *Materials Today* **18**
[6] Julien C M, Mauger A, Zaghib K and Groult H 2014 *Inorganics Inorganics* **2**
[7] Doeff M M 2013 *Cathode Materials. Batteries for Sustainability: Selected Entries from the Encyclopedia of Sustainability Science and Technology*
[8] Pillot C 2014 *The Rechargeable Battery Market and Main Trends*
[9] Hou Y, Li J, Gao X, Wen Z, Yuan C and Chen J 2016 *Nanoscale* **8**
[10] Xu D W, Xin S, You Y, Li Y, Cong H P and Yu S H 2016 *Chem Nano Mat* **2**
[11] Yu L, Hu H, Wu H B and Lou X W 2017 *Adv. Mater* **29**
[12] Kitsyuk E P 2017 *Research and development of electrochemical energy storage devices nanostructured electrodes formation processes. Dis. Cand. Sci.* MIET
[13] Yu Liu and Xinsheng Peng 2017 *Applied Materials Today* **8** 104–15
[14] Lebedev E F 2017 *Development of formation processes and study of properties of heat generation and energy storage elements of thermoelectric batteries. Dis. Cand. Sci.* MIET

Investigation of the deposition of metal nanoclusters on the hidden surface of porous electrode materials by electrophoresis

V V Sleptsov, A V Savkin, D Y Kukushkin and A O Diteleva

Moscow Aviation Institute (National Research University), 125993, GSP-3, A-80, Moscow, Volokolamskoe shosse 4, Russia

E-mail: savkin@mati.ru

Abstract. The formation of a coating on the entire surface of a highly porous electrode material based on activated carbon fabric using magnetron sputtering and subsequent deposition of a colloidal solution of metals is considered. The experimental results show the possibility of metallization of the hidden surfaces of porous carbon materials by metal nanoclusters of 2-10 nm size by electrophoresis without reducing their initial porosity.

1. Introduction

Miniaturization of electronics has led to the fact that the main volume and weight of many devices is a power source, which are increasingly demanding in terms of battery life, the amount of stored energy, the number of recharge cycles and service life, which should be at least the service life of the electronic device itself. Thus, today there is a need for a sharp increase in the energy characteristics of current sources, which can be achieved by using microelectronic technologies and nanotechnologies in their production [1, 2].

Priority directions of improvement of the miniature sources of electrical energy become chemical current sources, primarily, lithium batteries and supercapacitors. Being inferior to chemical current sources in nominal energy, supercapacitors significantly exceed them in specific power and stability of charge-discharge characteristics in a wide temperature range.

Microelectronic technologies and nanotechnologies will be in demand to improve the characteristics of electrode materials as the main elements of chemical current sources and supercapacitor structures [3, 4].

Prospective electrode materials must meet several criteria: developed surface area, high electrical conductivity, availability of porous structure to wetting, chemical inertness to structural materials and so on. The most fully listed requirements are met by carbon materials of varying degrees of porosity [5-10]. One of the promising materials in this group are carbon cloths from activated carbon fibers – "busofit" (figure 1) [11]. The capacitance of the capacitor depends primarily on the surface area, which in carbon materials can increase in comparison with the geometric surface by 10^3–10^5 times. In addition to a highly developed surface, electrode materials must have a low electrical resistance, both along the layer and across, as this determines the internal resistance of the current source, limiting its output current. To reduce the electrical resistance of busofit and the creation of electrode material on its basis it is necessary to solve two problems.

First, apply a continuous layer of metal to the surface of the busophyte, which will be a current

IOP Conf. Series: Materials Science and Engineering **387** (2018) 012073 doi:10.1088/1757-899X/387/1/012073

collector and reduce the internal resistance. This operation can be successfully realized with the help of magnetron sputtering. Since, this is a well-known and studied process, within the framework of this article, we will not dwell on its consideration.

(a) **(b)**

Figure 1. Initial busofit (**a**) and a snapshot with increasing (**b**).

Secondly, it is necessary to deposit a layer of metal on each filament of the busophyte, which, on the one hand, will lower their electrical resistance, and on the other hand will increase the specific surface. To solve this problem, it is proposed to use the technology of depositing nanoclusters of metals on the surface of porous materials by electrophoresis. As is known, nanometer-sized particles have a number of unique properties. They carry a positive charge, and their electrophysical and physicochemical properties are determined by the size of the nanoclusters. Therefore, the main task is to obtain metal particles with the desired size range and then position them in the pore space by electrophoresis. In this case, the nanoparticles should have a uniform distribution over the surface.

2. Experimental

In the studies, the serial carbon cloth Busofit T-40 and similar carbon materials Busofit T-1, Busofit-T, Busofit-TM-4 were used. On the busofit threads, nanoclusters of nickel and silver were applied from the hydrosol. The preparation of metal nanoclusters was carried out by the method of electropulse dispersion due to an electric discharge between electrodes made of the metal immersed in distilled water [12]. The interelectrode distance was 100 µm, air gap of 1 mm, capacitor voltage up to 3 kV, capacitance of 1000 pF. Produced nanoclusters of metals are shown in figure 2.

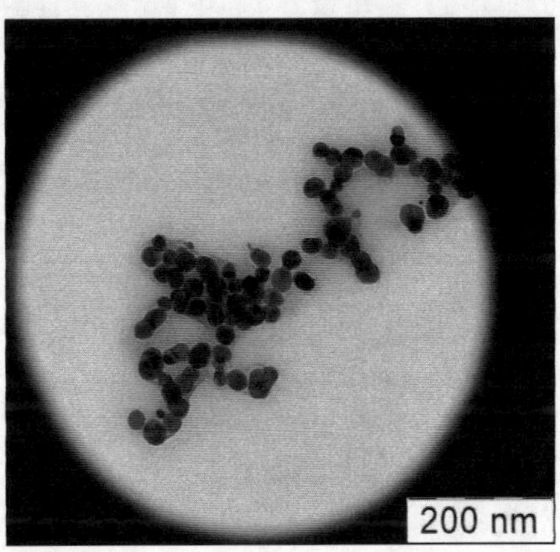

Figure 2. Produced silver nanoclusters of various sizes.

Under such a synthesis process, colloidal solutions of silver and nickel were obtained in which the initial sizes of the nanoclusters were 25–40 nm and 2–10 nm, respectively.

IOP Conf. Series: Materials Science and Engineering **387** (2018) 012073 doi:10.1088/1757-899X/387/1/012073

To metallize busofit with metal nanoclusters, an experimental laboratory reactor were developed using electrophoresis, its 3D model and a photograph of which is shown in figure 3.

(a) (b)

Figure 3. 3D model (**a**) and photo of the reactor for the application of metal nanoclusters (**b**): 1 - upper clamping cover, 2 - upper electrode, 3 - carbon cloth "busofit", 4 - base, 5 - lower electrode.

Busofit (3) was placed between the electrodes (2 and 5), after which the reactor was filled with a colloidal solution with silver or nickel nanoclusters and connected to an AC power source with a voltage of 40 V. Colloid particles have a charge, so they can move in an electric field. Reaching the destination electrode, the particles lose their charge and stick together - the colloidal solution coagulates on the surface of the material.

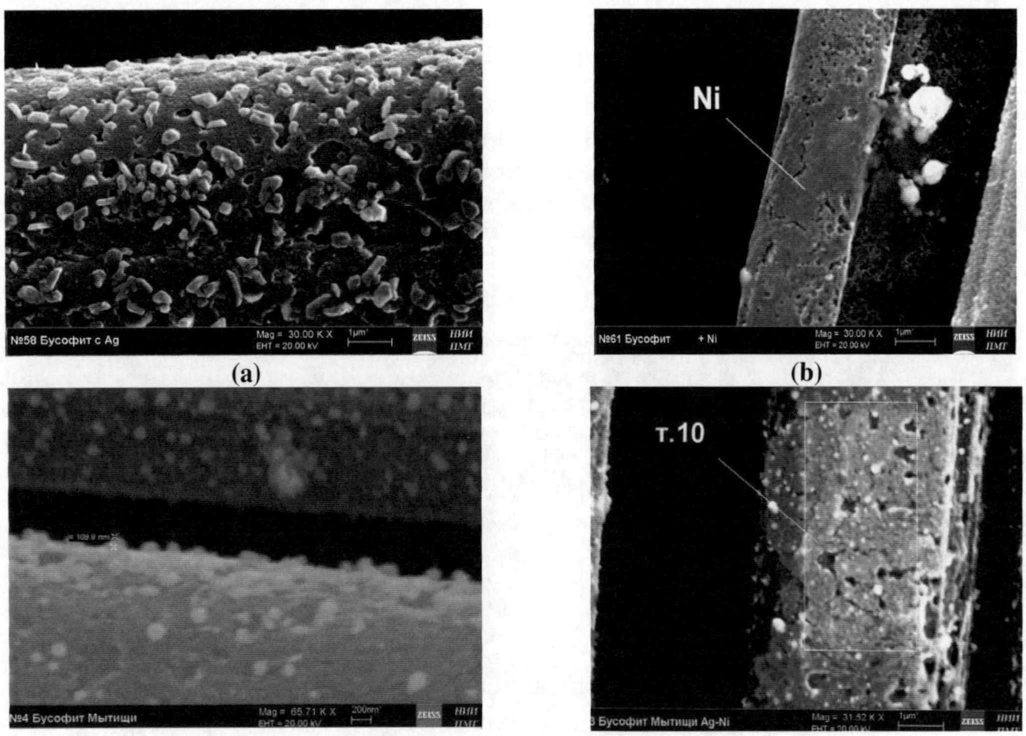

(a) (b)

(c) (d)

Figure 4. The busofit fibers metallized with silver (**a**), nickel (**b**), silver over nickel (**c**), (**d**).

Figure 4 shows shots of metallized busofit fibers obtained with an electron beam microscope. As can be seen, nickel is deposited on the busofit in a thin layer, and silver on the surface of the busofit fibers is formed in the form of large nanoparticles and crystallites, developing the surface of the fiber. The experiment of silver deposition on pre-nickel-plated busofit fibers also did not allow a single layer of silver to be deposited on the nickel sublayer as well as the initial busofit fibers. Apparently, this metallization result is associated with the initial size of metal nanoclusters, thin-film coatings are obtained from nanoclusters with sizes of 2–10 nm, and large nanoparticles and crystallites are formed from nanoclusters of a larger fraction.

3. Conclusions

The possibility of metallization of hidden surfaces of porous materials with metal nanoclusters is shown by electrophoresis without reducing their initial porosity. Such processing, combined with magnetron sputtering on the outer surfaces, allows the formation of a metal coating on the entire surface of a highly porous material, which reduces the electrical resistance and stabilizes the parameters of the electrode material.

To further study the metallization of highly porous materials, it is planned to use graduated colloidal solutions with a smaller spread of nanoclusters from different metals and alloys.

Acknowledgement

This work was supported by the grant № 8.7552.2017/BP from the Ministry of Education and Science of Russian Federation

Reference

[1] Arico A S, Bruce P, Scrosati B, Tarascon J M and Schalkwijk W 2005 *Nature Materials* **4** 366–77
[2] Huang X, Tan C, Yin Z and Zhang H 2014 *Advanced Materials* **14** 2185–204
[3] Bruno S 2007 *Nature Nanotechnology* **2** 598–9
[4] Lee S W, Yabuuchi N, Gallant B M, Chen S, Kim B, Hammond P T and Shao-Horn Y 2010 *Nature Nanotechnology* **5** 531–7
[5] Sleptsov V V, Elinson V M, Baranov A M and Tereshin S A 1995 *NATO ASI Series 3 High Technology* **1** 257–64
[6] Li X and Wei B 2013 *Nano Energy* **2/2** 159–73
[7] Pandolfo A G and Hollenkamp A F 2006 *Journal of Power Sources* **1** 11–27
[8] Blomquist N, Wells T, Andres B, Bäckström J, Forsberg S and Olin H 2017 *Scientific Reports* **7** 39836
[9] Weinstein L and Dash R 2013 *Materials today* **10** 356–7
[10] Dai L, Chang D W, Baek J and Lu W 2012 *Small* **8** 1130–66
[11] Astakhov M V, Galimzyanov R R, Klimont A A, Krechetov I S, Kundu M and Stakhanova S V 2016 *Current Nanoscience* **1** 83–9
[12] Ostroukhov N N, Tyanginskii A Y, Sleptsov V V, Tserulev M V 2014 *Inorganic Materials: Applied Research* **3** 284–8

Investigation of the synthesis of nanoparticles by pulsed electrical erosion in the conditions of overvoltage of the discharge gap

V V Sleptsov, A V Savkin, V I Berdnik, D Y Kukushkin and A O Diteleva

Moscow Aviation Institute (National Research University), 125993, GSP-3, A-80, Moscow, Volokolamskoe shosse 4, Russia

E-mail: savkin@mati.ru

Abstract. The process of obtaining silver nanoparticles by the method of electro erosion of metal electrodes when over-voltage discharge gap therebetween. For this purpose, a new scheme of an electric pulse erosion unit is proposed and investigated. Experimental results show the possibility of obtaining nanoparticles of conductive materials with a narrow size distribution by this method.

1. Introduction

The unique properties of nanoparticles have been the subject of intensive research for several decades. Modern science has shown the possibility of their use in many applications, including storage devices, creating new energy storage devices, MEMS, etc. [1–3]. To date, a number of methods for producing nanoparticles have been developed, which can be divided into two groups: chemical methods and physical methods [4–6]. The most common physical methods for producing nanoparticles are dispersion and laser ablation, which are carried out in vacuum or liquid media.

To obtain submicron and nanoscale particles of conductive materials (metals, alloys and semiconductors), a method of synthesis based on the electrospark erosion of raw materials in a liquid dielectric using the Electrical Discharge Machine (EDM), which is very attractive in its simplicity and accessibility, was proposed [7]. However, this method and a number of other known methods for synthesizing nanoparticles have the disadvantage of a large dispersion in the sizes (from 1 to 100 nm) of produced nanoparticles, which makes it extremely difficult to obtain or isolate nanoparticles with the required characteristics. However, it should be noted that such a disadvantage is typical for most physical high-performance methods for producing nanoparticles.

Further studies aimed at studying the method of electrospark erosion confirmed the possibility of obtaining nanoparticles of various conductive materials, including complex composition [8, 9]. Investigations of the influence of dielectric liquid media used in the synthesis of nanoparticles [10], including cryogenic liquids [11], have been carried out. A decrease in the size of the resulting metallic nanoparticles upon exposure to a liquid dielectric in the discharge zone by ultrasound due to cavitation and mixing has been established [12].

The results obtained by different groups of researchers made it possible to better understand the process of electrospark synthesis of nanoparticles, but could not eliminate the drawback associated with a large scatter of nanoparticle sizes. This can be due to several factors. In [10], it is proposed to use pulses with minimal energy for the synthesis of nanoparticles, which is difficult with the use of

IOP Conf. Series: Materials Science and Engineering **387** (2018) 012074 doi:10.1088/1757-899X/387/1/012074

standard pulse generators for EDM. In addition, as shown by the results of mathematical modeling of the behavior of vapor-gas bubbles formed in a liquid dielectric in breakdown [13], their collapse is in most cases accompanied by division into several parts whose dimensions depend on the condition of their formations, which subsequently leads to the formation of several nanoparticles.

On the basis of these results and arguments, we made the assumption that a decrease in the spread of the sizes of the produced nanoparticles is possible due to a change in the conditions for the occurrence of a spark breakdown, so that the process proceeds much faster. This can be achieved if a significant overvoltage is created on the EDM electrodes, which is not possible in any previously described scheme. To create an overvoltage, another switching element must be added to the circuit, which would commute the charged capacitor and the EDM electrodes to a high voltage.

2. Experimental

An experimental EDM was developed with the possibility of creating an overvoltage at the electrodes. The functional diagram of the experimental EDM is shown in figure 1.

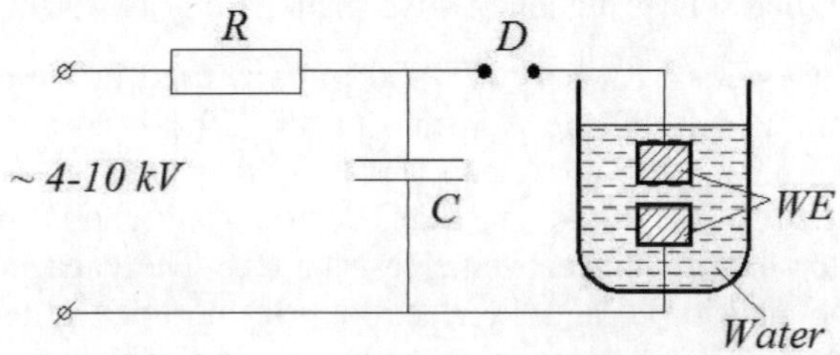

Figure 1. Functional diagram of the experimental EDM: C - high-voltage capacitor, R - limiting resistor, D - spark gap, WE - working electrodes.

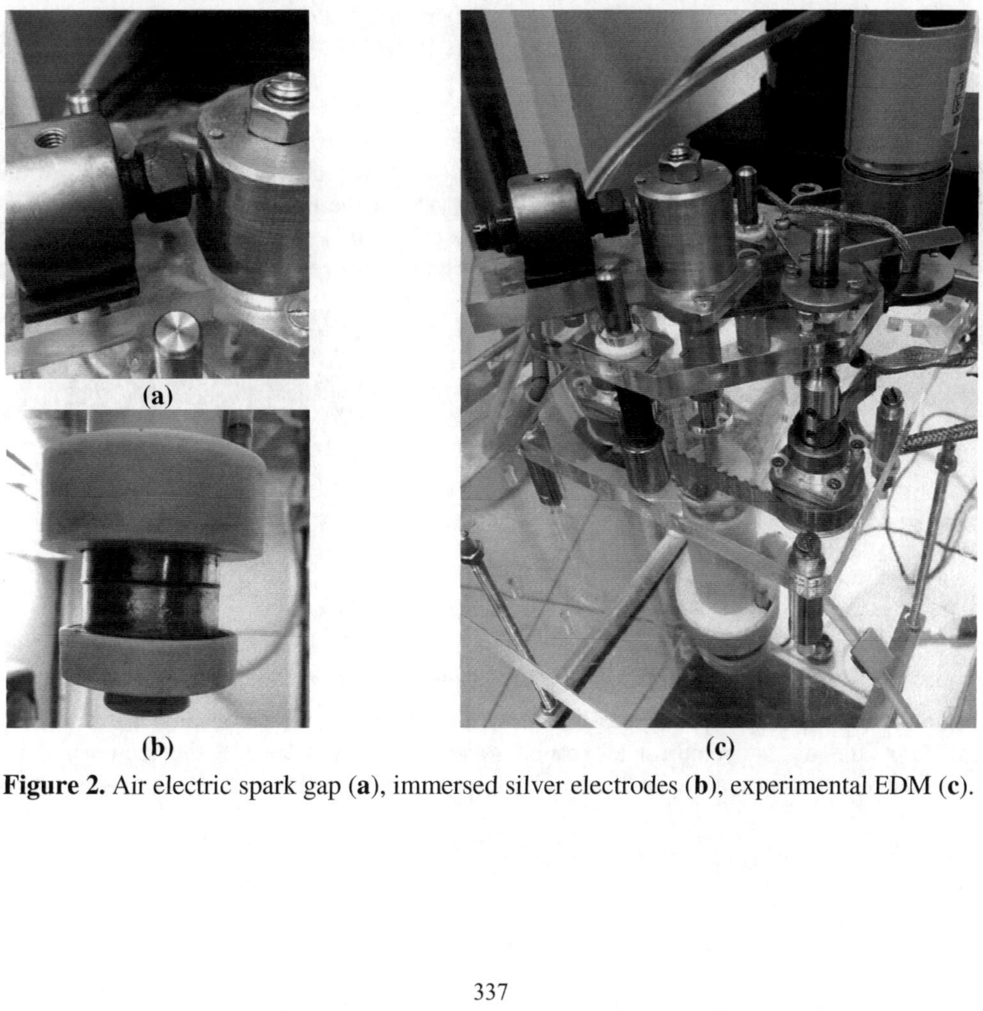

Figure 2. Air electric spark gap (**a**), immersed silver electrodes (**b**), experimental EDM (**c**).

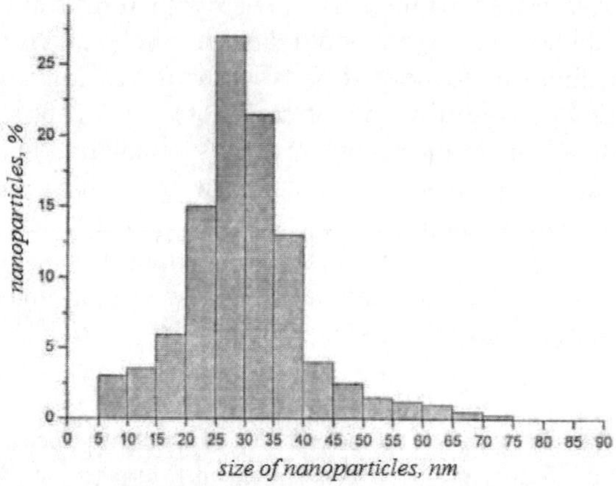

Figure 3. Distribution of silver nanoparticles by size.

A high-voltage capacitor (C) with a capacity of 1000 pF was used in the experimental EDM. The capacitor was charged from a high-voltage AC transformer (50 Hz) to a voltage of 1–10 kV. For capacitor switching (C), an air electric spark gap (D) is used, which is an air gap of 1 mm width between two 50 mm^2 electrodes. The photo of the air electric spark gap is shown in figure 2(a). The charge rate of the capacitor is determined by the resistor (R). The capacitor charged to a voltage sufficient to breakdown the air gap (D) and the electrode system (WE) is discharged through the air gap D to the working electrodes (WE) immersed in the liquid dielectric.

Distilled water was used as a liquid dielectric in the experiment. The working electrodes were made of silver, their photo is shown in figure 2(b). For the experiment, a gap of 70 μm was established between them. For uniform consumption of the electrode material during erosion, the rotation of the working electrode surfaces relative to each other was provided. A general view of the assembled experimental setup is shown in figure 2(b).

The experiment showed that at the indicated values of the parameters of the experimental setup, the breakdown voltage was of the order of 4 kV, and the discharge current in the pulses reached values of 1 ÷ 2 kA. The obtained samples of silver nanoparticles were subjected to the investigation of the size distribution by the method of dynamic light scattering (DLS), the measurement was carried out with the Photocor Compact-Z. The resulting particle size distribution is shown in figure 3. As can be seen from the figure, the distribution of the nanoparticles obtained in size is rather narrow. The diameter of more than 80 % of nanoparticles lies in the range of 20–45 nm.

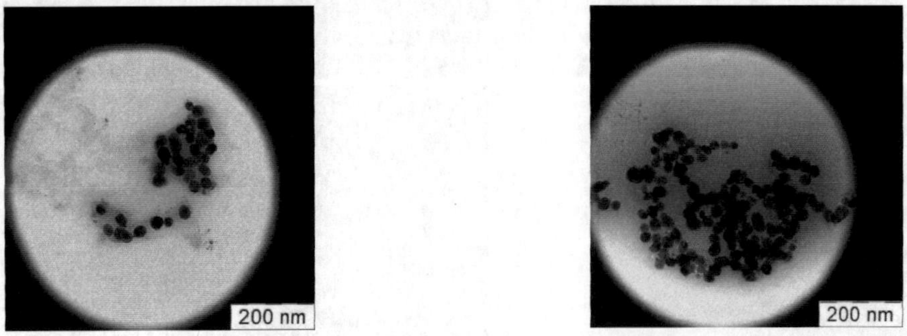

Figure 4. Typical photographs of the silver nanoparticles on an electron microscope.

In addition to the DLS method, images obtained with the transmission electron microscope LEO-912 AB OMEGA were used to confirm the nanoparticle size distribution. In the pictures, it is seen (figure 4) that most particles are spherical and have a size of 20 to 40 nm.

Electronograms samples obtained nanoparticles is shown in figure 5(a), electronograms sample of crystalline silver is shown in figure 5(b). The comparison showed that the obtained nanoparticles consist of crystalline silver, without a noticeable impurity of oxides or salts.

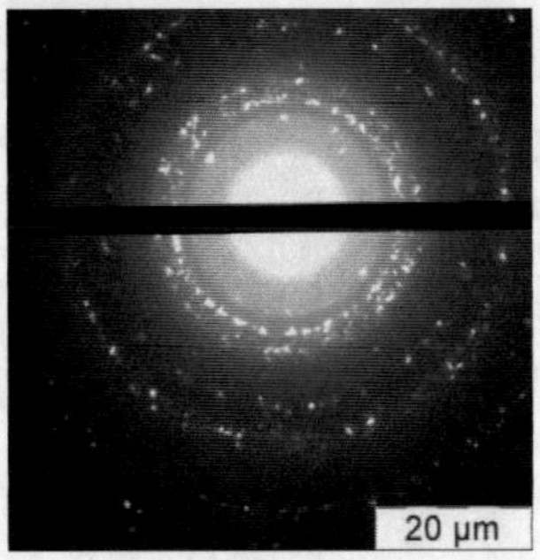

(a)

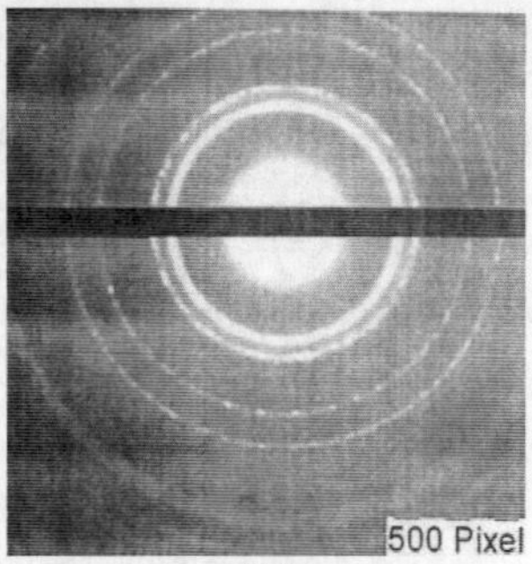

(b)

Figure 5. Electronogram of silver nanoparticles (**a**) and solid silver (**b**).

3. Conclusions

The possibility of producing nanoparticles by electrospark erosion under conditions of considerable overvoltage of the discharge gap was demonstrated. The resulting silver nanoparticles had a spherical shape and a rather narrow size distribution.

For a better understanding of the effect of overvoltage of the discharge gap on the production process of nanoparticles, it is planned to investigate this process for different ratios of discharge gaps and various conductive materials.

Acknowledgement

This work was supported by the grant № 8.7552.2017/BP from the Ministry of Education and Science of Russian Federation.

Reference

[1] Yan Y, Warren S C, Fuller P and Grzybowski B A 2016 *Nature Nanotechnology* **11** 603–8
[2] Karmakar S, Kumar S, Rinaldi R and Maruccio G 2011 *Journal of Physics: Conference Series* **292** 012002
[3] Theodorakos I, Zacharatos F, Geremia R, Karnakis D and Zergioti I 2015 *Applied Surface Science* **336** 157–62
[4] Gubin S P, Koksharov Y A, Khomutov G B and Yurkov G Y 2005 *Russian Chemical Reviews* **6** 489–520
[5] Lu A, Salabas E L and Schüth F 2007 *Angewandte Chemie* **8** 1222–44
[6] Rajput N 2015 *International Journal of Advances in Engineering & Technology* **4** 1806–11
[7] Berkowitz A E and Walter J L 1987 *Journal of Materials Research* **2** 277–88
[8] Hong J I, Parker F T, Solomon V C, Madras P, Smith D J and Berkowitz A E 2008 *J. Mater. Res.* **23** 1758–63
[9] Nguyen P K, Lee K H, Moon J, Kim S I, Ahn K A, Chen L H, Lee S M, Chen R K, Jin S and Berkowitz A E 2012 *Nanotechnology* **23** 415604
[10] Carrey J, Radousky H B and Berkowitz A E 2004 *Journal of applied physics* **3** 823–9
[11] Monastyrsky G 2015 *Nanoscale Research Letters* **10** 503
[12] Liu Y, Li X, Bai F, Chen J, Wang Y and Liu N 2014 *Particuology* **17** 36–41
[13] Shervani-Tabar M T and Mobadersany N 2013 *Theoretical and Computational Fluid Dynamics* **27** 701–19

IOP Conf. Series: Materials Science and Engineering **387** (2018) 012075 doi:10.1088/1757-899X/387/1/012075

Thin-film technologies for producing electrode materials for chemical sources of current (CSC) and ultra-high-volume capacitor structures (UCS)

V V Sleptsov, D Yu Kukushkin and A O Diteleva

[1] Moscow Aviation Institute (National Research University), 4 Volokolamskoe Shosse, 125310, Moscow, Russia

E-mail: 08fraktal@inbox.ru, anna.diteleva@mail.ru

Abstract. This paper describes analysis of thin-film technologies for promising CSC and UCS. It was shown that the stored energy amount is determined by the work to move charges in the preset direction. This relationship defines the CSC and UCS energy intensity and shows that the process of electric energy storage in them is the same and mainly determined by the polarization effects in the system.

1. Introduction

Electronics, instrumentation, power generation, transportation, medicine and numerous other areas of science and technology suggest a steady off-line supply sources share increase. Today, the best batteries samples are the secondary chemical sources of current (CSC) that provide energy intensities of around 250 W h/kg, and intensive research is under way to develop over the next 5 years energy storage units with the storage capacity of up to 450 to 500 W h/kg, and in the long term - up to the theoretically justified limit of 1000 to 1200 W h/kg. Energy intensity of ultra-high-volume capacitor structures (UCS) is 10 to 20 W h/kg, intensive research is under way to attain specific energy intensity of 20 to 40 0W h/kg and 220 to 250 W h/kg for the hybrid UCS. Conventional CSC and electrolytic UCS design is targeted at thick-film electrode materials formation, where chemically active CSC substance and ultra-porous UCS material are applied in paste form in a thick (30–60 μm) layer on aluminum or copper foil. This fairly cheap and productive technology is the most widely used for industrial purposes. However, recently, there developed a trend to reduce energy intensity of CSC produced by this technology [1–2]. In this way, there is a need to seek and develop new technologies that will provide required CSC and UCS energy intensity growth. The most perspective from this point of view are the thin-film technologies that have recently begun to develop which became actively developing lately [3–5].

2. Theoretical study of CSC and electrolytic UCS electrode materials thin-film formation technologies

The energy can be supplied to and extracted from the system in the form of heat or work [6–7]:

$$E = T \cdot S \pm \sum_{i=0}^{n} A_i \qquad (1)$$

where E is the energy supplied to the system; T is the system temperature; S is the system entropy; A_i is the work performed by the system (when energy is extracted) or $+A_i$, when energy is supplied to the system. For the real, irreversible processes the work, which can be supplied to the system or extracted from it, is written as the Product of two values - extensive (x) and intensive (y) ones. In this case, equation (1) should be written as follows:

$$dE = d(T \cdot S) + d(P \cdot V) + (\delta \cdot F) + d(\varphi \cdot q) + d\sum(\mu_i n_i) \quad (2)$$

where T – is the temperature; P – is the pressure; δ – is the surface energy; φ – is the electric potential; μ_i – is the chemical potential - extensive parameters; S – is the entropy; V – is the volume; q – is the charge; n_i – is the molal quantity - intensive parameters.

With the structural elements reduction to the nanodimensional values extensive parameters begin to depend on intensive ones. In this case, the energy intensity of nanostructured systems is much higher than the energy intensity of monolithics, and numerous experiments have unambiguously confirmed that fact [8–9]. By [10] shown that the thermodynamic and electrostatic accounting of polarization interactions of particles in the system lead to the same result - chemical potential of the molecule surrounded by its neighbors, producing an effective field E, is equal to the polarization work of the molecule under discussion. In this case, the chemical potential of the molecule (to an accuracy of up to the additive constant KT independent of the electric-field strength) can be represented as:

$$\mu = -PE = A_i \quad (3)$$

where P – is the dipole moment; E – is the electrical field strength; A_i – is the electric field work. Formula (3) as a result of a simple conversion can be written as:

$$dA_i = -ed\varphi \quad (4)$$

In this case, the Gibbs potential under the isobarometric conditions can be written as [10]:

$$dG_{PV} = \sum N_i d\mu_i = \sum N_i ed\varphi = \sum N_i Edp_i \quad (5)$$

Taking into account the electrocapillarity effects that occur during the two contiguous phases interfaces formation, the Gibbs potential is written as follows:

$$dG_S = \delta dF + Edp \quad (6)$$

where E – is the electrical field strength; P – is the dipole moment; δ – is the surface tension; F – is the surface area.

Using the Maxwell rearrangement and the Poisson's equation for equation (6), we obtain the fundamental electrocapillary equation [10–13]:

$$\frac{\partial\delta}{\partial x} = -\rho^{\text{free}} + \rho^{\text{bound}} = -q$$

(7)

where ρ^{free} and ρ^{bound} – are the free and bound charges surface density; ∂x – is the interface barrier jump. Based on equations (3, 4 and 7) [10, 13] molecule polarization work can be determined A_n.

$$A_n = pE = \mu \quad (8)$$

Equation (8) allows to draw the conclusion that the chemical potential of the molecule surrounded by its neighbors, producing an effective field E, is equal to the polarization work of this molecule. Differential equation (8) can be written as

$$dA_n = pdE + Edp = d\mu \quad (9)$$

In accordance with the equation of the first and second laws of Thermodynamics it can be written as [6, 10]

$$dG_{(dT=0,dp=0)} = \sum N_i d\mu_i = N_i ed\Delta\varphi \quad (10)$$

Ratio (10) is widely used in the theory of electrochemical cells that, first of all, convert the energy of a chemical reaction into electrical energy [10, 13]. The integrated form of equation can be written as follows:

$$\Delta G^{\circ}_{\text{ch}} = \sum \Delta G^{\circ}_{f\text{con}} - G^{\circ}_{f\text{init}} = \sum N_i \mu^{\circ}_{\text{con}} - \sum N_i \mu^{\circ}_{\text{init}} = z\varepsilon \quad (11)$$

where $z = \sum N_i e$ – is the amount of electricity (charge) that has passed through the external electrochemical cell circuit. This charge, as follows from the equation (11), is proportional to the amount of the cell substance that reacted, EMF = $\sum \Delta\varphi_i$ is the cell electromotive force (EMF) equal to the sum of the barrier jumps $\sum \Delta\varphi_i$. In this case, the cell EMF on a first approximation can be estimated as:

$$\text{EMF} = \frac{\Delta G^{\circ}_{\text{react.}}}{z} \tag{12}$$

Equations (11), (12) suggests that thin film structures less heat loss. This is due to the fact that in thin layers significantly less electrical resistance and, accordingly, reduced heat generation during the electron passes through them.

3. System without chemical reactions (big capacity capacitors electrolytic structures)

The major parameter of the systems accumulating electrical energy without chemical reactions is the electric capacitance. Differential capacity value can be found using electrocapillary curves [10, 13]

$$-\frac{d^2\delta}{d\varphi^2} = \frac{dq}{d\varphi} = C_d \tag{13}$$

By [10, 13, 14] was shown the influence of bound and free charges on the capacitance of DEL that were called dipole and electronic component respectively. Lippmann equation in this case can be converted to mean:

$$\frac{d\varphi}{d\rho} = \frac{d\varphi_e}{d\rho} + \frac{d\varphi_d}{d\rho} \text{ or } C^{-1} = C_e^{-1} + C_d^{-1} \tag{14}$$

where C_e^{-1} – is the electronic part of the DEL capacitance; C_d^{-1} – is the dipole part of the DEL capacitance.

Expression for the DEL electronic part capacitance can be represented as:

$$C_e = \frac{d\rho}{d\Delta x} = \frac{dp_e^V}{dz} \cdot \frac{dz}{d\Delta x} = \frac{p_e^V}{p_e^S} \cdot \frac{\varepsilon_S}{4\pi} \tag{15}$$

where and is the free charge volume and surface density respectively; z is the DEL thickness; is the interface barrier jump; is the DEL dielectric capacitance.

The conclusion could have been immediately drawn, that substances with high electrical conductivity should be used as the electrode material.

The second additive component in equation (14) is written in the following form:

$$\frac{d\rho}{d\varphi_d} = \frac{d\rho}{d\rho_d} \cdot \frac{d\rho d}{d\varphi d} = \bar{\varepsilon}\frac{d\rho d}{d\varphi} = C_d \tag{16}$$

where $\frac{d\rho}{d\rho d} = \bar{\varepsilon}$ – is the const, i.e., the effective dielectric constant, averaged throughout the dipole layer. Equation (16) suggests that the DEL_capacitance increases with the dipoles number and DEL dipole moment value increase, which ensures ε_S and factor $\frac{d\rho d}{d\varphi}$ increase.

4. Conclusions

1. It has been shown that the mechanism of CSC and UCS electrical energy accumulation is connected with the bound and free charges. The amount of accumulated energy is defined primarily by polarization effects in the system and is expressed by the formula: $A = PE$.

2. The interface barrier jump, which determines the operating voltage of the CSC and electrolytic UCS (supercondensers) to a precision of a certain constant is determined by the electron release into environment work - $\Delta\chi$ [10, 13]. CSC and electrolytic UCS energy intensity is determined by polarization processes and operating voltage, which depends on the barrier jumps on electrode materials.

3. To improve the CSC energy intensity, the chemically active substance should be nanostructured to increase the number of dipole moments and their size, and electrode materials should ensure stable maintaining of initial nanostructure during operation. One of the options is the use of a highly porous matrix when producing electrode materials. In this case, thin film technology allows the creation of hybrid UCS in which CSC and UCS properties are integrated. Lately this area is progressing actively and shows excellent results regarding energy intensity (energy intensity of 220 to 250 W h/kg).

4. Since CSC and UCS employ physically the same mechanism for electrical energy accumulation, it is thus possible to combine design solutions and develop a unified CSC and UCS electrode material based on the thin-film technology. In this case, it is possible to increase the specific energy intensity of the energy storage unit along with its specific capacity growth in the framework of a single design solution.

IOP Conf. Series: Materials Science and Engineering **387** (2018) 012075 doi:10.1088/1757-899X/387/1/012075

Acknowledgments
The study is carried out with the financial support from the State represented by the Ministry of Education and Science of the Russian Federation. Subsidies Provision Agreement No. RFMEFI57717X0275.

References
[1] Blomgren G E 2017 *Journal of the Electrochemical Society* **164**
[2] Myung S T 2017 *Nickel-Rich Layered Cathode Materials for Automotive Lithium-Ion Batteries: Achievements and Perspectives* **2** 196–223
[3] Kitsyuk E P 2017 *Research and development of the processes of formation of nanostructured electrochemical energy storage devices electrodes. Dis. Cand. Sci.* MIET
[4] Lebedev E F 2017 *Development of formation processes and study of properties of heat generation and energy storage elements of thermoelectric batteries. Dis. Cand. Sci.* MIET
[5] Vu Dyc Hoan 2017 *Research and development of thin-film multilayer electrolytic cells. Dis. Cand. Sci.* MAI
[6] Melvin-Hughes E A 1962 *Physical chemistry* (Moscow: publishing house l.L)
[7] Prigozhin M and Defei R 2010 *Chemical thermodynamics* (Publishing house Binom)
[8] Suzdalev I P 2009 *Nanotechnology: Physicochemistry of nanoclusters, nanostructures and nanomaterials* (Publishing House Librokom)
[9] Sleptsov V V 2015 *Physics and chemistry of of nanomaterials and nanotechnologies* (Sam Poligrafist)
[10] Salem R R 2010 *Physical chemistry. Introduction to the theoretical electrochemistry* (KomKniga)
[11] Damaskin B B, Petriy O A and Tsirlina G A 2010 *Electrochemistry* (Publishing House Khimia)
[12] Lewis J Rendil M 1936 *Chemical thermodynamics* (Publishing house ONTI Himteoret)
[13] Salem R R 2003 *Double layer theory* (Fizmatlit)
[14] Matveyev A N 2010 *Electricity and magnetism* (Lan)

IOP Conf. Series: Materials Science and Engineering **387** (2018) 012076 doi:10.1088/1757-899X/387/1/012076

Influence of acid-base and adsorptive characteristics of disperse aluminum modified by ammonium compounds on antifriction effect in lubricant

A G Syrkov[1], M O Silivanov[1], V R Kabirov[1] and I V Pleskunov[2]

[1]Department of general and technical physics, Saint Petersburg Mining University, 199106, Saint Petersburg, Russia
[2]Representative office of IMC Montan Company in the Republic of Belarus, IMC Montan, 220005, Minsk, Republic of Belarus

E-mail: syrkovandrey@mail.ru

Abstract. The article analyses the acid-base and adsorptive characteristics of the metal samples obtained using method of layering of different-sized molecules of quaternary ammonium compounds (QAC). The method allows to vary the hydrophilic-lipophilic, antifriction and other properties of commonly used disperse metals. The method includes the process of adsorption of reagents such as QAC based triamon (C_1-C_2) and alkamon (C_{16}-C_{18}), which have very different in length hydrocarbon radical at nitrogen atom. The acid-base properties of modified disperse aluminum were studied using Hammett indicator method and it has been established that there is a change in the quantity of the donor-acceptor centers of the powder caused by the adsorption of QAC. Spectrophotometric analysis of surface centers was performed using the Hammett indicators within the range of pKa values from -0.3 to 14.2. The distribution curves of the active adsorption centers of aluminum-based powders were plotted against the acid dissociation constant (acid strength - pK_a) of the indicators. A sharp decrease is observed in the number of centers in the Lewis acid sites region for the Al/(A+T) sample with co-adsorbed alkamone (A) and triamon (T). In addition, significant decrease in antifriction and water-repellent properties of the sample Al/T/A (consistent treatment of metal in vapor of T and A) in comparison with sample Al/(A+T) is explained by the increase of Bronsted base centers with pK_a = 9.5 and less intense interaction of QAC with Lewis acid centers with pK_a = 14.2.The following pattern was revealed: increase in the antifriction effect of oil based lubricants is accompanied by reduction of the Bronsted centers content and as a rule by the decrease of acidity function (H) of the surface of the QAC-modified Al-powder which was added to the lubricant.

1. Introduction

According to modern views,based on the classical works of Tanabe, the surface of solids consists of Lewis and Bronsted acid and base centers [1 - 4]. The spectrum of the distribution of adsorption centers (DAC) allows to predict the reactivity and sorptive characteristics of the surface and other physico-chemical properties of solids used as catalysts, fillers of composite and nanostructured materials [2, 4].

Using the method of adsorption of Hammett acid-base indicators [1], it is possible to determine the content of active centers, their distribution in terms of the ionization constant and the value of the Hammett acidity function. Previously conducted experimental research was mainly dedicated to the study of surface of solid oxides [1-3]. However, conducted scientific works have not sufficiently studied the stabilized disperse metals, such as aluminum powder, containing on the surface different-dimensional molecules of ammonium compounds, which possess electron-donating properties [5, 6]. At the same time, mentioned information is extremely important for studying the adsorption of modifiers on the surface of disperse metals, used as additives to the oil lubricant [7, 8].

The main goal of this work is to study the change in the acid-base properties of the surface of disperse aluminum caused by the adsorption of quaternary ammonium compounds, and to find corresponance of these properties with the antifriction effect of modified powders as additives to the oil lubricant.

2. Materials and methods

We have studied the distribution of donor-acceptor centers along the surface of the original aluminum powder (PAP-2, specific surface area = 2.6 m^2/g) and powder modified by various adsorption mechanisms in vapors of quaternary ammonium compounds (QAC). Mentioned methods of layering of QAC on metals (Al, Cu, Ni) from gas phase are fully described in the literature [5-8]. QAC-based surfactants were used as adsorbates: alkamon (A) and triamon (T). Composition of triamon corresponds to the chemical formula $[(HOC_2H_4)_3N^+CH_3][CH_3SO_3^-]$tris-(β-oxyethyl) methyl- ammonium-methyl sulfate; adsorbate A has a similar composition, but with a higher-molecular-weight radical (C_{16}-C_{18}) at the nitrogen atom [9]. The following types of samples have been obtained: with adsorbed T (Al/T), with adsorbed A (Al/A) and the sample with adsorbed A and T in mixture (Al/(A+T)). Powder Al/(A+T)previously has shown better hydrophobic and antifriction properties [5, 10, 11].

The measurement of the number of active center (q) was performed by spectrophotometric indicator method [3, 12] using acid-base indicators with pK_a value within the range from -0.3 to 14.2 and device PE5400UF. The indicators are listed with their values of ionization constants: Orthonitroaniline (-0.3), Brilliant green (1.5), Fuchsine (2.1), Xylenol orange (2.6), Methyl orange (3.5), Bromophenol blue (4.1), Methyl red (5.0), Bromocresol purple (6.4), Phenol red (8.0), Phenolphthalein (9.5), Alizarine Yellow (11), Indigo carmine (12.8), Thymolphthalein (14.2).

The methods of measuring solution density and calculation of the change in the optical density (ΔD) after adsorption of the indicators and the number of surface centers of a given strength (q_{pKa}) are described in the works [3, 12, 13]. The Hammett acidity function was calculated from equation: $H_0 = \sum pK_a \cdot q_{pKa} / \sum q_{pKa}$. Root mean square error of independent measurements of the optical density of one sample is $\Delta q_{pKa} = \pm 0.40\%$.

3. Results and discussion

Surface of the Al-based samples can be characterized as a set of Lewis and Bronsted acid and base centers (figure 1) which are able to interact with Hammett indicators.

The obtained experimental data shows that interaction of indicator and Al surface leads to the adsorption of certain indicator on the active centers with only corresponding acid strength. The adsorption of indicators provides changes in coloration of the solution, which can be observed spectrophotometrically. Such method is used to characterize the amount of certain active centers.

Indicators with lower dissociation constant are selectively adsorbed on the Lewis base centers. With the increase of acid strength of indicators, they are adsorbed on Bronsted acid ($0 < pK_a < 7$) and base ($7 < pK_a < 14$) centers. At the value of $pK_a > 14$ the adsorption occurs on the Lewis acid centers [3, 12].

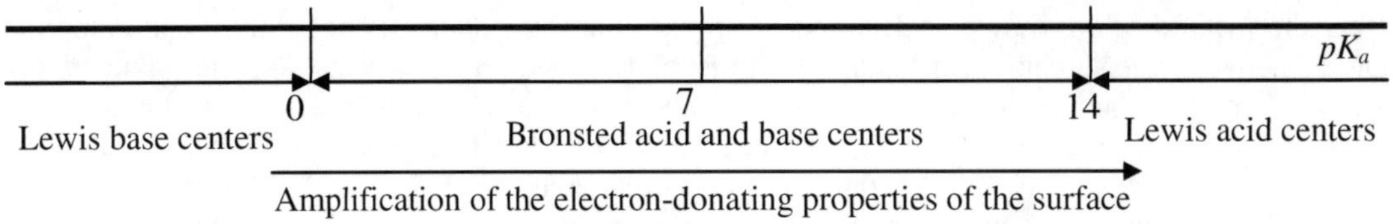

Figure 1. Formal layout of acid-base surface centers in respect to strength and type (typical for partially dehydrated amphoteric oxides, such as aluminum oxide).

Surface layer of the initial Al-powder PAP-2 consists of aluminum and aluminum oxide structures. This fact is proved by the following X-ray photoelectron spectrum peaks: for aluminum which appears at lower binding energies ($E_{binding}$ = 72.4eV – Al2p), for oxidized aluminum ($E_{binding}$ = 74.6eV – Al2p) and for oxygen ($E_{binding}$ = 531,3 eV – O1s), which is presented in the aluminum oxide structures [5]. Industrially manufactured powder PAP-2 contains stability agent – non-unifirm stearine nanofilm [14]. The thickness of Al-oxide layer may reach 0,002 μm [7, 14]. According to the research [15], stearine has low adsorption activity which is proved by the little adsorption (approximately 0,03 μmol/g) of Hammett indicators on the stearine in the experiments [7].

According to the characteristics of the DAC-spectrums (figure 2), all samples have 4 types of peaks on distribution curves: pK_a= -0.3; 5; 9.5; 14.2. Initial Al-powder contains mostly Lewis acid centers (pK_a= 14,2), which correspond to the centers with vacant atomic orbital at Al atom and capable to take an electron pair [2, 12]. Additionally, the composition surface of initial metal includes Bronsted base centers (pK_a= 9,5) which correspond to OH-groups. The peaks corresponding to Bronsted acid and base centers (pK_a= 5 and pK_a= 9,5) indicate the existence of hydrate-hydroxyl layer on the initial Al-powder surface [16,17]. Therefore, it has been concluded that initial powder is oxidized. In case of application of alkamon or triamon (curves II, III), there is a significant decrease in the number of Lewis acid centers with pK_a 14.2. However, when these substances are applied together from the mixture (A+T), the number of these centers decrease by more than an order of magnitude (curve IV). It is obvious that the adsorption of QAC occurs on Lewis acid centers. Moreover, when vapor mixture (A+T) is used the number of covered acid centers reaches the similar quantity to which could be provided by using vapours of triamon and alkamon separately taken together. This observation correlates with the concentration of QAC in the samples: Al/A – 0,13 mol %, Al/T – 0,22 mol %, Al/(A+T) – 0,32 mol % [6]. The most probable mechanism of adsorption of QAC on aluminum powder is the donor-acceptor interaction of the cation part $[N^+R_1R_2R_3R_4]$ of the adsorbate molecule with electron-accepting centers of the aluminum surface (R_1, R_2, R_3, R_4 – the radicals at the nitrogen atom in the compounds).

Adsorption of QAC occurs on the base OH-groups (pKa= 9,5), however in this case, the application of adsorbate (A+T), as seen from figure 2, does not provide the sum of the covered centers observed when applying A and T separately. The possibility of adsorption of hydrophobic QAC molecules on hydrophilic surface centers caused by the presence of polar groups ($CH_3SO_3^-$) in the structure of these molecules [9, 15].

The mechanism of changes in the peaks amplitude at pK_a= 5 and pK_a= -0,3 is far more complicated. Alkamon and triamon separately are capable to increase the amount of acid centers at pK_a= 5, however, at pK_a= 2,6 and pK_a= -0,3 only alkamon is capable to provide such results. Most probably, the mentioned increase is connected with addition of acid centers of each modifier (A or T) itself to the system. This hypothesis is proved by nearly constant value of specific surface area before and after the modification: (2.6 ± 0.1) m^2/g [5–7].

It is important to note that mixed adsorption of A and T (A+T, mass ratio A:T ≈ 1 [5]) provides significant results in coverage all types of centers listed, as shown in figure 2 (curve IV). The decrease in amount of acid centers was estimated by the change in amplitude of corresponding peaks.

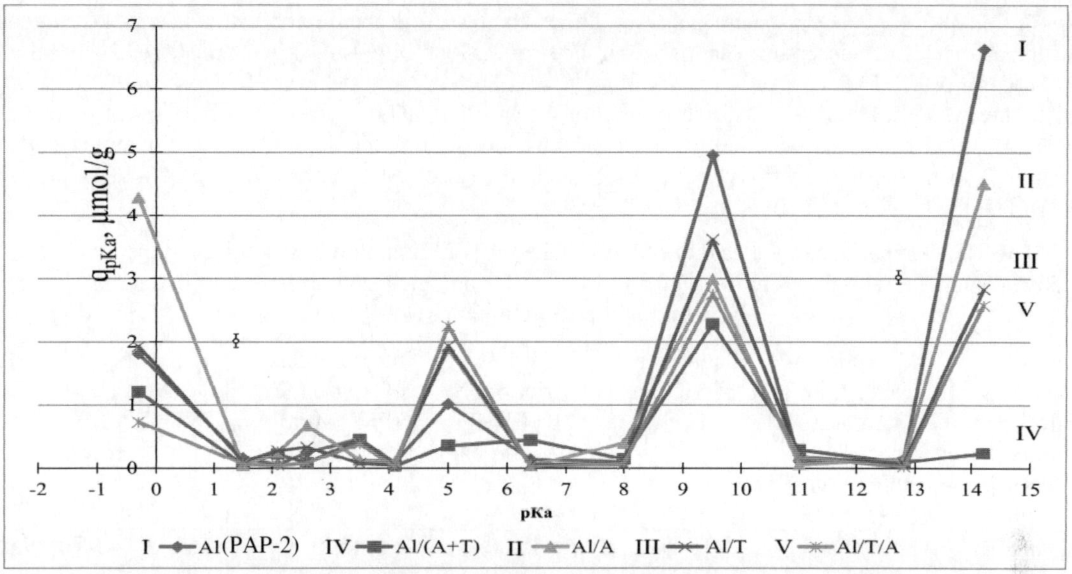

Figure 2. Distribution of acid-base centers on the surface of modified Al-powders.

According to the data presented in figure 2, total number of centers decreases depending on the adsorbate in the row: A, T, A+T. In addition, application of mentioned QAC leads to the decline of the acidity function, as shown in table 1 and according to the row: Al (PAP-2), Al/T/A, Al/T, Al/A, Al/(A+T).

Thus, the acid-base properties of disperse aluminum were studied using Hammett indicator method and it has been established that there is a change in the quantity of the donor-acceptor centers of the powder caused by the adsorption of QAC. The obtained data about the structure of the surface of the samples enables us to broaden the understanding about the evolution of the acid-base properties of the aluminum powder surface during the adsorption of ammonium compounds. These results provide new insights on the previously described [5 - 8, 11] mechanisms of enhancement of the hydrophobic and antifriction properties of the modified powder Al/(A+T), and on the increase of its chemical stability during atmospheric oxidation. The DAC-spectrums confirm that the adsorption of QAC occurs due to the formation of metal-nitrogen heteroatomic interaction with a shift of the electron density from metal to nitrogen, which is detected by X-ray photoelectron spectroscopy [5, 7, 8]. This data does not contradict the views that QAC, due to their diphylic nature, can be adsorbed on different active centers [5], such as Lewis acid and Bronsted basic groups of the surface (see peaks at pK_a 14.2 and 9.5 in figure 2). In addition, there is a trend towards a decrease in the total number of centers and surface acidity during mixed adsorption of A and T, which indicates greater possibility of occurrence of synergistic effects for Al/(A+T) sample rather than for Al/A and/or Al/T samples. These effects have been recorded for various properties and characteristics: hydrophobicity [5], decrease in the integral index, force and friction coefficient in lubricants with additives of Al/(A+T) [8, 11]. According to the table 1 the following trend is observed: the listed tribological characteristics decrease relatively to the acidity function H_0 and the number of active centers with pK_a 9.5 and 14.2 of the disperse additive of modified aluminum which is used in the oil lubricant.

It is interesting for the purpose of the research to compare the properties of Al/T/A (sequential layering of T and A) and Al/(A+T) samples, which contain both modifiers (A and T) and have different nitrogen content of chemisorpted QAC on the surface (0.55 and 0.32 mol% respectively). Interpretation and explanation of the substantially different properties of these samples, especially antifriction is very complicated [8]. New DAC-spectra analysis allows to relate the mentioned difference to less intense fulfillment of electron-accepting centers of surface (q=2.57 μmol/g – Al/T/A against q=0.24 μmol/g – Al/(T+A) for pKa = 14.2.

Another reason of increase of force of friction in case of Al/T/A sample can be a result of distinct hydrophilization of the surface. Truly, the concentration of hydrophilic centers, which represents the appearance of the hydrate-hydroxyl cover, is greater in case of Al/T/A than in case of Al/(A+T): 2.75 against 2.28 μmol/g for pKa = 9.5.

Table 1. Comparison of the acidity of the surface (H_0) of aluminum-based powders, the number of active centers (q) with different pK_a (9.5 and 14.2) and tribological characteristics of industrial oil I-20 with modified aluminum powder.

Disperse additive to the oil lubricant (1 mass %)	Adsorption of water from saturated vapors by an additive (g/g) [5, 8]	Coefficient of friction [11]	Force of friction, N (P=17 MPa, steel-bronze) [11]	Integral index of friction D (P=51MPa) [5, 7, 8]	q, μмоль/г		H_0
					pK_a14.2	pK_a 9.5	
Al (initial)	0.0023	0.0092	45.06	1690	6.63	4.95	9.65
Al/T	0.0017	0.0077	37.60	780	2.83	3.62	7.77
Al/A	0.0021	0.0076	37.03	910	4.50	2.99	7.06
Al/T/A	0.0016	0.0074	36.50	1000	2.57	2.75	8.54
Al/(A+T)	0.0013	0.0064	31.57	300	0.24	2.28	6.32

One of the possible reasons for the unusual properties of the Al/(A+T) sample in comparison with the Al/A and Al/T samples can be the formation of new adsorptive compound Al/T···A in the surface layer of aluminum due to structural similarity of the T and A molecules, which is accompanied by an experimentally recorded increase in the binding energy of electrons of the characteristic level N1s by approximately 2 eV [5, 7, 10].

4. Conclusion

Under other equal conditions reduction of friction force and coefficient of friction in the studied tribosystem occurs as a rule when the acidity H_0 and the content of the Brensted basic centers of modified aluminum decrease respectively.

References

[1] Tanabe K 1973 *Solid acids and bases* (Moscow: Mir) 156
[2] Sychov M M, Minakova T S, Slizhov Yu G and Shilova O A 2016 *Acid-base characteristics of the surface of solids and control of properties of materials and composites* (St. Petersburg: Himizdat) 276
[3] Nechiporenko A P 2017 *Donor-acceptor properties of surface of solid-phase systems. Indicator method* (St. Petersburg: Lan) 284
[4] Finlayson M F and Shah B A 2012 *Journal of Adhesion Science and Technology* **4** 431–9
[5] Beloglazov I N and Syrkov A G 2011 *Chemico-physical foundations and methods for obtaining surface-nanostructured metals* (St. Petersburg: St. Petersburg State University) 72
[6] Syrkov A G 2013 *Russian Journal of General Chemistry* **83** 1392–3
[7] Syrkov A G 2013 *Smart Nanoobjects. From Laboratory to Industry* ed K. Levine (New York: Nova Science Publishers, Inc.) 214

[8] Bystrov D S, Syrkov A G, Pantyushin I V and Vakhreneva T G 2009 *Chemical physics and mesoscopy* **11** 462–6
[9] Reference book 1984 *Surface phenomena and surface-active substances* ed Abramzon A A, Shchukin E D (L .: Chemistry) 392
[10] Syrkov A G 2015 *Russian Journal of General Chemistry* **85** 1538–9
[11] Taraban V V, Syrkov A G, Silivanov M O and Nazarova Ye A 2015 *Chemical Physics and Mesoscopy* **17** 557–64
[12] Nechiporenko A P, Burenina T A and Koltsov S I 1985 *J: Society. Chemistry* **55** 1907–12
[13] Zakharova N V, Sychov M M, Korsakov V G and Myakin S V 2011 *Condensed media and interphase boundaries* **13** 56–62
[14] Directory 1972 *Aluminum. Metallurgy, processing and application of aluminum alloys* (M .: Metallurgy) 664
[15] Dedkov G V 2000 *Achievements of Physical science* **170** 585–618
[16] Sorochkina E A, Smotraev R V Kalashnikov Yu V and Gruzdeva E V 2013 *Questions of chemistry and chemical technology* **6** 102–4
[17] Sitnikov P A, Ryabkov Yu I, Ryazanov M A, Belykh A G, Vaseneva I N, Fedoseev M S and Tereshatov V V 2013 *Komi Scientific Center of the Ural Branch of the Russian Academy of Sciences. Chemical sciences* **3** 19–26

Development of a device for automating the diagnostics of plasma parameters by probe method

A S Talanov, A A Lisenkov and D K Kostrin

Department of electronic instruments and devices, Saint Petersburg Electrotechnical University "LETI", 197376, Saint Petersburg, Russia

E-mail: dkkostrin@mail.ru

Abstract. This work is devoted to the improvement of the probe method of plasma diagnostics in order to obtain more accurate results, as well as to simplify this process by developing new schemes for obtaining and processing data acquired in the study.

One of the main methods of plasma diagnosis is the probe method. The advantage of this method over others [1–9] is its direct contact with the plasma, which allows to accurately determine its parameters. For example, in the analysis of plasma using the optical method, it is necessary to observe the same solid angle of the plasma tube with the reference sample, also this method has a large error in the identification of data using spectrometers. In the probe method, this error is much smaller, since modern ammeters allow quite accurate measurements.

Also, the probe method has its drawback. This method applies to active methods of investigation of plasma. In such measurements, sensors are used that come into contact with the plasma, violating its initial state. However, taking into account the perturbation introduced by the electrode into the plasma, it is possible to achieve high accuracy results.

Summing up, we can formulate some basic requirements for modern methods of diagnostics of plasma parameters: accuracy of parameters measurement; ability to work at different pressures; simple design, easy operation and low cost; ability to work at different temperatures; speed of the study.

In this method, a laboratory model is used as a source of gas-discharge plasma. Into the gas-discharge plasma is introduced a perturbing metal body, which can be of different shapes and from different materials. A metal body is supplied with some potential relative to the anode. Schematic implementation of this model equipment is shown in figure 1(a).

In this model, the main element is a vacuum lamp with a pressure of 10^1–10^2 Pa, in which 2 electrodes are placed. Using the power supply 1 between the electrodes the voltage for the ignition of the gas discharge is set. The discharge current is determined by the resistor connected in series to the circuit. The bias potential on the probe is set by the power supply 2 relative to the anode and is controlled by a voltmeter connected in parallel to the power supply.

With the help of an ammeter the current flowing on it is also controlled. According to the data obtained, a probe current-voltage characteristic is plotted, from which such important plasma parameters as the electron and ion currents, and the floating potential are obtained. However, the data obtained is not sufficient to determine other important parameters such as electron temperature, electron and ion concentrations, and plasma potential at this point. For these purposes, the data are processed, which provides a fairly complete picture of the parameters.

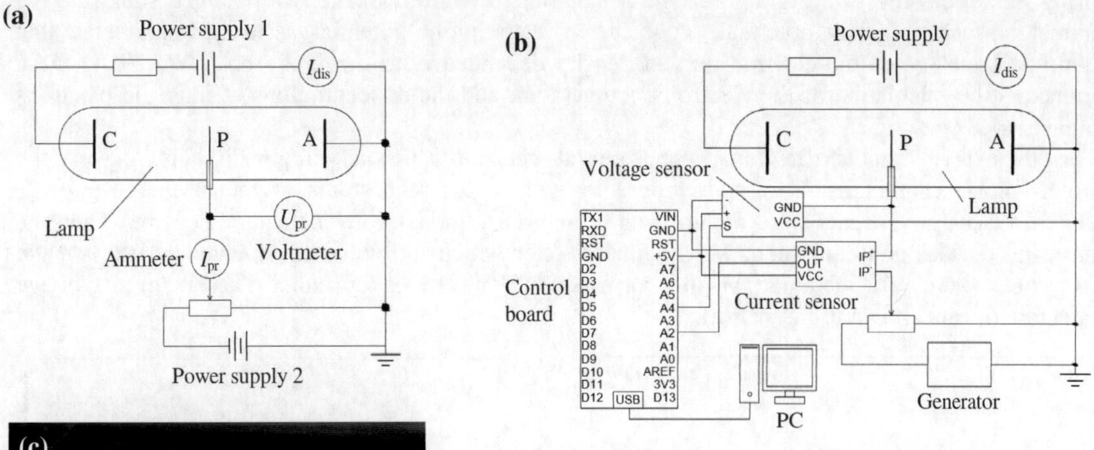

Figure 1. Schematic of the device for the probe method implementation (**a**); scheme for the automation of obtaining data (**b**); low-pressure gas-discharge lamp (**c**).

In order to eliminate the shortcomings of the system, it is necessary to develop a scheme for obtaining the probe current-voltage characteristic in the automatic mode. This will allow to capture data in a short period of time, thereby increasing the accuracy of the calculated plasma parameters.

The power supply 2 is replaced by a sawtooth pulse generator; the analog voltmeter and ammeter are replaced by a voltage sensor and a current sensor respectively. As the control board the Arduino scheme on the base of ATmega 328 microcontroller is used (figure 1(b)).

The basis of the laboratory model is a low-pressure gas-discharge lamp with soldered probes, shown in figure 1(c). Each probe has its own length, and therefore a different area of the receiving surface, which must be considered in the calculations. The data received from the probes is monitored using the port monitoring function in the Arduino software package. The monitored data is recorded and stored on the computer. For further data processing, the MagicPlot software package is used, in which the volt-ampere characteristics are built (figure 2).

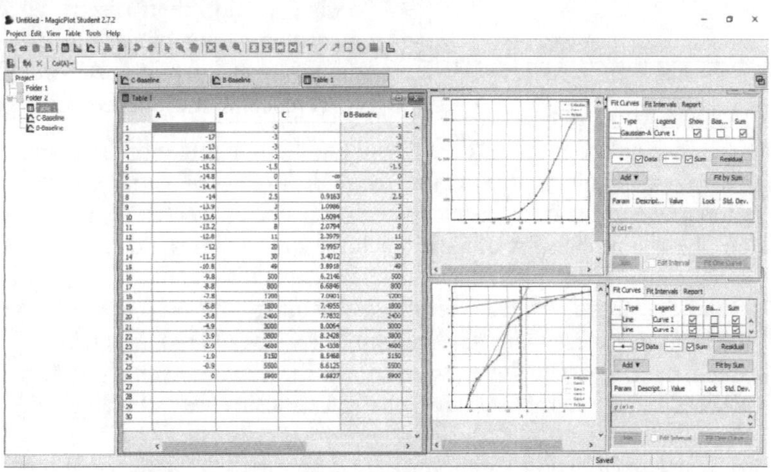

Figure 2. MagicPlot software package window.

During the processing of the graphs in the MagicPlot software package, we obtained some plasma parameters, such as plasma potential, probe current at the probe potential, as well as parameters that determine the slope of the electron current density dependence on the probe potential. Using these parameters the calculation of the electronic temperature and the concentrations of ions and electrons can be carried out.

For the experimental verification of the proposed scheme for the study of gas-discharge plasma, the current-voltage characteristics of two probes located at different points of the discharge gap for different discharge currents were obtained on a laboratory model with a standard scheme. Then the measuring devices and equipment were replaced, after which the volt-ampere characteristics of the two probes were also obtained at the corresponding discharge currents. These current-voltage characteristics are shown in figure 3(a).

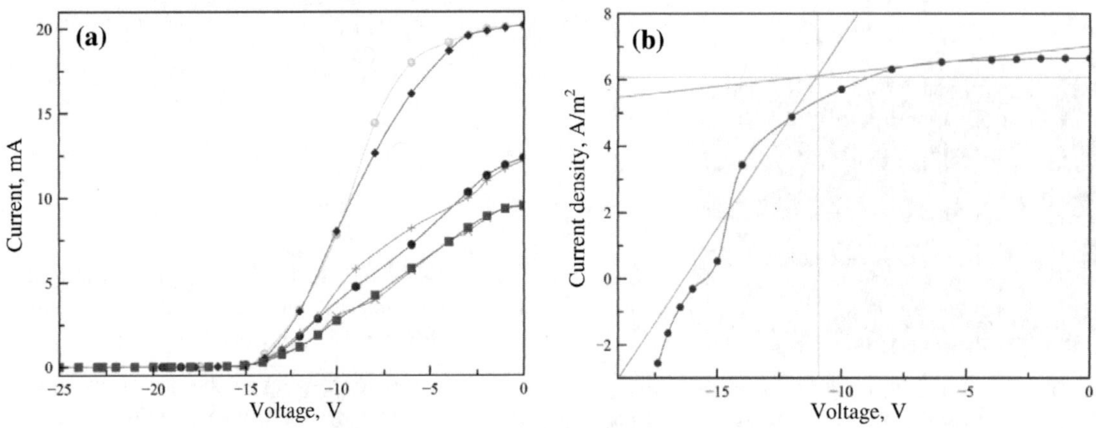

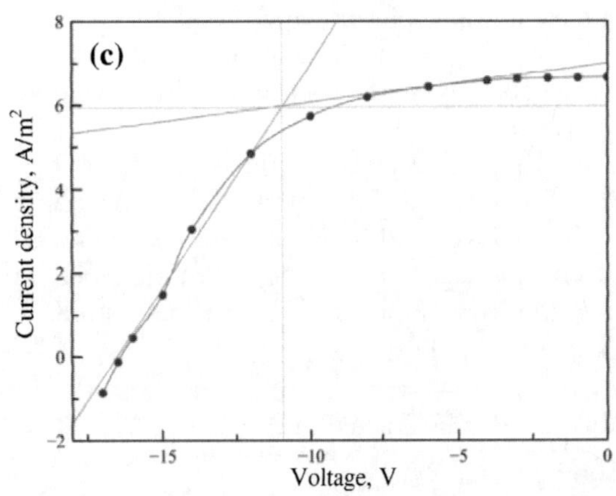

Figure 3. Volt-ampere characteristics of the probe at different discharge currents (**a**); logarithmic plot of current-voltage curve for the probe at I_{dis} = 40 mA before (**b**) and after (**c**) replacement of the measuring devices.

Analyzing the obtained dependences it can be noticed that the maximum probe current increases with increasing discharge current. This is due to the fact that as the potential between the electrodes increases, the emission of electrons from the cathode also increases. With a decrease in the modulus of the potential of the probe, the probability of electrons hitting it increases, due to this, an increase in the current is observed. When the probe potential value is greater than the plasma potential, an electronic layer is formed around the probe, which determines the maximum current on it. As the discharge current increases, the layer thickness increases, thereby increasing the maximum current on the probe. After that, the obtained characteristics were processed. In the software package MagicPlot were built logarithmic dependences of the current-voltage characteristics. These characteristics are shown in figures 3(a), (b).

After the experimental verification of the proposed scheme, the following conclusions can be made: the use of this scheme increases the accuracy of the plasma parameters measurement; the use of this scheme in the processing of data can uniquely determine the position of the tangents, which in turn reduces the divergence of possible options in obtaining the plasma parameters.

Based on the work done, it can be concluded that this scheme greatly simplifies the experimental work on obtaining plasma parameters, and also significantly reduces the processing time of the data. In accordance with the experimental work done, it can be said that it has led to satisfactory results. Despite this, there are great opportunities for further work to improve this diagnostic method.

References

[1] Uhov A A et al. 2014 *Journal of Physics: Conference Series* **567** 012039
[2] Kostrin D K et al. 2016 *Vakuum in Forschung und Praxis* **3** 34–7
[3] Gerasimov V A et al. 2016 *Journal of Physics: Conference Series* **729** 012019
[4] Kostrin D K et al. 2016 *Journal of Physics: Conference Series* **729** 012030
[5] Kostrin D K et al. *Journal of Physics: Conference Series* **735** 012055
[6] Kiselev A S et al. 2017 *Journal of Physics: Conference Series* **789** 012027
[7] Kostrin D K et al. 2017 *Journal of Physics: Conference Series* **857** 012022
[8] Kostrin D K et al. 2017 *Journal of Physics: Conference Series* **872** 012055
[9] Ramazanov A N et al. 2017 *Journal of Physics: Conference Series* **929** 012062

IOP Conf. Series: Materials Science and Engineering **387** (2018) 012078 doi:10.1088/1757-899X/387/1/012078

Finite element modeling of nanoindentation and scratching of Si, SiC, Ge crystals with anisotropic plasticity

R S Telyatnik

Institute of Problems of Mechanical Engineering, Russian Academy of Sciences, 199178, Saint Petersburg, Russia

E-mail: statphys@ya.ru

Abstract. Parameters of Hill's anisotropic elastoplasticity for diamond and Si, SiC, Ge semiconductors have been estimated from nonlinear stress-strain curves obtained by computational quantum chemistry. Using these parameters, finite element modeling of nanoindentation with Berkovich pyramid and frictionless scratching with spherical tip have been carried out. It has been shown, that the plastic anisotropy leads to the hardness anisotropy commonly observed by experimental indentation of (111), (011) and (001) crystallographic surfaces of the diamond-like crystals. The creep flow has been regarded in a quasi-static approach. Propagation of a median crack has been accounted by cohesive zone model.

1. Introduction

Indentation [1] is a common method of experimental determination of hardness and elastic properties of materials. In particular, nanoindentation and scratching experiments are widely used for testing thin film heterostructures. Since the indentation theory lacks analytical solutions, it is accompanied by computer modeling by finite element method (FEM), which requires data on plastic behavior of a material from other experiments, such as axial deformation of a rod beyond elastic linearity limit. However, a lack of such experimental data for materials of interest in turn forces researchers to calculate necessary properties of the materials by means of computational chemistry, especially when it comes to single crystals possessing all kinds of anisotropy. Indeed, for orthotropic materials (such as cubic diamond-like crystals C, SiC, Si, Ge), when compared to isotropic ones, not only the elastic anisotropy arises, when the shear modulus G becomes independent from the Young's modulus E and the Poisson's ratio ν, but the plastic anisotropy arises as well, when the elastic limit (yield point) for shear stress σ_Y^S becomes independent from the one for axial stress, e.g. compressive yield stress σ_Y^C, as we are going to consider an indentation process that generally comprises only shear and compressive strains. The independent pair of shear and compression elastic limits can be described by Hill's parameter h of plastic anisotropy [2]:

$$h^{s/c} = \left|\sqrt{3}\,\sigma_Y^S/\sigma_Y^C\right|. \quad (1)$$

If $h = 1$, the equation becomes the Mises' criterion of isotropic plasticity.

2. Parameters of anisotropic elastoplasticity

In our previous work [3], the nonlinear stress-strain curves up to creep or fracture limit for shear and axial deformations of C, SiC, Si, Ge crystal cells were calculated by computational quantum chemistry in ABINIT software for the ground state at $T = 0$ K by two methods: DFT LDA with TM pseudopotentials and DFT GGA with ONCV pseudopotentials. The averaged result for the current application is presented in the table 1. Limits of linear elasticity were estimated by bilinear approximations of the curves with logarithmic Hencky strain, so the resulting elastic constants don't necessarily coincide with the usual values for small strains near the equilibrium. The nonlinear parts of the curves represented only anharmonicity and didn't represent plastic irreversible deformations, since there were no corresponding crystal defects that appear at bigger scales and temperatures. So, the actual plastic yield points may differ from the linear limits obtained in this way, however there should be a correlation between them, which leads to the same Hill's relation, not to mention that computed fracture limits usually coincide with the experimental data, so moving only the yield point doesn't affect the bilinear curve very much. For FEM calculation beyond the ultimate strength, where creep or fracture occurs, we use a low-angle line of stress-strain dependency (1 or 5 % of the plastic line slope) to remain in equilibrium quasi-static approach (figure 1).

Table 1. Input data for FEM computation with multilinear hardening law exemplified by figure 1. Yield ($_Y$) and ultimate ($_U$) stresses σ for compression (^{c}) along [001] crystallographic axis and for shear (^{s}), accompanied by the corresponding axial ε and shear γ strains. Poisson coefficient ν actually depends on strain, a value at zero strain is used for stiff diamond C, while a representative value at compressive yield point is taken for other materials. $h^{s/c}$ is the Hill's anisotropic plasticity ratio between shear and compression, which is generally different from the one for tensile deformation [3].

	ε_Y^c	ε_U^c	γ_Y	γ_U	σ_Y^c(GPa)	σ_U^c(GPa)	σ_Y^s(GPa)	σ_U^s(GPa)	ν	$h^{s/c}$
Ge	–0.045	–0.15	0.055	0.21	–4.4	–9.5	3.7	12.1	0.307	1.46
Si	–0.065	–0.18	0.094	0.24	–7.3	–12.3	7.2	15.5	0.304	1.71
SiC	–0.107	–0.30	0.175	0.32	–27.9	–49.8	42.7	64.8	0.297	2.65
C	–0.161	–0.27	0.216	0.34	–161.8	–219.8	117.6	141.5	0.114	1.26

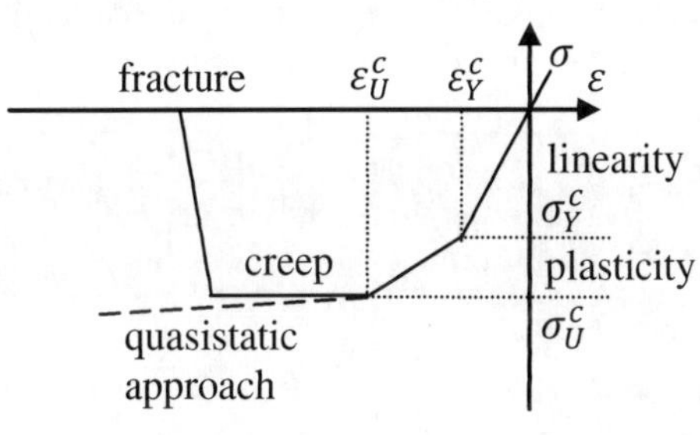

Figure 1. Schematic stress-strain curve by the example of compression. The quasistatic approach to the creep line is used in the FEM calculation beyond ε_U^c.

3. Indentation with Berkovich tip

Diamond three-sided Berkovich pyramid has its sides inclined at 65.27° angle to its geometrical axis aligned to the [001] crystallographic axis. The pyramid has been modelled with 100 nm tip curvature common for standard nanoindenters. FEM calculation of the 150 nm indentation process with account of Hill's anisotropic plasticity has been carried out in ANSYS software. Normal Lagrange formulation for frictionless contact between the bodies has been chosen to exclude penetration error, which has crucial importance in this kind of simulation. Combined symmetry of the indenter and the crystal sample has been taken into account to reduce the computational problem. Different rotational orientations of the pyramid in respect to the crystal surface may physically lead to a bit different results, but this difference happened to be within numerical error.

IOP Conf. Series: Materials Science and Engineering **387** (2018) 012078 doi:10.1088/1757-899X/387/1/012078

Figure 2 shows an example of the FEM model with residual plastic strain after indentation. Most of the plastic strain is accumulated under the pyramid edges, where cracks are usually observed in the experiments. Formation of a main median crack has been calculated by cohesive zone model (CZM), using data on the fracture stress limits from the table 1 and data on the crack surface energy (~2 J/m^2 for Si) calculated by computational quantum chemistry in [4]. The formation of crack hasn't impacted the value of the reaction force.

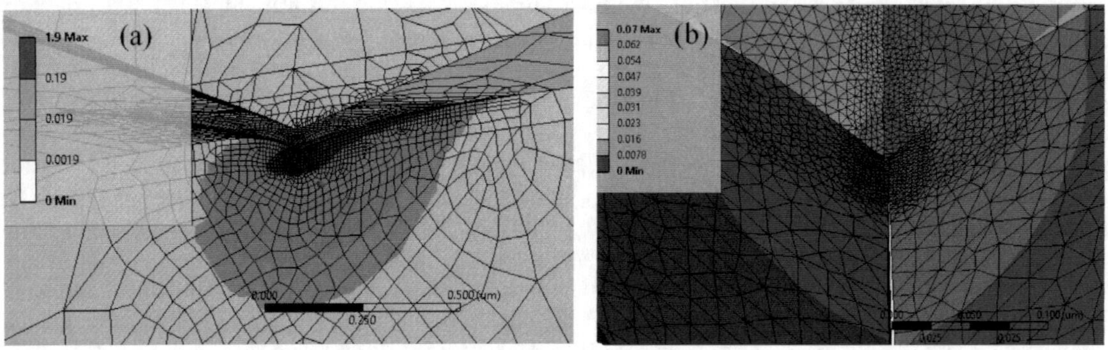

Figure 2. Section view of the finite element models of indentation of Si(111) surface: (a) after 150 nm indentation with residual equivalent plastic strain plotted in logarithmic colormap, (b) median crack (white color) propagation by cohesive zone model at 70 nm indentation with total deformation plotted in color. Vertical projection of the pyramid edge is aligned to the Si $[2\bar{1}\bar{1}]$ axis.

Figure 3 clearly shows by the example of silicon that the presence of plastic anisotropy leads to the different reaction forces for different orientation of the crystal sample, as it is commonly observed in the experiments. The Hill's parameter $h > 1$ reduces amount of plastic strain, thus narrowing the loading-unloading curves. Calculated hardness is 12.1 GPa, 11.8 GPa, and 10.9 GPa for Si(111), Si(011), and Si(001) surfaces correspondingly. Hardness is determined as a relation of the reaction force to the vertical projection of the contact area under conditions of a fully developed plastic zone.

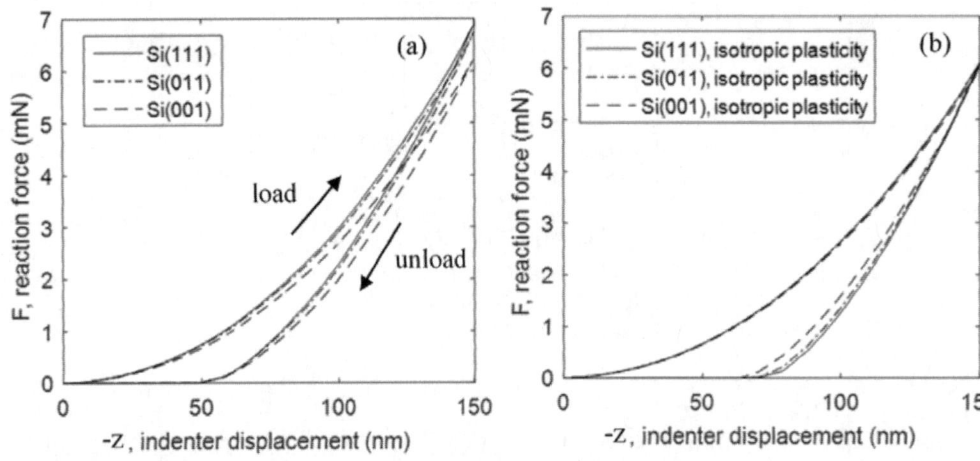

Figure 3. Loading-unloading curves of FEM-computed nanoindentation of different silicon surfaces for (a) anisotropic plasticity and (b) isotropic plasticity, Hill's parameter $h = 1$.

4. Indentation and scratching with spherical tip

A spherical tip, as the most suitable and safe for scratching tests, has been modelled with 25 μm radius of curvature. The spherical indenter exhibits the same hardness differentiation for various surfaces of Ge, Si [5], SiC materials as the Berkovich indenter. It is clear that nonzero component of the reaction force resistant to the direction of the scratching appears (figure 4(a)), but another lateral force may appear as well. In the case of anisotropic plasticity, even elastically isotropic (111) plane (stiffness tensor is invariant upon any rotation around the cubic [111] axis) loses its symmetry, so that different directions of the scratching produce different plastic strain zones (figure 4(b)). The difference in normal components of the reaction force between the scratch indentation and the usual one is small, but noticeable (figure 5).

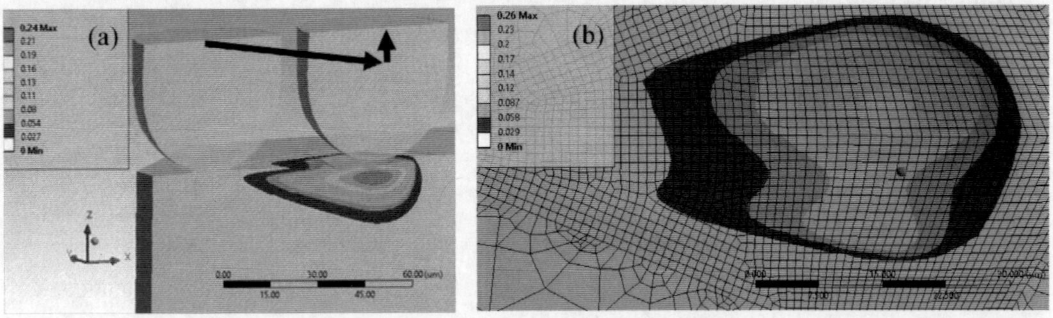

Figure 4. Residual plastic strain (in color) after indentation of Ge(111) surface by 6 μm with concurrent scratching by 60 μm (a) along the $[2\bar{1}\bar{1}]$ axis (section view across symmetry plane) (b) along the axis rotated by 30° from the $[2\bar{1}\bar{1}]$ axis (normal view showing asymmetry in the case of anisotropic plasticity).

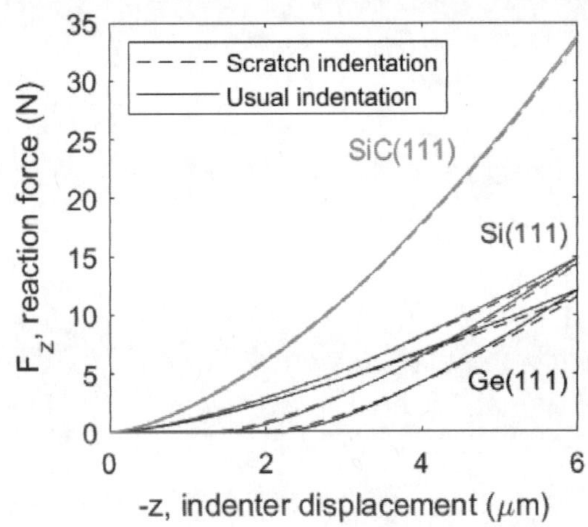

Figure 5. Z-component of the reaction force for indentation with a 25-μm spherical tip compared to the same indentation with concurrent scratching by 60 μm. The loading-unloading curve for SiC is almost fully elastic.

5. Conclusions

Introduction of anisotropic plasticity with the Hill's parameter $h > 1$ into finite-element calculation of indentation of diamond-like crystals Ge, Si, SiC has led to the correct hardness differentiation between the (111), (011) and (001) orientations of the crystals. Various aspects imposed by anisotropic plasticity on indentation and scratching tests have been graphically demonstrated.

Acknowledgments

This study was supported by the Russian Science Foundation, project no. 14-12-01102. The author acknowledges S. A. Kukushkin and A. S. Grashchenko for their engagement in the studied problem

IOP Conf. Series: Materials Science and Engineering **387** (2018) 012078 doi:10.1088/1757-899X/387/1/012078

with the work on the unique scientific installation "Physics, Chemistry and Mechanics of Crystals and Thin Films" (Institute of Problems of Mechanical Engineering RAS, St. Petersburg).

References

[1] Fisher-Crips A C 2011 *Nanoindentation* (Berlin: Springer)
[2] Hill R 1948 *Proc. Roy. Soc. London* **193** 281
[3] Telyatnik R S, Osipov A V and Kukushkin S A 2016 *Materials Physics and Mechanics* **29** 1
[4] Telyatnik R S, Osipov A V and Kukushkin S A 2015 *Physics of the Solid State* **57** 162
[5] Yu B and Qian L 2013 *Nanoscale Research Letters* **8** 137

IOP Conf. Series: Materials Science and Engineering **387** (2018) 012079 doi:10.1088/1757-899X/387/1/012079

Pulsed X-ray source based on photo X-ray tube

G Timofeev and N Potrakhov

Saint Petersburg Electrotechnical University "LETI", Saint Petersburg 197376, Russia

E-mail: gennady92@gmail.com

Abstract. The purpose of this paper is to present main parameters of high frequency X-ray source based on the photo X-tube with photocathode in compare with other tube types. Attention is paid to physical processes that influence such factors as minimum impulse duration and tube current. In the second part of the article the main parameters of designed pulsed X-ray source is given.

1. Introduction

X-ray tube is used for generating X-ray radiation. Such kind of radiation in conjunction with special screens, X-ray sensitive films and other detectors is used for non-destructive testing of objects structural defects or for fluorescent analysis. If object while it is being radiographed is stationary, the problem with studying of its volume does not arise. In cases when the investigated object is moving, its exposure time is directly related to the quality of the resulting image. In other cases, for example while the investigation of the scintillator "relax time" is carrying out, it is required to have a modulation of the exciting radiation faster than the object "relax time". Thus, there are tasks when the frequency characteristic of the X-ray source is very important.

An ordinary X-ray tube consists of an anode, a thermo cathode and a glass envelope. The heated filament emits electrons by thermionic emission. It means that if it is needed to change the tube current the cathode temperature has to be changed. That process takes some milliseconds. That's why such tube construction is not suitable for taking x-ray photos of movable objects.

2. Impulse X-ray tubes

As it was mentioned earlier, radiographing the moving object requires a source of pulsed X-ray radiation. The main rule is that the pulse time should be much shorter than the moving times of the object. To solve this problem, it began to look for a way to create impulse X-ray tubes. Nowadays, there are several major ones that should be considered.

The main type of impulse X-ray tube is an X-tube with cold cathode. The action principle of a tube with a cold cathode is based on the phenomenon of field emission and not on thermo-emission as in conventional tubes. The first samples of field emission tubes used a cathode in the form of a metal needle. A high potential between a cathode and anode produces a strong electric field at the tip of the needle sufficient to pull electrons out of the cold cathode. Then the electrons are accelerated towards an anode as in standard X-ray tubes. Generally, a vacuum of 10^{-8} Torr or better is required because bombardment of the cathode by positive ions of the rest gas deteriorates its sharp tip and reduces drastically the electron emission. The great advantage of such X-tube construction is opportunity to obtain impulses of tube current up to 10^{3} A and lasting several tens ns. But current vary greatly from

IOP Conf. Series: Materials Science and Engineering **387** (2018) 012079 doi:10.1088/1757-899X/387/1/012079

one impulse to other due to bombardment by positive ions "smooth" sharp tips and edges. The cathode lifetime limits by several thousands of impulse.

The latest developments use carbon-based cathodes in the form of graphite nanotubes or so-called "bucky paper". The main advantage of the carbon cathodes results from the fact that their bombardment by positive ions does not "smooth" sharp tips and edges. On the contrary, such a bombardment creates new sharp tips and edges due to the graphite structure. Therefore, the lifetime of carbon-based cathodes is significantly larger. Besides, one can work with lower vacuum (approx. 10^{-6} Torr). Operating frequencies of such tube constructions are in range up to several tens of kHz. Tube current may reach amount of several mA [2].

However, the standard x-ray tube with filament cathode can also provide a pulse X-ray. For example, for radiography in the pulsed mode periodically opening and closing the main circuit on the high voltage side are typically used. For this purpose it can be used either a three-electrode X-ray tube with a control grid or high-voltage triodes. The first variant used, for example, by Siemens is simpler. However, it requires a special X-ray tube, which able to limit the tube electron flow. In moment of fast limiting tube current the overvoltage is applying to the tube. But overvoltage is applying to the tube in moment of the fast current limiting. The second option is used, for example, by Phillips, is more complex. However, it allows conventional X-ray tube to be operated in conditions which are more relieved in comparison with the first embodiment. The opening and closing are performed here by HV triodes, x-ray tube does not suffer to overvoltage while current limiting. Such opening and closing the tube current circuit allows to reduce exposure to several ms. The tube current can reach several hundred of mA. Nowadays, a pulsed X-ray tube of the company Hamamatsu Photonics exists. The main difference is that the source of electrons is a photocathode. Such construction of X-ray tubes has a very small output intensity of X-ray radiation, but it is able to provide the modulation frequencies up to 10 MHz. It is used for determining fluorescence lifetimes between 20 ps and 1 ms.

3. Photo X-ray tube

Photo X-tube is the combination of photomultiplier tube and an ordinary X-ray tube. The schematic diagram of the photo X-tube is given in figure 1. The tube has a photo cathode to generate primary photo electron flow, which is proportional to the falling light intensity. But the amount of primary electrons is not enough for creating intensive current flow. A dynode system is used to multiply its quantity. Amplification ranging from 10^3 to as much as 10^6 provides output electron flow, which is enough for generating medium-intensity X-ray beam. After dynode system electrons get into the accelerating anode field, being accelerated, then collide with anode and generate an X-ray beam.

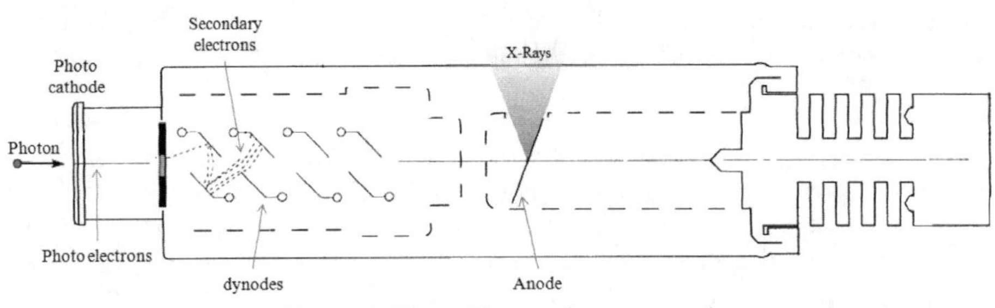

Figure 1. Photo X-ray tube construction.

It is considered, that the frequency characteristic of such kind systems depends only on electron flight time. The tube current control is carried out by modulating the intensity of the input optical radiation incident on the photocathode. For this purpose may be used either conventional LEDs or special lasers. It provides simplicity and stability of current control tube current from one pulse to the other, depending on the stability of the light emitter. According to the product datasheet [1], the tube is able to provide full frequency modulation in range up to 1 MHz with impulse current up to 1 mA.

Operating frequency range allows its use both as a source of modulated X-ray and to obtain a constant emission. Thus, the photo X-tube may be used for the similar purpose, which conventional

IOP Conf. Series: Materials Science and Engineering **387** (2018) 012079 doi:10.1088/1757-899X/387/1/012079

directly heated tubes are used for. However, the tube has enlarged size and is more expensive, that features limit its application area to range, where its frequency characteristic in couple with operating current will be crucial.

4. Pulsed X-ray source based on photo X-ray tube characteristics

To investigate the frequency response of the X-ray tube with the photocathode, a special detector based on PMT and a plastic scintillator was used. The source of optical radiation for the photocathode was LED and a laser diode. As can be seen from figure 2 LED allow to provide the rise time of the out radiation front of 50 ns. Using a laser diode, it was possible to achieve a time of 20 ns.

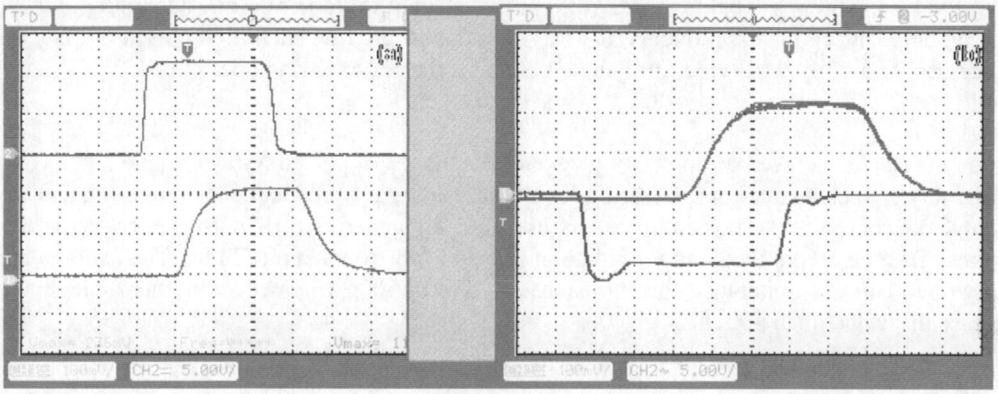

Figure 2. Oscillograms of the out radiation formed by LED (**a**) and the laser diode (**b**).

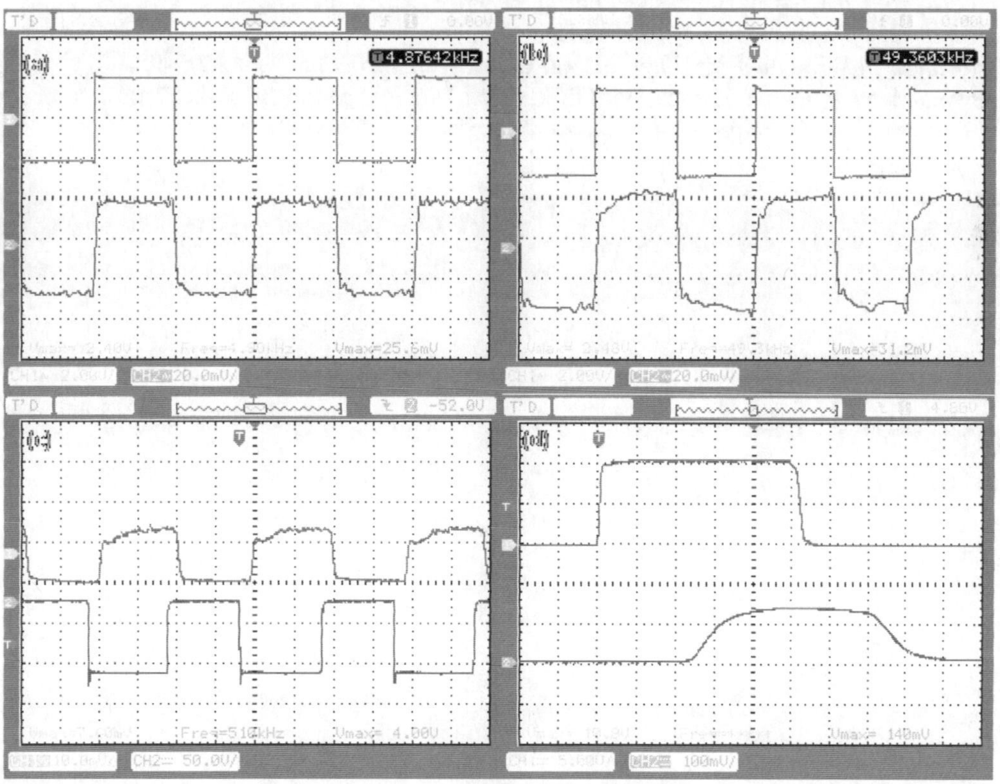

Figure 3. Oscillograms of the frequency response of the X-ray tube at frequencies of 5 kHz (**a**), 50 kHz (**b**) and 500 GHz (**c**), and also for single pulse with repetition 25 kHz (**d**).

IOP Conf. Series: Materials Science and Engineering **387** (2018) 012079 doi:10.1088/1757-899X/387/1/012079

The frequency response of the X-ray tube with a photocathode was estimated with a duty cycle of 50% to a frequency of 500 kHz. Further experiments were carried out with pulses duration of several hundred ns and a repetition frequency of several tens of kHz. The results are shown in figure 3. The oscillograms of Fig. 5a-c illustrate the operation of the out radiation under study at frequencies of 5, 50 and 500 kHz. The oscillograms in figures 3(a)–(c) illustrate the out X-ray pulses at frequencies of 5, 50 and 500 kHz. The oscillogram shown in figure 3(d), demonstrates the passage of a single pulse. The repetition frequency is 25 kHz, the average current is 100 μA. The peak current is 30 mA.

From figure 3(d), it is possible to measure the rise time of the output pulse equal to 20 ns. The impulse current in this experiment was 30 mA. The cathode assembly of the tube is designed on the basis of PMT-84, which is capable to provide a pulsed current of up to 500 mA according to technical documentation. During the experiments, it was found that to provide such high pulsed current values, it is necessary to modify the divider of dynode system of the photo X-ray tube.

5. Conclusion

Based on the results, achived during testing developed sources of pulsed X-ray radiation, one can state the suitability of such source for generating X-ray pulses over a wide frequency range. The advantage of such X-ray source is the possibility of its use both for generation of pulsed X-ray and for long exposures. The source provides output radiation rise and fall fronts about 20 ns. The average current can vary up to 1 mA. The impulse current reaches a value of 30 mA. However, in theory, it is possible to increase this value up to 500 mA.

References

[1] *X-Ray Tube with photo cathode datasheet.* URL: http://www.electron.spb.ru/data/%D0%A4%D0%A0%D0%A2%20datasheet.pdf

[2] *Handbook of Practical X-Ray Fluorescence Analysis*

[3] *Advances in X-ray Analysis, International Centre for Diffraction Data* 2005 **48**

[4] Reyes-Mena A, Jensen Charles, Bard Erik, Clark Turner D and Erdmann K G *Miniature X-Ray tubes utilizing carbon-nanotubebased cold cathodes* (MOXTEK Inc., Orem)

IOP Conf. Series: Materials Science and Engineering **387** (2018) 012080 doi:10.1088/1757-899X/387/1/012080

Obtaining of non-stoichiometric titanium oxide using reactive magnetron sputtering

S Y Udovichenko, A N Bobylev, D A Belotserkovtseva and D D Shpindyuk

Research and education center «Nanotechnology»,Tyumen State University , 625003, Tyumen, Russia

E-mail: udotgu@mail.ru

Abstract. A mode of obtaining of non-stoichiometric titanium oxide film in magnetron system with turbomolecular pump of low performance was found. The mode provides significantly lower volumetric flow rate of argon and reactive gas comparing with known mode. Dependencies of discharge voltage and film sputtering rate on reactive gas volumetric flow were investigated. Measurements of film obtained by the mode with low oxygen volumetric flow proved deviation of composition stoichiometry.

1. Introduction

Non-stoichiometric compounds (oxides, nitrides) of metals and semiconductors are actively used in electronics and nanotechnologies due to their special physical and chemical properties. There are research activities on light emission diodes (LEDs) creation on quantum luminescence dots that are silicon nanocrystals in a silicon nitride film of non-stoichiometric composition [1]. Thin films of transition metals oxides that change their electrical conductivity during current flow are used as an active layer of memristors [2] and memristor crossbars [3]. In addition, memristors can be made using reduced metal oxide films. Common method of obtaining such compounds is the atomic layer deposition. This method has significant lacks related to non-uniformity of elements distribution along the film thickness and inclusion of elements participating in chemical reactions during the metal oxide deposition from precursor.

The reactive magnetron sputtering method provides purity and excellent homogeneity of metal oxide films with high deposition rate and precise thickness control [1]. However, the atomic layer deposition (ALD) method is preferable for obtaining films of metal nitrides or dielectrics because it allows selection of desirable mass ratio of elements in precursor.

Control of film chemical composition is difficult during reactive magnetron sputtering. A sputtering of cathode target material on substrate occurs within inert gas presence and low flow rate of reactive gas. When flow of the gas is high enough to form a chemical compound, the target sputtering rate reducing and a stoichiometric compound is depositing onto the substrate. When the volumetric flow of the reactive gas decreases, sputtering mode remains stoichiometric. Further reducing of the gas flow leads to rapid change of the sputtering mode to deposition of target material. Observed hysteresis (figure 1) does not allow to describe sputtering process uniquely and to predict stoichiometry of a film.

An investigation of opportunity to obtain non-stoichiometric compound of titanium oxide within low flows of sputtering and reactive gases is the main aim of presented paper.

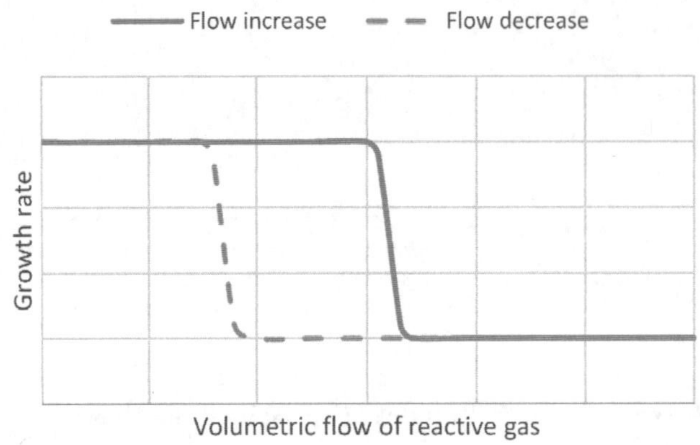

Figure 1. Simplified graph of dependence of the deposition rate on the reactive gas volumetric flow.

2. Stoichiometry control

S. Berg et al. [4] presented the model of reactive magnetron sputtering for the case of titanium sputtering within nitrogen atmosphere. The simulation results shows that nitrogen-titanium ratio directly depends on the reactive gas flow. This allows using reactive gas flow as state variable. The calculated data were proved during obtaining of stoichiometric TiN. This model was used primarily to increase the rate of stoichiometric compounds deposition.

B. Hoskins and D. Strukov [5] has modified Bergs model to obtain TiO_x compound of desired composition. As been shown, growth of non-stoichiometric films is possible when perturbative reactive gas flow combined with comparatively small discharge power and high argon to reactive gas partial pressures ratio. The state variable was normalized to 1, where 1 corresponds to stoichiometric composition and 0 corresponds to pure target material. The normalized state variable is called the oxidation factor. The experiment carried out at pressure of 0,08 Pa and discharge power of 150 W. Target diameter was 3 inches. Custom built high vacuum sputter chamber was pumped by Cryotorr 800 with a 1000 l/s pump rate. Argon flow rate was 190 sccm and oxygen flow rate was from 0 to 16 sccm. Experimental dependence of O/Ti ratio from oxidation factor F measured by Rutherford backscattering is approximated by hyperbolic function $O/Ti = a - 1/(bF)$, where a and b are constants.

3. Experimental

Actual experiment was carried out in compact magnetron unit of NT-MDT NanoFAB-100 complex. The work volume of the vacuum chamber is 50 liters. The unit is equipped with BOC Edwards STP-H451 low-performance turbine with a 450 l/s pump rate. A volumetric flow of gases in chamber adjusting by electromagnetic valves with accuracy of 0,1 sccm. The 99,99 % pure Ti target diameter is 78 mm. The distance between sputtering cathode target and substrate is equal 100 mm in the experiment. The vacuum chamber was pumped to remaining base pressure of $5 \cdot 10^{-5}$ Pa before the experiment.

Sputtering experiment was carried out in pulse mode of magnetron with constant discharge power of 115 W and constant chamber pressure of 0,1 Pa. Operating pressure was induced by 12 sccm flow of 99,99 % pure argon. Experimental dependences of TiO_x films growth rate on volumetric flow of O_2 (figure 2) and discharge voltage on volumetric flow of O_2 (figure 3) were obtained varying oxygen flow from 0 to 1,4 sccm.

Experimental plots (figure 2., figure 3.) show that chosen sputtering mode allows to avoid jumping behavior of sputtering curve. Nevertheless, relatively big oxygen flow and oxidation factor more than 0,5 lead to formation of hysteresis in S(Q) and especially in U(Q) curves. That circumstance leads us

IOP Conf. Series: Materials Science and Engineering **387** (2018) 012080 doi:10.1088/1757-899X/387/1/012080

to conclusion that composition of films grown by this sputtering mode is controllable only in plot section from F = 0 to F = 0,5 or from Q = 0 sccm to Q = 0,7 sccm.

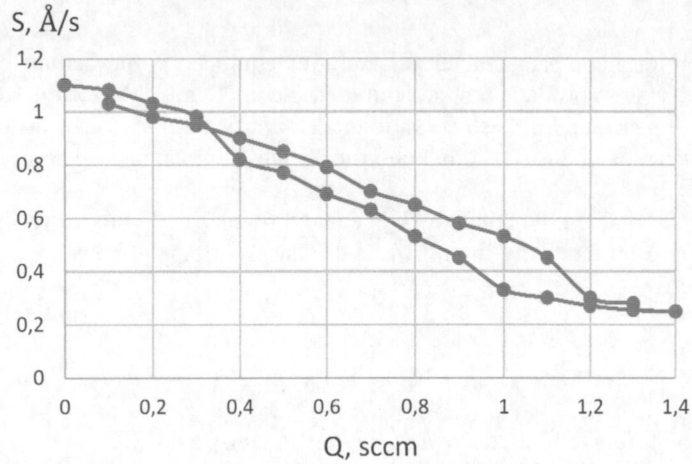

Figure 2. Experimental dependence of TiOx film growth rate on volumetric flow of O_2 at constant pressure and discharge power.

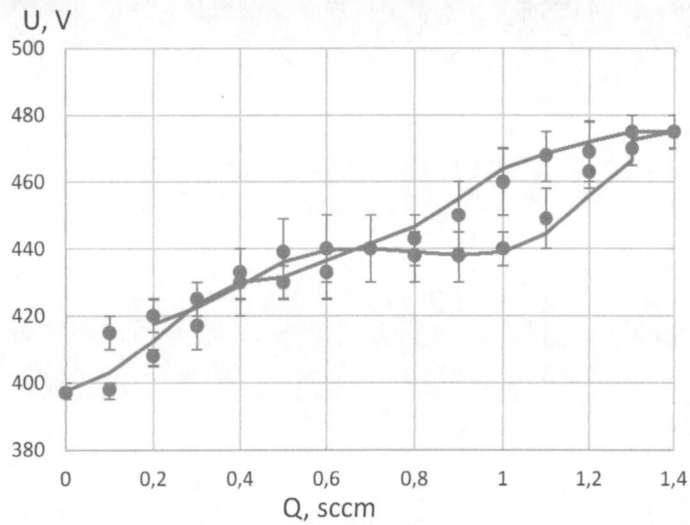

Figure 3. Experimental dependence of discharge voltage on volumetric flow of O_2 at constant pressure and discharge power. Error bars show range of voltage varying by feedback to maintain constant power of discharge.

To check stoichiometry of films obtaining by the mode, two points on plot were chosen. The first point corresponds F = 0,3 and Q = 0,42 sccm, the second point corresponds F = 1 and Q = 1,4 sccm. One hundred nanometer thick films were grown for each of points. Then films composition were measured by Oxford Instruments X-Max 20 mm^2 EDS device mounted on JEOL-6510 LV SEM. Measurement results are in table 1.

Table 1. EDS measurement results.

F	O, %	Ti, %	O/Ti
0,3	62,37	37,63	1,66
1	66,48	33,52	1,98

IOP Conf. Series: Materials Science and Engineering **387** (2018) 012080 doi:10.1088/1757-899X/387/1/012080

O/Ti rate at F = 1 actually should be equal 2. The observed deviation is caused by the error of measurement technique.

4. Conclusions

Obtained results show that compact magnetron facilities equipped with vacuum pumps with pump rates under 500 l/s are able to produce non-stoichiometric compounds. Moreover, volumetric flows of sputtering and reactive gases under these conditions are significantly less than in work [5]. That fact leads to cost reduction of industrial production of such compounds without decreasing of its efficiency.

O/Ti rate at F = 0,3 obtained in actual work is equal 1,66 and it slightly differs from O/Ti rate in [5]. Difference is caused by relatively high drift of discharge voltage at constant power maintenance and by EDS method error.

References

[1] Zhuravsky D V, Misiyuk K V, Udovichenko S Y et al. 2014 *Vacuum equipment and technology* **1** 65–6
[2] Lee J S, Lee S and Noh T W 2015 *Applied Physics Reviews* **3** 031303
[3] Bobylev A N and Udovichenko S Y 2016 *Russian Microelectronics* **6** 396–401
[4] Berg S, Larsson T, Nender C and Blom H-O 1988 *Journal of Applied Physics* **3** 887–91
[5] Hoskins B D and Strukov D B 2017 *Journal of Vacuum Science and Technology A: Vacuum, Surfaces and Films* **2** 020606

Investigation of physical and mechanical properties of coatings based on intermetallics of the Ti-Al system obtained in acetylene environment from vacuum-arc discharge plasma

E L Vardanyan, A Yu Nazarov, K N Ramazanov and R N Khusnimardanov

Ufa State Aviation Technical University, 12 K.Marx Street, Ufa, Russian Federation, 450000

E-mail: vardanyaned@gmail.com

Abstract. At the present time, protective coatings have found wide application in various branches of industry: mechanical engineering, aircraft engine building and also in the tool manufacture. In this work presented the results of the researches of the physico-mechanical properties of coatings based on the intermetallics of the Ti-Al system, obtained in the acetylene environment from vacuum-arc discharge plasma. The thickness, microhardness, and phase composition of the coatings were investigated when samples were placed at various distances from the table axis and rotation speeds.

1. Introduction

At the present time, protective coatings have found wide application in various branches of industry: mechanical engineering, aircraft engine building and also in the tool manufacture. Recently, much attention has been paid to the development and research of physico-mechanical properties of composite coatings, which have high physico-mechanical properties. Coatings based on multilayered compositions of the Ti-Al system have a great practical interest [1, 2]. In work [3] it is pointed out that the formation of intermetallic compounds is a perspective, because intermetallics have very high hardness and they are a good barrier for the diffusion processes to the flow, which should increase the efficiency of the cutting tool with coatings. In work [4] was developed a method for the preparation of Ti-Al intermetallics from vacuum-arc discharge plasma in the deposition of two single-component cathodes of titanium and aluminum in an argon inert gas environment. In work [5] was considered a method for obtaining Ti-Al intermetallides of their nitrides and carbides when vacuum-arc discharge from two single-component cathodes is deposited from the plasma, but their physico-mechanical properties are studied. In work [6] was considered the process of TiAlC deposition onto a silicon substrate with the help of a magnetron. The coatings were prepared by deposition from two targets of titanium and one from aluminum at different ratios of the reaction gas Ar and C_2H_2. The results showed that there are formed Ti_3AlC and TiC compounds. In work [7] was considered deposition of coatings from a vacuum-arc discharge plasma from Ti_3AlC_2 cathodes obtained by self-propagating high-temperature synthesis in the N_2 environment. The phase composition of coatings represents cubic two-element nitride of composition (Ti, Al) N, which occur a solid solution of aluminum in titanium nitride.

In these researches, the solid solutions, max phases of Ti-Al intermetallics, have been studied, however, the synthesis of intermetallic compounds in acetylene environment has not been considered and their effect on physical and mechanical properties and phase composition has not been investigated yet. And the connection with this, goal of this work – the synthesis of intermetallic carbides of the Ti-Al system from the vacuum-arc discharge plasma.

2. Methods of conducting experiments

The coating was applied to the mills of the modernized facility named NNV-6.6-I1. To explore the physico-mechanical properties of coatings based on intermetallic compounds Ti-Al synthesized in acetylene environment, 4 samples were established in the chamber, the scheme is shown in figure 1. The first sample was placed in the center, the second at a distance of 8 cm, the third – of 15 cm, the fourth – of 22 cm. In the first experiment, the table rotation speed was 1 rpm, in the second 14 rpm. Thus, coatings were deposited with a number of layers of 120 and 1680 pieces at the same processing time. The thickness of the coating was determined from the measurement of the well parameters, which was obtained with the CSM Calotest. The phase composition was determined in a Rigaku Ultima IV X-ray diffractometer.

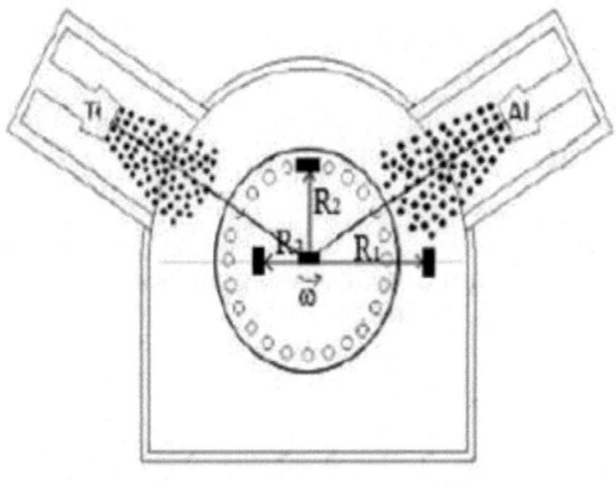

Figure 1. Scheme of the experiment.

3. Experimental results

The thickness of the coatings on the samples was 4.5–5.5 μm. A photo of the well for measuring the thickness of the coating is shown in figure 2.

Figure 2. Photo of a hole for measuring the thickness of coatings.

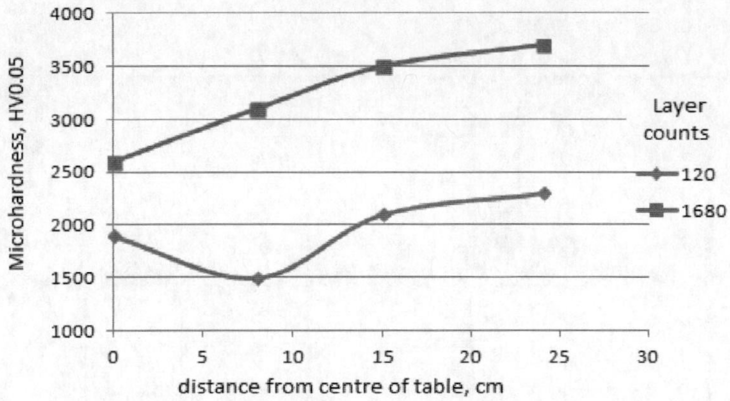

Figure 3. Graph of microhardness measurements.

The results of measuring the microhardness of the coatings obtained showed that as the distance from the axis and the number of layers increases, the microhardness increases from 1800 to 3700 $HV_{0.05}$. This can be explained by the fact that with increasing table rotation speed, the growth time of one layer decreases, and thus the thickness of the layer reduced from 300–500 nm to 20–50 nm and also by an increase in the amount of titanium carbide phase. The microhardness measurement schedule is shown in figure 3.

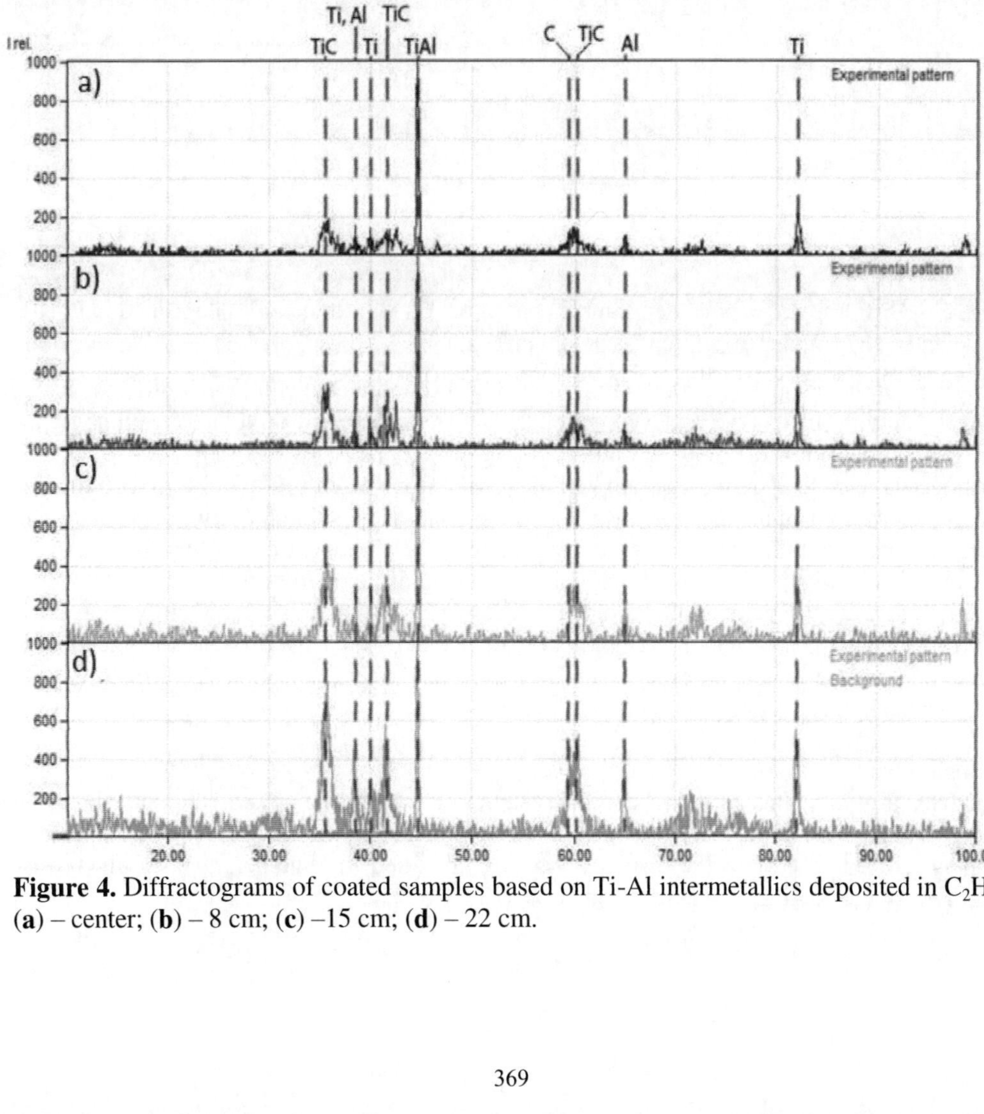

Figure 4. Diffractograms of coated samples based on Ti-Al intermetallics deposited in C_2H_2: (**a**) – center; (**b**) – 8 cm; (**c**) –15 cm; (**d**) – 22 cm.

IOP Conf. Series: Materials Science and Engineering **387** (2018) 012081 doi:10.1088/1757-899X/387/1/012081

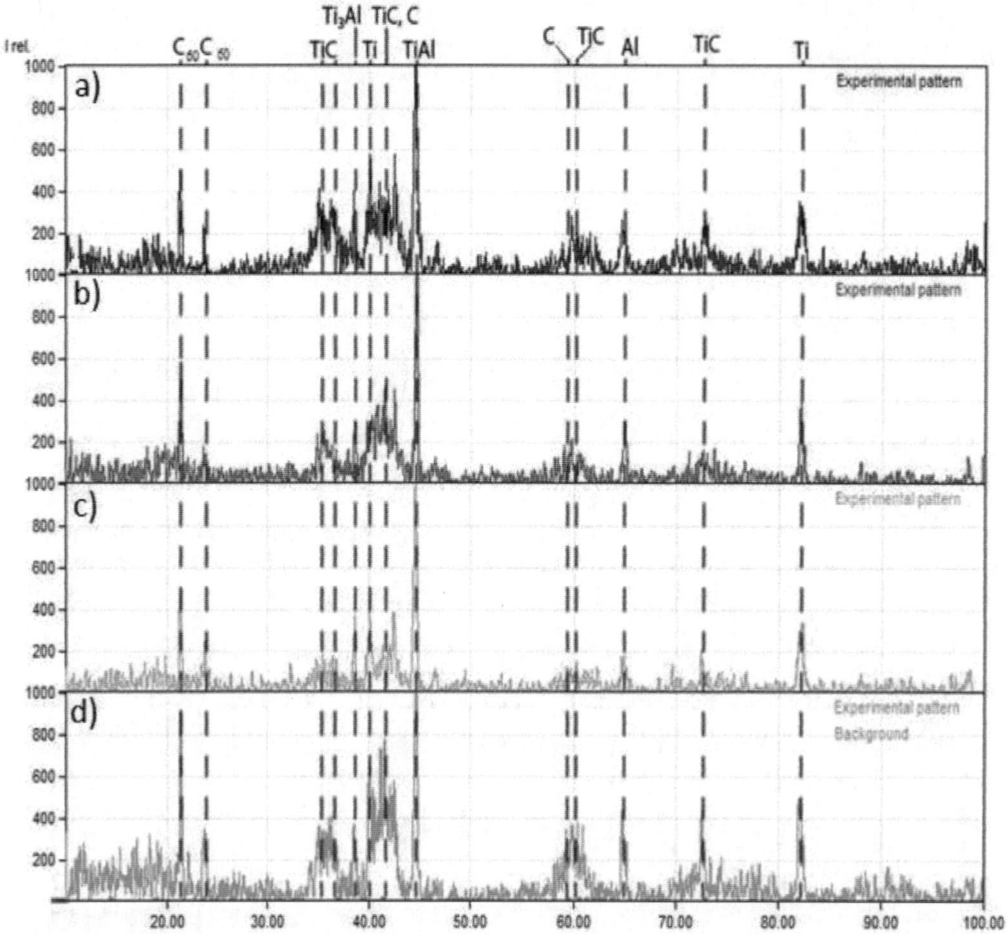

Figure 5. Diffractograms of coated samples based on Ti-Al intermetallics deposited in C_2H_2: (**a**) – center; (**b**) – 8 cm; (**c**) –15 cm; (**d**) – 22 cm.

Diffractograms of the samples are shown in figures 4, 5. From the results of X-ray diffraction analysis it is evident that with the number of layers 120 pieces there are formed TiAl intermetallics, Ti and Al in pure, titanium carbide TiC. With the number of layers 1680 pieces there are formed TiAl, $T_{i3}Al$ intermetallics, Ti and Al in pure, titanium carbide TiC, carbon C and fulurene C_{60}. The presence of fulurene in the coating can be explained by the fact that when depositing a coating with thinner layers in which C is contained, the process of the formation of fullurenes promotes the process. And also from the obtained results it can be seen that with increasing distance from the axis, the content of the intermetallic phase as well as titanium carbide increases. This is due to the fact that the increase in the distance from the axis is less than the sample under the Ti and Al plasma flux and the thickness of the layers deposited on the sample is less than in the center, and practically all of Al diffuses into Ti and forms intmetallide phases and titanium carbides.

4. Conclusion

The conducted investigations showed that TiAl system in the acetylene environment precipitates TiAl, Ti3Al, pure Ti and Al, TiC carbide, carbon C and fullerene C60 in the coating. The microthickness of the samples showed that when the distance from the axis and the number of layers increases, the microhardness of the coatings increases from 1800 to 3700 $HV_{0.05}$.

Acknowledgements
The financial support of Russian Foundation for Assistance to Innovative Small Innovative Enterprises #12979GU/2018

Reference

[1] Leyens C, Peters M and Kaysser WA 1997 *Surf Coat Tech.* **94–5** 34–40
[2] Pyachin S A and Burkov A A 2015 *Electronic material processing* **2** 16–23
[3] Verhoturov A D 1983 *Electrophysical and electrochemical methods of processing* **1** 3–6
[4] Vardanyan E L, Budilov V V 2016 *Surf. Invest: X-ray, Synchrotron and Neutron Techniques,* **10** 728–31
[5] Kireev R M 2008 *Vestnik USATU* **26** 96–9
[6] Harsh A Chaliyawala, Gaurav Gupta, Praveen Kumar, Srinivas G, Siju Harish C and Barshilia S 2015 *Surf Coat Tech.*
[7] Fedotov A F and Amosov A P 2013 *Hardening technology and coatings* **8** 33–8

Yttrium molybdates coating deposition of oxide-fibre/molybdenum-matrix composites

E L Vardanyan[1], A Yu Nazarov[1], S N Galyshev[2], R F Gallyamova[1] and S T Mileyko[2]

[1] Ufa State Aviation Technical University, Ufa, Russia
[2] Institute of Solid State Physics of Russian Academy of Sciences, Chernogolovka Moscow distr., Russia

E-mail: vardanyaned@gmail.com

Abstract. Shown a possibility to do a coating process on void-free surfaces which based on Yttrium molybdate on composite including molybdenum matrix by using ion-plasma deposition. Reported results of X-ray phase analysis of coating before and after annealing and also results of a microstructure analysis.

1. Introduction

The development of high temperature materials beyond both nickel superalloys and modern niobium alloys described in a series of the publications aims at molybdenum alloys in the Mo-Si-b systems [1, 2] and molybdenum-matrix fibrous composites [3, 4]. The behavior of fibrous metal matrix composites differs from that of metal alloys: fracture toughness and strength of the composites can go up simultaneously by increasing the volume fraction of strong brittle fibres [5, 6]. The alloys are characterized by a conflict between strength and fracture toughness. Molybdenum alloys in the Mo-Si-B system developed have high strength and creep rupture properties up to 1300 °C, sufficiently good oxidation resistance up to 1200 °C [5]. Coatings on the surface of specimens of such alloys to enhance the oxidation resistance at temperatures up to 1300 °C are also known [6]. However, fracture toughness of these alloys is too low to use them in heavily loaded structural elements [5].

Hence, molybdenum matrix fibrous composites should also be considered as a possible family of heat resistance materials. An interest of the present authors for oxide fibre/molybdenum-matrix composites was stimulated by (i) an effective way to produce such composites called the internal crystallization method (ICM) and (ii) an observation of a reduced oxidation rate of the molybdenum matrix provided it is reinforced with a fibre of a special chemical composition [5, 6]. Such composites are characterised by high strength and high creep resistance at temperatures up to about 1300 °C [5, 6]. In particular, if the reinforcing fibre contains Me (Me = yttrium or lanthanide) then molybdates of Me occur on the surface of a composite specimen that prevents rapid oxidation of molybdenum. The molybdates can be healing agents for defects in a coating that can arise in the service of the composite.

Therefore, the technology of coating the composites by molybdates is to be developed. In the present paper, yttrium molybdates coatings on molybdenum-matrix composites reinforced with the yttrium-aluminium perovskite (YAP) were deposited by ion-plasma method described in [3]. It is important to stress out that the information on yttrium molybdates is very scarce. There are a number of the compounds in the Y_2O_3 – MoO_3 phase diagram including $Y_2O_3 \cdot 3MoO_3$ ($Y_2Mo_3O_{12}$) with melting point 1310 °C, $Y_2O_3 \cdot MoO_3$ (Y_2MoO_6) and $3Y_2O_3 \cdot MoO_3$ (Y_6MoO_{12}) [6]. Note that the melting points of the last two compounds are not determined but they took to be much higher than 1310 °C.

2. Experimental methods

Preliminary experiments to determine coating parameters and regimes were conducted with a 18-10 stainless steel substrate. Then composite specimens with molybdenum matrix and fibres based on YAlO3 (YAP) were coated. The YAP-based-fibre/Mo-matrix composites were chosen as they are characterised by sufficiently high strength at temperatures up to 1400 °C and good fracture toughness properties. Coating was performed by using the ion-plasma technique, the corresponding equipment is shown schematically in figure 1. The apparatus allows coatings with either pure metals or compounds like nitrides, carbides, oxides of various metals.

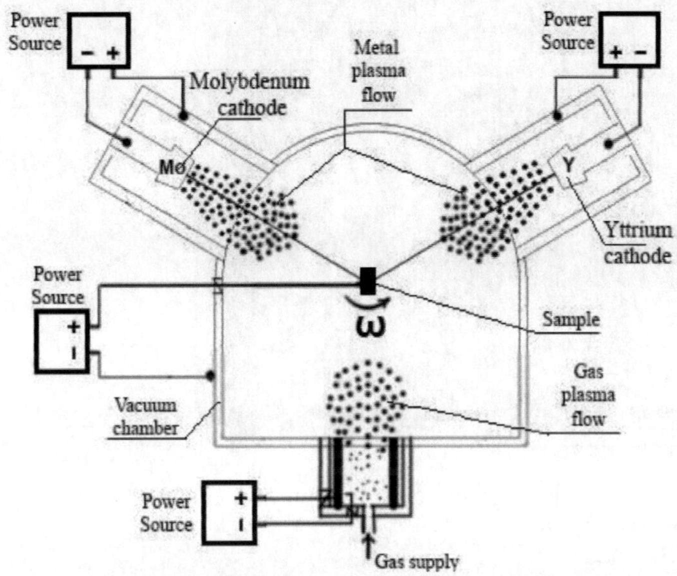

Figure 1. Schematics of the coating equipment.

The deposition of yttrium molybdates was carried out by using molybdenum and yttrium cathodes. To obtain yttrium molybdates of various stoichiometric compositions ($Y_2Mo_3O_{12}$, Y_2MoO_6 and Y_6MoO_{12}) three deposition regimes were developed. To provide an adhesion of the coating to molybdenum matrix the process started with depositing thin molybdenum sub-layer, which was done in argon gas atmosphere for 5 minutes. Then oxygen lets to the chamber and according regime switched on for 60 min to deposit one of the molybdates mentioned.

The phase analysis of the coatings was performed by X-ray diffractometer Rigaku Ultima equipped with the Cu-Kα radiation source; the range of angles 10–100°. The wavelength = 1,540562 Å, the survey step was 0,02°. Analysis of the X-ray spectra was carried out by using software Match! Crystal Impact.

As mentioned above, the first experiments to find appropriate deposition regimes were conducted by using a steel substrate. The X-ray phase analysis of coated steel samples after deposition (figure 3) reveal an amorphous structure of the coated layer. Hence, the specimens were annealed at a temperature of 950 °C for 1 hour in air atmosphere to crystallise the coating. The X-ray spectra (figure 3) show that crystallisation has been really occurred. The Y_2MoO_6 compound predominated in the coating obtained according to deposition regime #1, and the $Y_2Mo_4O_{15}$ molybdate exists in the coating obtained according to regime #3.

For the metallographic examination of the coatings on composite specimens, there were prepared oblique cross-sections with the 6° angle between the cut plane and the longitudinal direction of a specimen. The morphology of the coatings was observed by scanning electron microscope JEOL JSM-

6390 operated in the secondary electron mode. The microscope was equipped with INCA Energy that allowed analyzing the distribution of chemical elements across the coating thickness.

3. Results and discussion

Investigation of the structure showed that the coatings obtained can be characterized as continuous, without cracks and stratifications. The microstructure of the coated samples for all modes has a similar structure. An electronic image of the micro-texture of the coating obtained by mode № 1 is shown in figure 2(a). Points "Spectrum 1-4" indicate the areas of local energy-dispersive analysis: Spectrum 1 – oxide fiber, Spectrum 2 – molybdenum matrix, Spectrum 3 – undercoat coating, Spectrum 4 – coating. Similar studies were carried out for other regimes. Table 1 shows the results of an energy-dispersive analysis of the corresponding spectra for three samples.

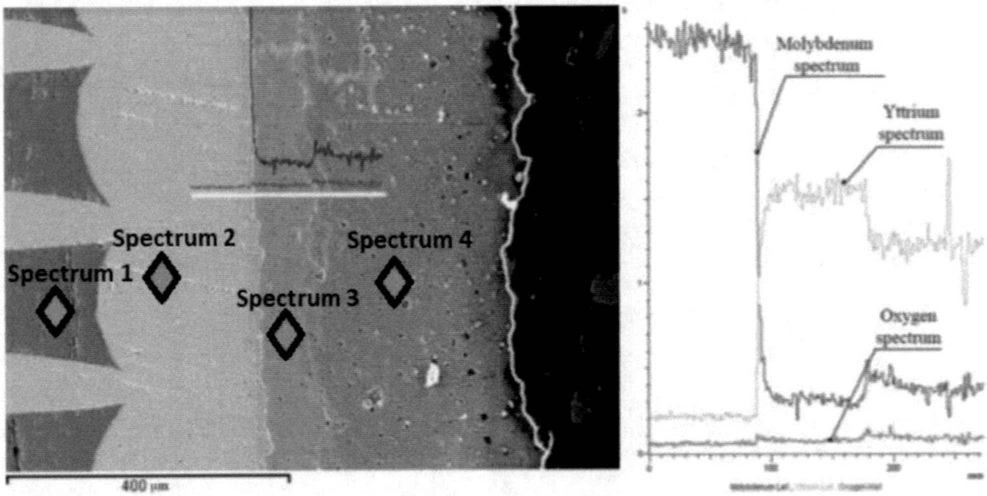

Figure 2. The SEM micrograph of the cross-section of a composite specimen with a coating obtained by regime #1.

Table 1. Chemical compositions of the composites and coatings obtained of EDS chemical micro analysis (at %).

Mode №1	O	Al	Y	Mo
Spectrum 1	48.34	22.86	28.80	
Spectrum 2				100.00
Spectrum 3	47.70		41.81	10.49
Spectrum 4	54.61		33.45	11.94
Mode №2	**O**	**Al**	**Y**	**Mo**
Spectrum 1	53.29	20.89	25.82	
Spectrum 2				100.00
Spectrum 3	47.86		41.03	11.11
Spectrum 4	50.71		41.70	7.59
Mode №3	**O**	**Al**	**Y**	**Mo**
Spectrum 1	53.41	20.84	25.75	
Spectrum 2				100.00
Spectrum 3	47.93		14.74	37.33
Spectrum 4	66.18		6.85	26.98

The composition of yttrium molybdates $Y_2Mo_3O_{12}$, Y_2MoO_6 and Y_6MoO_{12} is analyzed by the ratio of yttrium and molybdenum (the method does not allow determining the oxygen content with sufficient accuracy) (table 1, Spectrum 4 for all modes). The results of the analysis with indication of deviations from stoichiometric compositions are given in table 2.

Table 2. Correlation of coating compositions with stoichiometric compositions of yttrium molybdates.

№ Mode	Yttrium molybdates	Deviation, at. %
1	Y_2MoO_6	9,5
2	Y_6MoO_{12}	3,9
3	$Y_2Mo_3O_{12}$	11,2

The distribution of the elements along the line for mode № 1 is shown in figure 3. Comparison of electronic images and diagrams of distribution of the elements demonstrates the presence of a diffusion layer between the matrix material and the sublayer and the coating and the sublayer . So, for example, to cover the diffusion layer obtained by the mode №1 (figure 3(a), (b)), it is located between the 90 and 120 μm marks along the abscissa axis of the element distribution diagram; the sublayer is between 120 and 180 μm, followed by the main coating.

X-ray phase analysis of coated steel samples prior to annealing showed that the coating has an amorphous structure – only the substrate peaks (08H18N10 steel) and amorphous halos belonging to the coating material are present on the diffractograms (figure 3(a), (b)).

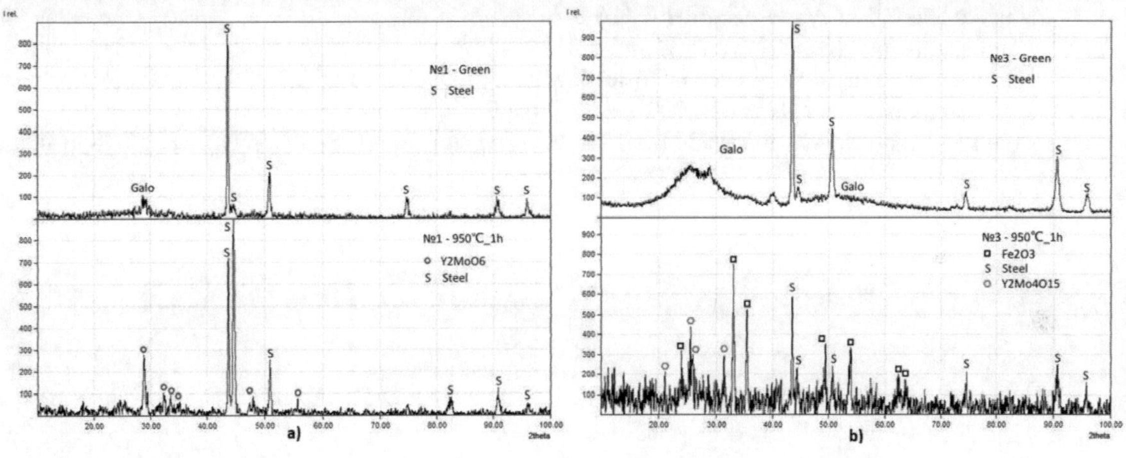

Figure 3. Diffractograms of coated steel samples before and after annealing obtained by mode: **(a)** – № 1; **(b)** – № 2.

To stabilize the structure of the coating, the samples were heated at 950 °C for 1 hour in an air atmosphere. The diffractograms of the coated steel samples after annealing are shown in figure 3.

Annealing for 1 hour resulted in the crystallization of yttrium molybdates in the coating, as evidenced by the presence of corresponding peaks. The Y_2MoO_6 compound predominates in the coating obtained according to the mode № 1, and $Y_2Mo_4O_{15}$ according to the mode №3. Peaks belonging to iron, chromium and nickel oxides in the first and second coatings were not detected. In the coating obtained by the mode № 2 there are peaks of iron oxide, which indicates the oxidation of the substrate, and therefore a shorter-term practical application of the coating obtained under this mode. The possibility of using coatings obtained by the modes № 1 and № 3 requires additional studies.

The X-ray phase analysis showed that when coatings are formed from two one-component cathodes made of Mo and Y in an oxygen atmosphere at a temperature of the processed sample not exceeding 500 °C, chemical compounds of yttrium molybdates of different phase composition are synthesized on the surface. Thus, there has been developed a fundamentally new method for forming coatings based on yttrium molybdates.

4. Conclusion

1. Based on the results of the research, it has been shown it is possible to apply continuous coatings on the basis of molybdates of yttrium by ion-plasma spraying on a composite with a molybdenum matrix. The thickness of the coatings obtained is about 35 μm.

2. There is developed a method for forming coatings based on yttrium molybdates from a vacuum-arc discharge plasma of the required stoichiometric composition.

3. Immediately after application, the coating material has an amorphous structure that acquires a crystalline structure after annealing at a temperature.

4. Based on the data of X-ray phase analysis, it can be concluded that the most prospected, from the point of view of practical application, are the coating obtained according to the modes № 1, № 2.

Acknowledgements

The financial support of Russian Science Foundation, project #16-19-10624 is gratefully acknowledged.

References

[1] Savickiy E M and Burhanov G S 1971 *Metallurgy of alloys of refractory and rare metals*
[2] Mileiko ST 1997 *Metal and Ceramic Based Composites* (Amsterdam: Elsevier)
[3] Mileiko S T 2017 *Composites Part B* **116** 1–6
[4] Jain P and Kumar K S 2010 *Acta Mater.* **6** 2124–42
[5] Heilmaier M, Krüger M, Saage H, Rösler J, Mukherji D, Glatzel U, Völkl R, Hüttner R, Eggeler G, Somsen Ch and Depka T 2009 *J. Met.* **7** 61–7
[6] Mileiko S T and Novokhatskaya N I 2015 *J. Materials Engineering and Performance* **7** 2836–40

Simulation of free-molecular and transition multicomponent gas flow through the system of filaments with different temperatures

A N Yakunchikov[1,2,3] **and V V Kosyanchuk**[1,2,3]

[1] Faculty of mathematics and mechanics, Lomonosov Moscow State University, Moscow, Russia
[2] Institute of Mechanics of Lomonosov Moscow State University, Moscow, Russia
[3] Mechanical Engineering Research Institute of the Russian Academy of Sciences, Moscow, Russia

E-mail: art-ya@mail.ru

Abstract. This work studies plane transition flow (Kn = 0.1–10) of a mixture of noble gases in the system, consisting of several rows of stretched filaments having different temperatures. Problem statement studying equilibrium state in closed system is considered. Calculations were performed using method of event-driven molecular dynamics (EDMD). It is shown that temperature difference leads to gas separation effect, which can be thereafter enhanced by increasing the number stages (rows of filaments).

1. Introduction

First studies of rarefied gas flows induced by temperature differences started more than a century ago [1, 2]. This effect (referred to in literature as "thermal transpiration") was utilized by Knudsen to construct his gas pump without moving parts [3, 4]. Nowadays, this method of creating pressure difference attracts the interest of researchers due to rapid development of technology of nano/ micro-electromechanical systems (NEMS / MEMS). Numerous versions of pump geometries were proposed [5–7], including multistage micropumps [8].

The flow of gaseous mixtures in such systems was studied numerically and experimentally [9, 10] and the effect of separation of the components with different molecular masses was shown. To date, prototypes of gas separation microdevices have been created on the basis of a nonisothermal membrane [11, 12] and on the basis of a lattice of plates with different temperatures [13]. In this paper, we consider the possibility of creating a gas separating device based on the lattice of threads (filaments) kept at different temperatures.

2. Problem statement

A plane problem about the flow of a mixture of five gases (He, Ne, Ar, Kr, Xe) in a system consisting of several rows of stretched filaments having different temperatures was considered. Equilibrium is studied state in the system (see figure 1), where one end is closed and another one is connected to a reservoir with constant parameters – pressure p_0, temperature T_0 and composition x_{0i} (x_{0i} – is the molar concentration of component in reservoir).

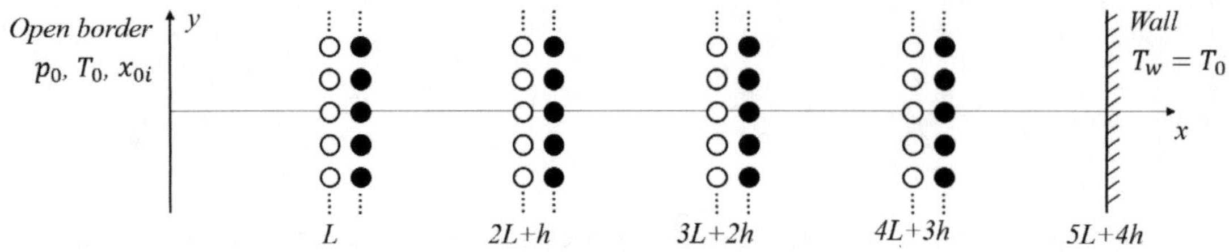

Figure 1. Geometry of system with four stages of filaments, closed at right end. White circles indicate cold filaments with temperature T_0, black ones indicate filaments with temperature T_1.

A system consisting of four stages of filaments was simulated. Each state contains two rows of filaments – cold one with temperature T_0 and hot one with temperature $T_1 = 2T_0$. Geometry of the problem is defined by h – distance between rows in one stage (see figure 2), L - distance between stages and d - diameter of a filament. Minimal distance between surfaces of neighboring filaments is denoted by $g = h - d$. Dimensions of the problem in directions perpendicular to x – axis were assumed much greater than listed parameters and thus symmetry conditions were used on y – bounds. Simulations were performed for 24 different configurations ($L/d = 5, 10, 15, 20, 25, 30, 35, 40$; $h/d = 1.25, 1.5, 2$) for Knudsen numbers $Kn_L = \lambda/L = 0.1 - 0.8$ и $Kn_g = \lambda/g = 4 - 15$. Molar concentrations at open end were set equal ($x_{0i} = 0.2, i = 1, \dots, 5$) in all simulations.

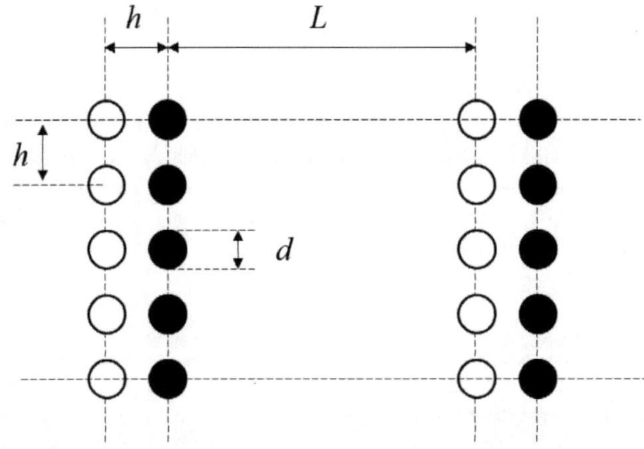

Figure 2. Geometry of one stage of filaments.

3. Calculations and results

Problem was solved using the method of event-driven molecular dynamics (EDMD) [14]. Strengths of the developed EDMD approach are meshless structure, deterministic (non-stochastic) implementation of intermolecular collisions (unlike in Direct Simulation Monte-Carlo method) and reduction of computational time compared to classical molecular dynamics simulations. Interaction between gas molecules and solid surface was described in terms of Maxwell scattering kernel with full energy and momentum accommodation.

Figure 3 illustrates simulation results for the case of geometry when ratio Kn_L/Kn_g of Knudsen numbers in chamber and between neighboring filaments has maximal value. Temperature between stages of filaments has linear profile with jump near surface of filaments and pressure is increasing with each next stage. Resulting pressure at the closed end of the system is approximately 2 times higher than the initial pressure in the reservoir. Number densities of the components increase as $1/T$ in the direction of the cold side of the chamber between the stages, and in the stages themselves concentrations decrease abruptly. The difference in molar concentrations is produced exactly in the chambers between the stages, and inside each stage (between neighboring rows) composition remains practically unchanged.

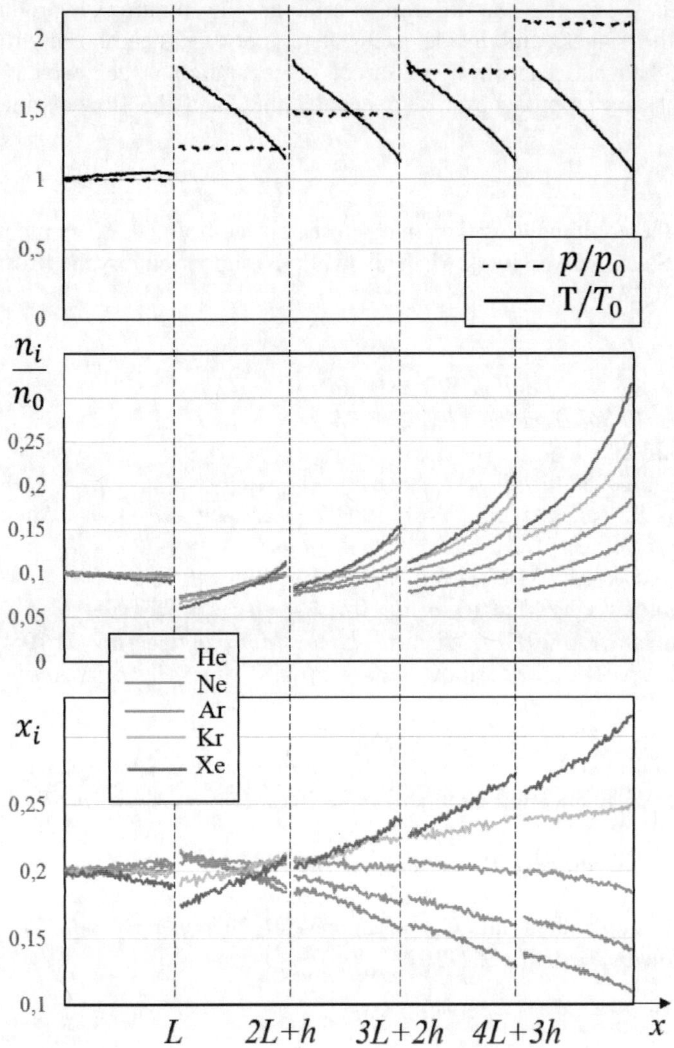

Figure 3. Temperature, pressure, number density and molar concentration of components for geometry $L/d = 40$, $g/d = 0.25$ for $Kn_L = 0.1$, $Kn_g = 15$.

The authors gave a qualitative physical explanation for this result by considering the stationary problem of heat transfer between two surfaces of different temperatures in two different regimes (free-molecular and continuum regime). That allowed to put forward a phenomenological hypothesis that the strongest separation effect is when the flow regime inside a stage approaches free-molecule limit, and the flow in the chamber between stages approaches continuum one. This hypothesis was completely confirmed by the results of EDMD simulations.

4. Conclusion

A plane problem about the transitional flow (Kn = 0.1–10) of a mixture of five noble gases in a system consisting of several rows of stretched filaments having different temperatures was studied. Pressure, temperature, density and molar concentration distributions were obtained. Temperature between stages of filaments has linear profile with jump near surface of filaments and pressure is increasing from stage to stage. As a result, pressure produced at the closed end of the system is approximately 2 times higher than the initial pressure in the reservoir. It is shown that under the influence of the temperature difference the effect of mixture separation arises. Number densities of the components increase in the direction of the cold side of the chamber between the stages, and in the stages themselves

IOP Conf. Series: Materials Science and Engineering **387** (2018) 012083 doi:10.1088/1757-899X/387/1/012083

concentrations decrease abruptly. It is shown that the difference in molar concentrations is produced exactly in the chambers between the stages, and inside each stage (between neighboring rows) composition remains practically unchanged. The strongest effect of separation is achieved under condition that flow regime inside a stage approaches free-molecule limit, and the flow regime in chambers is close continuum one.

Acknowledgments
The research was carried out using the equipment of the shared research facility of HPC computing resources at Lomonosov Moscow State University [15]. This work is supported by the Russian Science Foundation (grant № 17-71-10227).

References

[1] Maxwell J C 1879 *Philos. Trans. R. Soc. London* **170** 231–56
[2] Reynolds O 1879 *Philos. Trans. R. Soc. London* **170** 727–845
[3] Knudsen M 1909 *Ann. Phys.* **336** 205–29
[4] Knudsen M 1910 *Ann. Phys.* **338** 1435–48
[5] Aoki K, Degond P, Mieussens L, Takata S and Yoshida H 2008 *Multiscale Model. Simul.* **6** 1281–316
[6] Bond D M, Wheatley V and Goldsworthy M 2014 *Int. J. Heat Mass Transf.* **76** 1–15
[7] Bond D M, Wheatley V and Goldsworthy M 2016 *Int. J. Heat Mass Transf.* **93** 1038–58
[8] Gupta N K, An S and Gianchandani Y B 2012 *J. Micromechanics Microengineering* **22** 105026
[9] Naris S, Valougeorgis D, Kalempa D and Sharipov F 2004 *Phys. A Stat. Mech. its Appl.* **336** 294–318
[10] Kosuge S and Takata S 2008 *Eur. J. Mech. B/Fluids* **27** 444–65
[11] Matsumoto M, Nakaye S and Sugimoto H 2016 *AIP Conference Proceedings* **1786** 080011
[12] Nakaye S, Sugimoto H, Gupta N K and Gianchandani Y B 2015 *Eur. J. Mech. B/Fluids* **49** 36–49
[13] Sugimoto H and Abe T 2008 *AIP Conference Proceedings* (AIP) **1084** 1123–8
[14] Yakunchikov A and Kosyanchuk V 2018 *Comput. Fluids* **170** 121–7
[15] Sadovnichy V, Tikhonravov A, Voevodin V and Opanasenko V 2013 *Contemp. High Perform. Comput. From Petascale Towar. Exascale* 283–307

IOP Conf. Series: Materials Science and Engineering **387** (2018) 012084 doi:10.1088/1757-899X/387/1/012084

Investigation of the correlation between gas-sensitive properties and fractal dimension of nanostructured ZnO/ZnO<Cu, Fe> films obtained by the sol-gel method

N D Yakushova[1], I A Pronin[1], I A Averin[1], A A Karmanov[1], V A Moshnikov[1,2] and D Tz Dimitrov[3]

[1] Department of Nano- and Microelectronics, Penza State University, 440026, Penza, Russia

[2] Department of Micro- and Nanoelectronics, Saint-Petersburg Electrotechnical University "LETI", 197376, Saint-Petersburg, Russia

[3] Department of General and Inorganic Chemistry, University of Sofia, Sofia 1164, Bulgaria

E-mail: pronin_i90@mail.ru

Abstract. Correlations between gas sensitive properties and fractal dimension of nanostructured ZnO/ZnO <Cu, Fe> films obtained by the sol-gel method are established. The thickness of the films and their fractal dimension were determined by the number of dives in the sol of the upper zinc oxide film, alloyed with metal cations. It was found out that in all cases larger values of the fractal dimension correspond to lower values of the gas sensitivity.

1. Introduction

Gas-sensitive sensors appeared in 1970s as ceramic pressed samples of *n*-type oxide powders with a built-in heater. Then, sensors created by microelectronics technology on the basis of thin films were developed. As an alternative to the traditional methods of manufacturing sensor elements, a sol-gel method is used, the essential advantage of which is the possibility for achieving high homogeneity of synthesized materials, which greatly simplified the technology for obtaining nanostructured composites based on semiconductor oxides. The films obtained by sol-gel method have a developed relief, which leads to a significant increase of actual area of interaction with the gas phase, as well as the contact area between the lower and upper layers of the structure. This fact leads to a change of the characteristics of the gas sensor, which are not always explainable with the help of simple model representations of a thin continuous layer that do not take into account the nanorelief of contact surfaces, as well as features of transference phenomena. Therefore, it is important to evaluate the properties of the nanostructured surface of the films for the purpose of predicting the instrument parameters. The difficulty of numerical estimation of the nanorelief is in the strong dependence of the results on the scale of the measurements; therefore, a fractal approach is applied in the work, the results of which are invariant with respect to the scale. Thus, the purpose of the research was to investigate the correlation dependences between the fractal dimension of the surface of ZnO/ZnO<Cu, Fe> samples and their gas sensitive properties. The methods for manufacturing and study of samples are considered in detail in [1].

2. Calculation and discussion of results

The fractal dimension of the material was calculated using film surface images obtained by a scanning electron microscope. There are many ways of extracting the value of fractal dimension D from the image: "perimeter – area"; "area – volume of ratios"; variation, etc. [2]. Accordingly, the values of the fractal dimensions obtained by different methods also differ. In the research, D calculation for each sample was carried out by two methods: "perimeter – area" and approximations of a two-dimensional mass fractal.

The algorithm for calculating the fractal dimension of films by AFM and SEM images by the "perimeter – area" method is reduced to the following (figure 1):

1. Forming a contour image of the relief by cutting the film with the plane parallel to the plane of sample. Three contour images were obtained for each image, with subsequent averaging of the fractal dimension.
2. Covering the contour image of the surface with squares of $\delta \times \delta$ size of the variable side length, with subsequent counting the number of squares $N_\square$, proportional to the area S_c of the contours. Dimensions of δ in all cases are chosen in such a way that the surface behaves as a fractal object, which corresponds to a local approximation in the definition of fractal dimension.
3. Selecting the perimeter of each contour in the image.
4. Covering the resulting image with squares of $\delta \times \delta$ size with a variable side length and subsequent calculation of the number of squares $N_\square$, proportional to the perimeter P_c of the contours.
5. Replication of operations 1 – 4 for other δ values.

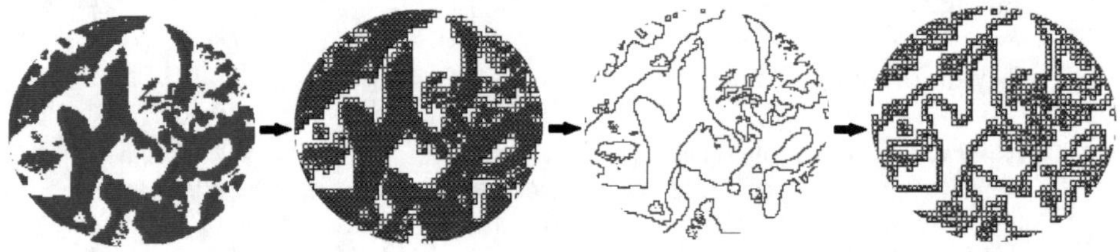

Figure 1. An algorithm for fractal dimension calculation by the "perimeter – area" method.

Since there is a relation $S_c(\delta) \sim [P_c(\delta)]^{2/D}$, then in the double logarithmic coordinates the tangent of the slope angle of this function characterizes the fractal dimension D [3]. The calculated values of the fractal dimension of the contour are related with the fractal dimension of the surface D_f by the following simple relation [4]: $D_f = D + 1$.

The method for fractal dimension calculation in terms of the mass two-dimensional fractal approximation is based on the approximation of a two-dimensional mass fractal aggregate. Since the technology for producing gas-sensitive films has layer-by-bivariate symmetry, then the scaling factors along the axis perpendicular to the substrate plane are strongly smoothed. This leads to the applicability of the two-dimensional mass fractal approximation. In this case, the fractal dimension is close to the Hausdorff-Besicovitch dimensionality. In this method, the calculation of the fractal dimensionality D_{fc} was carried out by slope ratio of the direct function $\ln(N_\square) = f(\ln(\delta))$.

To calculate the fractal dimension, ZnO-*Me* samples were made, the upper layer of which was formed by one, two and three dives in the sol. Figure 2, for example, shows morphology images of ZnO-Cu and ZnO-Fe film surfaces obtained by scanning electron microscopy. It is seen from the Figure, the results obtained by the two methods correlate with each other. In all cases, there is a decrease in fractal dimension with the increasing number of layers during the formation of the film. In general, this indicates that the surface becomes smoother. The obtained result can be related with improvement of the crystalline structure of the material, which is most likely due to the fact that the substrate for each subsequent layer is the previous one, which is substantially more porous and developed than the crystalline substrate Al_2O_3. The effect of improving the crystalline structure of ZnO using porous substrates is described in [5]. Figure 3 shows the obtained results for ZnO-Cu samples.

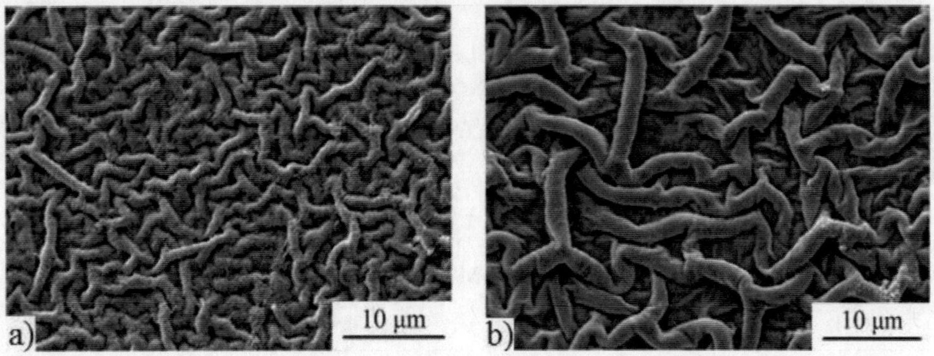

Figure 2. SEM images of ZnO-Cu (a) and ZnO-Fe (b) film surfaces formed by one dive in the sol.

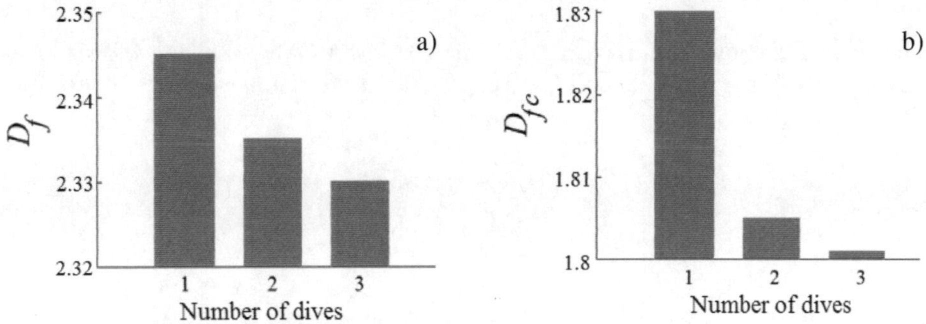

Figure 3. Dependence of the surface fractal dimension of ZnO-Cu films on the dives number in the sol, calculated by the "perimeter – area" method (**a**) and in the mass surface fractal approximation (**b**).

Figure 4 shows the results for calculating the fractal dimension for samples of ZnO-Fe series. Based on the results shown in figures 3 and 4, the following conclusions can be drawn: the fractal dimension of ZnO-Fe composition films in all cases is larger than the analogous composition of ZnO-Cu. This conclusion is in accordance with the difference in the radii of modifier cations and Zn^{2+}. In the case of ZnO-Fe, the difference in radii is large, which leads to substantial distortions of the crystal lattice, the appearance of mechanical stresses and, as a consequence, to more developed and defective surface; as in the case of ZnO-Cu samples, the surface fractal dimension in ZnO-Fe decreases with increasing number of film layers. This effect, apparently, is also due to an improvement in the crystallinity of the material during the formation of films on a porous substrate.

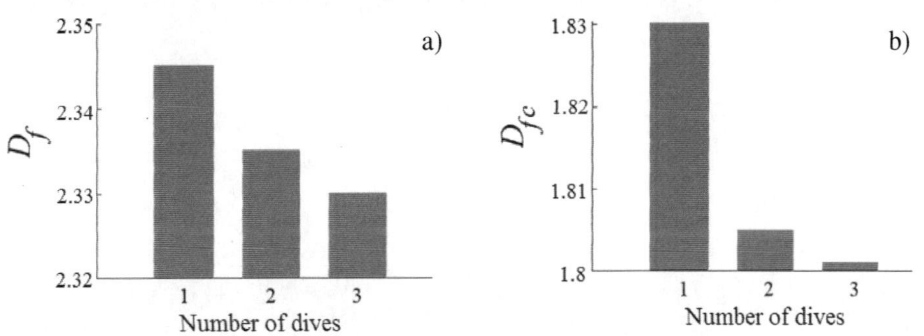

Figure 4. Dependence of the surface fractal dimension of ZnO-Fe films on the number of dives in the sol calculated by the "perimeter – area" method (**a**) and in the mass surface fractal approximation (**b**).

IOP Conf. Series: Materials Science and Engineering **387** (2018) 012084 doi:10.1088/1757-899X/387/1/012084

3. Conclusions

Thus, the correlations between gas sensitive properties and fractal dimension of nanostructured ZnO/ZnO <Cu, Fe> films obtained by the sol-gel method are established. It was found out that in all cases larger values of the fractal dimension correspond to lower values of the gas sensitivity.

Acknowledgments

The work was carried out with the financial support of the Ministry of Education and Science of the Russian Federation (project No. 16.897.2017/4.6), as well as in the framework of the Presidential Grant and the Presidential Scholarship (project No. MK-1882.2018.8 and SP-84.2018.1).

References

[1] Pronin I A 2015 *Nano- i mikrosistemnaya tekhnika* **8** 46–64
[2] Risovic D, Mahovic Poljacek S, Furic K and Gojo M 2008 *Appl. Surf. Sci.* **255** 3063–70
[3] Guzman-Castaneda J I, Garcıa-Borquez A and Arizabalo-Salas R D 2012 *Phys. Stat. Sol. B* **249** 1224–8
[4] Torkhov N A, Bozhkova V G, Ivonin I V and Novikov V A 2009 *Semiconductors* **43** 33–41
[5] Caia H, Shen H, Yin Y, Lu L, Shen J and Tang Z 2009 *J. Phys. Chem. Solids* **70** 967–71

Diagnosing system on the principle of high voltage pulses measurements

R V Yashkardin

St. Petersburg Electrotechnical Univeristy (ETU), 197376, Saint Petersburg, Russia

E-mail: yashkardin.rv@ya.ru

Abstract. The main aspects related to the development of measuring systems working with the high voltage pulses (from 1500 to 6000 V with current reaching 10^5A) are presented in the paper. The necessity of screening and noise protection for such systems is explained. The measures to ensure the electrical safety of the operating person while working with the system.

1. Introduction

Choosing parameters to control the functioning of the technological device is always complicated task. And it's even more complicated when the device is working in pulse mode with high density energy releases. Such devices usually use high power pulse discharges which is related with high amplitude currents and high voltages. One of the ways to control the processes taking place in the system is local measurements of the parameters of plasma radiation [1]. But it's not always feasible and sometimes can't provide the adequate data on the processes. It's especially common in the non-equilibrium systems and systems where plasma might move [2]. So, the main way to control if the electric technological system functioning correctly is still measuring the current and voltage parameters in the system. As an example of such system might be a technological device for production the ultradispersed particles in pulse arc discharge (high voltage (up to 5kV) short pulses (up to 200 us) at high currents (up to 100 kA)). A detailed description of the device functioning principles can be found in [3].

2. The processes taking place in a device

The following parameters were commonly used in the presented device: power 10^6–10^8 W, maximal current value 10^5 A, voltage – up to 5 kV, pulse duration 40–200 us (depends on the particular conditions).

Processes taking place in a device are described by massive system of integral-differential equations, including Maxwell equations describing electric and magnetic fields amplitudes, thermal conductivity equations (Stefan problem) describing thermal processes on the electrodes, gas dynamic problem considering inner heat sources and electromagnetic processes taking place in plasma discharge, etc. Moreover, all the processes are complicatedly related to each other and the properties of the mediums at such temperatures and currents are poorly studied. At all causes problems for numerical modeling making it literally impossible.

That's why such processes should be studied in a combined way, using integral parameters. In terms of this combined method the preparatory experimental researches of voltage dynamics in different technological modes. Comparing received data to the results of numerical modeling the processes in a matching invariant electric circuit including the elements of the device sepa-rately allowed to distinguish and connect the oscilloscope curve components with the processes taking place in a discharge. As a result, we could figure the particular oscilloscope curves of the voltages on electrodes when the device is functioning "correctly". They are presented in figure 1.

We used Tektronix TDS 3012c oscilloscope connected to the electrodes through the resistive divider with impedances 12.6 MΩ and 100 kΩ. The input impedance of the oscilloscope was set 100 MΩ. The signal was recorded as 10000 points with the measuring time from 40 to 100 us.

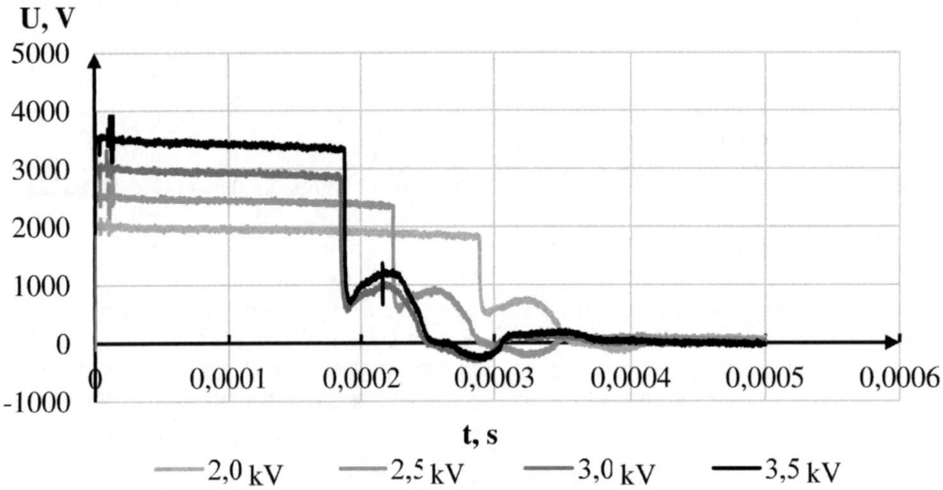

Figure 1. Voltage dynamics on the main electrodes at different initial voltage in capacitive 500uF battery.

Devices existing nowadays which are capable of measuring such voltages (kilovoltmeter spectral digital "KVC-120", and digital oscilloscope Tektronix MSO3012) can't satisfy all the requirements. And the major disadvantage of these devices is their high cost, which makes the whole technological device very expensive. Another disadvantage is the necessity for external computer signal processing. To overcome these difficulties, the system for diagnosing the technological device measuring the voltage pulses and comparing them to the standard signal curves was developed at the laboratory of "Electric Impulse Technologies".

3. Development of the technological device diagnosing system

The measurements are made through two parallel channels to provide control on two detection points of technological device. The device is controlled via signals sent from PC. The functional diagram is presented in figure 2.

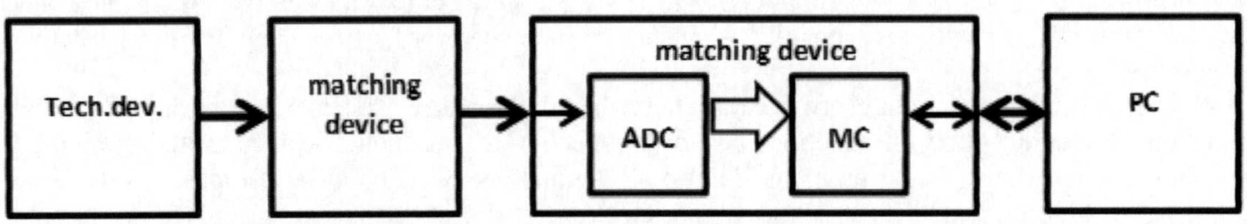

Figure 2. Functional diagram.

The functioning process of diagnosis system is the following. Measured voltage is sent to the input of matching device (resistive current bridge). The bridge contains metal oxide resistors connected in series (C2-23) 2 W, 300 Ω, 5 %, with total resistance 6 MΩ. The current bridge limits the currents in the system on 1 mA at maximal voltage 6 kV.

The signal from the input is sent to the galvanic splitter being optical isolator. Galvanic isola-tor divides the device into a number of independent circuits, with electric autonomous signals. After passing the galvanic isolation the signal is amplified and sent to the input of the analog-digital convertor (ADC). ADC sends digital signal to a microcontroller, which records a signal and sends it further to a computer. The circuit diagram is presented in figure 3.

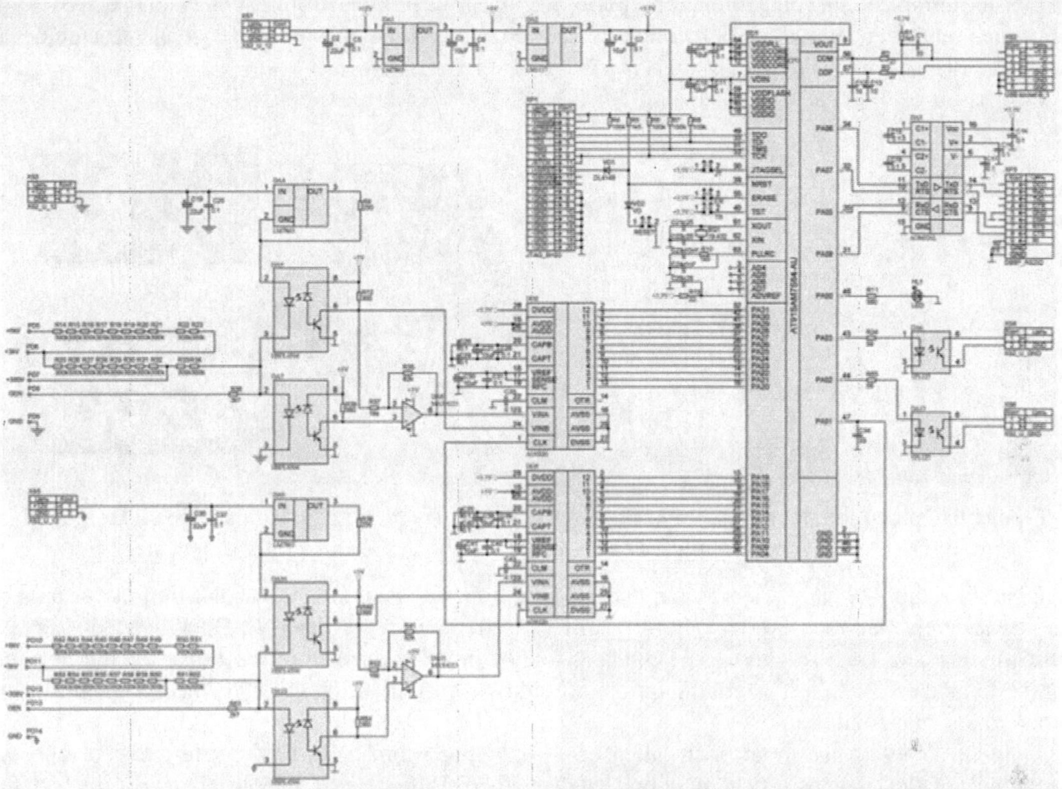

Figure 3. Electric circuit diagram.

As the system deals with the voltage pulses from 1500 up to 6000 V at currents up to 10^6 A, the systems of protection the circuits from noises and short circuit damaging. That's why for each channel of diagnosis system a galvanic isolation is envisaged. For the isolation HCPL-4504 optocoupler was chosen.

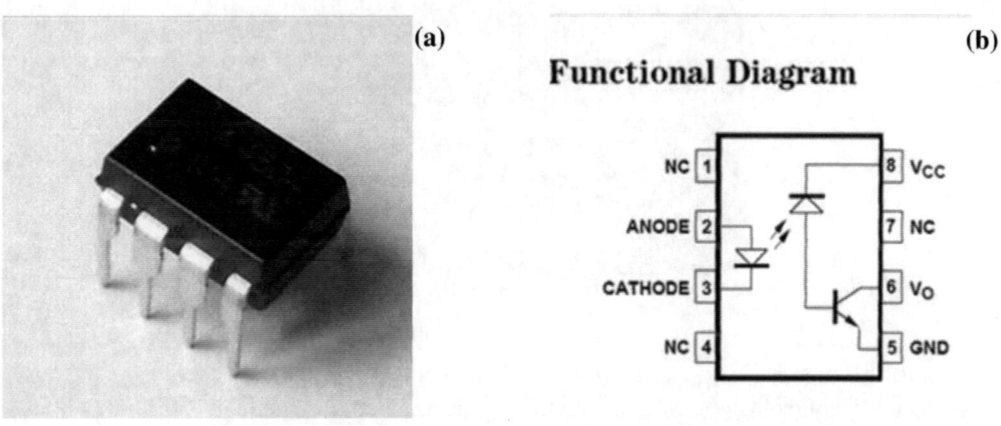

Figure 4. A photograph of the optocoupler HCPL-4504 (**a**) and functional scheme (**b**).

The optocouple HCPL-4504 has got high current transfer ratio and low transition latency. The isolating layer between photodiode and photodetector separates galvanically input and output. These optocouples have photodiode bias connectors and the collector of the output transistor figure 4(b). It decreases the capacity between base and collector improving the performance of the optocouples significantly. As the optocouples are semi-conductive devices, they have non-linear transformation characteristic. In order to compensate errors of the transformation two optocouples were used differentiately connected to the operational amplifier input. One optocouple was reference and the second – for actual measurements. The circuit diagram of one of the channels is presented at figure 5(a).

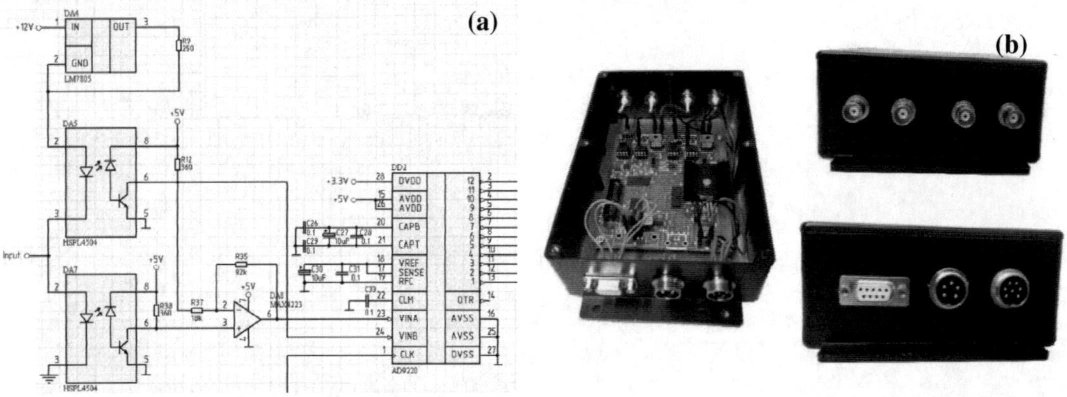

Figure 5. Optocouples commutation diagram (**a**), diagnosis system appearance (**b**) DD2 – ADC circuit; DA5, DA7 – optocouples; DA4 – voltage stabilizer; DA8 – operational amplifier.

Such an optocouple commutation principle allows minimizing the nonlinearity of the trans-formations and significantly decrease the temperature drift caused by currents flow. Another way to decrease transformation errors is to choose the optical isolator breakpoint on the linear part of the current-voltage characteristic. Such a commutation provides a current through refer-once and measuring optocouplers 20 mA current.

A signal comes to the input of the measuring optocouple and adds to the reference current. An appearance of the diagnosis system is presented at figure 5(b). Because both channels are isolated from each other the device contains three independent electric circuits. It required also three channel power supply system. A special power source (figure 6) was developed and construtted for these purposes.

Figure 6. Power supply.

In the electric circuit of each channel the transformer Transled TPG-32 was installed. The voltage from the secondary winding comes to the rectifying bridge. Rectified voltage is further applying to the input of the voltage stabilizer LM317. Using different additive resistive circuits on the stabilizer

IOP Conf. Series: Materials Science and Engineering **387** (2018) 012085 doi:10.1088/1757-899X/387/1/012085

provides creating a number of impedances options on the power source output, making the developed supply more versatile and functional.

Technological device functioning is provided by the pulse discharge. The noise currents appear in the electric circuits of the device and the devices connected to it while processing. Such currents also might appear in the circuits galvanically isolated from each other. Consequently, the diagnosis system has to be screened very carefully. The major type of noises in this case are nanoseconds long. Cancelling their impact is the most factor for device reliability improvement.

The standard IEC 61000-4-4 (GOST R 51317.4.4-99) allows to pick two main components of noise immunity for the device under nanoseconds long noises:

- Inner device grounding.
- Capacitive connections.

When nanoseconds noises pass through the circuit, the potential difference appears between different sections of the ground. And the major effort of the potential difference appearance is brought by the inductive component of the grounding circuit, not resistive one. As such short pulses are steep at their fronts, even minor inductiveness might cause device failure. Such noises are also very hard to capture even using digital recording oscilloscope.

Hence a consistent grounding layer causes the appearance the potential difference when very short noisy pulses pass through a circuit. Potential difference might reach up to couple volts, which might affect the gear.

The main method of protection diagnosis systems from the nanoseconds noises is using optocoules in the measurement links. However, using optical isolator didn't provide the required level of noise immunity though it increased 1,5–2 times. But this level wasn't enough to make the system work stable. For the further improvements a number of actions like using special wiring layouts on the device board. The example of such hint might be increasing a section of the ground wires on a layout. Empty spaces on the board were covered with grounding polygons on the opposite side of the board. Wiring and tracing the board considered the optimal points of grounding different circuits. The areas with "clear ground" were done.

One of the effective ways of high frequency noises influence decrease in a circuit is adding ferrite filters. Ferrite filters are damping the high frequency components in the conductors. It also provides the nanosecond noises significant reduction

The dynamics of the voltage on the main electrodes is processed in a special software on a PC. The software interface is presented at figure 7.

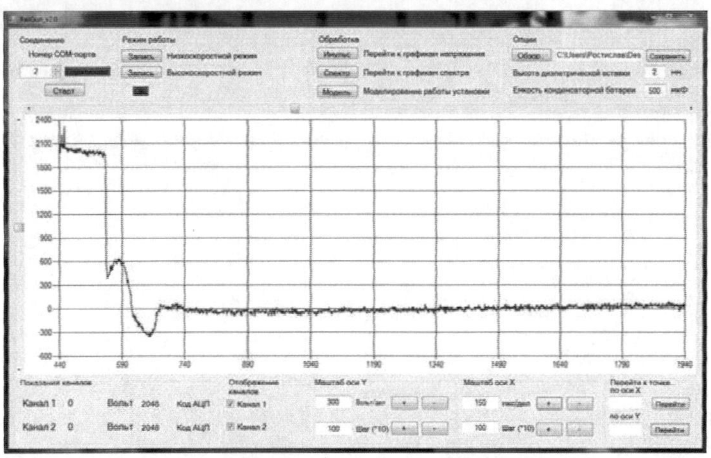

Figure 7. Software interface.

The voltage curve changes show the state of different parts of the technological device and its results. Comparing the voltage dynamics measurements and the pictures of substrates surfaces with sputtered

IOP Conf. Series: Materials Science and Engineering **387** (2018) 012085 doi:10.1088/1757-899X/387/1/012085

particles measured with atomic-forced microscopy would help the technological process properties and studying the results of particles production.

4. Conclusions

The developed system of diagnosis of the technological device provides registration the pulses up to 6 kV and beginning with 10 us long. The developed device includes complicated external noise reduction system. Such system includes screening, galvanic isolations and special hints on beard layouts. Moreover, the system includes a protective element for an operating person.

References

[1] Kostrin D K et al. 2016 *Journal of Physics: Conference Series* **729** 012030

[2] Goncharov V D and Samsonov D S 2015 *Journal of Technical Physics* **5** 37–42

[3] Yaskardin R V and Sorokin K S *2016 IEEE NW Russia Young Researchers in the Electrical and Electronic Engineering Conference (2016 ElConRusNW)*

Features of the methane clusters formation in supersonic jets of methane and its mixtures with monatomic gases

A E Zarvin, V E Khudozhitkov and V V Kalyada

Department of applied physics, Novosibirsk State University, 630090, Novosibirsk, Russia

E-mail: zarvin@phys.nsu.ru

Abstract. This work demonstrates the methane clusters formation in supersonic free jets of methane mixtures with helium and argon. The dependence of the clusterization process properties on the carrier gas, the volume fraction of methane in the mixture at the selected of gas jet parameters is shown. Impact of the methane dissociative ionization, and methane clusters fragmentation in the detector of the mass spectrometer on the process of condensation was demonstrated. The presence of protonated cluster ions of methane and the increase in their proportion with growing size of clusters was established. The formation of cluster ions with an attached methyl, methylene or methylidene was discovered.

Methane is widely used for the films deposition, because the plasma-chemical decomposition of methane allows the growth of diamond-like carbon, amorphous carbon films, polycrystalline carbon films, and carbon nanotubes. The use of methane clusters as a raw material for application to the surface may provide the creation of new nanomaterials with unique properties. The conventional method of obtaining atomic and molecular clusters is an adiabatic expansion of a gas with the desired composition from the stagnation chamber at known pressure and temperature using sonic or supersonic nozzle with a given geometry and at a sufficient rarefaction [1]. There have been reported several attempts to obtain large clusters of methane in streams of pure methane (see e.g. [2–3]). However, in contrast to the monosilane [4], at a supersonic expansion of pure methane the registration of large clusters can be achieved only at high stagnation pressures (tens of atmospheres) and/or low temperatures [5]. At the same time, clusters of methane and mixed argon-methane complexes are registered in a mixture with a carrier gas (argon) under moderate stagnation parameters as well [6]. Among the reasons of such clusterization peculiarities in the methane jets with carrier gas is primarily the entrainment of excess energy by argon during the formation of the oligomeric molecular gas cluster (triple collisions) [4–5]. The molecular hydrogen was also tested as such a “cold” reservoir [2].

The active use of argon as a carrier gas leaves a number of unresolved problems. Firstly, argon is easily condensed. Its condensation can also be the reason of oligomers nucleation, which can initiate the formation of binary supramolecular structures. Secondly, argon is a heavy gas compared to methane. This leads to the displacement of methane to the periphery of the supersonic jet. Therefore, it seems preferable to use lighter gas as the carrier gas, which condenses at much lower temperatures than methane. It is obvious that hydrogen or helium should be used as carrier gas. In our opinion, helium is more preferable, firstly, because with the expansion of a monatomic gas the temperature drops faster, and, secondly, it is possible to directly compare the processes taking place in a supersonic

IOP Conf. Series: Materials Science and Engineering **387** (2018) 012086 doi:10.1088/1757-899X/387/1/012086

jet of the various mixture. Investigations of supersonic expansion of binary mixtures with the possibility of methane condensation with a carrier gas are of the interest both as a prospect for obtaining pure one-component large-scale methane clusters and as an object of studying the formation processes of binary clusters.

In this paper, the analysis of the formed clusters was carried out using the method of molecular beam mass spectrometry at the gas-dynamic LEMPUS-2 complex of Novosibirsk State University [7]. A quadrupole mass spectrometer Hiden Analytical EPIC 1000 with a dynamic range of 2–1000 a.m.u. was used. Mass spectra were recorded at the electron energy 70 eV in the ionizer of a mass spectrometer. A supersonic jet was formed as a result of the expansion through a supersonic nozzle with a critical section diameter $d_* = 0.17$ mm, a diffuser length $L_d = 8.2$ mm and an output diameter $D_a = 2.2$ mm. The following parameters of the stagnation in the nozzle pre-chamber were maintained: pressure $P_0 = 0.3$ MPa, temperature $T_0 = 298$ K. We have chosen the entrance diameter of the molecular beam system skimmer ds = 0.44 mm. As a result, a sufficiently wide range of nozzle-skimmer distances was provided, in which both external and internal skimmer interaction could be neglected [8]. The pumping system provided a vacuum in the chamber of the mass spectrometric detector at the level no worse than 0.02 mPa. The sensitivity range of the mass spectrometer was about 7 orders of magnitude, from 10^0 to 10^7, but the amplitudes at a level less than $2 \cdot 10^1$, as a rule, were neglected because of the high noise of the useful signal.

In our work, experimental studies of the supersonic jets of helium-methane mixtures were conducted with the following component ratios (in volume percents): 90 – 10, 80 – 20, 60 – 40, 40 – 60, 20 – 80, 10 – 90, and also in pure methane and in an argon-methane mixture with a component ratio of 80 to 20.

It is established that clusters of methane in a mixture with helium are formed much more efficiently than in a mixture with argon. Earlier we found out [6] that oligomers Ar_n^+, $(CH_4)_m^+$, and also $((CH_4)_lAr_k)^+$ (where k, l, m, n – are integers) are registered at a volumetric methane content in argon of 10, 5 and 2.5 percents in the supersonic flow of such mixtures. The presence of clusters of both components in the mixture, as well as mixed clusters, confirmed the existence of parallel nucleation processes in argon and methane.

Figure 1 shows the mass spectra of helium-methane and argon-methane mixtures at the same volume concentration of methane (20 %). As noted above, the data are compared under the same stagnation conditions, the nozzle geometry and the molecular beam system at a fixed distance from nozzle to skimmer (40 mm).

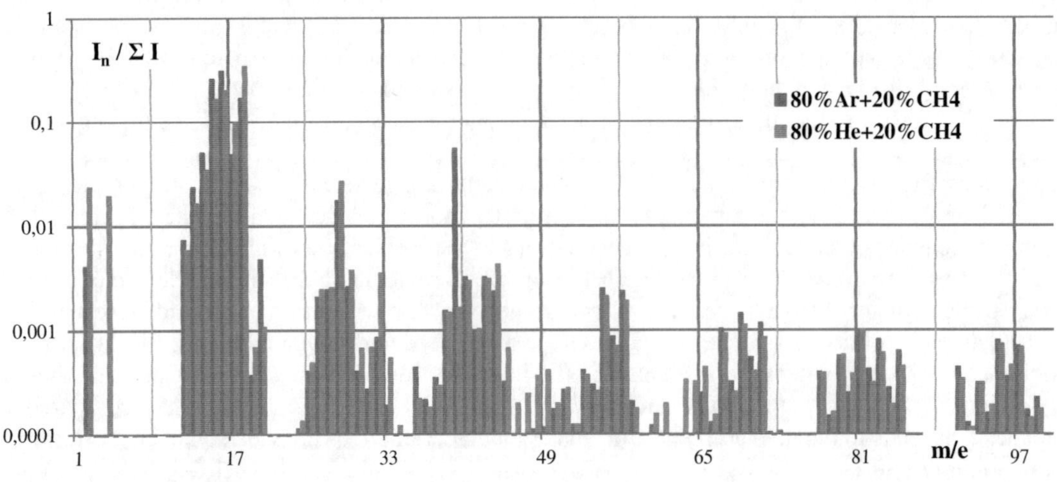

Figure 1. Histogram of mass spectra with two carrier gases, Ar (red lines) and He (green lines) within the range up to m/e = 100.

The amplitude of the mass peaks is normalized to the sum of the peaks of all masses. Naturally, this normalization cannot give a completely objective comparison of different mixtures, however it provides an opportunity to draw certain conclusions.

In both mixtures, there is a traditional distribution of the amplitudes of CH_x^+, where x = 0, 1, 2, 3, 4. Also, in both mixtures the amplitudes of methane monomers predominate over the carrier gas amplitudes. This indicates the intensive formation of methane oligomers with effective fragmentation of such clusters, mainly up to monomers. The intense peaks m/e = 18 and 17 are related to the water vapor background presence, not eliminated by the pumping system of the molecular beam detector chamber. The presence of an intense hydrogen peak, as well as a set of $C_nH_m^+$, peaks characteristic for mass spectra of the hydrocarbon cycle (m/e = 27, 29, 39, 41, 43, 55, 57, etc.) can be explained by the dissociative ionization of methane and the destruction of methane clusters during the electron impact and the existence of fast ion-cluster reactions during the detection period of registered ions in the detector of the mass spectrometer, which is $2\cdot10^{-4}$–$2\cdot10^{-5}$ sec. It should be noted that such peaks in the range up to m/e ~ 100 have larger amplitude than the cluster peaks, except for the dimer (in mixture with helium). With further growth of the mass number, the intensity of the hydrocarbon cycle peaks drops sharply, while the cluster peaks are observed throughout the recorded mass range.

In the given range, the differences between the two spectra are insignificant except for the masses, which should be attributed to He and Ar. One can only note the presence of more intense peaks of the hydrocarbon series radical fragments in a mixture of methane and helium. In contrast to the data [8] obtained in a mixture with argon at a lower concentration of methane, in our measurements no appreciable peaks related to argon oligomers or mixed argon-methane clusters were recorded.

The mass-spectra of larger clusters, N > 6 (where N is the cluster size), depending on the mass number m /e, are shown in figure 2. For convenience, the ordinate is given on a linear scale. The maximum values of the amplitudes of neighboring peaks are connected by straight lines.

As can be seen from the graph, the methane clusters are much larger in amplitude in the mixture with helium (green lines) than with argon (red lines). This confirms the assumptions about the efficiency of heat removal by light gas, as well as the possible parallel process of argon condensation in mixtures with methane that retards methane cluster formation.

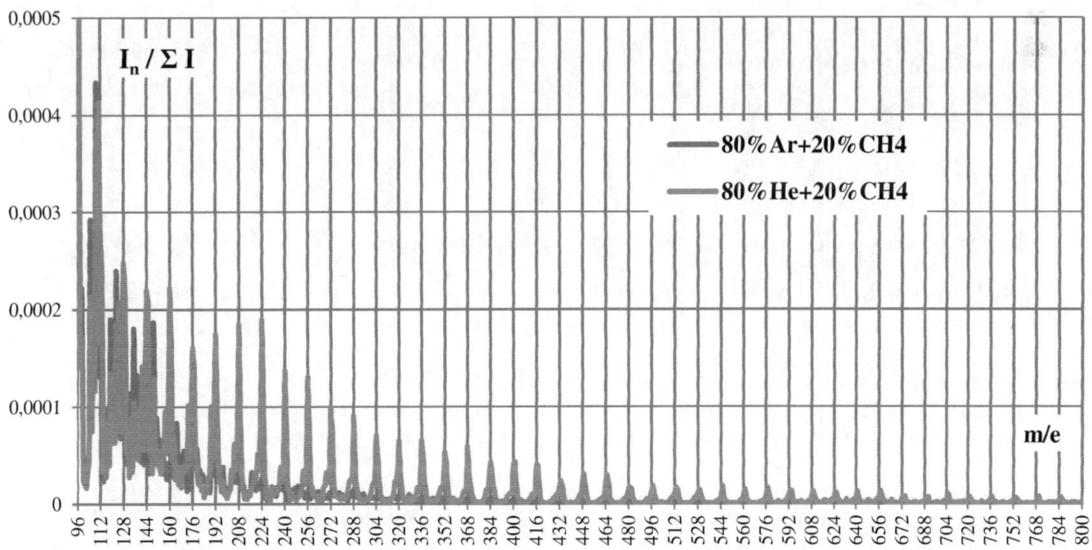

Figure 2. Graph of changes in the amplitudes of mass peaks I_n in mixtures of methane with argon (red lines) and helium (green lines) with normalization to the sum of the amplitudes of all peaks, depending on the mass number.

Measurements with other compositions of mixtures showed that with increasing percentage of methane in the mixture, the fraction of large clusters decreases, i.e. the process of cluster formation is weakens. In the mixtures where the volume concentration of methane exceeds 40 %, methane clusters have not been detected with the exception of a small number of dimers and trimers.

The mass spectrum of a mixture of 20 % CH_4 + 80 % He with the same gas dynamic parameters in the cluster size range from N = 6 (m/e = 96) to N = 34 (m/e = 544) can be considered in more detail in figure 3. The mass numbers m/e are plotted along the abscissa axis, and the amplitude of the ion peaks in the logarithmic scale is plotted along the ordinate. For convenience, the mass spectrum is divided into two ranges, $6 \leq N \leq 20$ and $20 \leq N \leq 34$, listed consecutively, one under the other. Noteworthy, that with a further increase in the mass number, cluster peaks are also present, but in the above mentioned mode of gas dynamic parameters, the signal level at N > 40 is insufficient for the reliability of the results.

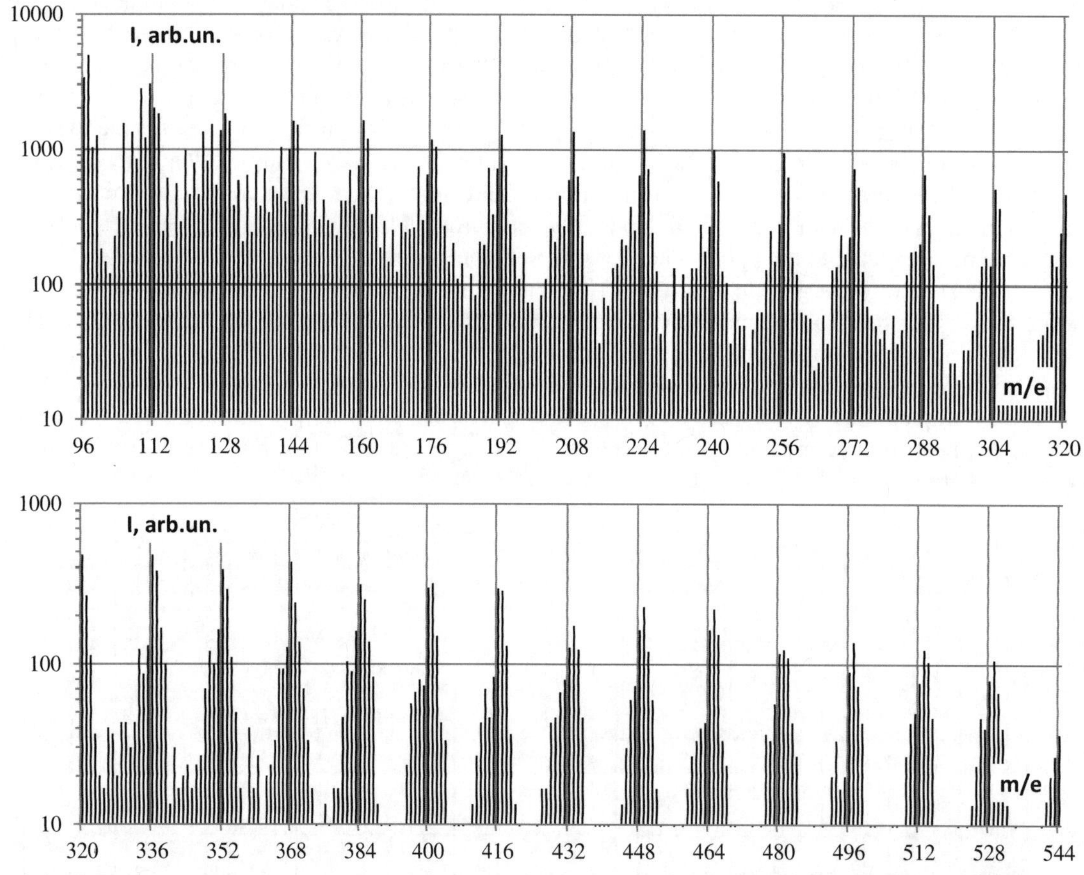

Figure 3. Mass spectrum of the gas mixture 20 % CH_4 + 80 % He. The nozzle-skimmer distance – X = 40 mm, the stagnation pressure – P_0 = 0.3 MPa, the background pressure in the main chamber – P_h = 4.4 Pa, in the post-skimmer chamber – $P_{ps} = 1\cdot10^{-3}$ Pa, in the detector chamber – $P_d = 4\cdot10^{-5}$ Pa.

It can be stated that in a mixture with a small content of methane in helium, methane clusters reach a size of at least N > 50. Local maximums of amplitudes in the cluster peaks are detected, with m/e = 96, 224, 368, 576, 704, which corresponds to the size of the clusters N = 6, 14, 23, 36, 44. These maxima can be identified with the "magic" numbers [13] of cluster methane ions.

In addition to cluster ions of the $(CH_4)_n^+$ type, the protonated cluster ions of the $(CH_4)_n\cdot H^+$ type are observed. This feature was also found in a mixture of methane and argon [6]. Many studies have been devoted to the formation of protonated molecules. In the literature on mass spectrometry, the results of protonation are described for some clusters [9–12]. However, the formation of protonated methane clusters requires a separate study. In this paper, we note that when the cluster size grows, the amplitudes of the protonated particles increase. Starting with cluster ions of size n > 25, the amplitude of such peaks becomes predominant. At the same time, noticeable amplitudes of twice protonated cluster ions are detected, the amplitude of which increases with the further growth of cluster size. The occurrence of protonated clusters indicates a possible presence of a substantially larger sized clusters in the investigated mixture flow that is inaccessible for the used mass spectrometric system.

It should be noted that in addition to the cluster peaks of the protonated series, we found in the spectrum of the mixture a cluster sequence of the type $(CH4)_{n-1}\cdot(CH_m)^+$, where m = 1, 2, 3. Clusters with m = 1 and 3 have the highest intensity. The presence of clusters of this type confirms the process of ion-cluster reactions in the detector of the mass spectrometer and determines the lower bound of the rate of such reactions.

Acknowledgment
The work was performed using the equipment of the center for collective use "Applied physics" of NSU with the financial support of grants of the Ministry of Education and Science of the Russian Federation 3.5918.2017/ITR and 3.5920.2017/ITR.

References
[1] Haberland H 1994 *Springer Series in Chemical Physics* **52**
[2] Yang S, Philippe L and Chatelet M 2007 *International Journal of Mass Spectrometry* **263** 190–4
[3] Zarvin A E, Korobeishchikov N G, Khodakov M D and Kalyada V V 2012 *Advances in Materials Physics and Chemistry* **2** 146–9
[4] Korobeishchikov N G, Zarvin A E, Madirbaev V Gh and Sharafutdinov R G 2005 *Plasma Chemistry and Plasma Processing* **25** 319–49
[5] Abu-samha M, Ryding M J, Uggerud E, Sæthre L J and Børve K J 2017 *The Journal of Chemical Physics* **147** 014301
[6] Zarvin A E, Kalyada V V, Korobeishchikov N G and Khodakov M D 2014 *The Monthly Scientific and Production Journal «Nano Ingineering»* **1** 3–6
[7] Zarvin A E, Kalyada V V, Madirbaev V Gh, Korobeishchikov N G, Khodakov M D, Yaskin A S, Khudozhitkov V E and Gimelshein S F 2017 *IEEE Transactions on Plasma Science* **45** 819–27
[8] Zarvin A E and Sharafutdinov R G 1979 *Journal of Applied Mechanics and Technical Physics* **20** 744–9
[9] Smirnov B M 2003 *Physics – Uspekhi* **46** 589–628
[10] Makarov G N *Physics – Uspekhi* **51** 319–53
[11] Wyslouzil B E and Wölk J 2016 *J. Chem. Phys.* **145** 211702
[12] Khodorkovskii M A, Artamonova T O, Murashov S V, Michael D, Rakcheeva L P, Belyaeva A A, Timofeev N A, Mel'nikov A S, Shakhmin A L and Dement'ev I A 2009 *Tech. Phys.* **54** 1–6
[13] Eletskii A V and Smirnov B M 1989 *Sov. Phys. Usp.* **32** 763–82